MW00592136

Springer Series on

Atoms+Plasmas

8

Editor: Günter Ecker

Springer
Berlin
Heidelberg
New York
Barcelona
Hong Kong
London
Milan
Paris
Singapore
Tokyo

Springer Series on

Atoms+Plasmas

Editors: G. Ecker P. Lambropoulos J. Mlynek H. Walther

1 **Polarized Electrons** 2nd Edition
By J. Kessler

2 **Multiphoton Processes**
Editors: P. Lambropoulos and S. J. Smith

3 **Atomic Many-Body Theory**
2nd Edition
By I. Lindgren and J. Morrison

4 **Elementary Processes in Hydrogen-Helium Plasmas**
Cross Sections and Reaction Rate Coefficients
By R. K. Janev, W. D. Langer, K. Evans Jr., and D. E. Post Jr.

5 **Pulsed Electrical Discharge in Vacuum**
By G. A. Mesyats and D. I. Proskurovsky

6 **Atomic and Molecular Spectroscopy**
2nd Edition
Basic Aspects and Practical Applications
By S. Svanberg

7 **Interference of Atomic States**
By E. B. Alexandrov, M. P. Chaika and G. I. Khvostenko

8 **Plasma Physics** 3rd Edition
Basic Theory with Fusion Applications
By K. Nishikawa and M. Wakatani

9 **Plasma Spectroscopy**
The Influence of Microwave and Laser Fields
By E. Oks

10 **Film Deposition by Plasma Techniques**
By M. Konuma

11 **Resonance Phenomena in Electron-Atom Collisions**
By V. I. Lengyel, V. T. Navrotsky and E. P. Sabad

12 **Atomic Spectra and Radiative Transitions** 2nd Edition
By I. I. Sobel'man

13 **Multiphoton Processes in Atoms** 2nd Edition
By N. B. Delone and V. P. Krainov

14 **Atoms in Plasmas**
By V. S. Lisitsa

15 **Excitation of Atoms and Broadening of Spectral Lines**
2nd Edition
By I. I. Sobel'man, L. Vainshtein, and E. Yukov

16 **Reference Data on Multicharged Ions**
By V. G. Pal'chikov and V. P. Shevelko

17 **Lectures on Non-linear Plasm Kinetics**
By V. N. Tsytovich

18 **Atoms and Their Spectroscopic Properties**
By V. P. Shevelko

19 **X-Ray Radiation of Highly Charged Ions**
By H. F. Beyer, H.-J. Kluge, and V. P. Shevelko

20 **Electron Emission in Heavy Ion-Atom Collision**
By N. Stolterfoht, R. D. DuBois, and R. D. Rivarola

21 **Molecules and Their Spectroscopic Properties**
By S. V. Khristenko, A. I. Maslov, and V. P. Shevelko

22 **Physics of Highly Excited Atoms and Ions**
By V. S. Lebedev and I. L. Beigman

23 **Atomic Multielectron Processes**
By V. P. Shevelko and H. Tawara

24 **Guided Wave Produced Plasmas**
By H. Schlüter, Y. M. Aliev, and A. Shivarova

25 **Quantum Statistics of Strongly Coupled Plasmas**
By D. Kremp, W. Kraeft, and M. Schlanges

26 **Atomic Physics with Heavy Ions**
By H. F. Beyer and V. P. Shevelko

K. Nishikawa   M. Wakatani

Plasma Physics

Basic Theory with Fusion Applications

Third, Revised Edition

With 80 Figures

Professor Kyoji Nishikawa
Kinki University, Faculty of Engineering
1 Umenobe, Takaya, Higashihiroshima,
Hiroshima 739-2166, Japan
e-mail: nisikawa@hiro.kindai.ac.jp

Professor Masahiro Wakatani
Kyoto University,
Graduate School of Energy Science
Gakasho, Uji, Kyoto 611-0011, Japan
e-mail: wakatani@energy.kyoto-u.ac.jp

Series Editors:

Professor Dr. Günter Ecker
Ruhr-Universität Bochum, Fakultät für Physik und Astronomie,
Lehrstuhl Theoretische Physik I, Universitätsstrasse 150,
D-44801 Bochum, Germany

Professor Peter Lambropoulos, Ph. D.
Max-Planck-Institut für Quantenoptik,
D-85748 Garching, Germany, and
Foundation for Research and Technology – Hellas (FO.R.T.H.),
Institute of Electronic Structure & Laser (IESL),
University of Crete, PO Box 1527, Heraklion, Crete 71110, Greece

Professor Jürgen Mlynek
Universität Konstanz, Universitätsstr. 10
D-78434 Konstanz, Germany

Professor Dr. Herbert Walther
Sektion Physik der Universität München, Am Coulombwall 1,
D-85748 Garching/München, Germany

Library of Congress Cataloging-in-Publication Data
Nishikawa, K. (Kyoji) Plasma physics: basic theory with fusion applications / K. Nishikawa, M. Wakatani. – 3rd, rev. ed. p. cm. – (Springer series on atoms + plasmas, ISSN 0177-6495; 8) Includes bibliographical references and index. ISBN 3-540-65285-X (hardcover : alk. paper)
1. Plasma (Ionized gases) 2. Controlled fusion. I. Wakatani, Masahiro. II. Title. III. Series
QC718.N57 1999 530'.4'4–dc21

ISSN 0177-6495
ISBN 3-540-65285-X 3rd. Ed. Springer-Verlag Berlin Heidelberg New York
ISBN 3-540-56854-9 2nd. Ed. Springer-Verlag Berlin Heidelber New York

Production: PRO EDIT GmbH, D-69126 Heidelberg
Cover concept by eStudio Calmar Steinen
Cover design: *design & production* GmbH, D-69121 Heidelberg
Typesetting: Satztechnik Steingräber, D-69126 Heidelberg
SPIN 10698009 57/3144 - 5 4 3 2 1 0 - Printed on acid-free paper

Preface to the Third Edition

After the first edition was published in 1990, many English textbooks were published on plasma physics. Some of them are more introductory, and some of them treat specific subjects and are more advanced than our book. This edition contains both fundamentals of plasma physics useful for studying laboratory research plasmas, industrial processing plasmas, space and solar plasmas, and applications to various types of experimental devices from magnetic confinement to inertia fusion. Since significant progress has been achieved in fusion research in the 1990s, new important results not included in the first edition have been added to the original text of Chap. 14. Other chapters and the appendix remain basically as in the second edition with a few corrections of misprints. We hope the refreshed book will serve sufficient knowledge to students in the coming 21st century.

July 1999

Kyoji Nishikawa
Masahiro Wakatani

Preface to the First Edition

The physics of fully ionized high temperature plasmas has made remarkable progress in the past two decades. This progress has been achieved mainly in connection with research into methods of controlling high temperature plasmas in large tokamak machines.

"Tokamak" is a generic name for the axisymmetric, toroidal, magnetic confinement devices used to produce high temperature plasmas and to stably confine them by means of a strong magnetic field. The purpose of such confinement is to yield a substantial amount of nuclear fusion inside the plasma. Investigations of high temperature plasmas using these machines, as well as others built for more fundamental research, have revealed numerous interesting properties characteristic of collision-free systems with collective electromagnetic interactions. The basic theory of these properties was established at an early stage of plasma physics research, and several excellent introductory textbooks have been published on these subjects. Books on more advanced topics, such as those dealing with the various nonlinear responses of high temperature plasmas to an external disturbance, also abound. More recently, some useful books devoted to specific applications of plasma physics, such as tokamak plasmas, laser-produced plasmas, and space plasmas, have been published. To our knowledge, however, no standard books are available which start from the very basic theory of high temperature plasmas and subsequently deal with its applications to present-day nuclear fusion plasma research in a self-contained manner. As a result, many students who want to study plasma physics and/or nuclear fusion research appear to encounter difficulties in understanding the relation between the emphasis in basic plasma physics and actual fusion plasma physics. This book is written with the aim of filling this gap by consolidating basic plasma physics and its applications to present-day nuclear fusion research.

We originally intended to intertwine the basic plasma physics with the applications to nuclear fusion research. However, on the advice of the editors, we finally decided to separate the basic plasma physics from the applications. This was done in consideration of the fact that there may be many readers who are interested in plasma physics itself, but not so much in nuclear fusion research. The book is therefore divided into two parts: Part I, Basic Theory, and Part II, Applications to Fusion Plasmas.

Since the plasma physics of nuclear fusion research is making daily progress, it is not an easy task to describe the most recent advances in this field. We thus decided to present only those subjects or theories which are more-or-less established and to avoid those which are still controversial. Some of the hot topics in the field are intentionally omitted since they are described in other literature (e.g. review articles). We intend the present book to be an introduction which will enable the reader to understand literature on the more advanced topics.

The main body of Part I was written by Nishikawa, the part dealing with magnetohydrodynamics by Wakatani. In Part II, the sections dealing with magnetic confinement were written by Wakatani and those dealing with inertial confinement are by Nishikawa. The entire text was then integrated by Nishikawa.

We are grateful to the staff of Springer-Verlag, particularly Dr. H. Lotsch and Dr. A.M. Lahee, for suggesting to us that we write this book and patiently assisting us in the writing of the manuscript. We are highly obliged to Dr. N. Aristov, the copy-editor of Springer-Verlag, who carefully read the manuscript and corrected our English. Moreover, she gave us much advice on how to present the material. Our secretaries, Miss Yuko Utsumi of the Plasma Physics Laboratory, Kyoto University, and Miss Saeko Fujioka of the Faculty of Science, Hiroshima University, have generously spent their precious time in typing our manuscript. Without their devoted effort our book would not have been completed.

Hiroshima, Kyoto
February 1990

K. Nishikawa
M. Wakatani

Contents

Part II Applications to Fusion Plasmas

1. Introduction

The three states of matter, solid, liquid and gas, are well-known to us. As the temperature is elevated, solid is liquefied and liquid is evaporated to form a gaseous state. If we further increase the temperature, the molecules constituting the gas are decomposed into atoms and the atoms are then decomposed into electrons and positively charged ions. The degree of ionization increases as the temperature rises. For the case of hydrogen gas at normal pressure, the ionization becomes almost complete at about $(2 \sim 3) \times 10^4$ K. The ionized gas formed in this way is called *high-temperature plasma.* It consists of a large number of negatively charged light electrons and positively charged heavy ions, both electrons and ions moving with high speed corresponding to high temperature. The net negative charge of the electrons cancels the net positive charge of the ions in the plasma. This is called the *overall charge neutrality* of the plasma.

There are other types of plasmas, such as the electron gas in metals and the plasma inside stars, which are formed by the quantum effect due to high densities. In this book we shall not deal with such plasmas, but restrict ourselves to the fully ionized high temperature plasmas which will be simply referred to as *plasmas*; these are the plasmas which are of relevance to present mainstream research on controlled thermonuclear fusion.

Historically, plasma research was initiated by studies of gas discharges. In 1929, *Tonks* and *Langmuir* observed an electric oscillation in a rarefied gas discharge [1.1]. They referred to this oscillation as "plasma oscillation". Since then the word "plasma" has been used to represent a conducting gas. Because it has properties which are quite different from those of ordinary neutral gas, the plasma state is often called the fourth state of matter, distinct from the previously known three states of matter, solid, liquid, and gas.

Experimental research on plasmas had made little progress over more than a quarter of this century because of difficulties in controlling them. They often show unexpected behavior which is hardly reproducible due to the presence of complicated atomic processes and the governing boundary conditions. On the other hand, the plasma state has received considerable attention from theoreticians. A gaseous system consisting of a large number of charged particles cannot be treated by the standard theory of gases based on the expansion in powers of the density. The difficulties are due to the long range character of the Coulomb interactions which intrinsically requires many-body correlation effects. A number of theoretical efforts have been devoted to a formulation which properly takes into account the many-body effects. The works by *Vlasov*, *Landau*, *Mayer*, *Balescu*,

Lenard and *Guernsey* [1.2–7] will be specifically mentioned here. A theoretical model, treating plasma in a strong magnetic field as a continuous electromagnetic fluid (magnetohydrodynamics or MHD), has also been established by *Alfven* [1.8] and others [1.9, 10].

Progress in plasma research has revealed that plasma is an active medium which exhibits a wide variety of nonlinear phenomena. Response to an external disturbance in a plasma is typically represented by excitation of a wave or a collective motion of charged particles which is a result of the long range character of the Coulomb interaction. Since collisional dissipation effects are weak in high temperature plasmas, the excited waves can easily grow to a high level and exhibit a variety of nonlinear behavior. Thus plasma physics is a field of active research on nonlinear physics [1.11–15].

Progress in plasma physics emerged in the second half of this century, motivated by two very important applications: space exploration and thermonuclear fusion. In the 1940s, radio-astronomers discovered that more than 99% of the universe is in a plasma state. Since the first Sputnik spacecraft was launched in 1954 into an orbit circling the earth, a number of space probes have been used to collect data on the earth's magnetosphere. This information has disclosed that the space outside the earth is not just a void but is filled with an active dilute plasma of energetic charged particles. This plasma exhibits many interesting physical phenomena such as the interaction of the solar wind plasma with the earth's magnetic field, particle acceleration and trapping, wave excitation and propagation, etc. The solar system is now considered to be the primary laboratory in which a rich variety of plasma processes can be studied [1.16, 17]. Controlled thermonuclear fusion research was started in the early 1950s. In the beginning, it was conducted as classified research. Declassification took place in 1958 at the second Geneva Conference on the Peaceful Use of Atomic Energy [1.18]. Since then thermonuclear fusion research has been carried out with active international collaboration. Various ideas for controlling high temperature plasmas had been proposed by 1960, but the experiments showed unexpected difficulties in controlling plasmas until the results of the USSR T-3 tokamak were reported in 1968 [1.19]. They showed a dramatic improvement in plasma confinement and promising scaling properties. Since then a large number of tokamaks has been constructed all over the world and a substantial amount of experimental data have been accumulated. Concurrently, extensive theoretical and computational efforts have been devoted to plasma equilibrium, stability, transport, and heating with specific application to tokamak devices. Thus the recent progress in fusion-oriented plasma physics has largely depended on tokamak research. Another field of plasma physics that has also evolved is that of inertial confinement fusion which requires the compression of a plasma to very high densities, as high as those in the center of the sun.

The present book aims to be a comprehensive textbook for a graduate level plasma physics course relevant to thermonuclear fusion research. The prerequisites are a knowledge of mechanics, electromagnetism, fluid dynamics, and applied mathematics, which are normally taught at the undergraduate level for majors in physics and engineering. This work is also intended to be an introduction to fusion theory which should allow the reader to understand review articles on specific subjects of fusion research.

The book consists of two parts: Part I, *Basic Theory*, and Part II, *Applications to Fusion Plasmas*. Part I is organized as follows. After an introductory description of plasma properties and charged particle motion in Chaps. 2, 3, a concise presentation of the basic formulation of plasma theory is given in Chap. 4. In Chap. 5, the two fluid model is used to describe the fundamental collective responses of plasma. Chapter 6 deals with the kinetic theory and Chap. 7 with the general theory of linear response in plasmas. Examples of nonlinear response are described in Chap. 8 which contains somewhat advanced topics in nonlinear plasma theory. Part II starts with an introduction to controlled thermonuclear fusion research in Chap. 9. Then Chaps. 10, 11 are devoted to magnetohydrodynamics and its application to equilibrium and stability of fusion plasmas. Brief descriptions of the wave-plasma interaction and transport processes, with specific reference to current fusion plasma research, are given in Chaps. 12, 13. Finally in Chap. 14, recent progress in fusion plasma research is surveyed.

Although the plasmas treated in this book are restricted to the classical fully ionized neutral plasmas, most of the important properties described in this book, specifically the motion of charged particles, collective motion, and nonlinear phenomena, are also applicable to other types of plasmas such as low temperature weakly ionized plasmas, high-density quantum mechanical plasmas, and non-neutral plasmas.

Part I

Basic Theory

2. Basic Properties of Plasma

As mentioned in Chap. 1, plasma is a conducting gas and, as the fourth state of matter, has properties which are distinct from ordinary neutral gas or nonconducting fluid. In this chapter, we shall describe the basic properties of plasma from a statistical mechanical viewpoint. First, in Sects. 2.1, 2, we show the intrinsically *collisionless* and *continuous* nature of plasma. In Sect. 2.3 the two fundamental aspects of the plasma motion, the individual and the collective aspects, are shown as being characteristic of a collisionless continuum. Various plasmas are described in Sect. 2.4.

2.1 Plasma Condition

Plasma is formed under the condition that the average kinetic energy of an electron substantially exceeds the average Coulomb energy needed for an ion to bind the electron. The average kinetic energy of an electron is represented by the electron temperature T_e. Here, and throughout this book, we shall use temperature as an energy unit in which the Boltzmann constant is taken to be unity. The average Coulomb energy needed for an ion to bind the electron is represented by $e^2/4\pi\varepsilon_0 d$, where d is the average distance between the nearest electron and ion, e is the elementary charge and ε_0 is the permittivity of vacuum. For simplicity, we assume that the ions are singly ionized. Let n_0 be the average number of electrons or ions per unit volume, then we can estimate d by the relation $(4\pi d^3/3)n_0 = 1$. The condition that plasma is formed can then be written as

$$\left(\frac{4\pi}{3}n_0\right)^{1/3}\frac{e^2}{4\pi\varepsilon_0 T_e} \ll 1\ . \tag{2.1.1}$$

Taking the 3/2 power of the left-hand side and omitting the numerical factor $4\pi\sqrt{3}$, we have the relation

$$n_0\lambda_D^3 \gg 1\ , \tag{2.1.2}$$

where λ_D is defined by

$$\lambda_D = \sqrt{\varepsilon_0 T_e/n_0 e^2}\ . \tag{2.1.3}$$

This parameter has the dimension of length and is called the *Debye length*. As we shall see later, it is one of the most fundamental properties that characterize the plasma. The condition (2.1.1) or (2.1.2) is called the *plasma condition*. The inverse of the left hand side of (2.1.2),

$$g \equiv \frac{1}{n_0 \lambda_D^3}, \tag{2.1.4}$$

is often called the *plasma parameter*. Since $g \ll 1$, it is used as one of the basic expansion parameters in plasma theory.

We note here that although in the following the plasma condition (2.1.1) or (2.1.2) will be assumed to be true, it is neither a necessary nor a sufficient condition for a plasma to exist. The degree of ionization of an atom at thermal equilibrium is determined by the ionization equilibrium condition in statistical mechanics. If we can treat the neutral atoms as being independent of each other (i.e., as an ideal gas), then the ionization equilibrium formula is given by [2.1]

$$\frac{n_0^2}{N} = \frac{G_e G_i}{G} \left(\frac{2\pi m_e T}{h^2} \right)^{3/2} \exp\left(-\frac{w}{T}\right), \tag{2.1.5}$$

where N is the number of neutral atoms in unit volume, G_e, G_i and G are, respectively, the partition functions for the internal motion of an electron, an ion, and a neutral atom, m_e is the electron mass, h is the Planck constant and w is the ionization potential energy. This formula is called the *Saha ionization formula* [2.1]. For the case $G_e G_i / G = 1$, (2.1.5) gives

$$\frac{n_0}{N} = 2^{3/2} \frac{n_0 \lambda_D^3}{\sqrt{n_0 a_0^3}} \exp\left(-\frac{w}{T}\right), \tag{2.1.6}$$

where $a_0 (= h^2 \varepsilon_0 / m_e e^2 \pi)$ is the Bohr radius. It is clear from this formula that condition (2.1.1) or (2.1.2) is neither necessary nor sufficient for n_0/N to become sufficiently large. At low temperatures ($T \ll w$), a high degree of ionization is possible only at an extremely low density. At high temperatures, $n_0 \lambda_D^3$ is not necessarily large if $n_0 a_0^3 \ll 1$. It should be noted, however, that the condition $w < T$ is not necessary for high ionization. As was mentioned earlier, a good hydrogen plasma can be formed at $T = (2 \sim 3) \times 10^4$ K, although $w = 13.6$ eV which is about 1.5×10^5 K. This is due to the large factor $n_0^{-1} (2\pi m_e T / h^2)^{3/2}$ in front of the factor $\exp(-w/T)$. Physically, this large factor comes from the increase of the entropy of the electron due to the ionization.

2.2 Ideal Plasma

The starting point for the description of gaseous behavior in the theory of gases is the approximation of an ideal gas, that is, the limiting case of low particle density and high temperature. Similarly, in the theory of plasmas one can define a limiting case of an "ideal plasma". In contrast to an ideal gas, an ideal plasma is defined by the condition

$$n_0\lambda_D^3 \equiv N_D \to \infty \quad (\lambda_D \simeq 1) , \tag{2.2.1}$$

that is, the particle density goes to infinity. This can be understood when we examine the physical meaning of the Debye length λ_D. As is well known, the Coulomb potential is of *long range* in the sense that it extends over an infinite distance in vacuum or in an unpolarized medium. In a plasma, however, the Coulomb potential of any charged particle (to be referred to as the *test particle*) induces an electric polarization because it attracts (or repels) the charge of the opposite (or same) sign. The polarization charge tends to cancel the charge of the test particle and restricts the effect of its Coulomb potential to a finite range. As we shall show in Sect. 5.1, it is the Debye length that gives this range in the case of static test charge. The restriction of the effective range of the Coulomb potential by the polarization charge is called the *Debye shielding*. In an ideal plasma, the Debye shielding is caused by an infinite number (of order N_D) of charged particles, because every charged particle suffers only an infinitesimal (of order g) amount of effect of the test particle.

Two basic properties emerge from the limit in (2.2.1). First, it implies (although $n_0 \to \infty$!) that the plasma is *collisionless*. This comes from the condition that the particle kinetic energy be infinitely large in comparison to the Coulomb energy of the nearest particle, which is inversely proportional to one third of the particle density, as stated in Sect. 2.1. Thus, the motion of a particle in an ideal plasma is free from two-body Coulomb interactions or collisions. This collisionless aspect of an ideal plasma can be seen more quantitatively by estimating the *mean free path* l of a particle. When r_0 be the collision diameter, then it is found from the standard formula

$$l \sim \frac{1}{n_0 \pi r_0^2} . \tag{2.2.2}$$

The collision diameter for the electron can be estimated as the distance at which the electron orbit is strongly modified by a scatterer which may be either an electron or an ion. A strong orbit modification occurs when the electron kinetic energy $m_e v^2/2$ becomes comparable to the Coulomb energy $e^2/4\pi\varepsilon_0 r_0$. On average, $m_e v^2/2$ can be replaced by T_e, so that we have

$$r_0 \sim \frac{e^2}{4\pi\varepsilon_0 T_e} . \tag{2.2.3}$$

Substitution of (2.2.3) into (2.2.2) yields

$$l \sim \frac{16\pi\varepsilon_0^2 T_e^2}{n_0 e^4} = 16\pi N_D \lambda_D . \tag{2.2.4}$$

Actually (2.2.4) is not quite correct. As we shall show in Sect. 3.5, an important effect arises from an accumulation of many *small angle scattering processes* due to distant encounters. If cumulative small angle scattering is taken into account, the electron mean free path should be written, instead of as in (2.2.4), as

$$l_e \sim \frac{16\pi N_D \lambda_D}{\log \Lambda}\,, \tag{2.2.5}$$

where Λ is of order N_D. Essentially the same result can be obtained for the ion mean free path, as we can see from the mass independence of the formula. In any case, the mean free path becomes infinite in the limit of (2.2.1).

The second basic property of an ideal plasma also arises from the fact that the limit in (2.2.1) implies that the density is infinite. The ideal plasma can therefore be regarded as a *continuum*. In this sense, the ideal plasma has a similarity to a neutral fluid described by the usual fluid dynamics. The property of the ideal plasma is, however, fundamentally different from that of the neutral fluid in two aspects.

First, whereas the neutral fluid can be treated as a continuum in real physical space, the ideal plasma is to be treated as a continuum in the combined physical and velocity space (phase space). This is due to the difference between the ratios of the particle mean free path to the typical macroscopic scale length in a neutral fluid and an ideal plasma. In a neutral fluid, the mean free path is much shorter than the scale length of the macroscopic observation L, and hence the particle velocity suffers frequent changes by collisions when observed on the macroscopic scale (Fig. 2.1), except in the case of a Knudsen gas. The individual particle velocity is immaterial in this case, therefore the neutral fluid can be treated as a continuum in real space. On the other hand, since in the ideal plasma the mean free path is infinitely long, each particle keeps the memory of its initial velocity, so that the notion of the velocity space distribution becomes important. This is the characteristic feature of the *collisionless continuum*.

Second, since the plasma consists of charged particles, it shows a strong response to electromagnetic fields. The response often appears as an electric current and/or space charge and modifies the original electromagnetic fields. Thus the plasma fluid should be treated as an *electromagnetic fluid* which is described by a coupled system of fluid equations and Maxwell's equations for electromag-

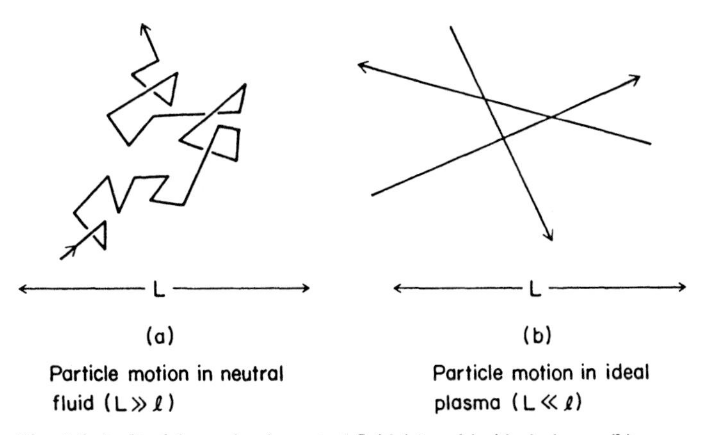

Fig. 2.1a,b. Particle motion in neutral fluid (a) and in ideal plasma (b)

netism. It is primarily for this reason that plasma is often called the fourth state of matter, distinct from an ordinary gas.

2.3 Individual Motion and Collective Motion

As mentioned in Sect. 2.2, the ideal plasma is characterized as a collisionless continuum. The properties of real plasmas are close to those of an ideal plasma and give rise to two characteristic types of particle motion.

The first one is a result of the collisionless nature of the ideal plasma. Each particle moves over the macroscopic scale length without being disturbed by collisions with other particles, keeping the initial memory of the motion. This type of motion is called the *individual motion* of the particles. It does not necessarily imply a motion at constant velocity. Since plasma is an electromagnetic fluid, there often exists a large-scale electromagnetic field which affects the individual particle motion. However, because the electromagnetic field is macroscopic, the particle orbit is determined by solving the equations of motion under a given initial condition. No stochastic nature characteristic of collisions comes into play in the individual motion of the particle.

The second type of motion arises from the continuous nature of the ideal plasma. The characteristic motion in a continuous medium is a *collective motion* which is a result of a coherent motion of a large number of particles.

At first glance, the individual and collective natures of the particle motion appear to be inconsistent with each other. This apparent paradox can be resolved by the following consideration. Although in a plasma each particle moves without being disturbed by collisions with other particles, at every encounter the particle orbit is modified by an infinitesimal amount of order g, which in turn causes a similar amount of orbit modification to the other particle in the encounter. There are approximately N_{D} electrons and ions in the unit volume and every one of them suffers such an *orbit modification*. Its direction is either attractive or repulsive depending on the sign of the charge. As a result, an electric polarization charge (charge separation) or current of order $gN_{\mathrm{D}} \sim 1$ is induced. This induced charge or current is observed as a collective motion of the plasma.

Thus, in addition to the shielding length of the Coulomb potential of a static test charge described above, the characteristic length of this collective motion of charged particles can also be parametrized by the Debye length.

As we can see from the above argument, the individual motion and the collective motion of plasma particles are not mutually independent, but are closely related to each other. The individual particle motion is determined by the macroscopic electromagnetic field which is produced or modified by the collective motion of the plasma, and the collective motion is produced by the sum of a large number of very small orbit modifications of the particle individual motion. In the remainder of this book we will explore the various physical phenomena which arise from these two characteristic types of motion and their interplay.

2.4 Types of Plasmas

If one wants to produce a good plasma in a laboratory, one usually needs to construct a fairly large apparatus. If one goes outside the earth, however, the plasma state is the most abundant state of matter. It is thought that more than 99.9% of matter in the universe is in plasma. Indeed, plasmic matter begins at about 50 km above the earth's surface in the *ionosphere*. There are various types of plasmas in the universe, ranging from very high density ($n \sim 10^{36}/\mathrm{m}^3$) inside a white dwarf to very low density ($n \sim 10^6/\mathrm{m}^3$) in interstellar space. Figure 2.2 shows various plasmas as a function of temperature and density. In this figure the various states are classified into four groups. The very high temperature state where the electrons must be treated relativistically is called *relativistic plasma*, whereas the very high density state where the electrons must be treated as a quantum–mechanical degenerate Fermi gas is called *degenerate plasma*. The remaining region is divided into *classical plasma* and *neutral gas* depending on whether the plasma condition (2.1.2) is satisfied or not. As can be seen from this figure, the neutral gas state is restricted to only a very narrow region. In this book

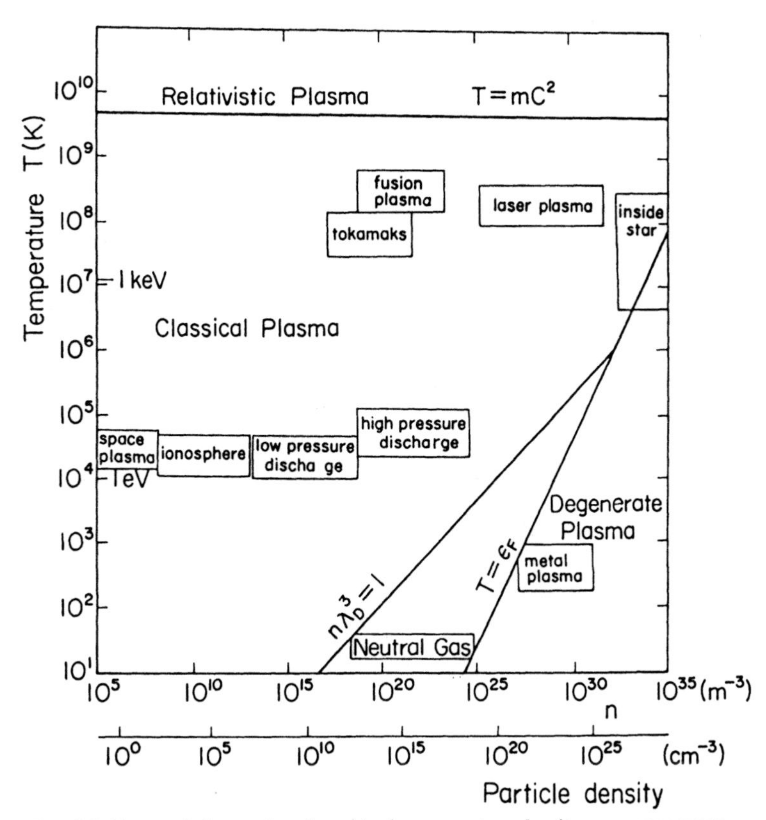

Fig. 2.2. Types of plasmas in a logarithmic temperature-density parameter space

we shall be interested in the region of the classical plasma, specifically, in those plasmas which have been studied in relation to thermonuclear fusion research. In this region, charged particle motion can be treated by nonrelativistic classical mechanics. However, as we shall show later, the plasmas relevant to inertial confinement fusion research are often at very high density and the electrons have to be treated quantum mechanically as a degenerate Fermi gas. Moreover, when a plasma is heated by an electromagnetic wave, for example, by electron cyclotron resonance heating, a fraction of the electrons are selectively heated to a very high energy so that they do have to be treated relativistically. The relativistic effect is certainly important for special problems, but to avoid unnecessary complications, we shall neglect it unless otherwise stated.

As mentioned above, the plasma is a collisionless medium and, hence, is not necessarily in thermal equilibrium. For instance, a typical discharge plasma in a vacuum tube has an electron temperature which is much higher ($T_e \sim 1\,\mathrm{eV}$) than the ion temperature ($T_i \sim 0.1\,\mathrm{eV}$) and the degree of ionization in such a plasma is typically 10^{-3}–10^{-5}, which is between the value predicted by the Saha formula of the electron temperature and that of the ion temperature. A relatively highly ionized plasma at low temperature ($\sim 0.3\,\mathrm{eV}$) and low density ($\sim 10^{16}/\mathrm{m}^3$) can be produced by contact ionization of an alkali beam, by using a device called the Q-machine. Such a low density, low temperature plasma is useful for studying the basic properties of wave propagation in a plasma. Various methods for producing laboratory plasmas are described in [2.2].

In a low ionization plasma, various atomic processes, such as ionization and recombination, excitation and radiation, play important roles in determining its properties. These processes are also important in fusion plasmas, particularly in connection with spectroscopic diagnostics of plasmas. These problems are described in [2.3, 4]. In this book we shall restrict ourselves to the problems in fully ionized plasmas and do not discuss these atomic processes.

Table 2.1. Density, temperature, Debye length, plasma parameter, mean free path, and typical size of various plasmas

$n(\mathrm{m}^{-3})$	Te(eV)	λ_D (m)	$n\lambda_D^3$	[1] L(m)	[2] l(m)	
10^6	1	7.4	4×10^8	10^7	1.0×10^{10}	Interstellar gas[3]
10^{12}	10^2	7.4×10^{-2}	4×10^8	10^6	1.0×10^8	Solar corona
10^{18}	10^2	7.4×10^{-5}	4×10^5	1	1.5×10^2	
10^{19}	10^3	7.4×10^{-5}	4×10^6	1	1.3×10^3	Hot plasma
10^{20}	10^4	7.4×10^{-5}	4×10^7	1	1.2×10^4	Thermonuclear plasma
10^{26}	10^4	7.4×10^{-8}	4×10^4	10^{-3}	3.4×10^{-2}	Laser plasma

1) system size

2) electron mean free path

3) system size for interstellar gas is chosen to be the radius of the typical orbit of a spaceprobe around the earth

In Table 2.1, we show the density, the temperature, the Debye length, the plasma parameter, the system size, and the mean free path of various classical plasmas. We can see from this table that the Debye length is a good measure for the observation of plasma properties while the mean free path is as large as or greater than the plasma size. Thus the real plasmas shown in this table can be treated as being close to the ideal plasma.

PROBLEMS

2.1. Calculate the left-hand side of (2.1.1) for the following cases:
a) $n_0 = 10^{20}/\mathrm{m}^3$, $T_e = 10\,\mathrm{keV}$ (typical magnetically confined fusion core plasma)
b) $n_0 = 10^{31}/\mathrm{m}^3$, $T_e = 1\,\mathrm{keV}$ (typical inertially confined fusion core plasma)
c) $n_0 = 10^6/\mathrm{m}^3$, $T_e = 10\,\mathrm{eV}$ (typical space plasma near the earth)
d) $n_0 = 10^{17}/\mathrm{m}^3$,$T_e = 1\,\mathrm{eV}$ (typical laboratory plasma)

2.2. Draw the $n_0 = N$ line as given by (2.1.6) with $w = 13.6\,\mathrm{eV}$ in Fig. 2.2.

2.3. The collision diameter of a typical neutral gas in the standard state is $5 \times 10^{-10}\,\mathrm{m}$. Estimate the mean free path.

2.4. If the gravitational force acting on a proton exceeds the Coulomb force acting on the proton by an electron, how far away from the proton must the electron be?

2.5. There is no shielding effect similar to Debye shielding in universal gravitation. Consider a half sphere of radius R composed of an equal number of electrons and protons distributed uniformly at a constant temperature T. If the universal gravitation acting on a proton at the center of the sphere equals the Coulomb force of an electron a Debye length away from the proton, then how large is the radius R?

3. Individual Particle Motion

In this chapter, we consider motion of a single particle in a given electromagnetic field. First in Sect. 3.1 we briefly describe the basic equations. Motion of a charged particle in a static magnetic field is described in some detail in Sects. 3.2, 3. Then effects of an oscillating electric field on the charged particle motion are discussed in Sect. 3.4. Finally collisional effects are described in Sect. 3.5.

3.1 General Relations

We use classical mechanics to consider the motion of a point particle of mass m and charge q in an electromagnetic field. As we can see from Problems 2.4, 5, the gravitational force is negligible in comparison to the electromagnetic force. The force acting on the particle is, therefore, the Lorentz force due to the electric field $\boldsymbol{E}$ and the magnetic field $\boldsymbol{B}$ (strictly speaking, it is the magnetic induction). The relativistic effect is neglected, so the particle mass is treated as constant. The equation of motion for the particle of velocity $\boldsymbol{v}$ is then given by

$$m\frac{d\boldsymbol{v}}{dt} = q(\boldsymbol{E} + \boldsymbol{v} \times \boldsymbol{B}) , \qquad (3.1.1)$$

where $\boldsymbol{v} \times \boldsymbol{B}$ is the cross product of these two vectors. Although the equation itself is simple, its solutions are quite diverse and complicated since the electromagnetic fields $\boldsymbol{E}$ and $\boldsymbol{B}$ in general depend on both space and time:

$$\boldsymbol{E} = \boldsymbol{E}(\boldsymbol{r}, t), \quad \boldsymbol{B} = \boldsymbol{B}(\boldsymbol{r}, t) . \qquad (3.1.2)$$

In (3.1.1) we put $\boldsymbol{r} = \boldsymbol{r}(t)$, the position of the particle which is given by the relation

$$\frac{d\boldsymbol{r}}{dt} = \boldsymbol{v}(t) . \qquad (3.1.3)$$

In the following, we consider only those simple cases in which the magnetic field is static, i.e., $\partial B/\partial t = 0$ or $\boldsymbol{B} = \boldsymbol{B}(\boldsymbol{r})$, and its spatial dependence is small. Due to the absence of the magnetic monopole, we have the scalar product

$$\nabla \cdot \boldsymbol{B} = 0 . \qquad (3.1.4)$$

3.2 Cyclotron Motion

We consider the case of no electric field, i.e., $\boldsymbol{E} = 0$. In this case, since the force acting on the particle is perpendicular to the particle velocity, the Lorentz force does no work on the particle. The particle kinetic energy is therefore conserved,

$$\tfrac{1}{2}mv^2 = \text{const}\,, \tag{3.2.1}$$

where $v = |\boldsymbol{v}|$.

If the magnetic field is uniform, i.e., $\nabla\boldsymbol{B} = \overleftrightarrow{0}$ (the simple product of two vectors $\boldsymbol{XY}$ stands for the dyadic tensor), then the acceleration parallel to the magnetic field vanishes, or

$$v_{\|} = \text{const}\,, \tag{3.2.2}$$

where $v_{\|} = \boldsymbol{b}\cdot\boldsymbol{v}$, $\boldsymbol{b}$ being the unit vector along $\boldsymbol{B}$; $\boldsymbol{b} = \boldsymbol{B}/B$. Along the magnetic field, the particle moves at constant speed. From (3.2.1,2), the kinetic energy of the particle motion perpendicular to $\boldsymbol{B}$ is also conserved, i.e.,

$$\tfrac{1}{2}mv_{\perp}^2 = \text{const}\,, \tag{3.2.3}$$

where $v_{\perp}$ is the modulus of the perpendicular velocity, $v_{\perp} = \sqrt{v^2 - v_{\|}^2}$. The particle therefore undergoes a circular motion at constant speed $v_{\perp}$ around the magnetic field. This motion is called the *cyclotron motion* or *gyration*. Its angular frequency, to be denoted by ω_c, can be readily obtained from (3.1.1) as

$$\omega_c = \frac{|q|B}{m} \tag{3.2.4}$$

and is called the *cyclotron* or the *Larmor frequency*. The radius of the cyclotron motion, to be denoted by ϱ_L, is given by

$$\varrho_L = \frac{v_{\perp}}{\omega_c} \tag{3.2.5}$$

and is referred to as the *cyclotron* or the *Larmor radius*. The direction of the cyclotron motion depends on the charge of the particle and is shown in Fig. 3.1.

The values of ω_c and ϱ_L for the electron and the proton are shown in Fig. 3.2, where the abscissa is the magnetic field strength B and the parameter shown is at constant perpendicular energy $mv_{\perp}^2/2$. For a given magnetic field, the cyclotron

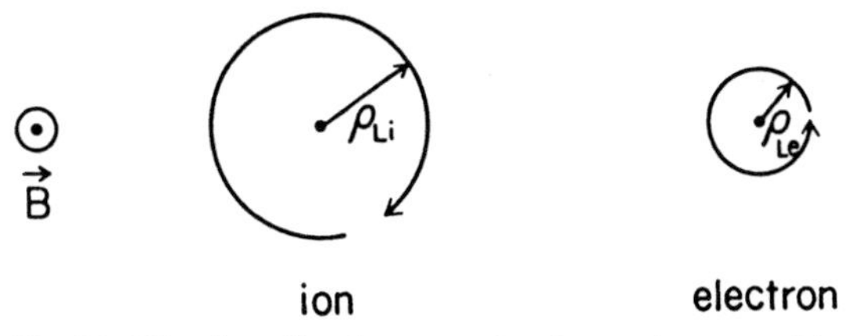

Fig. 3.1. Direction of cyclotron motion for an electron and ion

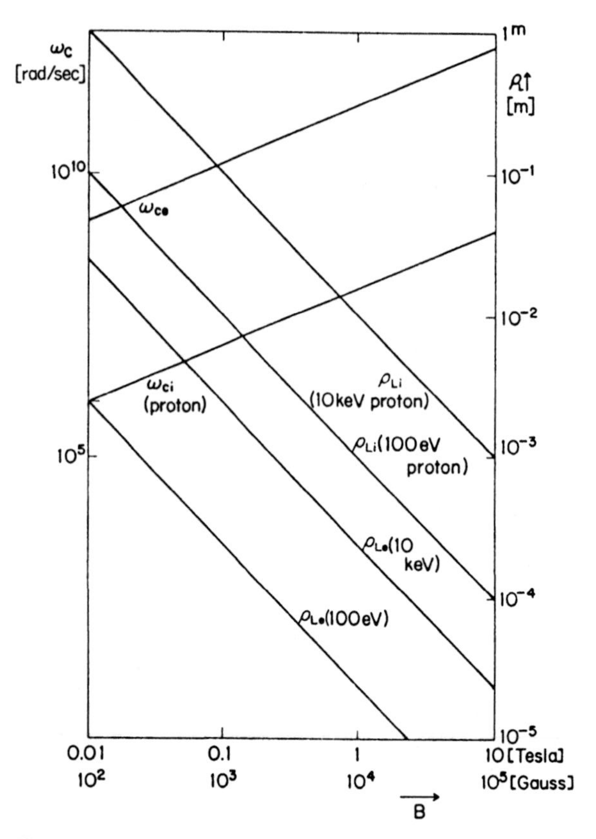

Fig. 3.2. Values of cyclotron frequency and Larmor radius versus magnetic field strength for various plasma temperatures

frequency is much greater for the electron than for the ion, while the Larmor radius, at a given perpendicular energy, is much smaller for the electron than for the ion.

Superposition of the parallel motion and the cyclotron motion yields a spiral motion as shown in Fig. 3.3. If we choose the z-axis to be along the magnetic field, the particle motion can be expressed as

$$\begin{aligned} v_z &= v_\parallel = \text{const}, \\ v_x &= v_\perp \cos(\Omega t + \theta_0), \\ v_y &= -v_\perp \sin(\Omega t + \theta_0), \end{aligned} \tag{3.2.6}$$

and

$$\begin{aligned} z &= z_0 + v_z t, \\ x &= x_0 + (v_\perp/\Omega)\sin(\Omega t + \theta_0), \\ y &= y_0 + (v_\perp/\Omega)\cos(\Omega t + \theta_0), \end{aligned} \tag{3.2.7}$$

where $\Omega = qB/m$, which is equal to $+\omega_c$ for the ion and $-\omega_c$ for the electron, θ_0 is

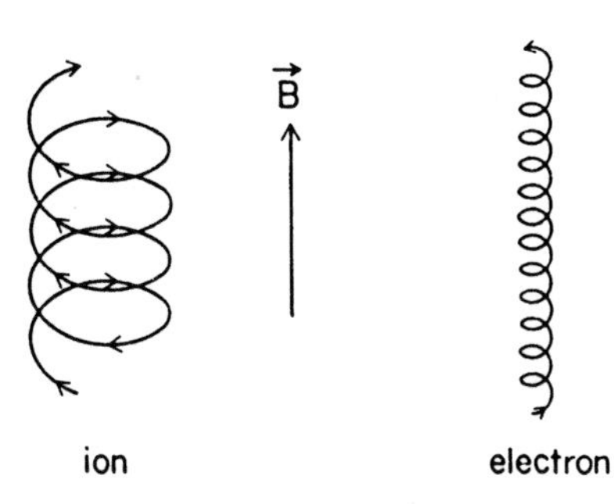

Fig. 3.3. Spiral motions of an electron and an ion in a magnetic field

the initial phase of the cyclotron motion, and x_0, y_0, z_0 are the initial coordinates of the particle.

3.2.1 Guiding Center

Since the cyclotron motion is usually rapidly oscillating, it is more convenient to consider the motion averaged over the cyclotron period $2\pi/\omega_c$. The motion obtained in this way is called the *guiding center motion* and the corresponding coordinates of the particle are called the *guiding center coordinates*. In the present case of a uniform magnetic field, the guiding center motion becomes a motion of constant speed along the magnetic field,

$$\bar{v}_z = v_{\parallel} = \text{const}, \quad \bar{v}_x = \bar{v}_y = 0 \tag{3.2.8}$$

and the guiding center coordinates are

$$\bar{x} = x_0, \quad \bar{y} = y_0, \quad \bar{z} = z_0 + v_{\parallel} t \ , \tag{3.2.9}$$

where the bar denotes the average over the cyclotron period. The frame of reference which moves with the guiding center is called the *guiding center frame*. In this frame, the particle motion is a mere cyclotron motion which we denote by $\tilde{\boldsymbol{v}}_{\perp}(t)$ and $\tilde{\boldsymbol{r}}_{\perp}(t)$, i.e.,

$$\begin{aligned} \tilde{\boldsymbol{v}}_{\perp}(t) &= [v_{\perp}\cos(\Omega t + \theta_0), \ -v_{\perp}\sin(\Omega t + \theta_0)] \\ \tilde{\boldsymbol{r}}_{\perp}(t) &= \left[\frac{v_{\perp}}{\Omega}\sin(\Omega t + \theta_0), \ +\frac{v_{\perp}}{\Omega}\cos(\Omega t + \theta_0)\right] \ . \end{aligned} \tag{3.2.10}$$

The motion in the rest frame is given by

$$\boldsymbol{v} = \bar{\boldsymbol{v}} + \tilde{\boldsymbol{v}}_{\perp}, \quad \boldsymbol{r} = \bar{\boldsymbol{r}} + \tilde{\boldsymbol{r}}_{\perp} \ . \tag{3.2.11}$$

3.2.2 Adiabatic Invariant

Consider the case in which the magnetic field slowly varies in space. The variation is assumed to be sufficiently slow that the magnetic field at the particle position hardly changes during the cyclotron motion. For such a quasi-periodic motion, there exists an adiabatic invariant given by

$$J_1 = \oint p_\theta dq_\theta = m \int_0^{2\pi} v_\theta r \, d\theta = 2\pi \varrho_L m \varrho_L \Omega = 2\pi m v_\perp^2 / \Omega \, , \tag{3.2.12}$$

where p_θ and q_θ are the canonical momentum and angle variable representing the cyclotron motion. Note that $\varrho_L \Omega$ is the gyration speed and $2\pi\varrho_L$ is the circumference of the cyclotron orbit.

3.2.3 Magnetic Moment

We note that $mv_\perp^2/\Omega$ is the angular momentum of the gyrating particle and that it is proportional to the *magnetic moment* μ associated with the circular current due to the cyclotron motion

$$\mu = \frac{q\Omega}{2\pi} \pi \varrho_L^2 = \frac{mv_\perp^2}{2B} \, . \tag{3.2.13}$$

Here $q\Omega/2\pi$ is the current due to the cyclotron motion and $\pi\varrho_L^2$ is the area surrounded by the circular current. The result implies that the magnetic moment is an adiabatic invariant for the case of slow variation of the magnetic field.

3.2.4 Magnetic Mirror

Consider the case that the magnetic field strength varies slowly along the field line. As the guiding center of the particle moves along the field line, the magnetic field strength at the particle position changes. Since the magnetic moment is conserved, $v_\perp^2$ varies proportionally to the magnetic field strength B. Therefore the kinetic energy of the parallel motion varies according to the relation

$$\frac{1}{2} m v_\parallel^2 + \mu B = K = \text{const} \, . \tag{3.2.14}$$

This relation implies that the guiding center motion along the field line behaves like a particle motion in a magnetic potential energy μB. When the particle comes to the point where μB equals K, the parallel velocity vanishes and the particle is reflected. Thus, a magnetic field configuration varying along the field line can act as a mirror for the particle guiding center motion. Such a configuration is called the *magnetic mirror*.

3.2.5 Magnetic Mirror Trap

A magnetic field configuration which forms a potential well for the guiding center motion can trap a charged particle in the well. Such a configuration is called the *magnetic mirror trap* or *magnetic mirror confinement system*, and is used for confinement of particles along the magnetic field line. The dipolar magnetic field of the earth is a well-known example of a magnetic mirror trap. It should be noted, however, that the magnetic mirror trap is effective only for those particles whose parallel energy $mv_\parallel^2/2$ is less than $\mu(B_{max} - B)$, where B_{max} is the maximum strength of the magnetic field.

3.2.6 Pitch Angle

The angle given by

$$\theta = \tan^{-1} \frac{v_\perp}{v_\parallel} \tag{3.2.15}$$

is called the *pitch angle*. The magnetic mirror configuration can trap only those particles whose pitch angles are greater than the critical angle

$$\theta_c = \tan^{-1} \sqrt{B/(B_{max} - B)} \; . \tag{3.2.16}$$

Particles having pitch angles smaller than θ_c escape from the magnetic mirror.

3.2.7 Longitudinal Invariant

A particle trapped in a magnetic mirror undergoes a periodic motion along the field line. The period of this oscillating motion may change for various reasons. For instance, if the mirror field is not axisymmetric, the reflection point changes with time due to the drift across the field line (curvature drift and gradient B drift, see Sect. 3.3). However, if the deviation from the axisymmetry is sufficiently small, the trapped particle motion becomes quasi-periodic and there appears an adiabatic invariant,

$$J_2 = m \oint v_\parallel ds \; , \tag{3.2.17}$$

where ds is the line element along the parallel motion and the integral is over one oscillation period. This quantity is called the *longitudinal (adiabatic) invariant* or the *second adiabatic invariant*.

3.3 Drift Across a Magnetic Field

When an effective force $\boldsymbol{F}$ is present in the equation of motion for the guiding center of a charged particle in a direction perpendicular to the static magnetic field $\boldsymbol{B}$, a guiding center motion is also induced in the direction perpendicular to both $\boldsymbol{B}$ and $\boldsymbol{F}$. Such a guiding center motion is called the *particle drift*.

3.3.1 Drift Velocity

We write the guiding center equation of motion corresponding to this situation as

$$m\frac{d\bar{\boldsymbol{v}}}{dt} = \boldsymbol{F} + q\bar{\boldsymbol{v}}_\perp \times \boldsymbol{B} \; . \tag{3.3.1}$$

At steady state the left-hand side vanishes. Taking the vector product of the resulting equation with $\boldsymbol{b} = \boldsymbol{B}/B$, we have

$$\boldsymbol{b} \times (\bar{\boldsymbol{v}} \times \boldsymbol{B}) = \boldsymbol{F} \times \boldsymbol{b}/q \,. \tag{3.3.2}$$

The vectorial identity

$$\boldsymbol{a} \times (\boldsymbol{b} \times \boldsymbol{c}) = (\boldsymbol{a} \cdot \boldsymbol{c})\boldsymbol{b} - (\boldsymbol{a} \cdot \boldsymbol{b})\boldsymbol{c} \tag{3.3.3}$$

gives the left-hand side of (3.3.2) as

$$\boldsymbol{b} \times (\bar{\boldsymbol{v}} \times \boldsymbol{B}) = B\bar{\boldsymbol{v}} - \bar{v}_{\|}\boldsymbol{B} = B(\bar{\boldsymbol{v}} - \bar{v}_{\|}\boldsymbol{b}) = B\bar{\boldsymbol{v}}_{\perp} \,.$$

Use of this relation in (3.2.2) yields

$$\bar{\boldsymbol{v}}_{\perp} = \frac{\boldsymbol{F} \times \boldsymbol{b}}{qB} = \frac{\boldsymbol{F} \times \boldsymbol{B}}{qB^2} \,. \tag{3.3.4}$$

This is the drift velocity due to the force $\boldsymbol{F}$ in the guiding center equation of motion.

The physical mechanism of this drift motion can be seen from the following considerations. Consider a geometry as shown in Fig. 3.4. The magnetic field is directed toward the reader out of the plane of the page (i.e., in the z-direction) and the force $\boldsymbol{F}$ is upward within the plane (i.e., in the y-direction). The particle cyclotron motion is in the plane (i.e., in the xy-plane). Due to the force $\boldsymbol{F}$ the particle is accelerated in the positive y-direction. As a result, the local gyrating speed of the particle changes such that it is greater at $y > \bar{y}$ than at $y < \bar{y}$, where $\bar{y}$ is the y-component of the guiding center coordinates. The local Larmor radius $\varrho_{\mathrm{L}}(y)$ also changes such that $\varrho_{\mathrm{L}}(y > \bar{y}) > \varrho_{\mathrm{L}}(y < \bar{y})$. Variation of the local gyration speed or the local Larmor radius results in a net drift of the guiding center along the x-axis as shown in Fig. 3.4.

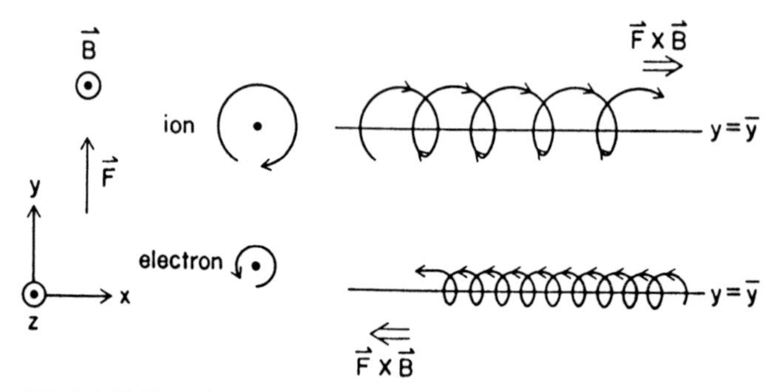

Fig. 3.4. Drift motions of an electron and an ion in the presence of an external force $\boldsymbol{F}$

3.3.2 $\boldsymbol{E} \times \boldsymbol{B}$ Drift

The most well-known example of the particle drift is the $\boldsymbol{E} \times \boldsymbol{B}$ *drift* which arises when there is a static electric field $\boldsymbol{E}$ with a component perpendicular to the magnetic field. In this case, $\boldsymbol{F} = q\boldsymbol{E}$, hence (3.3.4) gives the drift velocity as

$$\bar{\boldsymbol{v}}_{\perp} = \frac{\boldsymbol{E} \times \boldsymbol{B}}{B^2} \equiv \bar{\boldsymbol{v}}_E \,. \tag{3.3.5}$$

This result shows that the $\boldsymbol{E} \times \boldsymbol{B}$ drift velocity is the same for all particles independent of their charge, mass, and velocity. Thus the $\boldsymbol{E} \times \boldsymbol{B}$ drift constitutes a massive plasma flow across the magnetic field.

3.3.3 Curvature Drift

Consider a curved magnetic field line as in the case of the magnetic mirror configuraion. The particle guiding center motion that follows the field line is deflected along the curve and as a result the particle undergoes an inertial force or a centrifugal force perpendicular to the field line. This force can be derived from the inertial term for the particle motion along the field line

$$m\frac{d}{dt}(v_{\parallel}\boldsymbol{b}) = m\frac{dv_{\parallel}}{dt}\boldsymbol{b} + mv_{\parallel}\frac{d\boldsymbol{b}}{dt} . \tag{3.3.6}$$

The force across the magnetic field comes from the second term on the right-hand side. The first term is parallel to $\boldsymbol{B}$ and is responsible for the magnetic mirror effect.

Here the curvature of the field line is represented in terms of the $\boldsymbol{r}$-dependence of $\boldsymbol{b}, \boldsymbol{b} = \boldsymbol{b}(\boldsymbol{r})$. We substitute the particle coordinate $\boldsymbol{r}(t)$ into $\boldsymbol{r}$. Then

$$\frac{d\boldsymbol{b}}{dt} = \frac{d\boldsymbol{r}}{dt} \cdot \nabla\boldsymbol{b} = \boldsymbol{v} \cdot \nabla\boldsymbol{b}(\boldsymbol{r}) .$$

We now take the time average over the cyclotron period. Then the second term on the right-hand side of (3.3.6) gives

$$\overline{mv_{\parallel}\frac{d\boldsymbol{b}}{dt}} = \overline{mv_{\parallel}\boldsymbol{v} \cdot \nabla\boldsymbol{b}(\boldsymbol{r})} .$$

To lowest order we can use (3.2.6) or (3.2.7) which is obtained for the case of straight field line. Since $v_{\parallel}$ is constant while $\boldsymbol{v}_{\perp}$ is rapidly oscillating, only the parallel component of $\overline{v_{\parallel}\boldsymbol{v} \cdot \nabla}$ remains nonvanishing. We can therefore write

$$\overline{\boldsymbol{b} \cdot \boldsymbol{v}\frac{d\boldsymbol{b}}{dt}} \doteq v_{\parallel}^2\boldsymbol{b} \cdot \nabla\boldsymbol{b} \equiv -\frac{v_{\parallel}^2\boldsymbol{R}_c}{R_c^2} , \tag{3.3.7}$$

where R_c is the radius of local curvature of the magnetic field line and $\boldsymbol{R}_c/R_c$ is the unit vector in the direction of the centrifugal force.

Coming back to the original equation of motion, we can write

$$m\frac{d\boldsymbol{v}}{dt} = m\frac{d\boldsymbol{v}_{\perp}}{dt} + m\frac{d}{dt}(v_{\parallel}\boldsymbol{b}) = q\boldsymbol{v} \times \boldsymbol{B}$$

from which we obtain for the component perpendicular to $\boldsymbol{B}$

$$m\frac{d\boldsymbol{v}_{\perp}}{dt} = -mv_{\parallel}\frac{d\boldsymbol{b}}{dt} + q\boldsymbol{v} \times \boldsymbol{B} .$$

The time average of this equation yields the guiding center equation of motion as

$$m\frac{d\bar{\boldsymbol{v}}_{\perp}}{dt} = \frac{mv_{\parallel}^2\boldsymbol{R}_c}{R_c^2} + q\bar{\boldsymbol{v}}_{\perp} \times \boldsymbol{B} .$$

The first term on the right hand side is the effective force, so that the particle drift velocity $\bar{\boldsymbol{v}}_\perp$ is obtained from (3.3.4) as

$$\bar{\boldsymbol{v}}_\perp = \frac{mv_\parallel^2}{R_c^2}\frac{\boldsymbol{R}_c \times \boldsymbol{B}}{qB^2} \equiv \bar{\boldsymbol{v}}_c . \tag{3.3.8}$$

This drift is called the *curvature drift* or the *centrifugal drift*. Contrary to the case of $\boldsymbol{E}\times\boldsymbol{B}$ drift, the curvature drift depends both on charge and the particle velocity $\boldsymbol{v}$. In particular, the electrons and ions drift in opposite directions. Therefore the curvature drift yields a plasma current across the magnetic field.

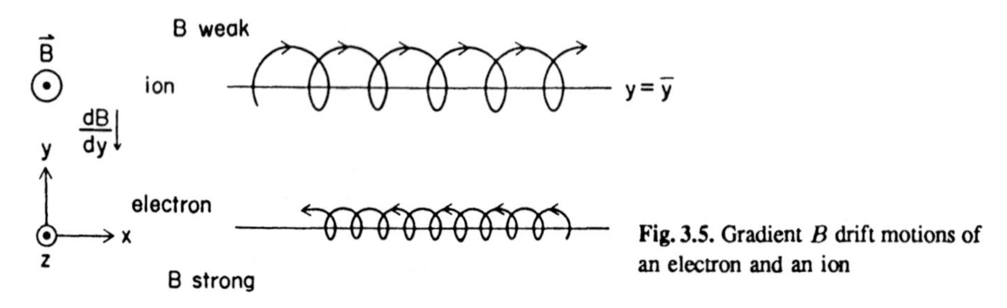

Fig. 3.5. Gradient B drift motions of an electron and an ion

3.3.4 Gradient B Drift

A similar drift occurs when the magnetic field is straight, but has a gradient ∇B across the field. This is due to the variation of the local Larmor radius, which is inversely proportional to B, during the cyclotron motion. In Fig. 3.5, the magnetic field is stronger in the region $y < \bar{y}$ than in the region $y > \bar{y}$, which causes the local Larmor radius $\varrho_L(y)$ to vary such that $\varrho_L(y > \bar{y}) > \varrho_L(y < \bar{y})$ and yields a drift along the x-axis. The effective force for this case is obtained from the equation of motion perpendicular to $\boldsymbol{B}$:

$$m\frac{d\boldsymbol{v}_\perp}{dt} = q[\boldsymbol{v}_\perp \times \boldsymbol{B}(\boldsymbol{r}(t))] . \tag{3.3.9}$$

We write $\boldsymbol{r}(t) = \bar{\boldsymbol{r}} + \tilde{\boldsymbol{r}}_\perp(t)$ following (3.2.11) and approximate $\boldsymbol{B}(\boldsymbol{r}(t))$ as

$$\boldsymbol{B}(\boldsymbol{r}(t)) \doteq \boldsymbol{B}(\bar{\boldsymbol{r}}) + \tilde{\boldsymbol{r}}_\perp(t) \cdot \nabla\boldsymbol{B}(\bar{\boldsymbol{r}}) .$$

Substituting this expression into (3.3.9) and taking the time average over the cyclotron period, we get for the right-hand side of (3.3.9)

$$q[\bar{\boldsymbol{v}} \times \boldsymbol{B}(\bar{\boldsymbol{r}})] + q\{\overline{\tilde{\boldsymbol{v}}_\perp(t) \times [\tilde{\boldsymbol{r}}_\perp(t)\cdot\nabla]\boldsymbol{B}(\bar{\boldsymbol{r}})}\} .$$

We calculate the second term using (3.2.10) as follows:

$$\begin{aligned} &q\{\overline{\tilde{\boldsymbol{v}}_\perp(t) \times [\tilde{\boldsymbol{r}}_\perp(t)\cdot\nabla]\boldsymbol{B}(\bar{\boldsymbol{r}})}\} \\ &= q\overline{\tilde{v}_y(t)\tilde{x}(t)}\frac{\partial}{\partial\bar{x}}B_z(\bar{\boldsymbol{r}})\hat{\boldsymbol{e}}_x - q\overline{\tilde{v}_x(t)\tilde{y}(t)}\frac{\partial}{\partial\bar{y}}B_z(\bar{\boldsymbol{r}})\hat{\boldsymbol{e}}_y \\ &= -\frac{q}{2}\frac{v_\perp^2}{\Omega}\nabla B_z(\bar{\boldsymbol{r}}) \doteq -\frac{mv_\perp^2}{2B}\nabla B(\bar{\boldsymbol{r}}) , \end{aligned} \tag{3.3.10}$$

where we used the trigonometric formulas

$$\int_0^T \cos^2(wt)dt = \int_0^T \sin^2(wt)dt = \frac{1}{2}, \quad \int_0^T \cos(wt)\sin(wt)dt = 0 , \qquad (3.3.11)$$

where $T = 2\pi/\omega$ and in the last expression of (3.3.10) we approximated $B_z(\bar{\boldsymbol{r}})$ by $B(\bar{\boldsymbol{r}})$. The drift velocity due to the force in (3.3.10) can be calculated from (3.3.4) as

$$\bar{\boldsymbol{v}}_\perp = \frac{mv_\perp^2}{2qB^3}\boldsymbol{B} \times \nabla B(\bar{\boldsymbol{r}}) \equiv \bar{\boldsymbol{v}}_{\mathrm{G}} \qquad (3.3.12)$$

which is called the *gradient B drift.*

3.4 Motion in an Oscillating Electric Field

Consider an electric field $\boldsymbol{E}(t)$ which oscillates at angular frequency ω_0:

$$\boldsymbol{E}(t) = \boldsymbol{E}_0 \cos\ \omega_0 t\ . \qquad (3.4.1)$$

3.4.1 Polarization Drift

We start with the case of no magnetic field and constant amplitude $\boldsymbol{E}_0$. In this case the solution of the equation of motion can be written in the form $\boldsymbol{v}(t) = \boldsymbol{v}_0 + \Delta\boldsymbol{v}(t), \boldsymbol{r}(t) = \boldsymbol{r}_0 + \boldsymbol{v}_0 t + \Delta\boldsymbol{r}(t)$, where $\boldsymbol{v}_0, \boldsymbol{r}_0$ are constant and $\Delta\boldsymbol{v}(t), \Delta\boldsymbol{r}(t)$ represent the motion induced by the oscillating electric field and are given by

$$\Delta\boldsymbol{v}(t) = \frac{q\boldsymbol{E}_0}{m\omega_0}\sin\omega_0 t, \quad \Delta\boldsymbol{r}(t) = -\frac{q}{m\omega_0^2}\boldsymbol{E}(t)\ . \qquad (3.4.2)$$

Both $\Delta\boldsymbol{v}(t)$ and $\Delta\boldsymbol{r}(t)$ depend on charge q, such that the electrons and ions oscillate in opposite directions, inducing an oscillating charge and current. This motion is called the *polarization drift.*

3.4.2 Magnetic Field Effect

We next consider the case in which a static and uniform magnetic field $\boldsymbol{B}$ is present. We choose it to be directed in the z-direction and the direction of $\boldsymbol{E}_0$ lies in the xz-plane. Since the motion in the z-direction is unaffected by the magnetic field, we consider the motion in the xy-plane. It is described by the equations of motion

$$\frac{d}{dt}v_x - \Omega v_y = \frac{q}{m}E_{0x}\cos\omega_0 t\ , \qquad \frac{d}{dt}v_y + \Omega v_x = 0\ . \qquad (3.4.3)$$

The solution of these equations can be written in the form $\boldsymbol{v}_\perp(t) = \boldsymbol{v}_{\perp 0}(t) + \Delta\boldsymbol{v}_\perp(t)$

where $\boldsymbol{v}_{\perp 0}(t)$ is the solution in the absence of the electric field $\boldsymbol{E}(t)$ and $\Delta \boldsymbol{v}_\perp(t)$ is the part induced by $\boldsymbol{E}(t)$. We can write $\Delta \boldsymbol{v}_\perp(t)$ as

$$\Delta v_x(t) = \Delta \bar{v}_x \sin \omega_0 t \ , \quad \Delta v_y(t) = \Delta \bar{v}_y \cos \omega_0 t \ . \tag{3.4.4}$$

Substitution of (3.4.4) into (3.4.3) yields for the case $\omega_0^2 \neq \Omega^2$

$$\Delta \bar{v}_x = \frac{\omega_0}{\omega_0^2 - \Omega^2} \frac{q}{m} E_{0x} \ , \quad \Delta \bar{v}_y = \frac{\Omega}{\omega_0^2 - \Omega^2} \frac{q}{m} E_{0x} \ . \tag{3.4.5}$$

In the special case of $\omega_0 \to 0, \Delta \boldsymbol{v}_\perp \to \bar{\boldsymbol{v}}_E$ as given by (3.3.5), therefore, the component of $\Delta \boldsymbol{v}_\perp$ perpendicular to $\boldsymbol{E}$ [i.e., $\Delta v_y(t)$] is called the $\boldsymbol{E} \times \boldsymbol{B}$ drift and is denoted by $\Delta \boldsymbol{v}_E$. Its general expression is given by

$$\Delta \boldsymbol{v}_E(t) = \frac{\Omega^2}{\Omega^2 - \omega_0^2} \frac{\boldsymbol{E}(t) \times \boldsymbol{B}}{B^2} \ . \tag{3.4.6}$$

The component parallel to $\boldsymbol{E}$ (i.e., Δv_x) is reduced to (3.4.2) in the limit $\Omega \to 0$, namely the *polarization drift* and its velocity is denoted by $\Delta \boldsymbol{v}_\mathrm{p}(t)$ which is given by

$$\Delta \boldsymbol{v}_\mathrm{p}(t) = \frac{q}{m} \frac{1}{\Omega^2 - \omega_0^2} \frac{\boldsymbol{B} \times [(d\boldsymbol{E}/dt) \times \boldsymbol{B}]}{B^2} \ . \tag{3.4.7}$$

3.4.3 Cyclotron Resonance

An important feature of (3.4.5–7) is that when ω_0^2 approaches Ω^2 they become infinitely large. This is called the *cyclotron resonance*. Under this condition, the particle gyrating motion is in phase with the electric field oscillation, namely, the gyrating particle is continuously accelerated by the oscillating electric field. Stated in other words, the oscillating electric field is observed as a dc field in the frame of reference moving with the gyrating motion of the particle. Of course, in the limit of $\omega_0^2 \to \Omega^2$, the above solution is no longer valid. In this case we have to take into account a collisional drag force on the particle motion. Since the cyclotron resonance is very effective in accelerating charged particles, it is widely used for the production and heating of plasmas.

3.4.4 Ponderomotive Force

We now consider the case of a spatially varying oscillating electric field

$$\boldsymbol{E} = \boldsymbol{E}_0(\boldsymbol{r}) \cos \omega_0 t \ , \tag{3.4.1$'$}$$

where $\boldsymbol{E}_0(\boldsymbol{r})$ now depends on $\boldsymbol{r}$. The dependence is assumed to be sufficiently slow that in the lowest order we can neglect it. In this case an effective force, called the *ponderomotive force*, arises along the gradient of $\boldsymbol{E}_0(\boldsymbol{r})$.

We first describe a physical picture of the ponderomotive force by considering the case of no magnetic field. We assume that the electric field is in the x-direction and consider the electron displacement $\Delta x(t)$ which oscillates with time in phase

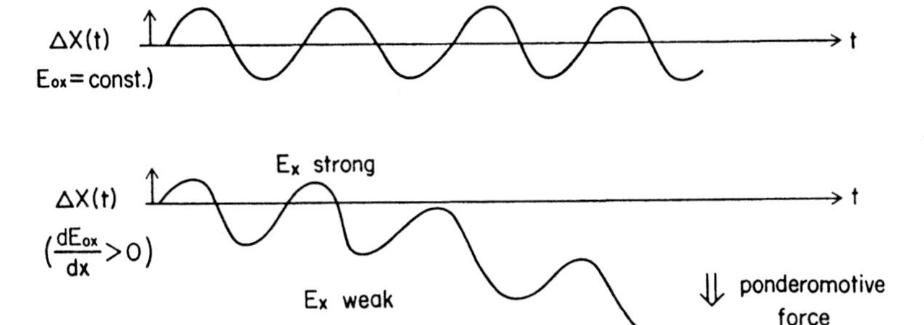

Fig. 3.6. Particle motion in the presence of a ponderomotive force

with the electric field as shown in (3.4.2) (Fig. 3.6). The electron is accelerated to the negative x-direction when $E_x > 0$, and to the positive x-direction when $E_x < 0$. Suppose that the amplitude E_{0x} increases with x, then the force acting on the electron changes its strength depending on the displacement Δx. The force is stronger when $\Delta x > 0$ than when $\Delta x < 0$. As a result a net acceleration occurs to the negative x-direction as seen from Fig. 3.6. To derive the expression for the ponderomotive force we denote the particle coordinate as $\boldsymbol{r}(t) = \bar{\boldsymbol{r}} + \Delta\boldsymbol{r}(t)$ where $\bar{\boldsymbol{r}}$ is now the time average over the oscillation period $2\pi/\omega_0$ and $\Delta\boldsymbol{r}(t)$ is the oscillating part which in the lowest order is given by (3.4.2). We then approximate the field amplitude $\boldsymbol{E}_0(\boldsymbol{r})$ by

$$\boldsymbol{E}_0(\boldsymbol{r}) = \boldsymbol{E}_0(\bar{\boldsymbol{r}}) + \Delta\boldsymbol{r}(t) \cdot \nabla \boldsymbol{E}_0(\bar{\boldsymbol{r}}) . \tag{3.4.8}$$

The equation of motion can then be written as follows

$$m\frac{d\boldsymbol{v}}{dt} = q\boldsymbol{E}_0(\boldsymbol{r})\cos\omega_0 t + \Delta\boldsymbol{v} \times \boldsymbol{B}(\boldsymbol{r}, t) .$$

Substituting (3.4.2) into the above equation and averaging over the time period $2\pi/\omega_0$, we get

$$m\overline{\frac{d\boldsymbol{v}}{dt}} = -\frac{m}{2}\nabla\overline{|\Delta\boldsymbol{v}|^2} = -\frac{q^2}{4m\omega_0^2}\nabla|\boldsymbol{E}_0(\bar{\boldsymbol{r}})|^2 , \tag{3.4.9}$$

where we used the trigonometric formulas in (3.3.11). We see that the ponderomotive force as given by (3.4.9) always acts to expel the particles from high field region and that it is stronger at lower frequencies and for lighter particles. We also emphasize that the ponderomotive force is a *nonlinear force* since it is proportional to the square of the field amplitude.

3.4.5 Magnetic Field Effect

The magnetic field effect on the ponderomotive force is important when the oscillating electric field $\boldsymbol{E}(t)$ has a component perpendicular to the magnetic field $\boldsymbol{B}$. In order to illustrate this effect, we consider the case when $\boldsymbol{B}$ is in the

z-direction and $\boldsymbol{E}(t)$ is in the x-direction with the amplitude E_{0x} varying slowly along the x. The equations of motion in the xy–plane are given by (3.4.3) with $E_{0x} = E_{0x}(x)$. As before we denote $x(t) = \bar{x} + \Delta x(t)$ where $\bar{x}$ is the average of $x(t)$ over period $2\pi/\omega_0$ and $\Delta x(t)$ is the oscillating part, and approximate E_{0x} by

$$E_{0x}(x) \doteq E_{0x}(\bar{x}) + \Delta x(t)\frac{\partial}{\partial \bar{x}}E_{0x}(\bar{x}) \, . \tag{3.4.10}$$

From (3.4.4) $\Delta x(t)$ is given by

$$\Delta x(t) = \int_0^t \Delta v_x(t')dt' = -\frac{\Delta \bar{v}_x}{\omega_0}(\cos\omega_0 t - 1) \, , \tag{3.4.11}$$

$\Delta \bar{v}_x$ being given by (3.4.5) with $E_{0x} = E_{0x}(\bar{x})$. Substituting (3.4.10) into the right-hand side of (3.4.3) and using (3.4.11) we obtain the average force in the x-direction as

$$m\overline{\frac{dv_x}{dt}} = \overline{q\Delta x(t)\frac{\partial}{\partial \bar{x}}E_{0x}(\bar{x})\cos\omega_0 t} = -\frac{q}{2\omega_0}\Delta \bar{v}_x\frac{\partial}{\partial \bar{x}}E_{0x}(\bar{x}) \, ,$$

where the trigonometric formula given in (3.3.11) was used. Use of (3.4.5) for Δv_x yields

$$m\overline{\frac{dv_x}{dt}} = -\frac{q^2}{2m}\frac{1}{\omega_0^2 - \Omega^2}E_{0x}\frac{\partial}{\partial \bar{x}}E_{0x} = -\frac{q^2}{4m}\frac{1}{\omega_0^2 - \Omega^2}\frac{\partial E_{0x}^2}{\partial \bar{x}} \, . \tag{3.4.12}$$

This is the ponderomotive force. The only difference from (3.4.9) is that ω_0^{-2} is now replaced by $(\omega_0^2 - \Omega^2)^{-1}$. As a result, the direction of the ponderomotive force changes depending on whether $\omega_0^2 > \Omega^2$ or $\omega_0^2 < \Omega^2$. Since Ω^2 is proportional to m^{-2}, the oscillating electric field of frequency ω_0 either attracts the particles (when $\omega_0^2 < \Omega^2$) or repels them (when $\omega_0^2 > \Omega^2$) depending on the mass of the particles. Moreover, the ponderomotive force is resonantly enhanced when ω_0^2 approaches Ω^2 due to the cyclotron resonance. These features can be used for selective acceleration of particles that resonate with the oscillating field, in a chosen direction.

3.5 Coulomb Collisions

Although collisions are rare in a high temperature plasma, they are of vital importance in relaxation phenomena, such as approach to equilibrium, transport processes (diffusion, heat transport, and electrical conduction) and plasma heating, etc.

3.5.1 Differential Cross Section

The standard formula for the integral cross sectional area is $\sigma = \pi r_0^2$, where r_0 is an effective collision radius and after the collision the particles are scattered into all angles. In our system we can estimate the length of this radius by noting that the Coulombic energy existing between two particles is equal to their relative kinetic energy:

$$\frac{q_1 q_2}{4\pi\varepsilon_0 r_0} \simeq \frac{1}{2} m^* v^2 \,. \qquad (3.5.1)$$

Here m^* is the reduced mass of the colliding particles $1/m^* = \sum_j (1/m_j)$ is their charge, with j being a particle index, and v is their relative velocity. This yields for the cross section of scattering by a Coulomb potential, also known as *Rutherford scattering*

$$\sigma(v) \sim \pi \left(\frac{q_1 q_2}{2\pi\varepsilon_0 m^* v^2} \right)^2 \,. \qquad (3.5.2)$$

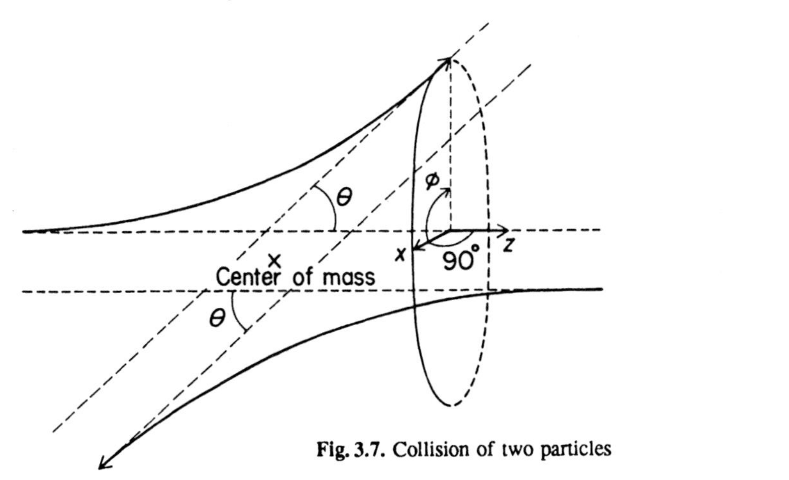

Fig. 3.7. Collision of two particles

The differential cross section, where only the particles scattered into a given deflection angle θ and azimuthal angle ϕ (Fig. 3.7) are counted is given by

$$\sigma(v, \theta, \phi) = \left(\frac{q_1 q_2}{8\pi\varepsilon_0 m^* v^2} \right)^2 \frac{1}{\sin^4 \theta/2} \,. \qquad (3.5.3)$$

The differential cross section can be defined as follows. Consider a particle of species 1 (incident particle) moving in the medium consisting of particles of species 2 (to be referred to as the field particles). Let $\nu_{12}(v, \theta, \phi) dv \sin\theta \, d\theta \, d\phi$ be the average number of those scattering events per unit time in which particle 1 is scattered by a particle of species 2 moving with speed in the range $v \sim v + dv$ relative to the incident particle through an angle in the range $\theta \sim \theta + d\theta$, $\phi \sim \phi + d\phi$ as measured in the frame where the center of mass of the two colliding particles is at rest. Then

$$\nu_{12}(v,\theta,\phi) = f_2(v) v \sigma(v,\theta,\phi) \tag{3.5.4}$$

where $f_2(v)dv$ is the number density of the field particles with relative speed in the range $v \sim v + dv$. Since $\sigma(v,\theta,\phi)$ is independent of ϕ, so is $\nu_{12}(v,\theta,\phi)$, thus it shall hereafter be neglected.

The differential cross section as given by (3.5.3) has two important properties. First, the cross section decreases rapidly with increasing relative speed v. Second, the cross section becomes markedly large as the scattering angle decreases. Both properties are the results of the long range character of the Coulomb interaction.

3.5.2 Cumulative Small-Angle Scattering

As mentioned in the preceding paragraph, small angle collisions have a significant effect on $\sigma(v,\theta)$. In particular, the dominant contribution to a large-angle deflection of a particle orbit arises from a cumulative effect of many small angle scattering processes rather than a single large angle scattering process.

We examine this effect by considering the case when the incident particle 1 is an electron and particle 2 (a field particle) is an ion. We assume that the particles are moving with a speed on the order of their thermal speed $v_T = (T/m)^{1/2}$ and that the electron and ion temperatures are about the same. Then the ion speed is much smaller than the electron speed, $v_i \ll v_e$, so that the relative speed v can be approximated by v_e. On the other hand, the ion momentum is much greater than the electron momentum, $m_i v_i \gg m_e v_e$, so that the center of mass, which is at rest, is practically on the ion, $v_i = 0$. We can therefore regard the electron as moving in a medium consisting of ions at rest. Since the angular direction of the scattering is expected to be random, the process can be regarded as an angular diffusion. The scattering angles θ, ϕ can be represented by the velocity components after collisions v_x and v_y, where $v_x = v \sin\theta \cos\phi$ and $v_y = v \sin\theta \sin\phi$. Obviously, the average rate of change of v_x and $v_y \langle \Delta v_x / \Delta t \rangle$ and $\langle \Delta v_y / \Delta t \rangle$ vanishes. The angular bracket denotes the average taken with respect to the collision rate $\nu_{ei}(v,\theta,\phi)$. We therefore calculate $\langle \Delta v_\perp^2 / \Delta t \rangle$, where $v_\perp^2 = v_x^2 + v_y^2$, which is given by

$$\left\langle \frac{\Delta v_\perp^2}{\Delta t} \right\rangle \equiv \int_0^{2\pi} d\phi \int_0^{\pi} d\theta \sin\theta (v_x^2 + v_y^2) \nu_{ei}(v,\theta,\phi)$$

$$= 2\pi v^2 \int_0^{\pi} d\theta \sin^3\theta \nu_{ei}(v,\theta) \, . \tag{3.5.5}$$

Then the quantity defined by

$$\frac{1}{v_\perp^2} \left\langle \frac{\Delta v_\perp^2}{\Delta t} \right\rangle = 2\pi \frac{v^2}{v_\perp^2} \int_0^{\pi} d\theta \sin^3\theta \nu_{ei}(v,\theta) \equiv \bar{\nu}_{ei}(v, \Delta\theta) \tag{3.5.6}$$

gives the average frequency for the particle orbit being deflected by an angle of order $\Delta\theta = \sin^{-1}(v_\perp/v)$. For $\Delta\theta = \frac{\pi}{2}$, it is the effective collision frequency for the 90 degree deflection of the electron orbit by static ions. If we substitute (3.5.3, 4) into (3.5.6), we find the integral diverges at $\theta \to 0$. This itself implies the importance of the small angle scattering effects. The divergence can be avoided if we note that the small angle scattering corresponds to an encounter of two particles a large distance apart. Because of the Debye screening, however, the Coulomb potential cannot extend to such a large distance in a plasma. Therefore there must be a natural lower limit for the value of θ. The integral in (3.5.6) can then be carried out by introducing a lower limit, say at $\theta = \theta_{\min}$, which is related to the Debye shielding. Substituting (3.5.3) into (3.5.4) and then into (3.5.5) and noting that in the present case of electron scattering by ions, $m^* = m_e$ and $f_i(v) = n_i$ which is the ion number density (all the ions can be treated as being at rest), we obtain

$$\bar{\nu}_{ei}(v, \Delta\theta) = 32\pi \left(\frac{Ze^2}{8\pi\varepsilon_0 m_e v^2}\right)^2 \frac{n_i v}{\sin^2 \Delta\theta} \log\left[\frac{1}{\sin(\theta_{\min}/2)}\right] \tag{3.5.7}$$

where $Z = q_i/e$ and we have dropped the $\frac{1}{2}\sin^2(\theta_{\min}/2)$ term since it is small compared to the logarithmic term.

For electrons moving with thermal speed $v_{T_e} = \sqrt{T_e/m_e}$, the collision frequency for the 90 degree deflection becomes

$$\bar{\nu}_{ei}\left(v, \frac{\pi}{2}\right) \equiv \bar{\nu}_{ei\perp} = \frac{1}{2\pi}\frac{Z^2 e^4}{\varepsilon_0^2}\frac{n_i}{\sqrt{m_e} T_e^{3/2}} \log\left[\frac{1}{\sin(\theta_{\min}/2)}\right] . \tag{3.5.8}$$

In addition to the angular deflection, the incident electron is subject to a frictional force. Indeed, accompanied by the angular scattering of angle θ, the electron's velocity component parallel to the incident velocity decreases by an amount $\Delta v_\parallel \equiv v[1 - \cos\theta] = 2v\sin^2\theta/2$. The average rate of change of $\Delta v_\parallel$ can then be calculated in the same way as (3.5.7) as

$$\begin{aligned}
\left\langle\frac{\Delta v_\parallel}{\Delta t}\right\rangle &= 4\pi v \int_{\theta_{\min}}^{\pi} d\theta \sin\theta \sin^2\frac{\theta}{2} \nu_{ei}(v, \theta) \\
&= 16\pi v^2 \left(\frac{Ze^2}{8\pi\varepsilon_0 m_e v^2}\right)^2 n_i \log\left[\frac{1}{\sin(\theta_{\min}/2)}\right] \\
&= \frac{Z^2 e^4 n_i}{4\pi\varepsilon_0^2 m_e^2 v^2} \log\left[\frac{1}{\sin(\theta_{\min}/2)}\right] .
\end{aligned} \tag{3.5.9}$$

Substituting the thermal speed into v, we obtain the collision frequency for the deceleration process $\bar{\nu}_{ei\parallel}$ as

$$\bar{\nu}_{ei\parallel} = \left[\frac{1}{v}\left\langle\frac{\Delta v_\parallel}{\Delta t}\right\rangle\right]_{v=v_e} = \frac{Z^2 e^4 n_i}{4\pi\varepsilon_0^2 \sqrt{m_e} T_e^{3/2}} \log\left[\frac{1}{\sin(\theta_{\min}/2)}\right] . \tag{3.5.10}$$

This frequency determines the electrical resistivity η through the relation

$$\eta = \frac{c m_e}{n_e e^2} \bar{\nu}_{ei\parallel} . \tag{3.5.11}$$

The appropriate numerical factor c was calculated by *Spitzer* and the resulting resistivity is called the *Spitzer resistivity* [3.1].

Equation (3.5.7) becomes infinitely large for $v \to 0$ and $\Delta\theta \to 0$. For sufficiently small v the above formula is invalid since the ions can no longer be treated as static. Naturally, $\Delta\theta$ has to be larger than θ_{min}. We also note that in the present approximation of static ions (i.e., the infinite ion mass approximation), there is no energy exchange between the electron and the ion: the scattering is purely elastic.

3.5.3 Coulomb Logarithm

The collision frequencies given in (3.5.7–10) depend on θ_{min}, which remains to be determined. Since it appears only through the logarithmic term, its uncertainty has little effect on the collision frequency, so that we do not need a precise expression. It can be evaluated by the relation

$$\theta_{\text{min}} \sim \frac{\Delta v_\perp}{v}$$

where $\Delta v_\perp$ is the perpendicular velocity component acquired by the scattering at a distance of λ_{D}, the Debye length. The velocity component $\Delta v_\perp$ is produced by the Coulomb force of order $e^2/4\pi\varepsilon_0\lambda_{\text{D}}^2$ which acts during the period of order λ_{D}/v,

$$\Delta v_\perp \sim \frac{e^2}{4\pi\varepsilon_0\lambda_{\text{D}}^2 m_{\text{e}}}\frac{\lambda_{\text{D}}}{v} = \frac{e^2}{4\pi\varepsilon_0\lambda_{\text{D}} m_{\text{e}} v} .$$

Therefore we find

$$\theta_{\text{min}} \sim \frac{e^2}{4\pi\varepsilon_0\lambda_{\text{D}} m_{\text{e}} v^2} \sim g \sim \frac{1}{N_{\text{D}}} ,$$

where in the last expression we put $v = v_{\text{e}}$. The logarithmic factor then becomes

$$\log\left[\frac{1}{\sin(\theta_{\text{min}}/2)}\right] = \log\Lambda, \quad \Lambda \sim 2n\lambda_{\text{D}}^3 . \tag{3.5.12}$$

This logarithmic factor is called the *Coulomb logarithm* and represents the relative importance of the cumulative small-angle scattering effects. It is typically on the order of 10–20.

3.5.4 Collision Frequencies and Relaxation Times

For other types of collisions, we have to take into account the difference in velocity between the center of mass frame and the laboratory or rest frame. In addition, one has to take into account the velocity distribution of the field particles. Let $\boldsymbol{v}_1$ and $\boldsymbol{v}_2$ be the incident velocities of the colliding particles as measured in the rest frame. Then the change of the relative velocity due to the collision $\Delta\boldsymbol{v} = \Delta(\boldsymbol{v}_1 - \boldsymbol{v}_2)$ is related to that of the change in velocity of each particle $\Delta\boldsymbol{v}_j$ by the relation

$$\Delta \boldsymbol{v} = \frac{m_1}{m^*}\Delta \boldsymbol{v}_1 = -\frac{m_2}{m^*}\Delta \boldsymbol{v}_2 \; ; \tag{3.5.13}$$

as seen from the momentum conservation, $m_1 \Delta \boldsymbol{v}_1 + m_2 \Delta \boldsymbol{v}_2 = \mathbf{o}$. We calculate $\langle |\Delta \boldsymbol{v}_\perp|^2/\Delta t\rangle$ and $\langle \Delta \boldsymbol{v}_\parallel/\Delta t\rangle$ in the same way as in Sect. 3.5.2, where $\Delta \boldsymbol{v}_\perp$ and $\Delta \boldsymbol{v}_\parallel$ are, respectively, the components of $\Delta \boldsymbol{v}$ perpendicular and parallel to $\boldsymbol{v}$. Noting that $v = |\boldsymbol{v}| = |\boldsymbol{v}_1 - \boldsymbol{v}_2|$, we have

$$\left\langle \frac{|\Delta \boldsymbol{v}_\perp|^2}{\Delta t} \right\rangle = 32\pi \left(\frac{q_1 q_2}{8\pi\varepsilon_0 m^*}\right)^2 \log \Lambda \int d^3\boldsymbol{v}_2 \frac{f_2(\boldsymbol{v}_2)}{|\boldsymbol{v}_1 - \boldsymbol{v}_2|} \tag{3.5.14}$$

$$\begin{aligned}\left\langle \frac{\Delta \boldsymbol{v}_\parallel}{\Delta t} \right\rangle &= 16\pi \left(\frac{q_1 q_2}{8\pi\varepsilon_0 m^*}\right)^2 \log \Lambda \int d^3\boldsymbol{v}_2 \frac{(\boldsymbol{v}_1 - \boldsymbol{v}_2) f_2(\boldsymbol{v}_2)}{|\boldsymbol{v}_1 - \boldsymbol{v}_2|^3} \\ &= -16\pi \left(\frac{q_1 q_2}{8\pi\varepsilon_0 m^*}\right)^2 \log \Lambda \frac{d}{d\boldsymbol{v}_1} \int d^3\boldsymbol{v}_2 \frac{f_2(\boldsymbol{v}_2)}{|\boldsymbol{v}_1 - \boldsymbol{v}_2|} ,\end{aligned} \tag{3.5.15}$$

where $f_2(\boldsymbol{v}_2) d^3\boldsymbol{v}_2$ is the average number of the field particles in the velocity range $\boldsymbol{v}_2 \sim \boldsymbol{v}_2 + d\boldsymbol{v}_2$. The collision frequencies of angular deflection and deceleration of particle 1 in collisions with the field particles of species 2, $\bar{\nu}_{12\perp}$ and $\bar{\nu}_{12\parallel}$, can then be defined by the relations

$$\left\langle \frac{|\Delta \boldsymbol{v}_{1\perp}|^2}{\Delta t} \right\rangle = \left(\frac{m^*}{m_1}\right)^2 \left\langle \frac{|\Delta \boldsymbol{v}_\perp|^2}{\Delta t} \right\rangle \equiv \bar{\nu}_{12\perp} |v_1|^2 \tag{3.5.16}$$

$$\left\langle \frac{\Delta \boldsymbol{v}_{1\parallel}}{\Delta t} \right\rangle = \left(\frac{m^*}{m_1}\right) \left\langle \frac{\Delta \boldsymbol{v}_\parallel}{\Delta t} \right\rangle \equiv -\bar{\nu}_{12\parallel} v_1 \, . \tag{3.5.17}$$

The inverses of $\bar{\nu}_{12\perp}$ and $\bar{\nu}_{12\parallel}$ correspond to the momentum relaxation times,

$$\bar{\nu}_{12\perp}^{-1} = \tau_{12\perp}, \quad \bar{\nu}_{12\parallel}^{-1} = \tau_{12\parallel}, \tag{3.5.18}$$

and are called the *angular deflection time* and *the slowing-down time*, respectively.

To obtain a more explicit expression, we calculate the velocity integral in (3.5.14, 15) by expanding $|\boldsymbol{v}_1 - \boldsymbol{v}_2|^{-1}$ in terms of the Legendre polynomials $P_n(z)$ as

$$\frac{1}{|\boldsymbol{v}_1 - \boldsymbol{v}_2|} = \begin{cases} (1/v_1) \sum_{n=0}^{\infty} (v_2/v_1)^n P_n(\cos\theta_{12}) & (v_1 > v_2) \\ (1/v_2) \sum_{n=0}^{\infty} (v_1/v_2)^n P_n(\cos\theta_{12}) & (v_2 > v_1) \end{cases} \tag{3.5.19}$$

where $v_j = |\boldsymbol{v}_j|$ and θ_{12} is the angle between the two vectors $\boldsymbol{v}_1$ and $\boldsymbol{v}_2$. For the case where $f_2(\boldsymbol{v}_2)$ is isotropic, i.e., $f_2(\boldsymbol{v}_2) = f_2(v_2)$, we then get

$$\int d^3\boldsymbol{v}_2 \frac{f_2(\boldsymbol{v}_2)}{|\boldsymbol{v}_1 - \boldsymbol{v}_2|} = 4\pi \left\{ \frac{1}{v_1} \int_0^{v_1} dv_2 v_2^2 f_2(v_2) + \int_{v_1}^{\infty} dv_2 v_2 f_2(v_2) \right\} , \tag{3.5.20}$$

$$\frac{d}{d\boldsymbol{v}_1} \int d^3\boldsymbol{v}_2 \frac{f_2(\boldsymbol{v}_2)}{|\boldsymbol{v}_1 - \boldsymbol{v}_2|} = -\frac{\boldsymbol{v}_1}{v_1^3} 4\pi \int_0^{v_1} dv_2 v_2^2 f_2(v_2) \, . \tag{3.5.21}$$

In particular, when $f_2(v_2)$ is the Maxwell-Boltzmann distribution, i.e.,

$$f_2(v_2) = \left(\frac{1}{2\pi v_{T2}^2}\right)^{3/2} n_2 \exp\left(-\frac{v_2^2}{2v_{T2}^2}\right) , \tag{3.5.22}$$

where $v_{T2}^2 = T_2/m_2$, the above integral becomes

$$\int d^3\boldsymbol{v}_2 \frac{f_2(v_2)}{|\boldsymbol{v}_1 - \boldsymbol{v}_2|} = \frac{n_2}{v_1}\phi\left(\frac{v_1}{v_{T2}\sqrt{2}}\right) \tag{3.5.23}$$

$$\frac{d}{d\boldsymbol{v}_1}\int d^3\boldsymbol{v}_2 \frac{f_2(v_2)}{|\boldsymbol{v}_1 - \boldsymbol{v}_2|} = -\frac{\boldsymbol{v}_1}{v_1}\frac{n_2}{v_{T2}^2} G\left(\frac{v_1}{v_{T2}\sqrt{2}}\right) , \tag{3.5.24}$$

where

$$\phi(x) = \frac{2}{\sqrt{\pi}}\int_0^x dy e^{-y^2}, \tag{3.5.25}$$

$$G(x) = \frac{1}{2x^2}\left[\phi(x) - x\frac{d\phi(x)}{dx}\right] . \tag{3.5.26}$$

Substituting these expressions into (3.5.14–17), we finally obtain

$$\tau_{12\|}^{-1} = \frac{q_1^2 q_2^2 n_2}{2\pi\varepsilon_0^2 m_1^2} \log\Lambda \frac{1}{v_1^3}\phi\left(\frac{v_1}{v_{T2}\sqrt{2}}\right) , \tag{3.5.27}$$

$$\tau_{12\|}^{-1} = \frac{q_1^2 q_2^2 n_2}{2\pi\varepsilon_0^2 m_1^2} \log\Lambda \left(1 + \frac{m_1}{m_2}\right)\frac{1}{v_{T2}^2}\frac{1}{v_1} G\left(\frac{v_1}{v_{T2}\sqrt{2}}\right) . \tag{3.5.28}$$

If we note

$$\begin{array}{ll} \text{for } x \to 0: & \phi(x)/x \to 2/\sqrt{\pi},\ G(x)/x \to 2/3\sqrt{\pi} \\ \text{for } x \to \infty: & \phi(x) \to 1,\ 2x^2G(x) \to 1, \\ \text{for } x \sim 1: & \phi(x),\ G(x) \sim \mathrm{O}(1), \end{array} \tag{3.5.29}$$

we find the following relations for the particles moving with the thermal speed ($v_1 \sim v_{T1}$):

$$\tau_{\mathrm{ee}\perp}^{-1} \sim \tau_{\mathrm{ee}\|}^{-1} \sim \frac{e^4}{2\pi\varepsilon_0^2}\frac{n_\mathrm{e}}{\sqrt{m_\mathrm{e}}T_\mathrm{e}^{3/2}}\log\Lambda \equiv \tau_\mathrm{e}^{-1} \tag{3.5.30a}$$

$$\tau_{\mathrm{ei}\perp}^{-1} \sim 2\tau_{\mathrm{ei}\|}^{-1} \sim Z\tau_\mathrm{e}^{-1} \tag{3.5.30b}$$

$$\tau_{\mathrm{ii}\perp}^{-1} \sim \tau_{\mathrm{ii}\|}^{-1} \sim \sqrt{\frac{m_\mathrm{e}}{m_\mathrm{i}}}(\frac{T_\mathrm{e}}{T_\mathrm{i}})^{3/2} Z^2\tau_\mathrm{e}^{-1} \equiv \tau_\mathrm{i}^{-1} \tag{3.5.30c}$$

$$\tau_{\mathrm{ie}\perp}^{-1} \sim 2\frac{T_\mathrm{e}}{T_\mathrm{i}}\tau_{\mathrm{ie}\|}^{-1} \sim \frac{m_\mathrm{e}}{m_\mathrm{i}}Z^2\tau_\mathrm{e}^{-1} \equiv \tau_{\mathrm{ie}}^{-1} . \tag{3.5.30d}$$

For $T_\mathrm{e} \sim T_\mathrm{i}$ and $Z = 1$, we then find

$$\frac{\tau_e^{-1}}{\tau_i^{-1}} \sim \frac{\tau_i^{-1}}{\tau_{ie}^{-1}} \sim \sqrt{\frac{m_i}{m_e}} \gg 1 \,. \tag{3.5.31}$$

Now, the like particle collisions contribute to the approach of the system towards the Maxwell-Boltzmann (equilibrium) distribution (3.5.22). We then find from (3.5.31) that the electrons approach equilibrium much faster than the ions.

We next consider the relaxation time needed for the electrons and ions to approach a mutual equilibrium at the same temperature. This relaxation time is called the *energy equilibration time* and is denoted by τ_{eq}. Since the energy change of the incident particle 1 due to the collision with a field particle is given by

$$\Delta E = \frac{m_1}{2}(|\boldsymbol{v}_1 + \Delta\boldsymbol{v}_2|^2 - v_1^2) = m_1\left(\boldsymbol{v}_1 \cdot \Delta\boldsymbol{v}_1 + \frac{|\Delta\boldsymbol{v}_1|^2}{2}\right), \tag{3.5.32}$$

its average rate of change is written as

$$\begin{aligned}\left\langle\frac{\Delta E}{\Delta t}\right\rangle &= m_1\left(\boldsymbol{v}_1\left\langle\frac{\Delta\boldsymbol{v}_1}{\Delta t}\right\rangle + \frac{1}{2}\left\langle\frac{|\Delta\boldsymbol{v}_1|^2}{\Delta t}\right\rangle\right)\\ &= \left(\bar{\nu}_{12\perp} - 2\bar{\nu}_{12\parallel}\right)\frac{1}{2}m_1|\boldsymbol{v}_1|^2 \,,\end{aligned} \tag{3.5.33}$$

where we used the relations (3.5.16, 17) and have ignored $\langle|\Delta\boldsymbol{v}_\parallel|^2/\Delta t\rangle$ since it is much smaller [on the order of $(\log\Lambda)^{-1}$] than the other terms. As we shall later show aposteriori, the energy equilibration time is longer than the time needed for each species of particle to approach its own equilibrium, i.e., $\tau_{eq} \gg \tau_{ee}, \tau_{ii}$, so that we can assume that each species of particle is in a Maxwell-Boltzmann distribution specified by its own temperatures T_1 or T_2. Then we can use (3.5.27, 28) to obtain

$$\begin{aligned}\left\langle\frac{\Delta E}{\Delta t}\right\rangle &= \frac{q_1^2 q_2^2 n_2}{4\pi\varepsilon_0^2 m_1}\log\Lambda\left[\frac{1}{v_1}\phi\left(\frac{v_1}{v_{T2}\sqrt{2}}\right) - \left(1+\frac{m_1}{m_2}\right)\frac{v_1}{v_{T2}^2}G\left(\frac{v_1}{v_{T2}\sqrt{2}}\right)\right]\\ &= -\frac{q_1^2 q_2^2 n_2}{4\pi\varepsilon_0^2 m_2}\log\Lambda\left[\left(1+\frac{m_2}{m_1}\right)\sqrt{\frac{2}{\pi}}\frac{1}{v_{T2}}\exp\left(-\frac{v_1^2}{2v_{T2}^2}\right)\right.\\ &\qquad \left. - \frac{1}{v_1}\phi\left(\frac{v_1}{v_{T2}\sqrt{2}}\right)\right] ,\end{aligned} \tag{3.5.34}$$

where (3.5.26) was used. We then average (3.5.34) over the Maxwell-Boltzmann distribution for species 1:

$$\left\langle\!\!\left\langle\frac{\Delta E}{\Delta t}\right\rangle\!\!\right\rangle = \frac{n_1}{(2\pi)^{3/2}v_{T1}^2}\int d^3\boldsymbol{v}_1 \exp\left(-\frac{v_1^2}{2v_{T1}^2}\right)\left\langle\frac{\Delta E}{\Delta t}\right\rangle . \tag{3.5.35}$$

Noting the relations

$$\frac{1}{(2\pi)^{3/2}v_{T1}^3}\int d^3\boldsymbol{v}_1 \exp\left(-\frac{v_1^2}{2v_{T1}^2}\right)\exp\left(-\frac{v_1^2}{2v_{T2}^2}\right)$$

$$= \frac{1}{(1+\beta^{-2})^{3/2}} \frac{1}{(2\pi)^{3/2} v_{T1}^3} \int d^3 \boldsymbol{v}_1 \exp\left(-\frac{v_1^2}{2v_{T1}^2}\right) \frac{1}{v_1} \phi' \left(\frac{v_1}{v_{T2}\sqrt{2}}\right)$$

$$= \frac{4\sqrt{2}\beta^2}{\pi v_{T2}} \int_0^\infty dx x^2 e^{-\beta x^2} \int_0^x dy e^{y^2} = \sqrt{\frac{2}{\pi}} \frac{1}{v_{T2}} \frac{1}{\sqrt{1+\beta^{-2}}} ,$$

where $\beta = v_{T2}/v_{T1}$, and the last expression was integrated by parts, we obtain

$$\left\langle\!\left\langle \frac{\Delta E}{\Delta t} \right\rangle\!\right\rangle = -\frac{q_1^2 q_2^2 n_1 n_2}{4\pi\varepsilon_0^2 m_2} \sqrt{\frac{2}{\pi}} \log \Lambda \frac{1}{v_{T2}} \left[\left(1 + \frac{m_2}{m_1}\right) \frac{1}{(1+\beta^{-2})^{3/2}} - \frac{1}{(1+\beta^{-2})^{1/2}}\right] \tag{3.5.36}$$

with

$$\left(1 + \frac{m_2}{m_1}\right) = (1+\beta^{-2}) - \frac{T_1 - T_2}{T_2} \frac{m_2}{m_1} .$$

Finally, setting (3.5.36) equal to $(3n_1/2) dT_1/dt$, we arrive at the temperature relaxation equation

$$\frac{dT_1}{dt} = -\frac{T_1 - T_2}{\tau_{eq}} \tag{3.5.37}$$

with

$$\tau_{eq} = \frac{3\pi\sqrt{2\pi}\varepsilon_0^2 m_1 m_2}{q_1^2 q_2^2 n_2 \log \Lambda} \left(\frac{T_1}{m_1} + \frac{T_2}{m_2}\right)^{3/2} . \tag{3.5.38}$$

By setting $T_1 = T_e$, $T_2 = T_i$, $m_1 = m_e$, $m_2 = m_i$, $q_1 = -e$, $q_2 = Ze$, and $n_2 = n_i$, the electron-ion energy equilibration time τ_{eq}^{ei} is then given as

$$\tau_{eq}^{ei} \doteq \frac{3\pi\sqrt{2\pi}\varepsilon_0^2 m_i}{Z^2 e^4 n_i \log \Lambda} \frac{T_e^{3/2}}{\sqrt{m_e}} . \tag{3.5.39}$$

Clearly it is of the same order as τ_{ie} and hence the assumption of $\tau_{eq} \gg \tau_{ee}, \tau_{ii}$ is justified.

3.5.5 Effective Collision Frequency

The collision frequencies (3.5.30) are for particles moving with thermal speed and scattering at 90 degrees. Let us reexamine the parameter dependence of the general expression for the collision frequency (3.5.7), i.e.,

$$\nu_{ei}(v, \Delta\theta) \propto \frac{1}{v^3 \sin^2 \Delta\theta} . \tag{3.5.40}$$

This relation shows that the collision frequency is smaller for faster particles and that it is large for small-angle scattering.

We mentioned before that the electrical resistivity is determined by $\bar{\nu}_{ei\|}$ through the relation (3.5.11). This is true only at thermal equilibrium. However, if there is a special class of electrons which move at a much faster speed than the thermal electron speed, their collision frequency is low, so that these electrons can be accelelated by an electric field for a much longer time than the time $\nu_{ei\|}^{-1}$. The more they are accelerated, the fewer collisional events do they suffer. Thus these particles can be accelerated to a very high speed, close to the relativistic limit. Such electrons are called the *run-away electrons*. Similarly, the relaxation times of fast particles are much longer than the average relaxation times given by (3.5.30). In many real plasmas of interest, there exist high energy components. These components constitute a *high energy tail* in the velocity distribution and are sometimes called the *supra-thermal particles*.

Let us examine the momentum relaxation of supra-thermal ions, such as those produced by energetic neutral beam injection (NBI) and the α-particles produced by nuclear reactions. As seen from (3.5.27–29), their relaxation times are proportional to v_1^3. In particular, because of the mass ratio in (3.5.28), the relaxation time of the ions with $v_1 > v_{Te}$ is determined by the slowing-down process due to the electrons,

$$\tau_{ie\|}(v_1 > v_{Te}) \doteq \frac{4\pi\varepsilon_0^2 m_e m_i}{Z^2 e^2 n_e \log \Lambda}\, v_1^3 \,. \tag{3.5.41}$$

In many cases of interest, in order to significantly decrease the velocity we do not need 90 degree deflection in collisions; only a small-angle deflection is sufficient. For instance, particles trapped in a local magnetic mirror potential can be detrapped by relatively small pitch angle scattering. If the critical pitch angle for trapping is θ_c, the *effective collision frequency* ν_{eff} needed for detrapping is given by

$$\nu_{eff} \sim \left(\frac{\pi}{2\theta_c}\right)^2 \nu_\perp \,, \tag{3.5.42}$$

where $\nu_\perp = 1/\tau_\perp$ is the 90 degree deflection frequency.

3.6 Relativistic Effects

The equation of motion is written as

$$\frac{d\boldsymbol{v}}{dt} = \boldsymbol{v} \times \boldsymbol{\omega}_c \,, \tag{3.6.1}$$

where $\boldsymbol{\omega}_c = \omega_c \boldsymbol{b}$ and

$$\omega_c = \frac{eB}{\gamma m} \tag{3.6.2}$$

is the cyclotron frequency and $\gamma = [1 - (v/c)^2]^{-1/2}$. The fact that the cyclotron frequency depends on γ means that the relativistic effect significantly changes the electron cyclotron resonance $\omega = \omega_{ce}$ in an inhomogeneous magnetic field. This implies that the relativistic effect is important in the analysis of electron cyclotron wave absorption for high electron temperature plasmas where $T_e \gtrsim 20\,\mathrm{keV}$.

We now consider a charged particle moving in a combination of electric and magnetic fields $\boldsymbol{E}$ and $\boldsymbol{B}$, where both are uniform and static. As a special case, $\boldsymbol{E} \perp \boldsymbol{B}$ will be treated first. Consider a Lorentz transformation to a coordinate frame K' moving with a velocity $\boldsymbol{u}$ with respect to the original coordinate frame. The equation for the charged particle in K' is

$$\frac{d\boldsymbol{p}'}{dt'} = e(\boldsymbol{E}' + \boldsymbol{v} \times \boldsymbol{B}') , \tag{3.6.3}$$

where the primed variables are in the K' system. The electromagnetic field becomes

$$\boldsymbol{E}' = \gamma(\boldsymbol{E} + \boldsymbol{u} \times \boldsymbol{B}) - \frac{\gamma^2}{\gamma+1}\frac{\boldsymbol{u}}{c}\left(\frac{\boldsymbol{u}}{c} \cdot \boldsymbol{E}\right) \quad \text{and} \tag{3.6.4}$$

$$\boldsymbol{B}' = \gamma\left(\boldsymbol{B} - \frac{1}{c^2}\boldsymbol{u} \times \boldsymbol{E}\right) - \frac{\gamma^2}{\gamma+1}\frac{\boldsymbol{u}}{c}\left(\frac{\boldsymbol{u}}{c} \cdot \boldsymbol{B}\right) . \tag{3.6.5}$$

Let us consider the case of $|\boldsymbol{E}| < c|\boldsymbol{B}|$. If $\boldsymbol{u}$ is chosen as the $\boldsymbol{E} \times \boldsymbol{B}$ drift velocity, then

$$\boldsymbol{u} = \frac{\boldsymbol{E} \times \boldsymbol{B}}{B^2} . \tag{3.6.6}$$

Then from (3.6.4, 5) we find the fields in K' to be $\boldsymbol{E}'_{\parallel} = 0, \boldsymbol{E}'_{\perp} = 0$ and $\boldsymbol{B}'_{\parallel} = 0, \boldsymbol{B}'_{\perp} = \boldsymbol{B}/\gamma$, where $\gamma = [1 - (u/c)^2]^{-1/2}$. Here $\parallel$ and $\perp$ refer to the direction of $\boldsymbol{u}$. In the K' frame the only field acting is the static magnetic field $\boldsymbol{B}'_{\perp}$ which points in the same direction as $\boldsymbol{B}$, but it is weaker than $\boldsymbol{B}$ by a factor $1/\gamma$. Thus, the motion in K' is a spiraling around the line of force.

The drift velocity $\boldsymbol{u}$ has physical meaning only if it is less than the velocity of light, i.e., for $|\boldsymbol{E}| < c|\boldsymbol{B}|$. If $|\boldsymbol{E}| > c|\boldsymbol{B}|$, the particle is continuously accelerated in the direction of $\boldsymbol{E}$ by the strong electric field and it is considered that its average energy continues to increase with time. To understand this we consider a Lorentz transformation from the original frame to a system K'' moving with a velocity

$$\boldsymbol{u}' = c^2\frac{\boldsymbol{E} \times \boldsymbol{B}}{E^2} . \tag{3.6.7}$$

In this frame, from (3.6.4, 5) with $\boldsymbol{E}' \to \boldsymbol{E}'', \boldsymbol{B}' \to \boldsymbol{B}'', \gamma \to \gamma' = [1 - (u'/c)^2]^{-1/2}$ the electric and magnetic fields are $\boldsymbol{E}''_{\parallel} = 0$, $\boldsymbol{E}''_{\perp} = \boldsymbol{E}/\gamma'$ and $\boldsymbol{B}''_{\parallel} = 0, \boldsymbol{B}'_{\perp} = 0$. Thus in the K'' system the particle moves in a purely electrostatic field.

If $\boldsymbol{E}$ has a component parallel to $\boldsymbol{B}$, the behavior of the particle is not straightforward to understand. The scalar product $\boldsymbol{E} \cdot \boldsymbol{B}$ is a Lorentz invariant. If $\boldsymbol{E} \cdot \boldsymbol{B} \neq 0$, both $\boldsymbol{E}$ and $\boldsymbol{B}$ will exist simultaneously in all-Lorentz frames.

For a charged particle in a uniform static magnetic field B and $E = 0$, the transverse motion is periodic. The action integral for this transverse motion is

$$J = \oint \boldsymbol{p}_\perp \cdot d\boldsymbol{l} \tag{3.6.8}$$

where $\boldsymbol{p}_\perp = \gamma m \boldsymbol{v}_\perp + e\boldsymbol{A}$ is the transverse component of the canonical momentum and $d\boldsymbol{l}$ is a line element along the circular path of the particle. This gives

$$J = \gamma m \omega_c \pi \varrho_L^2 = e(B\pi\varrho_L^2) \, . \tag{3.6.9}$$

The quantity $B\pi\varrho_L^2$ is the flux through the particle orbit. This means $\gamma\mu$ is an adiabatic invariant, where $\mu = e\omega_c\varrho_L^2/2$ is the magnetic moment. In time varying fields, μ is an adiabatic invariant only in the non-relativistic limit.

PROBLEMS

3.1. Calculate the cyclotron frequency and the Larmor radius of the following particles: (i) an electron of 1 eV in the earth's magnetic field ($\sim$ 0.3 G); (ii) an alpha particle of 3.6 MeV in a magnetic field of 20 kG; (iii) a boron atom B of 200 keV in a magnetic field of 50 kG.

3.2. The magnetic mirror force arises from the Lorentz force component,

$$m dv_\parallel / dt = -q v_\theta B_r .$$

Show that it is given by $-\mu \partial B / \partial z$.

3.3. Show that in vacuum the sum of the curvature drift and the gradient B drift is given by

$$\boldsymbol{v}_c + \boldsymbol{v}_G = \frac{m}{q}\left(v_\parallel^2 + \frac{v_\perp^2}{2}\right)\frac{\boldsymbol{B} \times \nabla \mathrm{B}}{B^3} \, .$$

3.4. Discuss the guiding center motion when a static electric field is present in the direction of the gradient B drift as shown in Fig. 3.8.

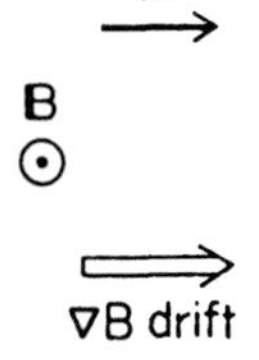

E

B

∇B drift

Fig. 3.8. Directions of static electric field and gradient B drift, for Problem 3.4

3.5. Estimate the $\boldsymbol{E} \times \boldsymbol{B}$ drift velocity, the curvature drift velocity and the gradient B drift velocity, in a magnetic field of 10^4 G. Use the units kV/m for the electric field, m^{-1} for the gradient and keV for the energy.

3.6. Discuss the particle motion in the presence of a weak oscillating magnetic field $B_\perp \cos \omega_0 t$ perpendicular to a static magnetic field B_0.

3.7. Show that the ponderomotive force (3.4.9) can be written as

$$-\nabla\overline{(m/2|\Delta v|^2)} = -\nabla(q/2\overline{\Delta r \cdot E)}\,.$$

3.8. Estimate the collision frequencies ν_{ee} and ν_{ei} for the following plasmas: (i) $T_e = 10\,\mathrm{eV}$ and $n = 10^{18}\mathrm{m}^{-3}$; (ii) $T_e = 10\,\mathrm{keV}$ and $n = 10^{20}\mathrm{m}^{-3}$; (iii) $T_e = 10\,\mathrm{keV}$ and $n = 10^{31}\mathrm{m}^{-3}$.

4. Formulation of Plasma Theory

In this chapter, we begin the exposition of the theory of plasmas. We shall restrict ourselves primarily to plasmas consisting of charged particles which can be represented by point charge and point mass. Each particle is assumed to obey the classical nonrelativistic equations of motion and the plasma has no net charge, namely, there are equal number of negative charges (electrons) and positive charges (ions).

The basic equations describing the behavior of such a plasma can roughly be divided into three classes depending on the level of detail of the description. The most detailed description is to follow the motion of each constituent particle, including the mutual interactions with other particles. In this description, the plasma is represented by a distribution function in Γ-space (the phase space of all the particles) and the electromagnetic fields produced by all the charged particles. The system of equations corresponding to such a description is called the *Klimontovich system of equations*. Although these equations can completely describe the plasma, the direct information gained by such description is often useless because it contains too much detail.

A more useful description which treats the plasma as a phase space continuum, is the μ-space description where a single particle phase space distribution is used to describe the behavior in the presence of an averaged electromagnetic field consistent with the continuum particle distribution. In this case we can either include the collisional effect (*Boltzmann equation*) or neglect it (*Vlasov* or *collisionless Boltzmann equation*). In the presence of a strong magnetic field, we can further simplify the equation by considering only the time-averaged guiding center motion (*drift-kinetic equation*).

The most widely used description is similar to that of an ordinary macroscopic fluid. Here the plasma is represented by a continuum in real space and only the macroscopic quantities such as density, fluid velocity, temperature, and the electromagnetic fields consistent with these macroscopic variables are used to describe the plasma behavior. Although this description neglects the phase space continuity of plasma, it is a good description of an electromagnetic fluid. There are two types of macroscopic descriptions: one is to treat the plasma as a two-component fluid consisting of an electron fluid and an ion fluid, and the other is to treat it as a one-component fluid by assuming local charge neutrality. The former is useful to derive the electrostatic response of a plasma, whereas the latter, usually called the magnetohydrodynamic (MHD) model, can describe a plasma in the presence of a strong inhomogeneous magnetic field. In fact, it can be derived from the more fundamental equations in the limit of small Larmor radius and large cyclotron frequency.

4.1 The Γ-Space Description (Klimontovich System of Equations)

Consider an sth species of particles (s = e for the electron and i for the ion) and let the number of such particles in the spatial region $\boldsymbol{r} \sim \boldsymbol{r}+d\boldsymbol{r}$ and the velocity region $\boldsymbol{v} \sim \boldsymbol{v}+d\boldsymbol{v}$ at time t be $\hat{F}_s(\boldsymbol{r}, \boldsymbol{v}, t) d^3\boldsymbol{r} d^3\boldsymbol{v}$.

If we choose the volume element $d^3\boldsymbol{r}d^3\boldsymbol{v}$ to be sufficiently small such that only one particle or less can exist inside it, then $\hat{F}_s$ will become either zero (if no particle exists there) or one, that is, it is a δ-function. Let there be a total of N_s particles and let the position and velocity of the jth particle at time t be $\boldsymbol{r}_j(t)$ and $\boldsymbol{v}_j(t)$ ($j = 1, 2, \cdots, N_s$), then the distribution function $\hat{F}_s$ can be written as

$$\hat{F}_s(\boldsymbol{r}, \boldsymbol{v}, t) = \sum_{j=1}^{N_s} \delta[\boldsymbol{r} - \boldsymbol{r}_j(t)]\ \delta\ [\boldsymbol{v} - \boldsymbol{v}_j(t)]\ , \tag{4.1.1}$$

where $\delta[\boldsymbol{a}]$ for a vector $\boldsymbol{a}$ represents the product of the three δ-functions for each vector component:

$$\delta[\boldsymbol{a}] = \delta[a_x]\ \delta[a_y]\ \delta[a_z]\ . \tag{4.1.2}$$

Differentiating (4.1.1) with respect to time yields

$$\frac{\partial \hat{F}_s}{\partial t} = \sum_{j=1}^{N_s} \left(\frac{\partial \hat{F}_s}{\partial \boldsymbol{r}_j} \cdot \frac{d\boldsymbol{r}_j}{dt} + \frac{\partial \hat{F}_s}{\partial \boldsymbol{v}_j} \cdot \frac{d\boldsymbol{v}_j}{dt} \right)\ . \tag{4.1.3}$$

Here, $d\boldsymbol{r}_j/dt = \boldsymbol{v}_j$ and $d\boldsymbol{v}_j/dt$ is given from the equation of motion as

$$\frac{d\boldsymbol{v}_j}{dt} = \frac{q_s}{m_s}[\hat{\boldsymbol{E}}(\boldsymbol{r}_j, t) + \boldsymbol{v}_j \times \hat{\boldsymbol{B}}(\boldsymbol{r}_j, t)] \equiv \hat{\boldsymbol{K}}_s(\boldsymbol{r}_j, \boldsymbol{v}_j, t)\ , \tag{4.1.4}$$

where q_s, m_s are the charge and mass of the sth species of particle and the hat stands for microscopic quantities, implying that the fields produced by point charges, as well as the external fields, are included.

Equation (4.1.3) then becomes

$$\begin{aligned} \frac{\partial \hat{F}_s}{\partial t} &= \sum_{j=1}^{N_s} \left[\frac{\partial \hat{F}_s}{\partial \boldsymbol{r}_j} \cdot \boldsymbol{v}_j + \frac{\partial \hat{F}_s}{\partial \boldsymbol{v}_j} \cdot \hat{\boldsymbol{K}}_s(\boldsymbol{r}_j, \boldsymbol{v}_j, t) \right] \\ &= \sum_{j=1}^{N_s} \left[\frac{\partial \hat{F}_s}{\partial \boldsymbol{r}_j} \cdot \boldsymbol{v} + \frac{\partial \hat{F}_s}{\partial \boldsymbol{v}_j} \cdot \hat{\boldsymbol{K}}_s(\boldsymbol{r}, \boldsymbol{v}, t) \right]\ , \end{aligned}$$

where we used the fact that $\hat{F}_s$ is a sum of δ-functions. We can write from (4.1.1)

$$\sum_{j=1}^{N_s} \frac{\partial \hat{F}_s}{\partial \boldsymbol{r}_j} = -\frac{\partial \hat{F}_s}{\partial \boldsymbol{r}}, \quad \sum_{j=1}^{N_s} \frac{\partial \hat{F}_s}{\partial \boldsymbol{v}_j} = -\frac{\partial \hat{F}_s}{\partial \boldsymbol{v}}\ ,$$

from which we finally obtain

$$\frac{\partial \hat{F}_s}{\partial t} + \boldsymbol{v} \cdot \frac{\partial \hat{F}_s}{\partial \boldsymbol{r}} + \hat{\boldsymbol{K}}_s(\boldsymbol{r}, \boldsymbol{v}, t) \cdot \frac{\partial \hat{F}_s}{\partial \boldsymbol{v}} = 0 . \tag{4.1.5}$$

This equation is called the *Klimontovich equation.* The acceleration $\hat{\boldsymbol{K}}_s$ or the electromagnetic fields $\hat{\boldsymbol{E}}$ and $\hat{\boldsymbol{B}}$ satisfy Maxwell's equations:

$$\begin{aligned}
\nabla \times \hat{\boldsymbol{E}}(\boldsymbol{r}, t) &= -\frac{\partial}{\partial t}\hat{\boldsymbol{B}}(\boldsymbol{r}, t) \\
\nabla \times \hat{\boldsymbol{B}}(\boldsymbol{r}, t) &= \frac{1}{c^2}\frac{\partial}{\partial t}\hat{\boldsymbol{E}}(\boldsymbol{r}, t) + \mu_0 \hat{\boldsymbol{J}}(\boldsymbol{r}, t) \\
\nabla \cdot \hat{\boldsymbol{B}}(\boldsymbol{r}, t) &= 0 \\
\nabla \cdot \hat{\boldsymbol{E}}(\boldsymbol{r}, t) &= \frac{1}{\varepsilon_0}\hat{\sigma}_{\mathrm{e}}(\boldsymbol{r}, t)
\end{aligned} \tag{4.1.6}$$

where $\hat{\boldsymbol{J}}$ and $\hat{\sigma}_{\mathrm{e}}$ are the microscopic current density and charge density and are given in terms of $\hat{F}_s$ as

$$\hat{\boldsymbol{J}}(\boldsymbol{r}, t) = \sum_s q_s \int \boldsymbol{v} \hat{F}_s(\boldsymbol{r}, \boldsymbol{v}, t) d^3\boldsymbol{v} , \tag{4.1.7}$$

$$\hat{\sigma}_{\mathrm{e}}(\boldsymbol{r}, t) = \sum_s q_s \int \hat{F}_s(\boldsymbol{r}, \boldsymbol{v}, t) d^3\boldsymbol{v} . \tag{4.1.8}$$

We note that in (4.1.6), the current and charge produced by the plasma particles are all included in $\hat{\boldsymbol{J}}$ and $\hat{\sigma}_{\mathrm{e}}$, so that the relations between $\hat{\boldsymbol{B}}$ and $\hat{\boldsymbol{H}}$ and $\hat{\boldsymbol{E}}$ and $\hat{\boldsymbol{D}}$ are the same as those in vacuum.

Equations (4.1.5–8) constitute the Klimontovich system of equations. The structure of this system is the following: given the particle distribution function $\hat{F}_s$, it determines $\hat{\boldsymbol{J}}$ and $\hat{\sigma}_{\mathrm{e}}$ through (4.1.7, 8) which in turn determine $\hat{\boldsymbol{E}}$ and $\hat{\boldsymbol{B}}$ through (4.1.6); the acceleration $\hat{\boldsymbol{K}}_s$ is then expressed in terms of $\hat{F}_s$ which has to satisfy (4.1.5) (self-consistency).

Since the Klimontovich system of equations are partial differential equations, their solution is not unique unless initial and boundary conditions are prescribed. The boundary conditions could be determined by assuming an idealized vessel wall and vacuum electromagnetic fields which are produced by external means. On the other hand, specifying the initial conditions requires a knowledge of the positions and velocities $\{\boldsymbol{r}_j, \boldsymbol{v}_j\}$ of all the constituent particles at $t = 0$, where $\{\boldsymbol{r}_j, \boldsymbol{v}_j\} = \{\boldsymbol{r}_1(t = 0), \boldsymbol{v}_1(t = 0), \cdots, \boldsymbol{r}_N(t = 0), \boldsymbol{v}_N(t = 0)\}$ and N is the total number of particles ($N = \sum_s N_s$). The solution of the Klimontovich system of equations therefore depends on the $6N$ initial parameters $\{\boldsymbol{r}_j, \boldsymbol{v}_j\}$. We denote this by

$$\hat{F}_s(\boldsymbol{r}, \boldsymbol{v}, t) = \hat{F}_s(\boldsymbol{r}, \boldsymbol{v}, t | \{\boldsymbol{r}_j, \boldsymbol{v}_j\}) , \quad \hat{\boldsymbol{E}}(\boldsymbol{r}, t) = \hat{\boldsymbol{E}}(\boldsymbol{r}, t | \{\boldsymbol{r}_j, \boldsymbol{v}_j\}) \text{ etc.} \tag{4.1.9}$$

Naturally it is impossible to know all $6N$ initial parameters, so that in practice we cannot solve the Klimontovich system of equations. Even if we knew the initial parameters and even if we succeeded in solving the equations, the results would contain too much information to be of any use!

4.2 The μ-Space Description (Boltzmann-Vlasov System of Equations)

The μ-space description can be obtained from the Γ-space Klimontovich system of equations by averaging their solutions over the ensemble of initial parameters. The resulting distribution function and the field variables then become smooth functions of $\boldsymbol{r}$ and $\boldsymbol{v}$. In the notation used here, this average is denoted by an angular bracket, and the averaged quantities are now without a hat as

$$\begin{aligned} \left\langle \hat{F}_s(\boldsymbol{r},\boldsymbol{v},t|\{\boldsymbol{r}_j,\boldsymbol{v}_j\})\right\rangle &\equiv F_s(\boldsymbol{r},\boldsymbol{v},t)\,, \\ \left\langle \hat{\boldsymbol{E}}(\boldsymbol{r},t|\{\boldsymbol{r}_j,\boldsymbol{v}_j\})\right\rangle &\equiv \boldsymbol{E}(\boldsymbol{r},t) \text{ etc.} \end{aligned} \tag{4.2.1}$$

The quantities defined by (4.1.6–8) can also be averaged as in (4.2.1) and the same equations as (4.1.6–8) hold for the averaged quantities, since these equations are linear with respect to the quantities with hats. However, because of its nonlinearity, taking the average of (4.1.5) we have

$$\frac{\partial F_s}{\partial t} + \boldsymbol{v}\cdot\frac{\partial}{\partial \boldsymbol{r}}F_s + \left\langle \hat{\boldsymbol{K}}_s\cdot\frac{\partial}{\partial \boldsymbol{v}}\hat{F}_s\right\rangle = 0\,. \tag{4.2.2}$$

The last term on the left-hand side differs from $\boldsymbol{K}_s\cdot\partial F_s/\partial\boldsymbol{v}$. We denote the difference by

$$\left\langle \hat{\boldsymbol{K}}_s\cdot\frac{\partial}{\partial \boldsymbol{v}}\hat{F}_s\right\rangle_{\text{corr}} \equiv \left\langle \hat{\boldsymbol{K}}_s\cdot\frac{\partial}{\partial \boldsymbol{v}}\hat{F}_s\right\rangle - \boldsymbol{K}_s\cdot\frac{\partial}{\partial \boldsymbol{v}}F_s\,. \tag{4.2.3}$$

This quantity is usually called the *two-body correlation term* or the term representing the *discreteness effect* of the plasma. Indeed, $\hat{\boldsymbol{K}}_s$ and $\hat{F}_s$ are expressed in terms of the sum of contributions of the discrete particles represented by δ-functions. Thus, $\left\langle \hat{K}_s\cdot(\partial/\partial\boldsymbol{v})\hat{F}_s\right\rangle$ consists of the terms like

$$\langle \delta[\boldsymbol{r}-\boldsymbol{r}_j(t)]\;\delta[\boldsymbol{v}-\boldsymbol{v}_j(t)]\;\delta[\boldsymbol{r}-\boldsymbol{r}_i(t)]\;\delta[\boldsymbol{v}-\boldsymbol{v}_i(t)]\rangle\,.$$

If the particles are mutually independent of each other, then for $j\neq i$ the above average is reduced to

$$\langle \delta[\boldsymbol{r}-\boldsymbol{r}_j(t)]\;\delta[\boldsymbol{v}-\boldsymbol{v}_j(t)]\rangle\,\langle \delta[\boldsymbol{r}-\boldsymbol{r}_i(t)]\;\delta[\boldsymbol{v}-\boldsymbol{v}_i(t)]\rangle$$

so that its contribution to (4.2.3) vanishes. Nonzero contributions to (4.2.3) can be expected only from the terms $j = i$. The physical process represented by such terms is the following. Consider a particle which we call the 'test particle'. The electromagnetic fields produced by this particle modify the orbits of the surrounding particles. This orbit modification in turn induces a field at the position of the test particle and modifies its orbit. It is clear that (4.2.3) always represents the two-body correlation effect or the collisional effect. The collision term can also be derived in a way similar to the Boltzmann collision integral in an ordinary gas, but by taking into account the collective screening effect of the surrounding particles (Sect. 6.8).

We note that in the limit of an ideal plasma, the two-body correlation term can completely be ignored. Indeed, the ideal plasma is a continuum in phase space and has no discrete properties, hence, the two-body correlation effect is totally negligible when compared with the effect of the average electromagnetic field. Of course, this argument is true only when we look at the plasma on a spatial scale greater than or equal to the Debye length. On sufficiently small spatial scales, the discreteness effects become very important. We also note that the definition of the average $\langle\ \rangle$ has not yet been clearly stated. For a system not in thermal equilibrium, the ensemble average cannot be precisely defined. We shall not dwell on this point, but restrict ourselves to mentioning that the average smears out the discreteness effects to an extent which depends on the given situation of interest.

In the discussion below, we shall consider only plasmas very close to an ideal plasma and neglect two-body correlation terms. The resulting system of equations can then be written as follows:

$$\frac{\partial F_s}{\partial t} + \boldsymbol{v} \cdot \frac{\partial}{\partial \boldsymbol{r}} F_s + \boldsymbol{K}_s \cdot \frac{\partial}{\partial \boldsymbol{v}} F_s = 0 , \tag{4.2.4}$$

$$K_s(\boldsymbol{r}, \boldsymbol{v}, t) = \frac{q_s}{m_s} [\boldsymbol{E}(\boldsymbol{r}, t) + \boldsymbol{v} \times \boldsymbol{B}(\boldsymbol{r}, t)] , \tag{4.2.5}$$

$$\begin{aligned} \nabla \times \boldsymbol{E} &= -\frac{\partial \boldsymbol{B}}{\partial t} , \quad \nabla \times \boldsymbol{B} = \frac{1}{c^2} \frac{\partial \boldsymbol{E}}{\partial t} + \mu_0 \boldsymbol{J} , \\ \nabla \cdot \boldsymbol{B} &= 0 , \qquad \nabla \cdot \boldsymbol{E} = \frac{1}{\varepsilon_0} \sigma_{\mathrm{e}} , \end{aligned} \tag{4.2.6}$$

$$\boldsymbol{J}(\boldsymbol{r}, t) = \sum_s q_s \int \boldsymbol{v} F_s d^3\boldsymbol{v} , \tag{4.2.7}$$

$$\sigma_{\mathrm{e}}(\boldsymbol{r}, t) = \sum_s q_s \int F_s d^3\boldsymbol{v} . \tag{4.2.8}$$

We refer to (4.2.4) as the *Vlasov equation* or the *collisionless Boltzmann equation*, and (4.2.4–8) as the *Vlasov system of equations*. Note that they are exactly of the same form as the Klimontovich system of equations. However, their components are different on the following two points. First, whereas the Klimontovich equations describe the *exact* behavior of the plasma, the Vlasov equations are approximate and valid only when two-body correlation effects are totally negligible. Second, every solution of the Klimontovich equations contains $6N$ initial parameters $\{\boldsymbol{r}_j, \boldsymbol{v}_j\}$ but the solution of the Vlasov equations do not have any such parameters. The initial condition for the latter is simply that the solution be smooth in phase space.

Both the Klimontovich system and the Vlasov system are invariant under time reversal, so that the physical processes described by these system of equations are essentially *reversible* processes. This can be seen by the invariance of equations by the transformations $t \to -t, v \to -v$ and $B \to -B$. This property of the equations is the fundamental difference from the Boltzmann equation which describes the irreversible approach to equilibrium by collisional processes. However, the Vlasov equation is *nonlinear* with respect to the electromagnetic fields, as shown by the last term on the left hand side of (4.2.4). This term is nonlinear because the distribution function F_s depends on the electromagnetic field. Because of this nonlinearity, the problem of solving the Vlasov equation is in general extremely difficult, and the solution describes a rich variety of interesting phenomena.

4.3 Drift Kinetic Theory

For plasma immersed in a strong magnetic field, the Larmor radii of the individual particles are much smaller than the characteristic length of the magnetic field, L. Since $\varrho_L/L \ll 1$, where ϱ_L is a representative Larmor radius, we can expand the Vlasov equations for such plasmas in this small parameter. The lowest order equation in the expansion is called *drift kinetic equation.* It is also possible to derive this equation directly from the guiding center equations of motion rather than by expanding the Vlasov equation. Then the guiding center orbits are exactly described by the resulting drift kinetic equation.

In the guiding center approximation of the particle motion, the cyclotron motion is averaged out and the magnetic moment and the parallel velocity are enough to describe the motion (Sect. 3.2). Hence we consider a five-dimensional phase space with coordinates α_i ($i = 1$ to 5), where $\{\alpha_i\} = \{\boldsymbol{x}, \mu, v_\parallel\}$; $\boldsymbol{x}$ is the guiding center position, μ is the magnetic moment, and $v_\parallel$ is the guiding center velocity component parallel to $\boldsymbol{B}$ at $\boldsymbol{x}$. The volume element in guiding center phase space is defined by

$$dV = J(\alpha_1, \cdots, \alpha_5, t) d\alpha_1 d\alpha_2 d\alpha_3 d\alpha_4 d\alpha_5 \,. \tag{4.3.1}$$

The collisionless guiding center drift kinetic equation follows from the requirement that the number

$$dN = f dV \tag{4.3.2}$$

of guiding centers in the co-moving volume element dV be constant in time,

$$\frac{d}{dt}(f\, dV) = 0 \,. \tag{4.3.3}$$

Here the total time derivative in phase space is defined by

$$\frac{d}{dt} \equiv \frac{\partial}{\partial t} + \sum_i \frac{d\alpha_i}{dt} \frac{\partial}{\partial \alpha_i} \,. \tag{4.3.4}$$

It is noted that

$$\frac{\partial \alpha_i}{\partial t} = 0\,, \qquad \frac{\partial \alpha_i}{\partial \alpha_j} = \delta_{ij} \text{ (Kronecker delta)}\,, \tag{4.3.5}$$

and $d\alpha_i/dt$ in (4.3.4) is a function of $\{\alpha_i\}$ and t.

After a short time dt, the volume element dV will become $dV+[d(dV)/dt]dt$, where

$$\left\{dV + \frac{d}{dt}(dV)dt\right\} \Big/ dV$$
$$= \frac{\partial(\alpha_1 + (d\alpha_1/dt)dt, \cdots, \alpha_5 + (d\alpha_5/dt)dt)}{\partial(\alpha_1, \cdots, \alpha_5)} + \frac{1}{J}\frac{dJ}{dt}dt \tag{4.3.6}$$

by the rule for transforming small elements of volume. The first term on the right of (4.3.6) is a determinant whose non-diagonal elements are all proportional to dt, and whose diagonal elements are

$$1 + \frac{\partial}{\partial \alpha_1}\left(\frac{d\alpha_1}{dt}\right)dt, \cdots, 1 + \frac{\partial}{\partial \alpha_5}\left(\frac{d\alpha_5}{dt}\right)dt\,. \tag{4.3.7}$$

Thus, neglecting squares and higher powers of dt,

$$\frac{d}{dt}(dV) = \left\{\sum_i \frac{\partial}{\partial \alpha_i}\left(\frac{d\alpha_i}{dt}\right) + \frac{1}{J}\frac{dJ}{dt}\right\}dV\,. \tag{4.3.8}$$

Equation (4.3.3) can be transformed to yield

$$\frac{df}{dt} + Sf = 0\,, \tag{4.3.9}$$

with S defined by

$$S \equiv \frac{1}{dV}\frac{d}{dt}(dV) = \sum_i \frac{\partial}{\partial \alpha_i}\left(\frac{d\alpha_i}{dt}\right) + \frac{1}{J}\frac{dJ}{dt}\,. \tag{4.3.10}$$

Here, (4.3.9) is the collisionless drift kinetic equation. The quantity S vanishes identically when the Vlasov description is applicable to the chosen phase space volume dV. The left-hand side of the drift kinetic equation is then a total time derivative in phase space, hence f = const along the phase space trajectories. If $S \neq 0$, then f varies along the phase space trajectories. It is important to note that (4.3.9) can be transformed to yield

$$\frac{\partial}{\partial t}(Jf) + \sum_i \frac{\partial}{\partial \alpha_i}\left(\frac{d\alpha_i}{dt}Jf\right) = 0 \tag{4.3.11}$$

as an alternative form of the drift kinetic equation and that S can be written in the form

$$S = \frac{1}{J}\left[\frac{\partial J}{\partial t} + \sum_i \frac{\partial}{\partial \alpha_i}\left(\frac{d\alpha_i}{dt}J\right)\right] \tag{4.3.12}$$

as an alternative to (4.3.10). The form of (4.3.11) is the same whether Vlasov description is satisfied ($S = 0$) or not ($S \neq 0$).

We will now turn to a guiding center drift theory valid for time-independent magnetic fields, which is consistent with the Vlasov description. The equations of motion are

$$\frac{d\boldsymbol{x}}{dt} \equiv \boldsymbol{v}_{\mathrm{D}} = v_{\|}\boldsymbol{b} + \boldsymbol{v}_E + \boldsymbol{v}_{\mathrm{G}} + \frac{v_{\|}^2}{\Omega}(\nabla \times \boldsymbol{b}) , \tag{4.3.13}$$

$$\frac{dv_{\|}}{dt} = \left(\frac{q}{m}\boldsymbol{E} - \frac{\mu}{m}\nabla B\right) \cdot \left[\boldsymbol{b} + \frac{v_{\|}}{\Omega}\nabla \times \boldsymbol{b}\right] , \tag{4.3.14}$$

$$\frac{d\mu}{dt} = 0 , \tag{4.3.15}$$

where $\Omega = qB/m$ and $\boldsymbol{v}_E$ and $\boldsymbol{v}_{\mathrm{G}}$ are $\boldsymbol{E} \times \boldsymbol{B}$ drift and gradient B drift, respectively, given in Sect. 3.2. The last term in (4.3.14) can be derived by using $\mu B + mv_{\|}^2/2 + q\phi = K = \mathrm{const}$:

$$\begin{aligned} \frac{dv_{\|}}{dt} &= \frac{d}{dt}\left[\left(\frac{2}{m}\right)^{1/2} \sqrt{K - \mu B - q\phi}\right] \\ &= -\frac{1}{\sqrt{2m}}\left(\mu\frac{dB}{dt} + q\frac{d\phi}{dt}\right)(K - \mu B - q\phi)^{-1/2} , \end{aligned} \tag{4.3.16}$$

$dB/dt = d\boldsymbol{x}/dt \cdot \nabla B$ and $d\phi/dt = d\boldsymbol{x}/dt \cdot \nabla\phi$. By substituting (4.3.13) into $d\boldsymbol{x}/dt$, and noting $(\boldsymbol{v}_E + \boldsymbol{v}_G) \cdot (\mu\nabla B - q\boldsymbol{E}) = 0$, we obtain

$$\frac{dv_{\|}}{dt} = -\left(\frac{\mu}{m}\nabla B + \frac{q}{m}\nabla\phi\right) \cdot \left[\boldsymbol{b} + \frac{v_{\|}}{\Omega}(\nabla \times \boldsymbol{b})\right] , \tag{4.3.17}$$

for $\boldsymbol{E} = -\nabla\phi$. A special property of this guiding center equation is that (4.3.14) contains a term for the parallel drift which is of higher order in ϱ_{L}/L, and that $v_{\|}$ agrees with the usual guiding center velocity along the magnetic field line only in the leading order in ϱ_{L}/L. It is noted that the last term of (4.3.13) is identical to the curvature drift in Sect. 3.2 since $\boldsymbol{b} \cdot \nabla\boldsymbol{b} = -\boldsymbol{b} \times (\nabla \times \boldsymbol{b})$ and $\nabla \times \boldsymbol{B} = 0$:

$$\boldsymbol{v}_{\mathrm{c}} = \frac{v_{\|}^2}{\Omega}\boldsymbol{b} \times (\boldsymbol{b} \cdot \nabla\boldsymbol{b}) = \frac{v_{\|}^2}{\Omega}(\nabla \times \boldsymbol{b}) . \tag{4.3.18}$$

The Vlasov description follows as an exact consequence of the preceding guiding center equations of motion. Use of the phase space volume element

$$dV = \frac{2\pi}{m}B\,d\boldsymbol{x}\,dv_{\|}\,d\mu \tag{4.3.19}$$

leads to $S = 0$ as defined in (4.3.12). By noting that B is assumed to be stationary, S reduces to

$$S = \frac{1}{B}\left[\nabla \cdot \left(\frac{d\boldsymbol{x}}{dt}B\right) + \frac{\partial}{\partial v_{\|}}\left(\frac{dv_{\|}}{dt}B\right)\right] . \tag{4.3.20}$$

The resultant drift kinetic equation has the form

$$\frac{\partial f}{\partial t} + \boldsymbol{v}_{\mathrm{D}} \cdot \nabla f + \frac{dv_{\|}}{dt}\frac{\partial}{\partial v_{\|}} f = 0 \tag{4.3.21}$$

or alternatively,

$$\frac{\partial}{\partial t}(Bf) + \nabla \cdot (\boldsymbol{v}_{\mathrm{D}} Bf) + \frac{\partial}{\partial v_{\|}}\left(\frac{dv_{\|}}{dt}Bf\right) = 0 , \tag{4.3.22}$$

corresponding to (4.3.11). Here the nabla operator means that the spatial derivative is performed with $v_{\|}$ and μ kept constant. This section is largely based on [4.2] which is by the author's mistake missing in the previous editions.

4.4 Two-Fluids Theory

Another approach to describe a plasma with no net charge is as a continuous medium in real space. In this approximation, information about the velocity distribution function is averaged out and kinetic effects relating to wave-particle interactions, which will be discussed in Chap. 6, are neglected. The mass, momentum, and energy conservation equations are used to obtain a macroscopic description of the plasma.

The distribution function describing the dynamics of charged particles in phase space (μ-space) is given by $F_s(\boldsymbol{r}, \boldsymbol{v}, t)$, where $\boldsymbol{r}$ denotes the real space and $\boldsymbol{v}$ denotes the velocity space, respectively. As before, the suffix s denotes the particle species; $s = \mathrm{i}$ for an ion and $s = \mathrm{e}$ for an electron. The particle number in the phase space volume element $d^3\boldsymbol{r}d^3\boldsymbol{v}$ at time t is given by $F_s(\boldsymbol{r}, \boldsymbol{v}, t)d^3\boldsymbol{r}d^3\boldsymbol{v}$. The volume element should be large enough to treat $F_s(\boldsymbol{r}, \boldsymbol{v}, t)$ as a continuous function.

The particle density $n_s(\boldsymbol{r}, t)$ at the position $\boldsymbol{r}$ at time t is given by the integral of the distribution function over the velocity space:

$$n_s(\boldsymbol{r}, t) = \int F_s(\boldsymbol{r}, \boldsymbol{v}, t) d^3\boldsymbol{v} . \tag{4.4.1}$$

Particle balance in the real space volume element $d^3\boldsymbol{r}$ is expressed in terms of the time evolution of $n_s(\boldsymbol{r}, t)$ and the divergence of particle flux $\boldsymbol{\Gamma}_s(\boldsymbol{r}, t)$ as

$$\frac{\partial n_s}{\partial t} + \nabla \cdot \boldsymbol{\Gamma}_s = 0 . \tag{4.4.2}$$

This is the mass conservation equation, since integration over the closed volume under the boundary condition $\boldsymbol{v}_s \cdot \boldsymbol{n} = 0$ gives a constant total number of particles or $\int n_s d^3\boldsymbol{r} = \mathrm{const.}$ Here, $\boldsymbol{n}$ is the unit vector normal to the surface of the closed volume. In (4.4.2) the particle source term due to ionization of neutral atoms and particle sink term due to recombination between an electron and an ion are neglected, since we are concerned with fully ionized plasmas. The particle flux

$\boldsymbol{\Gamma}_s$ is defined by

$$\boldsymbol{\Gamma}_s(\boldsymbol{r},t) = \int \boldsymbol{v} F_s(\boldsymbol{r},\boldsymbol{v},t) d^3\boldsymbol{v} = n_s(\boldsymbol{r},t)\bar{\boldsymbol{v}}_s(\boldsymbol{r},t) , \tag{4.4.3}$$

where the bar denotes the average with respect to the velocity distribution function. The *continuity equation* (4.4.2) is required for both electrons and ions.

Hereafter v^l represents the component of the velocity vector along the l-direction in cartesian coordinates. The momentum of the l-direction is given by

$$\Pi_s^l(\boldsymbol{r},t) = \int m_s v^l F_s(\boldsymbol{r},\boldsymbol{v},t) d^3\boldsymbol{v} = n_s(\boldsymbol{r},t) m_s \overline{\boldsymbol{v}_s^l}(\boldsymbol{r},t) . \tag{4.4.4}$$

The l-component of the momentum flow in the unit area perpendicular to the k-direction is given by

$$P_s^{lk}(\boldsymbol{r},t) = \int m v^l v^k F_s(\boldsymbol{r},\boldsymbol{v},t) d^3\boldsymbol{v} = n_s(\boldsymbol{r},t) m_s \overline{v_s^l v_s^k} . \tag{4.4.5}$$

This quantity is called the momentum transfer tensor and is related to the pressure tensor. The velocity vector can be separated into an average velocity vector $\bar{\boldsymbol{v}}_s$ and a fluctuating $\tilde{\boldsymbol{v}}_s$, i.e, $\boldsymbol{v}_s = \bar{\boldsymbol{v}}_s + \tilde{\boldsymbol{v}}_s$. By using this relation, (4.4.5) can be rewritten as

$$P_s^{lk}(\boldsymbol{r},t) = n_s(\boldsymbol{r},t) m_s \bar{v}_s^l(\boldsymbol{r},t) \bar{v}_s^k(\boldsymbol{r},t) + \tilde{P}_s^{lk}(\boldsymbol{r},t) , \tag{4.4.6}$$

where

$$\tilde{P}_s^{lk}(\boldsymbol{r},t) = m_s \int \tilde{v}^l \tilde{v}^k F_s(\boldsymbol{r},\boldsymbol{v},t) d^3\boldsymbol{v} . \tag{4.4.7}$$

The quantity $\tilde{P}_s^{lk}$ is called the pressure tensor. In deriving (4.4.6), we used the relation $\overline{\tilde{v}_s^l} = 0$.

Hereafter the distribution function is assumed to be Maxwellian with temperature T_s. This assumption is not essential to obtain the macroscopic equations in a closed form. By including the average velocity $\bar{\boldsymbol{v}}_s$, the Maxwell distribution is given by

$$F_s(\boldsymbol{r},\boldsymbol{v}) = n_s(\boldsymbol{r}) \left[\frac{m_s}{2\pi T_s(\boldsymbol{r})}\right]^{3/2} \exp\left\{-\frac{m_s[\boldsymbol{v} - \bar{\boldsymbol{v}}_s(\boldsymbol{r})]^2}{2T_s(\boldsymbol{r})}\right\} . \tag{4.4.8}$$

This is called the local Maxwellian distribution since the average quantities, n_s, T_s and $\bar{\boldsymbol{v}}_s$ are spatially dependent. By substitution of (4.4.8) into (4.4.7), the pressure tensor is given by

$$\tilde{P}_s^{lk} = \begin{pmatrix} n_s T_s & 0 & 0 \\ 0 & n_s T_s & 0 \\ 0 & 0 & n_s T_s \end{pmatrix} . \tag{4.4.9}$$

Thus for a Maxwellian distribution the pressure tensor becomes isotropic and corresponds to a scalar pressure $P_s = n_s T_s$.

Occasionally, small deviations from the Maxwell distribution become important. Then the pressure tensor is given by

$$\tilde{P}_s^{lk} = n_s T_s \delta_{lk} + \delta \tilde{P}_s^{lk} , \tag{4.4.10}$$

where δ_{lk} is the Kronecker delta (unity for $l = k$ and zero for $l \neq k$). The off-diagonal component of $\delta \tilde{P}_j^{lk}$ is given by

$$\delta \tilde{P}_s^{lk} = \int m_s(\tilde{v}^l \tilde{v}^k - (\tilde{v}^l)^2 \delta_{lk}) F_s(\boldsymbol{r}, \boldsymbol{v}, t) d^3\boldsymbol{v} . \tag{4.4.11}$$

This quantity is related to the viscosity tensor which we will not discuss further. In real problems, even when there is a deviation from a Maxwellian distribution, it is often neglected. This is because the description of macroscopic or averaged quantities is rather insensitive to the details of the distribution function.

Now we consider the momentum balance in the real space volume element $d^3\boldsymbol{r}$. When electron-ion collision effects are negligible, the momentum balance is

$$\frac{\partial \Pi_s^l}{\partial t} + \sum_k \frac{\partial}{\partial x_k} P_s^{kl} = F_s^l , \tag{4.4.12}$$

where the second term on the left hand side corresponds to the momentum flow. The term F_s^l is the electromagnetic force per unit volume. This force is found by taking the average of the force $q_s(\boldsymbol{E} + \boldsymbol{v}_s \times \boldsymbol{B})$ for electrons or ions over the velocity space:

$$F_s^l = q_s n_s (\boldsymbol{E} + \bar{\boldsymbol{v}}_s \times \boldsymbol{B})^l . \tag{4.4.13}$$

The gravitational force on the charged particles is neglected since it is much smaller than the electromagnetic force. Then (4.4.12) can be written as

$$\frac{\partial}{\partial t}(n_s m_s \bar{v}_s^l) + \sum_k \frac{\partial}{\partial x_k}(n_s m_s \bar{v}_s^l \bar{v}_s^k)$$
$$= -\sum_k \frac{\partial}{\partial x_k} \tilde{P}_s^{lk} + q_s n_s (\boldsymbol{E} + \bar{\boldsymbol{v}}_s \times \boldsymbol{B})^l . \tag{4.4.14}$$

By using the continuity equation (4.4.2) in (4.4.14), we have

$$n_s m_s \left(\frac{\partial \bar{v}_s^l}{\partial t} + \sum_k \bar{v}_s^k \frac{\partial}{\partial x_k} \bar{v}_s^l \right)$$
$$= -\sum_k \frac{\partial}{\partial x_k} \tilde{P}_s^{lk} + q_s n_s (\boldsymbol{E} + \bar{\boldsymbol{v}}_s \times \boldsymbol{B})^l . \tag{4.4.15}$$

When the pressure tensor is assumed to be a scalar by (4.4.9), (4.4.15) becomes

$$n_s m_s \frac{d\bar{\boldsymbol{v}}_s}{dt} = -\boldsymbol{\nabla} P_s + q_s n_s (\boldsymbol{E} + \bar{\boldsymbol{v}}_s \times \boldsymbol{B}) , \tag{4.4.16}$$

where the left hand side is expressed in terms of the convective derivative, $d/dt = \partial/\partial t + \bar{\boldsymbol{v}}_s \cdot \boldsymbol{\nabla}$.

Next we will consider the energy balance in the volume element $d^3\boldsymbol{r}$ around the position $\boldsymbol{r}$. The energy density is defined by

$$\epsilon_s(\boldsymbol{r},t) = \int \frac{1}{2} m_s v^2 F_s(\boldsymbol{r},\boldsymbol{v},t) d^3\boldsymbol{v} = \frac{1}{2} m_s n_s \overline{v_s^2} \,. \tag{4.4.17}$$

The heat flow vector is defined by

$$Q_s^l(\boldsymbol{r},t) = \int \frac{1}{2} m_s v^2 v^l F_s(\boldsymbol{r},\boldsymbol{v},t) d^3\boldsymbol{v} = \frac{1}{2} n_s m_s \overline{v_s^2 v_s^l} \,. \tag{4.4.18}$$

The charged particles will be accelerated or decelerated by an electric field $\boldsymbol{E}$ in the direction of the force along $\bar{\boldsymbol{v}}_s$. The magnetic Lorentz force $\bar{\boldsymbol{v}}_s \times \boldsymbol{B}$, however, will not contribute to the energy change. Then the energy balance is described by

$$\frac{\partial \epsilon_s}{\partial t} + \sum_k \frac{\partial}{\partial x_k} Q_s^k = n_s q_s \bar{\boldsymbol{v}}_s \cdot \boldsymbol{E} \,. \tag{4.4.19}$$

Just as the momentum tensor can be separated into two parts by introducing $\boldsymbol{v}_s = \bar{\boldsymbol{v}}_s + \tilde{\boldsymbol{v}}_s$, so can the energy density:

$$\epsilon_s = \frac{1}{2} n_s m_s \bar{v}_s^2 + \frac{1}{2} n_s m_s \overline{\tilde{v}_s^2} \,. \tag{4.4.20}$$

The first term is the energy related to the macroscopic plasma flow and the second term is related to the internal energy. The heat flow vector is then given by

$$\begin{aligned} Q_s^l &= \frac{1}{2} m_s \int \left\{ \sum_k (\bar{v}^k + \tilde{v}^k)^2 \right\} (\bar{v}^l + \tilde{v}^l) F_s(\boldsymbol{r},\boldsymbol{v},t) d^3\boldsymbol{v} \\ &= \frac{1}{2} n_s m_s \bar{v}_s^2 \bar{v}_s^l + \frac{1}{2} n_s m_s (\overline{\tilde{v}_s^2} \bar{v}_s^l + 2 \sum_k \bar{v}_s^k \overline{\tilde{v}_s^k \tilde{v}_s^l}) + \frac{1}{2} n_s m_s \overline{\tilde{v}_s^2 \tilde{v}_s^l} \,. \end{aligned} \tag{4.4.21}$$

Here $\overline{\tilde{v}_s^l} = 0$ is used to obtain the last expression. The first term in (4.4.21) is the kinetic energy flow related to the average velocity. The second term is the energy flow related to the internal energy and is also called the convection of energy. The third term is related to the thermal conduction. Under the assumption of a Maxwell distribution, the following relations are obtained

$$\begin{aligned} \overline{(\tilde{v}_s^l)^2} &= \frac{1}{3} \overline{(\tilde{v}_s)^2} = \frac{T_s}{m_s} \\ \overline{\tilde{v}_s^l \tilde{v}_s^k} &= 0 \ (k \neq l) \\ \overline{\tilde{v}_s^2 \tilde{v}_s^l} &= 0 \,. \end{aligned} \tag{4.4.22}$$

By using these results we can reduce the energy density and the heat flow vector to the form

$$\begin{aligned} \epsilon_s &= \frac{1}{2} n_s m_s \bar{v}_s^2 + \frac{3}{2} n_s T_s \,, \\ Q_s^l &= \frac{1}{2} n_s m_s \bar{v}_s^2 \bar{v}_s^l + \frac{5}{2} n_s T_s \bar{v}_s^l \,. \end{aligned} \tag{4.4.23}$$

Substitution of these expressions into (4.4.19) gives

$$\begin{aligned} n_s m_s \sum_k \bar{v}_s^k \frac{\partial \bar{v}_s^k}{\partial t} + \frac{1}{2} m_s \bar{v}_s^2 \frac{\partial n_s}{\partial t} + \frac{3}{2}\frac{\partial}{\partial t}(n_s T_s) \\ + n_s m_s \sum_{k,l} \bar{v}_s^k \bar{v}_s^l \frac{\partial \bar{v}_s^l}{\partial x_k} + \frac{1}{2} m_s \bar{v}_s^2 \sum_k \frac{\partial}{\partial x_k}\left(n_s \bar{v}_s^k\right) \\ + \frac{5}{2} \sum_k \bar{v}_s^k \frac{\partial}{\partial x_k}(n_s T_s) + \frac{5}{2} n_s T_s \sum_k \frac{\partial \bar{v}_s^k}{\partial x_k} \\ = n_s q_s \bar{\boldsymbol{v}}_s \cdot \boldsymbol{E} . \end{aligned} \tag{4.4.24}$$

Use of the continuity equation and the equation of motion for $\partial n_s / \partial t$ and $\partial \bar{v}_s^k / \partial t$ yields

$$\frac{3}{2}\frac{\partial}{\partial t}(n_s T_s) + \frac{3}{2} \sum_k \bar{v}_s^k \frac{\partial}{\partial x_k}(n_s T_s) + \frac{5}{2} n_s T_s \sum_k \frac{\partial \bar{v}_s^k}{\partial x_k} = 0 . \tag{4.4.25}$$

It should be noted that the continuity equation can be written in a different form

$$\frac{dn_s}{dt} + n_s \sum_k \frac{\partial \bar{v}_s^k}{\partial x_k} = 0 . \tag{4.4.26}$$

Noting that the sum of the first and the second terms of (4.4.25) can be written in terms of the convective derivative and, by using (4.4.26), (4.4.25) becomes

$$\frac{3}{2}\frac{d}{dt}(n_s T_s) - \frac{5}{2} T_s \frac{dn_s}{dt} = 0 . \tag{4.4.27}$$

or

$$\frac{dP_s}{dn_s} = \frac{5}{3}\frac{P_s}{n_s} . \tag{4.4.28}$$

This relation is called the adiabatic law with the specific heat ratio $\gamma = 5/3$. It is valid when collisions are frequent and energy equipartition is established in a short time. In some rare cases of plasmas immersed in magnetic fields, equipartition of the energy along directions perpendicular to the magnetic field line is sufficient for the adiabatic assumption, but the energy exchange between the perpendicular directions and the parallel direction is incomplete. In this case $\gamma = 2$. The general form of (4.4.28) is therefore

$$\frac{dP_s}{dn_s} = \gamma \frac{P_s}{n_s} , \tag{4.4.29}$$

with $\gamma = (d+2)/d$ and d is the number of degrees of freedom.

In summary, the following set of equations is obtained by combining the continuity equation, momentum conservation, and energy conservation with the Maxwell equations:

$$\frac{\partial n_s}{\partial t} + \boldsymbol{\nabla} \cdot (n_s \bar{\boldsymbol{v}}_s) = 0 , \tag{4.4.30}$$

$$m_s n_s \left(\frac{\partial \bar{\boldsymbol{v}}_s}{\partial t} + (\bar{\boldsymbol{v}}_s \cdot \boldsymbol{\nabla}) \bar{\boldsymbol{v}}_s \right) = -\boldsymbol{\nabla} P_s + n_s q_s (\boldsymbol{E} + \bar{\boldsymbol{v}}_s \times \boldsymbol{B}) , \tag{4.4.31}$$

$$\frac{dP_s}{dn_s} = \gamma \frac{P_s}{n_s} , \tag{4.4.32}$$

$$\nabla \times \boldsymbol{E} = -\frac{\partial B}{\partial t} , \tag{4.4.33}$$

$$\frac{1}{\mu_0} \boldsymbol{\nabla} \times \boldsymbol{B} = (n_i q_i \bar{\boldsymbol{v}}_i + n_e q_e \bar{\boldsymbol{v}}_e) + \varepsilon_0 \frac{\partial \boldsymbol{E}}{\partial t} , \tag{4.4.34}$$

$$\boldsymbol{\nabla} \cdot \boldsymbol{E} = \frac{1}{\varepsilon_0} (q_i n_i + q_e n_e) , \tag{4.4.35}$$

$$\boldsymbol{\nabla} \cdot \boldsymbol{B} = 0 . \tag{4.4.36}$$

When the ions have charge state specified by Z, $q_i = Ze$ and $q_e = -e$, and μ_0 and ε_0 are the vacuum permittivity and vacuum dielectric constant, respectively.

Equation (4.4.35) is the Poisson equation of an electric field due to space charges, (4.4.33) is the induction equation and (4.4.34) is Ampere's law determining the magnetic field due to conduction current and displacement current. These are the *two-fluid equations*, where one fluid is pure electrons and the other pure ions. There are sixteen unknown quantities, i.e., $n_e, n_i, P_e, P_i, \bar{\boldsymbol{v}}_e, \bar{\boldsymbol{v}}_i, \boldsymbol{E}$ and $\boldsymbol{B}$ and eighteen equations. However, it should be noted that the divergence in (4.4.33) corresponds to the time derivative of (4.4.36). That is, (4.4.33) describes the time evolution of the initial condition set by (4.4.36). Similarly, the divergence of (4.4.34) is equivalent to taking the time derivative of (4.4.35), when we include the continuity equation (4.4.30). Thus, the Poisson equation (4.4.35) also specifies an initial condition. That leaves 16 remaining independent equations, equal to the number of independent variables. This argument shows that the two-fluid equations are closed and that they describe the dynamics of both an electron fluid and an ion fluid in self-consistent electromagnetic fields.

4.5 Resistive Magnetohydrodynamics (MHD)

When the collisional momentum exchange is considerable, the equation of motion for an electron fluid becomes

$$m_e n_e \left[\frac{\partial \boldsymbol{u}_e}{\partial t} + (\boldsymbol{u}_e \cdot \boldsymbol{\nabla}) \boldsymbol{u}_e \right] = -\boldsymbol{\nabla} P_e + n_e q_e (\boldsymbol{E} + \boldsymbol{u}_e \times \boldsymbol{B}) + \boldsymbol{R} , \tag{4.5.1}$$

and for an ion fluid

$$m_i n_i \left[\frac{\partial \boldsymbol{u}_i}{\partial t} + (\boldsymbol{u}_i \cdot \boldsymbol{\nabla}) \boldsymbol{u}_i \right] = -\boldsymbol{\nabla} P_i + n_i q_i (\boldsymbol{E} + \boldsymbol{u}_i \times \boldsymbol{B}) - \boldsymbol{R} , \tag{4.5.2}$$

where

$$\boldsymbol{R} = -n_e m_e(\boldsymbol{u}_e - \boldsymbol{u}_i)/\tau_{ei} \,. \tag{4.5.3}$$

Hereafter $\bar{\boldsymbol{v}}_s$ is denoted by $\boldsymbol{u}_s$. The collision time between electron and ion is denoted by τ_{ei}.

In this section the resistive one-fluid MHD equations will be derived from the two-fluid equations. By taking the sum of the equations of motion (4.5.1, 2),

$$m_e n_e \frac{d\boldsymbol{u}_e}{dt} + m_i n_i \frac{d\boldsymbol{u}_i}{dt} = -\boldsymbol{\nabla} P + \boldsymbol{J} \times \boldsymbol{B} + \sigma_e \boldsymbol{E} \tag{4.5.4}$$

is given, where $\sigma_e = -en_e + Zen_i$, $\boldsymbol{J} = -en_e\boldsymbol{u}_e + Zen_i\boldsymbol{u}_i$ and $P = P_e + P_i$. It is noted that the mass density ϱ and the average velocity $\boldsymbol{u}$ can be approximated by

$$\begin{aligned} \varrho &= n_i m_i + n_e m_e = n_i m_i \\ \boldsymbol{u} &= (\boldsymbol{u}_i n_i m_i + \boldsymbol{u}_e n_e m_e)/\varrho = \boldsymbol{u}_i \,. \end{aligned} \tag{4.5.5}$$

In the left-hand side of (4.5.4), the electron inertia term is usually much smaller than the ion inertia term and is therefore negligible. Then we have the equation

$$\varrho\left(\frac{\partial \boldsymbol{u}}{\partial t} + (\boldsymbol{u} \cdot \nabla)\boldsymbol{u}\right) = -\boldsymbol{\nabla} P + \boldsymbol{J} \times \boldsymbol{B} + \sigma_e \boldsymbol{E} \,. \tag{4.5.6}$$

as the equation of motion for one fluid; the plasma is approximated by a single magneto-fluid.

Similarly, by taking the sum of the electron continuity equation and the ion continuity equation, we get

$$\frac{\partial \varrho}{\partial t} + \boldsymbol{\nabla} \cdot (\varrho \boldsymbol{u}) = 0 \tag{4.5.7}$$

as the one-fluid continuity equation.

It is noted that the energy conservation equation or the adiabatic law (4.4.29) can be written as

$$\frac{d}{dt}(P_s n_s^{-\gamma}) = 0 \,. \tag{4.5.8}$$

After introducing charge neutrality $n_e = Zn_i = n$ and taking the sum of the adiabatic laws for the electron fluid and the ion fluid, we get

$$\frac{d}{dt}(Pn^{-\gamma}) = 0 \,. \tag{4.5.9}$$

If we use the continuity equation for the density instead of ϱ, (4.5.9) can be rewritten as

$$\frac{\partial P}{\partial t} + (\boldsymbol{u} \cdot \nabla)P + \gamma P \nabla \cdot \boldsymbol{u} = 0 \,. \tag{4.5.10}$$

From the equation of motion for an electron fluid (4.5.1), an expression for the relation between the electric field $\boldsymbol{E}$ and the current density $\boldsymbol{J}$, i.e., Ohm's law, can be derived. When the electron inertia term is negligible, Ohm's law is written as

$$\boldsymbol{E} + \boldsymbol{u} \times \boldsymbol{B} = \frac{1}{en}(\boldsymbol{J} \times \boldsymbol{B} - \nabla P_e + \boldsymbol{R}) , \tag{4.5.11}$$

where charge neutrality is assumed. The frictional force given in (4.5.3) can also be expressed as

$$\boldsymbol{R} = \frac{m_e}{\tau_{ei} e} \boldsymbol{J} = en\eta \boldsymbol{J} , \tag{4.5.12}$$

where we introduce the resistivity $\eta = m_e/n_e e^2 \tau_{ei}$. When the current drift velocity $\boldsymbol{u}_d = -\boldsymbol{J}/en$ is much smaller than $\boldsymbol{u}$, and $\beta_e = nT_e/(B^2/2\mu_0) < 1$, the right-hand side of (4.5.11) is simply replaced by $\eta \boldsymbol{J}$. Then Ohm's law becomes

$$\boldsymbol{E} + \boldsymbol{u} \times \boldsymbol{B} = \eta \boldsymbol{J} . \tag{4.5.13}$$

By taking the difference of the continuity equations for an electron fluid and an ion fluid we have

$$\frac{\partial \sigma_e}{\partial t} + \nabla \cdot \boldsymbol{J} = 0 . \tag{4.5.14}$$

For the electromagnetic fields, the same equations as in the two-fluids model are employed:

$$\frac{\partial \boldsymbol{B}}{\partial t} = -\nabla \times \boldsymbol{E} , \tag{4.5.15}$$

$$\frac{1}{c^2} \frac{\partial \boldsymbol{E}}{\partial t} + \mu_0 \boldsymbol{J} - \nabla \times \boldsymbol{B} = 0 , \tag{4.5.16}$$

where $c^2 = (\varepsilon_0 \mu_0)^{-1}$.

Thus, the resistive one-fluid MHD equations are closed by (4.5.6, 7, 10, 13–16) for the fifteen variables $\{\varrho, \sigma_e, P, \boldsymbol{u}, \boldsymbol{B}, \boldsymbol{E}, \boldsymbol{J}\}$. As before, the equations

$$\nabla \cdot \boldsymbol{E} = \frac{\sigma_e}{\varepsilon_0} , \tag{4.5.17}$$

$$\nabla \cdot \boldsymbol{B} = 0 \tag{4.5.18}$$

are initial conditions. It should be noted that the resistive MHD equations are actually closed when the resistivity η is given independently. In Coulomb collisions the resistivity is proportional to $T_e^{-3/2}$. But the pressure P also includes T_e. In order to include the temperature dependence of the resistivity within the resistive MHD equations, we need an expression for the electron temperature evolution or the assumption $T_e = T_i = P/2n$.

4.1. According to the Liouville theorem in classical mechanics, the one-particle phase space distribution $F(\boldsymbol{r}, \boldsymbol{p}, t)$ satisfies the Liouville equation

$$\frac{\partial F}{\partial t} = \frac{\partial H}{\partial \boldsymbol{r}} \cdot \frac{\partial F}{\partial \boldsymbol{p}} - \frac{\partial H}{\partial \boldsymbol{p}} \cdot \frac{\partial F}{\partial \boldsymbol{r}}, \tag{4.P.1}$$

where H is the single particle Hamiltonian

$$H = \frac{1}{2m}|\boldsymbol{p} - q\boldsymbol{A}|^2 + q\phi \tag{4.P.2}$$

and $\boldsymbol{p}$ is the canonical momentum $\boldsymbol{p} = m\boldsymbol{v} + q\boldsymbol{A}$. Here $\boldsymbol{A}$ and ϕ are, respectively, the vector and the scalar potential. Using the equation of motion

$$\frac{d\boldsymbol{r}}{dt} = \boldsymbol{v} = \frac{\partial H}{\partial \boldsymbol{p}}, \quad \frac{d\boldsymbol{p}}{dt} = -\frac{\partial H}{\partial \boldsymbol{r}}, \tag{4.P.3}$$

derive the Vlasov equation (4.2.4) from the Liouville equation.

4.2. Confirm (4.3.18) by showing that $\boldsymbol{b} \cdot (\nabla \times \boldsymbol{b}) = 0$.

4.3. For the phase space volume element of (4.3.19), show that $S = 0$, where S is given by (4.3.20), by noting that $d\boldsymbol{x}/dt$ and $dv_{\|}/dt$ are given by (4.3.13, 14), respectively.

4.4. Taking the moment of the Vlasov equation (4.2.4), derive the continuity equation (4.4.2) and the momentum balance equation (4.4.12) with (4.4.13).

5. Electrostatic Response

In general, plasma motion induced by an electric field produces an electromagnetic field. However, in many cases of interest, the field produced by the plasma motion can be represented solely by an electric field, without being accompanied by magnetic field perturbations. Such is the case when a charge density fluctuation is produced in a plasma with no magnetic field. In this case, the induced plasma motion is parallel to the electric field $\boldsymbol{E}$ due to the charge fluctuation, therefore the modification of the field due to the plasma motion is also parallel to $\boldsymbol{E}$. As a result, the curl of $\boldsymbol{E}$ vanishes and no magnetic field results. A similar situation takes place when an electric field perturbation is along a static magnetic field. In these cases, the electric field can be expressed in terms of an electrostatic potential $\boldsymbol{E} = -\nabla\phi$. The plasma response that can be expressed by an electrostatic potential alone is called the *electrostatic response*. In this chapter, we shall consider only such responses.

As mentioned in Chap. 2, plasma has a tendency to keep local charge neutrality. This electrostatic response is due to the local electric polarization of the plasma, and is called the *Debye shielding*. Section 5.1 deals with this Debye shielding. The following three sections treat oscillatory electrostatic responses or *electrostatic waves* in plasma without magnetic field, namely, the *electron plasma oscillation* in Sect. 5.2, the *two-stream instability* in Sect. 5.3, and the *ion acoustic wave* in Sect. 5.4.

When the electric field perturbation is at an angle to the magnetic field, the induced plasma motion has a component perpendicular to both the electric and magnetic fields due to the $\boldsymbol{E} \times \boldsymbol{B}$ drift. As a result, a variation perpendicular to $\boldsymbol{E}$ is induced, or $\nabla \times \boldsymbol{E} \neq 0$, so that a magnetic perturbation is produced. Even in this case, one often encounters the situation where the electric field is, to a good approximation, expressible by an electrostatic potential alone (*electrostatic approximation*). As an example of such cases, we consider in Sect. 5.5 the *electrostatic drift waves* which are the waves that often cause enhanced transport of plasma across the magnetic field.

The electrostatic responses considered in Sect. 5.1–5 are cases of *linear response*, that is, the plasma response is linear with respect to the electric field amplitude, which is usually true when the latter is sufficiently small. When the amplitude becomes large, for example, through an instability, nonlinear effects have to be invoked. Naturally, the theoretical treatment of nonlinear effects is much more difficult than that of the linear response; we shall present some of the simplest examples of nonlinear waves in the last section.

The method used in this chapter is based on the two-fluid model described in Sect. 4.4. For more rigorous analysis, one has to use the Vlasov model described in Chap. 6. The most important effect that requires the Vlasov model is the resonant exchange of energy between the individual particle motion and the wave motion. Such effects become important when the phase velocity of the wave is comparable to the thermal speed of the particles. Conversely, if the phase velocity of the wave is either much larger or smaller than the thermal speed of the particles, the two-fluid model usually gives a good approximation.

5.1 Debye Shielding

5.1.1 Debye Potential

The electrostatic potential produced by a charge e located at distance r is

$$\phi(r) = \frac{e}{4\pi\varepsilon_0 r} \tag{5.1.1}$$

The plasma consists of many charged particles and the electrostatic potential inside the plasma is given by the sum of all the potentials, in the form given in (5.1.1). Since the potential produced by a positive (or negative) charge attracts negative (or positive) charge, the net potential around a specific charge becomes smaller than that given by (5.1.1), approaching zero much faster than $1/r$ as r increases. This is called the *Debye shielding* and is one of the most important properties of the plasma.

Consider the potential $\phi(\boldsymbol{r})$ near an ion of positive charge q placed at the origin. The Poisson equation for $\phi(\boldsymbol{r})$ can be written as

$$-\nabla^2\phi(\boldsymbol{r}) = \frac{q}{\varepsilon_0}\delta[\boldsymbol{r}] + \frac{e}{\varepsilon_0}\left[Zn_{\mathrm{i}}(\boldsymbol{r}) - n_{\mathrm{e}}(\boldsymbol{r})\right] \tag{5.1.2}$$

where $\delta[\boldsymbol{r}](=\delta[x]\delta[y]\delta[z])$ is the product of Dirac delta functions and as before n_{i} and n_{e} are the ion and electron particle density, respectively. We consider the time scale in which the ion at the origin can be treated at rest. In this time scale, most of the ions do not respond to the potential, so that the induced ion density perturbation can be ignored, that is, n_{i} can be treated as being constant. The electrons can quickly respond to the potential, however, and hence n_{e} is perturbed by $\phi(\boldsymbol{r})$. Neglecting the electron inertia [the left hand side of (4.4.16)] and assuming the isothermal response, $P_{\mathrm{e}}(\boldsymbol{r}) = n_{\mathrm{e}}(\boldsymbol{r})T_{\mathrm{e}}$, we obtain from (4.4.16) with $\boldsymbol{E} = -\nabla\phi$ and $\boldsymbol{B} = 0$ the relation

$$n_{\mathrm{e}}(\boldsymbol{r}) = n_0 \exp\left[e\phi(\boldsymbol{r})/T_{\mathrm{e}}\right] \tag{5.1.3}$$

where n_0 is the initial unperturbed electron density. This relation describes the static balance of the electrostatic potential force and the electron pressure gradient force and is called the *Boltzmann relation*. Note that in the presence of a static magnetic field, the same relation follows from the electron equation of motion along the magnetic field. If we assume for simplicity that the ions are singly ionized, then we have $n_{\mathrm{i}} = n_0$ and (5.1.2) becomes

$$-\nabla^2\phi = \frac{n_0 e}{\varepsilon_0}\left[1 - \exp(e\phi/T_e)\right] + \frac{q}{\varepsilon_0}\delta[\boldsymbol{r}] \,. \tag{5.1.4}$$

We further linearize this equation with respect to ϕ by considering the region $e\phi \ll T_e$ and look for a spatially symmetric solution, $\phi(\boldsymbol{r}) = \phi(r)$. Equation (5.1.4) then becomes

$$\frac{1}{r^2}\frac{d}{dr}r^2\frac{d\phi}{dr} - \frac{e^2 n_0}{\varepsilon_0 T_e}\phi = -\frac{q}{\varepsilon_0}\delta[r] \tag{5.1.5}$$

which can be solved under the boundary condition that $\phi(r) \to 0$ for $r \to \infty$ as

$$\phi(r) = \frac{q}{4\pi\varepsilon_0 r}\exp(-r/\lambda_D) \tag{5.1.6}$$

where λ_D is the Debye length defined by

$$\lambda_D = \sqrt{\varepsilon_0 T_e / n_0 e^2} \,. \tag{5.1.7}$$

The potential given by (5.1.6) is called the *Debye potential* and is shown in Fig. 5.1, together with the bare Coulomb potential given by (5.1.1). We can see that the Debye potential goes to zero much faster than the Coulomb potential for $r > \lambda_D$. In practice, the Debye potential is effective only in the region $r < \lambda_D$, being almost perfectly screened by the electrons in the region $r \gg \lambda_D$.

5.1.2 Physical Picture of Debye Shielding

As before, consider an ion of charge q at rest at the origin. An electron of speed v passing by at a distance R away from the ion feels an attractive force equal to $eq/4\pi\varepsilon_0 R^2$ during the transit time τ which is on the order of $\sqrt{2}R/v$. As

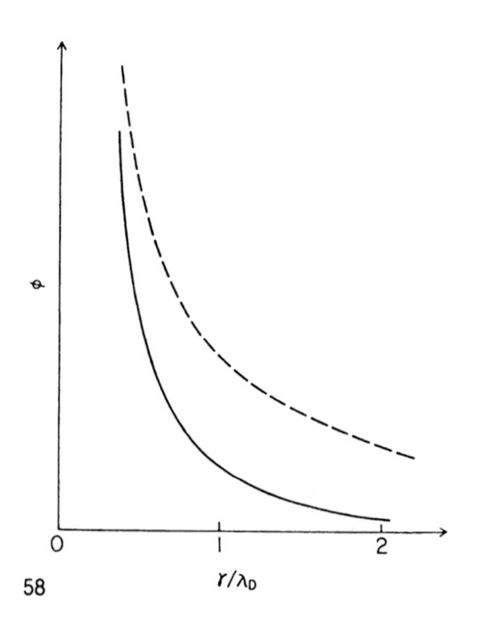

Fig. 5.1. Debye potential (*solid line*) and unshielded Coulomb potential (*dashed line*) as a function of a reduced length

a result, the electron orbit is displaced toward the origin by an amount $\Delta r \sim \tau^2 eq/8\pi\varepsilon_0 R^2 m \sim eq/4\pi\varepsilon_0 mv^2$. There are n_0 electrons in the unit volume, so that the charge density induced at the $r = R$ surface due to the displacement of the electron orbit is $-en_0 \langle \Delta r \rangle$, where the angled brackets $\langle\ \rangle$ denote an average. The potential due to the ion is completely screened outside the radius R if the net induced charge at $r = R$, $-4\pi R^2 en_0 \langle \Delta r \rangle$, equals $-q$. Using the above result for Δr, we obtain the value of R by

$$R^2 \sim \varepsilon_0 / \left[n_0 e^2 \left\langle 1/mv^2 \right\rangle \right] . \tag{5.1.8}$$

Approximating $\langle 1/mv^2 \rangle$ by $1/\langle mv^2 \rangle \sim 1/T_e$, we get $R^2 \sim \lambda_D^2$. We can see from this argument that low energy electrons contribute more efficiently to the Debye shielding than high energy electrons. This is because the orbit of a low energy electron has more time to be deflected than that of a high energy electron.

Three comments should be added. First, the linear approximation used to derive (5.1.5) breaks down when $r \to 0$. Equation (5.1.6) shows that the condition $e\phi \ll T_e$ is satisfied when

$$\frac{1}{n_0 \lambda_D^3} \frac{\lambda_D}{r} \exp\left(-r/\lambda_D\right) \ll 1$$

or unless

$$r < \lambda_D / n_0 \lambda_D^3 . \tag{5.1.9}$$

Since for such small values of r the potential due to the ion is close to the bare Coulomb potential, the Debye potential of (5.1.6) is, in fact, a good approximation for the entire range of r. Second, in the linear approximation the Boltzmann relation is reduced to the form

$$\frac{n_e - n_0}{n_0} = \frac{e\phi}{T_e} \tag{5.1.10}$$

which implies that the electron density perturbation is proportional to (or in phase with) the electrostatic potential. For a large spacial scale the local charge neutrality given by $n_i = n_e$ is satisfied due to the Debye shielding, so that in this case the ion density perturbation is also in phase with the electrostatic potential. Third, since the Boltzmann relation can be derived even in the presence of a static magnetic field, the Debye potential can also be derived irrespective of the presence of a static magnetic field, provided that the potential is isotropic.

5.2 Electron Plasma Oscillation

Consider a spatially uniform plasma free from magnetic fields. Suppose that in this plasma an electron density perturbation is produced which depends only on x and t. Such a perturbation can be made by a displacement in the x-direction of the electron fluid relative to the ion fluid. Then an electric field is produced and acts to bring the electron fluid back to the original position. Because of the electron inertia, however, the electron fluid overshoots the original position and must reverse itself again. This back and forth oscillation takes place at a high frequency called the *electron plasma frequency* which is a function of the electron mass. Thus, the response of the heavy ions to this oscillation can be ignored.

We examine this oscillation by using the two-fluid equations. We first consider the case in which the plasma pressure gradient force is negligible compared to the electric field force. This approximation is equivalent to neglecting the plasma temperature, and is referred to as the *cold plasma model* (Sect. 7.3). Since no magnetic field is involved, the independent variables are the density n_s, the fluid velocity $\boldsymbol{u}_s$, and the electric field $\boldsymbol{E}$, or equivalently, the electrostatic potential ϕ. The two-fluid equations for these variables are

$$\frac{\partial n_s}{\partial t} + \nabla \cdot (n_s \boldsymbol{u}_s) = 0 \tag{5.2.1}$$

$$m_s n_s \left[\frac{\partial \boldsymbol{u}_s}{\partial t} + (\boldsymbol{u}_s \cdot \nabla)\boldsymbol{u}_s \right] = -q_s n_s \nabla \phi \tag{5.2.2}$$

$$-\nabla^2 \phi = \frac{e}{\varepsilon_0}(n_\mathrm{i} - n_\mathrm{e}) \tag{5.2.3}$$

where in (5.2.3) we assumed that the ions are singly ionized and set $q_\mathrm{e} = -e$, $q_\mathrm{i} = e$. We denote the perturbations due to the electron displacement by superscript 1,

$$\begin{aligned} n_s(\boldsymbol{r}, t) &= n_0 + n_s^1(\boldsymbol{r}, t) \\ \boldsymbol{u}_s(\boldsymbol{r}, t) &= \boldsymbol{u}_s^1(\boldsymbol{r}, t) \\ \phi(\boldsymbol{r}, t) &= \phi^1(\boldsymbol{r}, t) \end{aligned} \tag{5.2.4}$$

where we assumed that $\boldsymbol{u}_s$ and ϕ are zero before the electron displacement. Substituting (5.2.4) into the electron continuity equation and neglecting the terms in second order with respect to the perturbation (i.e., using the linear approximation), we have

$$\frac{\partial n_\mathrm{e}^1}{\partial t} + n_0 \nabla \cdot \boldsymbol{u}_\mathrm{e}^1 = 0 \, . \tag{5.2.5}$$

Similarly, the electron equation of motion in the linear approximation becomes

$$m_\mathrm{e} \frac{\partial \boldsymbol{u}_\mathrm{e}^1}{\partial t} = e \nabla \phi^1 . \tag{5.2.6}$$

As mentioned above, the ion response is neglected, i.e., $n_\mathrm{i}^1 = 0$, $\boldsymbol{u}_\mathrm{i}^1 = 0$. Then the Poisson equation (5.2.3) in the linear approximation becomes

$$\nabla^2 \phi^1 = \frac{e}{\varepsilon_0} n_\mathrm{e}^1 \, . \tag{5.2.7}$$

Since we allow the perturbed quantities to depend only on x and t, we can express them in terms of a superposition of plane waves of the form $\exp(\mathrm{i}kx - \mathrm{i}\omega t)$ as

$$\begin{pmatrix} n_e^1 \\ \boldsymbol{u}_e^1 \\ \phi_e^1 \end{pmatrix} = \sum_k \int \frac{dw}{2\pi} \begin{pmatrix} \tilde{n}_e^1 \\ \tilde{\boldsymbol{u}}_e^1 \\ \tilde{\phi}^1 \end{pmatrix} \exp(\mathrm{i}kx - \mathrm{i}wt) , \tag{5.2.8}$$

where for simplicity, we suppressed the dependence of $\tilde{n}_e^1, \tilde{\boldsymbol{u}}_e^1$, and $\tilde{\phi}^1$ on k and ω. Equation (5.2.8) corresponds to the Fourier representation of the variables, k and ω corresponding to the wavenumber and frequency of the plane wave. Since the Fourier components of different sets of (k, ω) are independent of each other, (5.2.5–7) can be reduced to the form

$$\begin{aligned} \mathrm{i}\omega\tilde{n}_e^1 &= \mathrm{i}n_0k\tilde{u}_{ex}^1 \\ -\mathrm{i}\omega m_e\tilde{u}_{ex}^1 &= \mathrm{i}ke\tilde{\phi}^1 \\ -k^2\tilde{\phi}^1 &= (e/\varepsilon_0)\tilde{n}_e^1 . \end{aligned} \tag{5.2.9}$$

These equations can be derived from (5.2.5–7) by simply replacing $\partial/\partial t$ by $-\mathrm{i}\omega$ and ∇ by $\mathrm{i}\boldsymbol{k}$ with $\boldsymbol{k} = k\hat{\boldsymbol{x}}$, where $\hat{\boldsymbol{x}}$ is the unit vector in the x-direction. Such a replacement is always possible in the linear approximation for a spatially uniform and temporally stationary unperturbed system. (A more rigorous treatment is given in Chap. 7.) Solving (5.2.9), we get

$$(\omega^2 - \omega_{pe}^2)\tilde{u}_{ex}^1 = 0 , \tag{5.2.10}$$

where ω_{pe} is the electron plasma frequency defined by

$$\omega_{pe} = \sqrt{\frac{n_0e^2}{\varepsilon_0m_e}} . \tag{5.2.11}$$

Equation (5.2.10) has a nontrivial solution ($\tilde{u}_{ex}^1 \neq 0$) only when $\omega^2 = \omega_{pe}^2$. This is the *electron plasma oscillation*. The frequency depends only on the particle density and is independent of the wavenumber k. Therefore, the group velocity vanishes and the oscillation does not propagate inside the plasma. As shown in Chap. 7, the frequency and the wavenumber of a linear wave in a uniform, stationary medium satisfy a relation called the *dispersion relation*. In the present special case, however, ω is independent of k.

If, however, the finite electron temperature is taken into account, the plasma wave acquires a nonzero group velocity and propagates in the plasma. This effect can be included by adding the pressure gradient force $-\nabla P_e^1$ on the right-hand side of (5.2.6). Since the motion takes place only in the x-direction, we can treat it as a one-dimensional adiabatic motion for which the specific heat ratio is $\gamma = 3$. We then obtain

$$-\nabla P_e^1 = -3T_e\nabla n_{ex}^1 = -3T_e\frac{dn_{ex}^1}{dx}\hat{\boldsymbol{x}} . \tag{5.2.12}$$

Using the Fourier representation (5.2.8) and replacing d/dx in (5.2.12) by $\mathrm{i}k$, we obtain in the same way as before,

$$\left[\omega^2 - \left(\omega_{pe}^2 + 3T_ek^2/m_e\right)\right]\tilde{u}_{ex}^1 = 0 \tag{5.2.13}$$

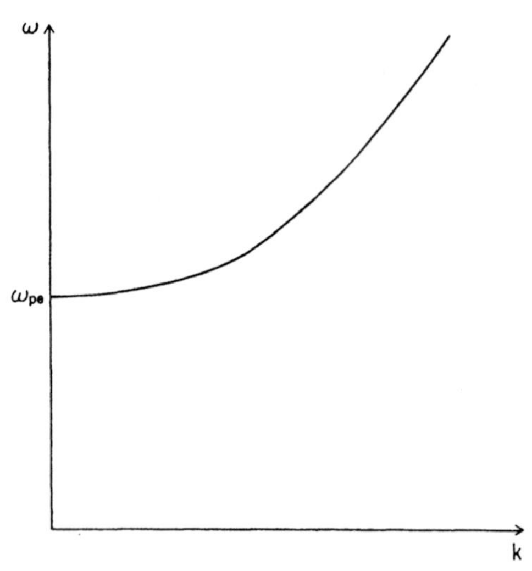

Fig. 5.2. Dispersion relation of an electron plasma wave

from which we obtain the dispersion relation

$$\omega^2 = \omega_{pe}^2 + 3k^2 v_{Te}^2 , \tag{5.2.14}$$

where $v_{Te} = \sqrt{T_e/m_e}$ is the thermal speed of the electrons. From (5.2.14), we have the group velocity,

$$v_g = \frac{d\omega}{dk} = \frac{3v_{Te}^2}{v_{ph}} , \tag{5.2.15}$$

where $v_{ph} = \omega/k$ is the *phase velocity* of the wave. The wave satisfying the dispersion relation (5.2.14), shown in Fig. 5.2, is called the *electron plasma wave* or the *Langmuir wave*.

Two remarks are in order. First, for small k or long wavelength ($2\pi/k$), the phase velocity is larger than the thermal speed of the electrons, but the group velocity is lower. This implies that as the wavelength increases the density gradient becomes smaller and hence, the energy transfer by the pressure gradient force is reduced. Second, as k becomes as large as $\lambda_D^{-1} = \omega_{pe}/v_{Te}$, the electron plasma wave suffers a strong damping due to the resonant wave-particle interaction. This will be discussed further in Chap. 6. Therefore, the dispersion relation (5.2.14) is valid only in the long wavelength region $k\lambda_{De} \ll 1$. There the pressure gradient term $3k^2 v_{Te}^2$ is small compared to the electric field term ω_{pe}^2 and the phase velocity (group velocity) is large (small) compared to the thermal speed of the electrons.

5.3 Two-Stream Instability

Consider the case in which the electron fluid is moving with speed $\boldsymbol{u}_e^0$ relative to the ion fluid in a uniform plasma with $n_e^0 = n_i^0 = n_0$. We use the coordinate system moving with the ion fluid and treat the plasma by the cold plasma model ($T_e = T_i = 0$). The magnetic field is assumed to be absent. The linear approximation for the perturbation (denoted by superscript 1 as before) applied to the two-fluid equations with the electron flow $\boldsymbol{u}_e^0$ yields

$$m_i n_0 \frac{\partial \boldsymbol{u}_i^1}{\partial t} = -e n_0 \nabla \phi^1 , \tag{5.3.1}$$

$$m_e n_0 \left(\frac{\partial \boldsymbol{u}_e^1}{\partial t} + \boldsymbol{u}_e^0 \cdot \nabla \boldsymbol{u}_e^1 \right) = e n_0 \nabla \phi^1 , \tag{5.3.2}$$

$$\frac{\partial n_i^1}{\partial t} + n_0 \nabla \cdot \boldsymbol{u}_i^1 = 0 , \tag{5.3.3}$$

$$\frac{\partial n_e^1}{\partial t} + n_0 \nabla \cdot \boldsymbol{u}_e^1 + \boldsymbol{u}_e^0 \cdot \nabla n_e^1 = 0 , \tag{5.3.4}$$

$$\nabla^2 \phi^1 = \frac{e}{\varepsilon_0} \left(n_i^1 - n_e^1 \right) , \tag{5.3.5}$$

which are a closed set of equations for the perturbed quantities. We choose the x-direction to be in the direction of $\boldsymbol{u}_e^0$ and consider the perturbation which depends only on x and t. Using the Fourier representation of (5.2.8), we obtain from (5.3.1, 2),

$$\tilde{u}_{ix}^1 = \frac{e}{m_i \omega} k \tilde{\phi}^1$$

$$\tilde{u}_{ex}^1 = -e k \tilde{\phi}^1 / m_e \left(\omega - k u_e^0 \right) \tag{5.3.6}$$

and from (5.3.3, 4),

$$\begin{aligned} \tilde{n}_i^1 &= \frac{k}{\omega} n_0 \tilde{u}_{ix}^1 = \frac{e n_0 k^2}{m_i \omega^2} \tilde{\phi}^1 \\ \tilde{n}_e^1 &= \frac{k n_0}{\omega - k u_e^0} \tilde{u}_{ex}^1 = -\frac{e n_0 k^2}{m_e (\omega - k u_e^0)^2} \tilde{\phi}^1 . \end{aligned} \tag{5.3.7}$$

Substitution of (5.3.7) into the Poisson equation (5.3.5) then yields

$$k^2 \left(1 - \frac{n_0 e^2}{\varepsilon_0} \left[\frac{1}{m_i \omega^2} + \frac{1}{m_e (\omega - k u_e^0)^2} \right] \right) \tilde{\phi}^1 = 0 \tag{5.3.8}$$

from which the following dispersion relation results:

$$\frac{\omega_{pi}^2}{\omega^2} + \frac{\omega_{pe}^2}{(\omega - k u_e^0)^2} = 1 , \tag{5.3.9}$$

where $\omega_{pi} = (n_0 e^2 / \varepsilon_0 m_i)^{1/2}$ is the ion plasma frequency.

Let us investigate the solution of this dispersion relation. We let k be real and solve for ω. Equation (5.3.9) is fourth order in ω, so there are four solutions for ω for a given real value of k. If they are all real, then all solutions correspond to purely oscillating waves. If, on the other hand, a complex solution is obtained, its complex conjugate must also be a solution. We denote such a pair of solutions by $\omega = \omega_r \pm i\gamma$, and they yield Fourier components varying in the form

$$\exp(ikx - i\omega_r t)\exp(\pm\gamma t) . \tag{5.3.10}$$

Without loss of generality, we can choose γ to be positive. Then, since the exp $(-\gamma t)$ solution decreases with time, but the exp $(+\gamma t)$ term grows in time, an instability ensues. This demonstrates that we can determine the stability of a wave by studying the reality of the solutions.

Noting that $\omega_{pe}^2/\omega_{pi}^2 = m_i/m_e$ and setting $\omega/\omega_{pe} = \xi, ku_e^0/\omega_{pe} = \eta$, we can write the left hand side of (5.3.9) as

$$\frac{m_e}{m_i}\frac{1}{\xi^2} + \frac{1}{(\xi - \eta)^2} \equiv F(\xi, \eta) . \tag{5.3.11}$$

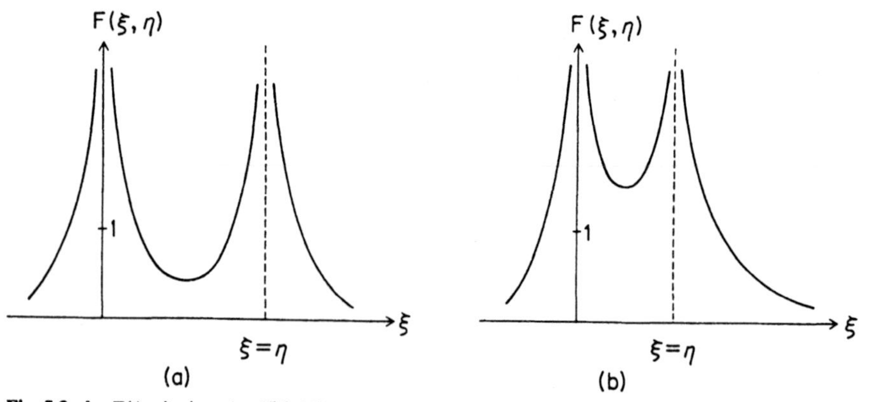

Fig. 5.3a,b. $F(\xi, \eta)$ given by (5.3.11) versus ξ. **(a)** large η; **(b)** small η

Our problem is to find the number of real solutions for $F(\xi, \eta) = 1$ for given real value of η. The function $F(\xi, \eta)$ is shown for large and small η in Fig. 5.3a and 5.3b, respectively. For large η, we have four real solutions for $F(\xi, \eta) = 1$. On the other hand, if η is small, we have only two real solutions, the two other solutions are complex. We therefore conclude that instabilities can occur only for small η. Such instabilities are called the *two-stream instabilities*. The critical value of η below which instability takes place can be obtained from the condition that the two real solutions in the region $0 < \xi < \eta$ coalesce,

$$\eta < \left[1 + (m_e/m_i)^{1/3}\right]^{3/2} . \tag{5.3.12}$$

For $m_e \ll m_i$, the instability arises when $ku_e^0 < \omega_{pe}$. This means that for a given k the wave becomes unstable when u_e^0 is less than ω_{pe}/k. This result sounds paradoxical, considering that the energy source of the instability is the kinetic energy of the electron fluid. In reality, when u_e^0 becomes very small, the unstable waves obtained from the dispersion relation (5.3.9) have large wavenumbers or small phase velocities, that is where the finite temperature of the electrons can no longer be ignored. Then the resonant wave-particle interaction, which will be discussed in Chap. 6, inhibits the instability. As will be shown in Sect. 6.3, for the plasma with $T_e = T_i$, instability can occur only when $u_e^0 > v_{Te}$.

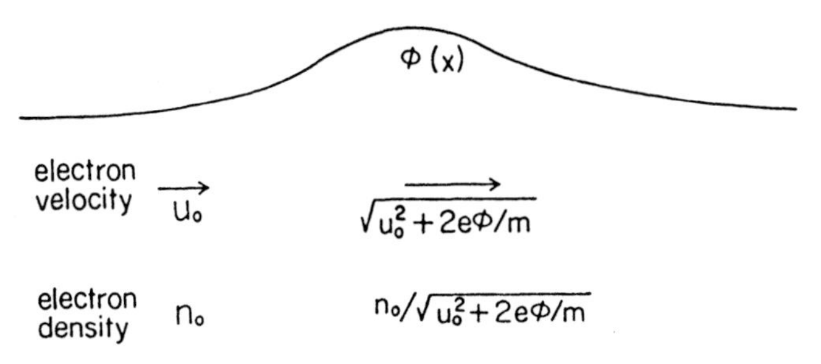

Fig. 5.4. Mechanism of two-stream instability. The electrons passing through the potential are accelerated. Since the electron flux $n_e u_e$ is everywhere constant, the number density n_e in this region must decrease

The physical mechanism of the two-stream instability can be explained as follows. Suppose a small potential hump $\phi(x)$ is produced as shown in Fig. 5.4. The electrons passing through the potential are accelerated by the force $-\partial\phi(x)/\partial x$ and become faster at the hump than in the region where $\phi = 0$. At steady state, the electron flux $n_e u_e$ must be constant at every point, so that at the hump n_e is lower than in the other region. Decrease of the electron density at the potential hump results in a further increase of the potential hump. In this way, the potential perturbation is amplified.

If the growth rate γ of the unstable wave is constant in time, the amplitude of the perturbation increases exponentially. Then the linear approximation used to derive the dispersion relation becomes invalid. Nonlinear effects associated with two-stream instabilities have received considerable attention in connection with plasma turbulence theory [5.1–3].

5.4 Ion Acoustic Waves

It is well known that a density perturbation in a neutral fluid propagates as a sound wave with phase velocity given by

$$v_{ph} = \sqrt{\frac{\gamma P}{\varrho}} = \sqrt{\frac{\gamma T}{M}} \, . \tag{5.4.1}$$

where P, ϱ, T and γ are, respectively, the pressure, mass density, temperature, and specific heat ratio of the fluid, and M is the particle mass. A similar wave can propagate in a plasma provided certain conditions are satisfied.

As before, we consider a spatially uniform and temporally stationary plasma without magnetic field. We use the two-fluid equations with electron temperature T_e and ion temperature T_i. We introduce a small low-frequency perturbation denoted by superscript 1 and linearize the equations of motion and continuity with respect to the perturbation:

$$m_i n_0 \frac{\partial \boldsymbol{u}_i^1}{\partial t} = -e n_0 \nabla \phi^1 - \gamma T_i \nabla n_i^1 , \tag{5.4.2}$$

$$m_e n_0 \frac{\partial \boldsymbol{u}_e^1}{\partial t} = e n_0 \nabla \phi^1 - T_e \nabla n_e^1 , \tag{5.4.3}$$

$$\frac{\partial n_i^1}{\partial t} + n_0 \nabla \cdot \boldsymbol{u}_i^1 = 0 , \tag{5.4.4}$$

$$\frac{\partial n_e^1}{\partial t} + n_0 \nabla \cdot \boldsymbol{u}_e^1 = 0 , \tag{5.4.5}$$

where we assumed that the ion pressure varies adiabatically while the electron pressure varies isothermally as the oscillation is sufficiently slow. We combine these equations with the Poisson equation

$$\Delta \phi^1 = \frac{e}{\varepsilon_0}(n_e^1 - n_i^1) . \tag{5.4.6}$$

For sufficiently slow spatial variation, we can assume efficient electron Debye shielding and use the local charge neutrality condition

$$n_e^1 = n_i^1 \tag{5.4.7}$$

in place of (5.4.6). Equations (5.4.2–5) and (5.4.6) or (5.4.7) are closed for the variables $n_e^1, n_i^1, \phi^1, \boldsymbol{u}_e^1$, and $\boldsymbol{u}_i^1$.

As in the previous two sections, we assume that the perturbation varies only in the x-direction, then these quantities can be expressed in the form similar to (5.2.8). Equations (5.4.2, 3) then become

$$-\omega m_i n_0 \tilde{u}_{ix}^1 = -e n_0 k \tilde{\phi}^1 - \gamma T_i k \tilde{n}_i^1 , \tag{5.4.8}$$

$$0 = e n_0 k \tilde{\phi}^1 - T_e k \tilde{n}_e^1 , \tag{5.4.9}$$

where in the second equation we neglected the electron inertia since it is small. From this relation we have

$$\tilde{n}_e^1 = n_0 \frac{e \tilde{\phi}^1}{T_e} , \tag{5.4.10}$$

which corresponds to (5.1.10). The ion continuity equation becomes

$$\omega \tilde{n}_i^1 = n_0 k \tilde{u}_{ix}^1 . \tag{5.4.11}$$

Using (5.4.7, 10, 11) to eliminate n_e^1, n_i^1, and $\tilde{\phi}^1$ in (5.4.8) we obtain

$$\left(\omega^2 - k^2 \frac{T_e + \gamma T_i}{m_i}\right) \tilde{u}_{ix}^1 = 0 \tag{5.4.12}$$

from which we get the following dispersion relation

$$\omega = \sqrt{\frac{T_e + \gamma T_i}{m_i}} k \equiv c_s k \ , \tag{5.4.13}$$

where c_s is called the *ion acoustic speed.* The wave satisfying this dispersion relation is called the *ion acoustic wave.*

We note that in deriving this dispersion relation the electron continuity equation was not used. The electron continuity equation gives u_e^1 or the oscillating current $J_x^1 = n_0^e(u_{ix}^1 - u_{ex}^1)$. For electrostatic waves, because there is no magnetic perturbation we do not have to calculate J_x^1. In general, in order to derive the dispersion relation for an electrostatic wave under the condition of static electron force balance (5.4.10), the electron continuity equation is unnecessary.

The assumption of isothermal variation of the electron pressure can be justified since the thermal speed of the electrons is much faster than the ion acoustic speed. The dispersion relation (5.4.13) shows that the phase velocity of the ion acoustic wave is equal to its group velocity.

Let us discuss the physical mechanism of the ion acoustic wave. Suppose a local ion density perturbation is produced, the electrons tend to shield the electric field due to the ion density perturbation. However, because of the thermal motion of the electrons the shielding is incomplete and an electrostatic potential on the order of T_e/e is left unshielded. This potential produces an electrostatic restoring force on the ion fluid. The resulting ion motion compresses the neighboring plasma because of the ion inertia, generating the propagating ion acoustic wave. The ion pressure gradient also acts as a restoring force, but it is not essential to this description. The ion acoustic wave can propagate in a plasma even in the absence of ion pressure ($T_i = 0$). In fact, because of the resonant wave-particle interaction which will be discussed in Chap. 6, the ion acoustic wave can propagate in a plasma only when $T_e \gg T_i$. In this case the ion acoustic speed can be written as $c_s = \sqrt{T_e/m_i}$ and the following inequalities hold

$$v_{Te} \gg \frac{\omega}{k} = c_s \gg v_{Ti} \ , \tag{5.4.14}$$

which are the condition that the wave propagates without significant damping.

At short wavelengths, the local charge neutrality (5.4.7) is no longer satisfied and we have to use the Poisson equation (5.4.6) which reads

$$k^2 \tilde{\phi}^1 = \frac{e}{\varepsilon_0}\left(\tilde{n}_i^1 - \tilde{n}_e^1\right) \ . \tag{5.4.15}$$

We substitute (5.4.10) into (5.4.15) and use (5.4.8, 11, 15) to eliminate $\tilde{n}_i^1, \tilde{n}_e^1$, and $\tilde{\phi}^1$ obtaining

$$\left[\omega^2 - \left(\frac{T_e}{m_i}\frac{1}{1+k^2\lambda_D^2} + \frac{\gamma T_i}{m_i}\right)k^2\right]\tilde{u}_{ix}^1 = 0\,.$$

The dispersion relation can now be written as

$$\omega = \sqrt{\frac{T_e}{m_i}\frac{1}{1+k^2\lambda_D^2} + \frac{\gamma T_i}{m_i}}\,k\,. \tag{5.4.16}$$

For long wavelength waves, we can neglect $k^2\lambda_D^2 \ll 1$ and we recover the dispersion relation (5.4.13). For $T_e \ll T_i$ and $1 \ll k^2\lambda_D^2 \ll T_e/T_i$, the dispersion relation (5.4.16) becomes

$$\omega \doteq \sqrt{\frac{n_0 e^2}{\varepsilon_0 m_i}} = \omega_{pi}\,, \tag{5.4.17}$$

which is the ion plasma frequency. In this case, the oscillation is called the *ion plasma oscillation*. Because of the short wavelength, the electron cannot shield the ion density perturbation and ion fluid oscillates in a uniform background sea of electrons. If the wavelength decreases to the point where the phase velocity of the wave becomes comparable to the thermal speed of the ions, the wave can no longer propagate because of strong wave-particle interactions.

5.5 Drift Waves

So far, we have considered waves in a uniform plasma. In this section, we consider a plasma which is spatially nonuniform in the density and/or temperature in a direction perpendicular to a static magnetic field. In some cases, spatial nonuniformity leads to waves with unique characteristics, an example of which are drift waves.

5.5.1 Diamagnetic Current

Spatial nonuniformity causes a *diamagnetic current* in a plasma. The mechanism of the diamagnetic current generation can be explained as follows: Charged particles in a magnetic field gyrate around a magnetic line of force as shown in Fig. 5.5. In the absence of Coulomb collisions, each charged particle gyrates around a fixed magnetic field line, that is, the guiding center is fixed in the plane perpendicular to the magnetic field. (We are assuming that the magnetic field is spatially uniform.) If we take into account the finite Larmor radius in a plasma of nonuniform density, then the number of particles crossing the plane $x = x_0$ in Fig. 5.5a is larger from the high density side than from the low density side. As a result, we get a net drift of plasma along the $\nabla n \times \boldsymbol{B}$ or y-direction as shown in

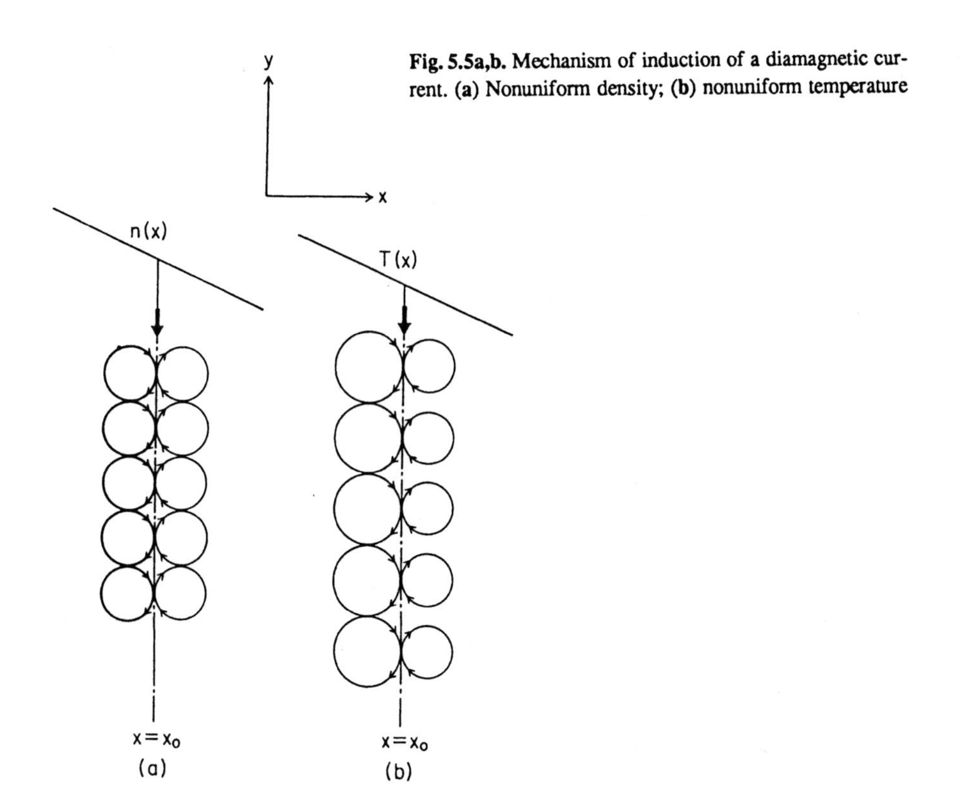

Fig. 5.5a,b. Mechanism of induction of a diamagnetic current. **(a)** Nonuniform density; **(b)** nonuniform temperature

Fig. 5.5a. Since the electrons and the ions gyrate in opposite directions, the ion drift and the electron drift also occur in opposite y-directions and a diamagnetic current is produced. For the case of uniform density but nonuniform temperature, particles crossing the plane $x = x_0$ from the high temperature side have larger Larmor radii or gyrating speeds than those crossing the plane $x = x_0$ from the low temperature side (Fig. 5.5b) which also results in a diamagnetic current in the $\nabla T \times \boldsymbol{B}$ or y-direction.

5.5.2 Diamagnetic Drift Velocity in the Slab Model

We consider a plasma of uniform temperature but nonuniform density with density gradient in the x-direction as shown in Fig. 5.5a. A uniform static magnetic field $\boldsymbol{B}$ exists in the z-direction and the diamagnetic current is in the y-direction. Such a model is called the *slab model* and is particularly useful in experiments where the plasma is confined in a cylindrical vessel with the magnetic field along the cylinder axis. The slab model can be used to simulate such a plasma if the spatial scale of interest across the magnetic field is much shorter than the radius of the cylinder. The cylindrical plasma is most naturally represented by the cylindrical coordinate system (r, θ, z), the corresponding coordinates in the slab model are $x \to r, y \to \theta, z \to z$.

We denote the plasma density and the magnetic field by $n_0(x)$ and $\boldsymbol{B} = B_0\hat{\boldsymbol{z}}$. We assume that the electron temperature is finite but the ion temperature is low ($T_\mathrm{i} \to 0$). In the unperturbed system, the electric field is absent ($\boldsymbol{E}_0 = 0$), and we further assume the absence of magnetic perturbations. A detailed analysis shows that this assumption is justified provided that the ratio of the plasma pressure to the magnetic pressure is very low; lower than the electron-to-ion mass ratio $m_\mathrm{e}/m_\mathrm{i}$ [5.4]. In this case, the electric field can be represented by the electrostatic potential ϕ. From the electron equation of motion in a stationary state, we have

$$0 = -en_\mathrm{e}(\boldsymbol{u}_\mathrm{e}^0 \times \boldsymbol{B}) - \nabla P_\mathrm{e} , \qquad (5.5.1)$$

from which we obtain

$$\boldsymbol{u}_\mathrm{e}^0 = -\frac{T_\mathrm{e}}{m_\mathrm{e}|\Omega_\mathrm{e}|}\frac{1}{n_0}\frac{dn_0}{dx}\hat{\boldsymbol{y}} , \qquad (5.5.2)$$

where $|\Omega_\mathrm{e}|$ is the electron cyclotron frequency. This velocity is called the electron *diamagnetic drift velocity*. A similar expression can be obtained for the ions, but since we are assuming $T_\mathrm{i} = 0$, $\boldsymbol{u}_\mathrm{i}^0 = 0$ and the diamagnetic current is determined by the electron diamagnetic drift velocity alone.

5.5.3 Drift Waves in the Slab Model

Drift waves are the quasielectrostatic waves which, in the coordinate system used in the previous section, propagate mainly in the y-direction with a phase velocity close to the electron diamagnetic drift velocity. They also propagate in the z-direction, but the wavelength in the z-direction is much longer than that in the y-direction. We denote the wavenumber vector by $\boldsymbol{k} = (0, k_y, k_z)$, then the parallel phase velocity ω/k_z is close to the thermal speed of the electrons while $1/k_y$ is of the same order or less than the inverse of the ion Larmor radius. As we shall show in Chap. 6, resonant electrons tend to excite the wave while resonant ions tend to damp the wave.

We shall derive the dispersion relation using the linearized two-fluid model. Taking into account the magnetic Lorentz force in (5.4.2, 3) and neglecting the ion temperature, we get

$$m_\mathrm{i} n_0 \frac{\partial \boldsymbol{u}_\mathrm{i}^1}{\partial t} = en_0(-\nabla\phi^1 + \boldsymbol{u}_\mathrm{i}^1 \times \boldsymbol{B}_0) \qquad (5.5.3)$$

$$m_\mathrm{e} n_0 \frac{\partial \boldsymbol{u}_\mathrm{e}^1}{\partial t} = -en_0(-\nabla\phi^1 + \boldsymbol{u}_\mathrm{e}^1 \times \boldsymbol{B}_0) - T_\mathrm{e}\nabla n_\mathrm{e}^1 . \qquad (5.5.4)$$

We combine these equations with the ion continuity equation

$$\frac{\partial n_\mathrm{i}^1}{\partial t} + n_0 \nabla \cdot \boldsymbol{u}_\mathrm{i}^1 + \boldsymbol{u}_\mathrm{i}^1 \cdot \nabla n_0 = 0 \qquad (5.5.5)$$

and the local charge neutrality condition (5.4.7) $n_\mathrm{i}^1 = n_\mathrm{e}^1$.

As in the case of ion acoustic waves we ignore the electron inertia in (5.5.4) whose z-component then yields the Boltzmann relation (5.4.10), while the perpendicular component yields the electron drift across the magnetic field. We use the Fourier representation

$$\begin{pmatrix} n^1 \\ \boldsymbol{u}^1 \\ \phi^1 \end{pmatrix} \sim \begin{pmatrix} \tilde{n}^1 \\ \tilde{\boldsymbol{u}}^1 \\ \tilde{\phi}^1 \end{pmatrix} \exp(\mathrm{i}\boldsymbol{k}\cdot\boldsymbol{r} - \mathrm{i}\omega t) , \tag{5.5.6}$$

where $\boldsymbol{r}$ stands for the position vector (x, y, z). Equations (5.5.3, 5) then become

$$-\mathrm{i}\omega m_\mathrm{i}\tilde{\boldsymbol{u}}_\mathrm{i}^1 = -\mathrm{i}\boldsymbol{k}e\tilde{\phi}^1 + e(\tilde{\boldsymbol{u}}_\mathrm{i}^1 \times \boldsymbol{B}_0) \tag{5.5.7}$$

$$-\mathrm{i}\omega\tilde{n}_\mathrm{i}^1 + \mathrm{i}n_0\boldsymbol{k}\cdot\tilde{\boldsymbol{u}}_\mathrm{i}^1 + \tilde{u}_{\mathrm{i}x}^1\frac{dn_0}{dx} = 0 . \tag{5.5.8}$$

We solve (5.5.7) for $\tilde{\boldsymbol{u}}_\mathrm{i}^1$ and substitute the result into (5.5.8). In the case $\omega \ll \Omega_\mathrm{i} = eB_0/m_\mathrm{i}$, (5.5.7) can be solved as

$$\tilde{\boldsymbol{u}}_{\mathrm{i}\perp}^1 = \mathrm{i}\frac{\hat{\boldsymbol{z}} \times \boldsymbol{k}\tilde{\phi}^1}{B_0} + \frac{\omega}{B_0\Omega_\mathrm{i}}k_y\hat{\boldsymbol{y}}\tilde{\phi}^1 . \tag{5.5.9}$$

The first term on the right hand side is the $\boldsymbol{E} \times \boldsymbol{B}$ drift while the second term is the polarization drift. The drift wave propagation is shown in Fig. 5.6. The parallel ion velocity is given by

$$\tilde{u}_{\mathrm{i}z}^1 = \frac{ek_z}{m_\mathrm{i}\omega}\tilde{\phi}^1 . \tag{5.5.10}$$

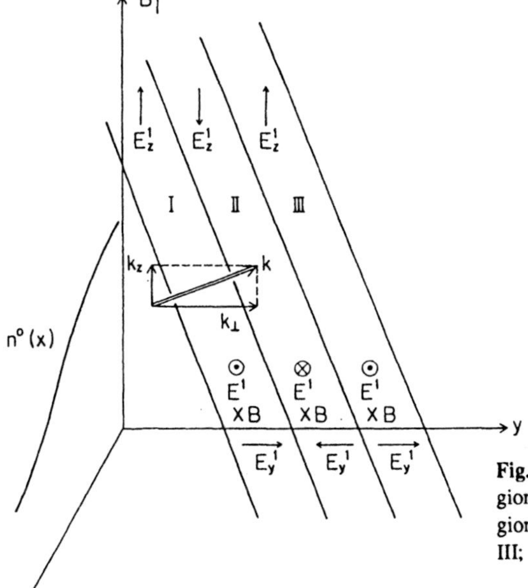

Fig. 5.6. Propagation of a drift wave. Regions I and III are in the same phase. Region II is out of phase with regions I and III; $k_z \ll k$

Substituting (5.5.9, 10) into (5.5.8), and assuming $\omega^2 \ll \Omega_i^2(k_z/k_y)^2$, we have

$$\frac{\tilde{n}_i^1}{n_0} = \left(\frac{k_z^2 T_e}{\omega^2 m_i} - \frac{k_y}{\omega}\frac{T_e}{eB_0}\frac{1}{n_0}\frac{dn_0}{dx}\right)\frac{e\tilde{\phi}^1}{T_e} . \tag{5.5.11}$$

The first term inside the parentheses can be written as $k_z^2 c_s^2/\omega^2$, where c_s is the ion acoustic speed. If we denote the second term inside the parentheses using the electron diamagnetic drift velocity (5.5.2) $\boldsymbol{u}_e^0 = u_{ey}^0\hat{y}$, then we can write (5.5.11) as

$$\frac{\tilde{n}_i^1}{n_0} = \left(\frac{k_z^2 c_s^2}{\omega^2} + \frac{k_y u_{ey}^0}{\omega}\right)\frac{e\tilde{\phi}^1}{T_e} . \tag{5.5.12}$$

Using the Boltzmann relation (5.4.10) and the local charge neutrality condition (5.4.7), we finally get

$$\left(\omega^2 - k_y u_{ey}^0 \omega - k_z^2 c_s^2\right)\frac{e\tilde{\phi}^1}{T_e} = 0 \tag{5.5.13}$$

from which the following dispersion relation is obtained

$$\omega^2 - \omega_* \omega - k_z^2 c_s^2 = 0 , \tag{5.5.14}$$

where $\omega_* = k_y u_{ey}^0$ and is called the *drift frequency*. The solution of (5.5.14) is shown in Fig. 5.7. In the absence of the density gradient, $u_{ey}^0 = 0$ and we get the ion acoustic waves propagating in the z-direction. For $k_z \to 0$, (5.5.14) yields two solutions $\omega_1 = \omega_*$ and $\omega_2 = -k_z^2\ c_s^2/\omega_*$. The phase velocity of the mode at $\omega = \omega_1$ equals the electron diamagnetic drift velocity.

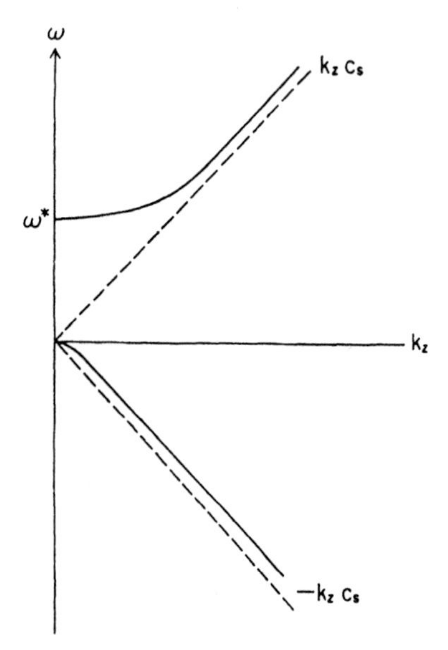

Fig. 5.7. Dispersion of drift wave given by (5.5.14)

5.5.4 Properties of the Drift Wave

The perturbed electric field $\boldsymbol{E}^1$ perpendicular to the magnetic field produces a $\boldsymbol{E}^1 \times \boldsymbol{B}$ plasma motion, $\boldsymbol{u}^1 = \boldsymbol{E}^1 \times \boldsymbol{B}_0/B_0^2$, as shown in Fig. 5.6. This motion satisfies the condition

$$\nabla \cdot \boldsymbol{u}^1 = \mathrm{i}\boldsymbol{k} \cdot \boldsymbol{u}^1 = \boldsymbol{k} \cdot \frac{\boldsymbol{k}\phi^1 \times \boldsymbol{B}_0}{B_0^2} = 0 , \tag{5.5.15}$$

and hence is an incompressible motion. Both electrons and ions move with the same $\boldsymbol{E} \times \boldsymbol{B}$ drift velocity $\boldsymbol{u}^1$ and the ion continuity equation becomes

$$\frac{\partial n_{\mathrm{i}}^1}{\partial t} = -\boldsymbol{u}_{\mathrm{i}}^1 \cdot \nabla n_0 = -u_{\mathrm{i}x}^1 \frac{dn_0}{dx} . \tag{5.5.16}$$

This relation indicates that in nonuniform plasma even an incompressible flow produces a density perturbation. This is due to the fact that the $\boldsymbol{E} \times \boldsymbol{B}$ drift along the density gradient carries the plasma from the high density side to the low density side causing a density variation. Such a density perturbation, or drift wave, propagates in the y-direction with a speed approximately equal to the diamagnetic drift velocity $u_{\mathrm{e}y}^0$.

If the parallel phase velocity ω/k_z lies in the region

$$v_{T\mathrm{i}} \ll \frac{\omega}{k_z} \ll v_{T\mathrm{e}} \tag{5.5.17}$$

then the resonant wave-particle interaction is small. However, as we shall show in an example in Sect. 6.5, drift waves become unstable for various reasons. In order to stabilize this *drift wave instability*, one can use a spatially varying y-component of the magnetic field represented by

$$\boldsymbol{B} = \frac{x}{L_s} B_0 \hat{\boldsymbol{y}} + B_0 \hat{\boldsymbol{z}} , \tag{5.5.18}$$

where L_s is a constant called the *shear length*. The magnetic field of (5.5.18) is shown in Fig. 5.8, and is directed along the z-direction at $x = 0$ and inclined toward the y-direction at $x \neq 0$. It depends on x and the declining angle becomes 45° at $x = L_s$.

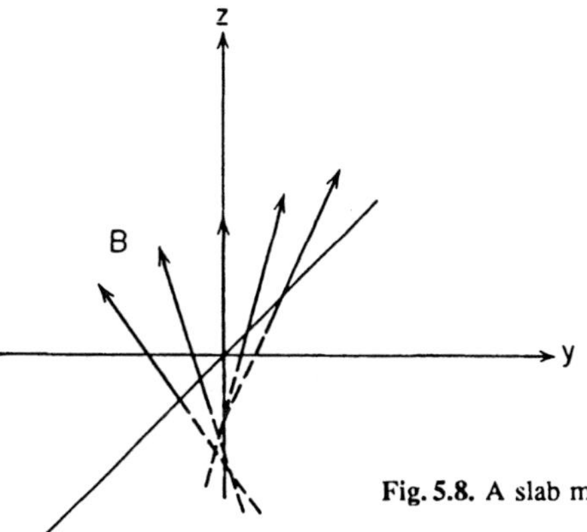

Fig. 5.8. A slab model representation for a sheared magnetic field

Such a magnetic field configuration is called a *sheared magnetic field*. By the transformation $x \to r, y \to \theta, z \to z$, a similar sheared magnetic field can be formed in a cylindrical plasma (Chap. 10). In a sheared magnetic field, the parallel wavenumber becomes $k_{\parallel} = k_z + k_y x/L_s$. As the parallel wavenumber varies in space, we represent the potential $\phi^1(x, y, z)$ by $\phi^1(x)\exp(ik_y y + ik_z z - i\omega t)$ and solve $\phi^1(x)$ as a linear eigenvalue problem in the x-direction. The result shows that in the presence of a sheared magnetic field the electrostatic drift wave is stable due to the resonant interaction with ions at the position where $\omega = kv_{Ti}$. [5.5–7].

In a more realistic geometry, drift waves can be destabilized due to the effect of magnetic field curvature, finite ion Larmor radius effect, etc.

Drift waves have been extensively studied in plasmas used for nuclear fusion research, because if we want to confine the plasma by a magnetic field, spatial nonuniformity across the magnetic field is inevitable and the unstable drift waves excite turbulence. This may cause an *anomalous transport* of plasma across the magnetic field (Chap. 13).

5.6 Nonlinear Waves

So far, we have restricted ourselves to very small amplitude waves and derived the dispersion relations for linear waves. However, waves in a plasma often acquire large amplitudes. Recently the focus of much research has shifted to such large amplitude waves which naturally involve nonlinear effects. Here we first discuss the various nonlinear effects and then derive the equations which describe nonlinear waves.

The first study of nonlinear waves was on shallow water waves. For waves of wavelength longer than the water depth, the fluid velocity u satisfies Korteweg-de Vries (K-dV) equation:

$$\frac{\partial u}{\partial t} + u\frac{\partial u}{\partial X} + \alpha\frac{\partial^3 u}{\partial X^3} = 0\,, \tag{5.6.1}$$

where $X = x - st$, s being the phase velocity of the water wave and α is a constant. We first derive the same form of the equation for ion acoustic wave propagation in the x-direction.

5.6.1 K-dV Equation for Ion Acoustic Waves

We first normalize the variables as follows:

$$\frac{n_i}{n_0} = n\,, \quad \frac{u_{ix}}{c_s} = u\,, \quad \frac{x}{\lambda_D} = \xi\,, \quad \omega_{pi}t = \tau\,, \quad \frac{e\phi}{T_e} = \varphi\,, \tag{5.6.2}$$

then the ion equation of motion and continuity equation are written as

$$\frac{\partial n}{\partial \tau} + \frac{\partial}{\partial \xi}(nu) = 0\,, \tag{5.6.3}$$

$$\frac{\partial u}{\partial \tau} + u\frac{\partial u}{\partial \xi} + \frac{\partial \varphi}{\partial \xi} = 0 \, . \tag{5.6.4}$$

For electrons we ignore the inertia term in the equation of motion, which yields the Boltzmann distribution

$$\frac{n_e}{n_0} = e^{\varphi} \, . \tag{5.6.5}$$

The Poisson equation becomes

$$\frac{\partial^2 \varphi}{\partial \xi^2} = e^{\varphi} - n \, . \tag{5.6.6}$$

We write $n = 1+n'$ where n' stands for the ion density perturbation, and transform to a frame moving with the ion acoustic speed $y = \xi - \tau$. Expanding e^{φ} in powers of φ, we get from (5.6.3, 4, 6) the following set of equations:

$$\frac{\partial n'}{\partial \tau} + \frac{\partial}{\partial y}(u - n' + n'u) = 0 \tag{5.6.7}$$

$$\frac{\partial u}{\partial \tau} + \frac{\partial}{\partial y}(\varphi - u) + u\frac{\partial u}{\partial y} = 0 \tag{5.6.8}$$

$$-\frac{\partial^2 \varphi}{\partial y^2} = n' - \varphi - \frac{\varphi^2}{2} - \cdots \, . \tag{5.6.9}$$

In the first order approximation, we ignore the nonlinear terms and the time derivative $\partial/\partial\tau$, then under the boundary condition that φ, u, and $n' \to 0$ at $|y| \to \infty$, we get

$$u = n' = \varphi \, . \tag{5.6.10}$$

In the second order approximation, we add (5.6.7–9) and use the lowest order relation (5.6.10) for the time derivative and nonlinear terms, obtaining

$$\frac{\partial u}{\partial \tau} + u\frac{\partial u}{\partial y} + \frac{1}{2}\frac{\partial^3 u}{\partial y^3} = 0 \, . \tag{5.6.11}$$

This is nothing but the K-dV equation with $\alpha = 1/2$. The second term in (5.6.11) corresponds to the nonlinear effect while the third term comes from the dispersion effect of the ion acoustic wave, i.e., the effect which arises from the $k^2\lambda_D^2$ correction in the dispersion relation (5.4.16).

We can easily show that (5.6.11) has a localized solution

$$u = u_\infty + B\,\mathrm{sech}^2\left[\left(\frac{x - ct}{\delta}\right)\right], \quad c = u_\infty + B/3 \, , \quad \delta = \sqrt{12\alpha/B} \, . \tag{5.6.12}$$

This solution is called the *solitary wave*, or simply *soliton*, and its properties have been extensively studied both numerically and analytically. The solitary wave solution for the ion acoustic wave is called the *ion acoustic soliton* and its basic properties have been experimentally verified by *Ikezi* et al. [5.8–10].

Let us discuss how such solitary wave solutions can be produced. In the long wavelength limit we can ignore the dispersion term $\partial^3 u/\partial y^3$. Then the solution of (5.6.1) has the form

$$u(y,t) = u(y - u(y,t), 0)$$

In the region $\partial u/\partial y < 0$ at $t = 0$, the gradient of u becomes steeper, implying the appearance of short wavelength (large k) components in that region. In the Fourier representation, this means that starting from the fundamental component (ω, k), higher harmonics $(2\omega, 2k), (3\omega, 3k)$ are produced by the nonlinear term. For higher harmonics, the dispersion term becomes important, which tends to suppress the short wavelength component. From the point of view of the linear dispersion relation, at long wavelengths, $\omega = kc_s$, so that $(2\omega, 2k), (3\omega, 3k), \ldots$ also satisfy the linear dispersion relation. As we go to higher harmonics, the linear dispersion relation $\omega = kc_s$ is no longer satisfied because of the $k^2\lambda_D^2$ term. In this way, the steepening due to the nonlinear term and the suppression effect of the higher harmonics due to the dispersion term balance each other resulting in a solitary wave.

5.6.2 Nonlinear Electron Plasma Waves

The electron plasma wave is a wave of high frequency, nearly equal to ω_{pe} for all possible wavenumbers that satisfy $k\lambda_D < 1$. Therefore, the higher harmonics of the linear electron plasma wave do not satisfy the linear dispersion relation. Harmonic generation, as in the case of ion acoustic waves, is unimportant for electron plasma waves. If we denote the linear wave by

$$A\cos(kx - \omega t) ,$$

then the lowest order nonlinear term is

$$A^2\cos^2(kx - \omega t) = \tfrac{1}{2}A^2[\cos(2kx - 2\omega t) + 1] \ . \qquad (5.6.13)$$

The first term inside the bracket is the second harmonic and is therefore unimportant. The second term denotes the space-time independent term. For the case of a strongly dispersive wave such as the electron plasma wave, nonlinear combination of the lowest order terms in two linear waves produces two terms similar to the two terms in (5.6.13). One term oscillates at frequency $2\omega_{pe}$ and is unimportant. The other is a term which depends only slowly on space and time. This means that the medium in which the wave propagates slowly changes its properties, and, as a result, the wave dispersion characteristics change both spatially and temporally. The principal nonlinear effect associated with electron plasma waves arises through such slow modulation of the dispersion characteristics. Such a description is valid only at long wavelengths (much longer than the Debye length). At shorter wavelengths resonant particles play important roles as will be shown in Chap. 6.

In the linear approximation, the wave equation for the electron plasma wave can be written as

$$\frac{\partial^2}{\partial t^2}u_{ex} + \omega_{pe}^2 u_{ex} - 3v_{Te}^2\frac{\partial^2}{\partial x^2}u_{ex} = 0 \,. \tag{5.6.14}$$

Setting $u_{ex} \sim \exp(ikx - i\omega t)$, we can immediately derive the dispersion relation (5.2.14). Equation (5.6.14) is characterized by two parameters, n_0 through ω_{pe}^2 and T_e through v_{Te}^2. Since at long wavelengths the thermal effect [the last term in (5.6.14)] is unimportant, the dominant nonlinear effect will arise from the modulation of the plasma density. Suppose the plasma density is modified nonlinearly from n_0 to $n_0 + \delta n_e(x,t)$. The time dependence of $\delta n_e(x,t)$ is naturally much slower than the plasma oscillation. Then (5.6.14) will be written as

$$\frac{\partial^2}{\partial t^2}u_{ex} - 3v_{Te}^2\frac{\partial^2}{\partial x^2}u_{ex} + \omega_{pe}^2\left[1 + \frac{\delta n_e(x,t)}{n_0}\right]u_{ex} = 0 \,. \tag{5.6.15}$$

We write

$$u_{ex}(x,t) = u(x,t)e^{-i\omega_{pe}t} + u^*(x,t)e^{i\omega_{pe}t} \,, \tag{5.6.16}$$

where u^* is the complex conjugate of u. We can assume that both u and u^* depend on time only slowly, and neglect the second derivative $\partial^2 u/\partial t^2$. Then substituting (5.6.16) into (5.6.15) and neglecting $\partial^2 u/\partial t^2$, we get

$$i\frac{\partial}{\partial t}u(x,t) + \frac{3}{2}\frac{v_{Te}^2}{\omega_{pe}^2}\frac{\partial^2}{\partial x^2}u(x,t) - \frac{\omega_{pe}}{2}\frac{\delta n_e(x,t)}{n_0}u(x,t) = 0 \,. \tag{5.6.17}$$

This equation has the form of the Schrödinger equation in quantum mechanics, in which the nonlinear density modification term [the last term in (5.6.17)] plays the role of the potential. Without this term, the solution is a plane wave of a free particle.

This plane wave solution represents the linear wave, and in this analogy, the nonlinear wave corresponds to the quantum mechanical motion of a particle in the presence of a potential. A particle trapped in a potential well corresponds to a localized wave trapped in a density depression. Solitary waves are also obtained from such localized solutions, but in this case, since the wave oscillates at high frequency, it is called an *envelope soliton.*

Let us express the density modification $\delta n_e(x,t)$ in terms of $u(x,t)$. We use the time average of the electron equation of motion to average out the oscillating terms at high frequencies on the order of ω_{pe} and retain only slowly varying terms such as $\delta n_e(x,t)$. The result is

$$m_e\frac{\partial \bar{u}_{ex}}{\partial t} + \overline{m_e u_{ex}\frac{\partial}{\partial x}u_{ex}} = -\overline{\frac{1}{n_e}\frac{\partial P_e}{\partial x}} + e\frac{\partial \bar{\phi}}{\partial x} \,, \tag{5.6.18}$$

where the bar denotes the time average. The first term on the left hand side can be ignored because of the low frequency and small electron mass. The second term on the left hand side is nonlinear. Substitution of (5.6.16) yields to lowest order

$$\frac{\partial}{\partial x}\left(\frac{m_e}{2}\overline{u_{ex}^2}\right)=\frac{\partial}{\partial x}\left(m_e|u(x,t)|^2\right) . \tag{5.6.19}$$

This is nothing but the ponderomotive force in the fluid representation. The first term on the right hand side of (5.6.18) also contains the nonlinear effect, but a detailed calculation using the electron continuity equation shows that this nonlinear term is on the order of $k^2\lambda_D^2$ compared to the ponderomotive force, and hence will be neglected. That is, we use the linear approximation for the pressure gradient term

$$-\overline{\frac{1}{n_e}\frac{\partial P_e}{\partial x}}=-\frac{T_e}{n_0}\frac{\partial \delta n_e}{\partial x} , \tag{5.6.20}$$

where we assumed that the slow variation is isothermal.

From these considerations, we get

$$\frac{\partial}{\partial x}(m_e|u(x,t)|^2)=-\frac{\partial}{\partial x}\left(\frac{T_e}{n_0}\delta n_e-e\bar{\phi}\right) . \tag{5.6.21}$$

Under the boundary condition that $\delta n_e, \bar{\phi}\to 0$ at $u(x,t)\to 0$, we get

$$\frac{\delta n_e}{n_0}=\frac{e\bar{\phi}-m_e|u(x,t)|^2}{T_e} . \tag{5.6.22}$$

Finally, in order to calculate $\bar{\phi}$, we use the ion equation of motion. Since the ion response to high frequency electron plasma oscillation is negligible, we can use the linear approximation:

$$m_i\frac{\partial u_i}{\partial t}=-e\frac{\partial\bar{\phi}}{\partial x}-\frac{1}{n_0}\frac{\partial}{\partial x}(n_iT_i) . \tag{5.6.23}$$

For sufficiently slow temporal variation, we neglect the ion inertia and get

$$e\bar{\phi}=-T_i\frac{\delta n_i}{n_0} , \tag{5.6.24}$$

where δn_i is the ion density perturbation. For slow variation we can assume charge neutrality, $\delta n_i=\delta n_e$, then from (5.6.22, 24)

$$\frac{\delta n_e}{n_0}=-\frac{m_e|u(x,t)|^2}{T_e+T_i} . \tag{5.6.25}$$

If we substitute this relation into (5.6.17), we get a closed equation for $u(x,t)$:

$$\mathrm{i}\frac{\partial}{\partial t}u(x,t)+\frac{3v_{Te}^2}{2\omega_{pe}^2}\frac{\partial^2}{\partial x^2}u(x,t)+\frac{\omega_{pe}}{2}\frac{m_e}{T_e+T_i}|u(x,t)|^2u(x,t)=0 . \tag{5.6.26}$$

This nonlinear wave equation is called the *nonlinear Schrödinger equation*. Its analytical properties have been extensively studied [5.11]. Relation (5.6.25) shows that the electron density depression is produced in the high field region. Such a density depression is produced by the ponderomotive force which expels the particle from the high amplitude region. As the depression is produced, the wave tends to localize in that region and as a result the ponderomotive force is further enhanced. In this way the localization of the high frequency wave, which can be expressed in terms of an instability, proceeds. This instability occurs in the modulation of the amplitude and phase of a large amplitude plane wave and is called the *modulational instability*. As the localization progresses due to this instability, the second term in (5.6.26) becomes important. This term corresponds to the diffraction term in quantum mechanics and spreads out the wave packet. Thus the nonlinear localization is suppressed by the diffraction and a stationary balance is obtained. This is the envelope soliton.

As a final remark, we note that we can derive the nonlinear Schrödinger equation of the form

$$\mathrm{i}\frac{\partial}{\partial t}u(x,t) + p\frac{\partial^2}{\partial x^2}u(x,t) + q|u(x,t)|^2 u(x,t) = 0 , \tag{5.6.27}$$

where q is a coefficient of the cubic nonlinear term and p is the *group dispersion*

$$p = \frac{1}{2}\frac{d^2\omega}{dk^2} \tag{5.6.28}$$

for a general class of one-dimensional propagation of a strongly dispersive wave in the frame moving with the group velocity of the wave. Equation (5.6.27) has a finite amplitude plane wave solution

$$u(x,t) = A\exp(-\mathrm{i}\nu t), \quad \nu = -qA^2, \tag{5.6.29}$$

but is unstable against a small perturbation of the amplitude and phase if $pq > 0$ (*modulational instability*, Problem 5.8) [5.12].

PROBLEMS

5.1. Show that the potential (5.1.6) satisfies (5.1.5).

5.2. Consider an electron plasma in a background sea of uniform neutralizing positive charge. The electrons consist of bulk cold electrons of density n_0 and a cold beam of density n_b and velocity u_0, with $n_b \ll n_0$. Derive the dispersion relation for this beam-plasma system and find the condition for the-two stream instability.

5.3. Setting $\omega = \omega_r + \mathrm{i}\gamma$ in (5.3.9), find the maximum growth rate.

5.4. (i) In the presence of a magnetic field, each charged particle has a magnetic moment $-\mu \boldsymbol{b}$. By summing up the magnetic moments of all the particles of a species s, show that the magnetization of the sth species of particle per unit volume is given by

$$\boldsymbol{M}_s = -\frac{nT}{B^2}\boldsymbol{B} . \tag{5.P.1}$$

(ii) Show that the drift current due to the pressure gradient force $\boldsymbol{J}_{d_s} = -(\nabla P \times \boldsymbol{b})/B$ is the sum of the curvature drift current, the ∇B drift current, and the diamagnetic current $\nabla \times \boldsymbol{M}_s$.

5.5. (i) Using the cold ion model and the Boltzmann relation (5.6.5) for the electron, derive the following nonlinear equation for the one-dimensional stationary wave propagation [set $\partial/\partial\tau = 0$ in (5.6.3, 4)]

$$\frac{d^2\varphi(\xi)}{d\xi^2} = -\frac{\partial U(\varphi)}{\partial\varphi}, \tag{5.P.2}$$

where $U(\varphi)$ is called the Sagdeev potential and is given by

$$U(\varphi) = -\left[e^{\varphi} + 2\varphi_0\sqrt{1 - \varphi/\varphi_0}\right] + 1 + 2\varphi_0 + U(0), \tag{5.P.3}$$

where φ_0 is a constant related to the energy: $u^2/2 + \varphi = \varphi_0$.

(ii) Discuss the solution of (5.P.2) using the analogy of the particle equation of motion $d^2x/dt^2 = -\partial U(x)/\partial x$ in a potential $U(x)$.

5.6. (i) Show that the transformation $\varphi - \varphi_m = \tilde{\varphi}$ and $\varphi_0 - \varphi_m = \tilde{\varphi}_0$ yields $U(\tilde{\varphi}) = [U(\varphi) - U(\varphi_m)]\exp(-\varphi_m)$, where φ_m is the solution of $\partial U(\varphi)/\partial\varphi = 0$, i.e.,

$$\exp(\varphi_m) = [1 - \varphi_m/\varphi_0]^{-1/2}. \tag{5.P.4}$$

(ii) Show that by the transformation $\xi \exp(\varphi_m/2) = \tilde{\xi}$, (5.P.2) is recovered for $\tilde{\varphi}$:

$$\frac{d^2\tilde{\varphi}(\tilde{\xi})}{d\tilde{\xi}^2} = -\frac{\partial U(\tilde{\varphi})}{\partial\tilde{\varphi}}. \tag{5.P.5}$$

(iii) Using the above result, prove that any problem (5.P.2) with $\varphi_0 < 1/2$ or $\varphi_m < 0$ can be transformed to an equivalent problem with $\varphi_0 > 1/2$ or $\varphi_m > 0$.

(iv) Show from the result in (iii) that the solitary wave solution (localized solution) of (5.P.2) is always supersonic if the charge neutrality condition is satisfied at $|\xi| \to \infty$.

5.7. (i) Introducing a small perturbation of the form

$$\Delta U = (X + iY)\exp(-i\nu t) \tag{5.P.6}$$

to the plane wave solution (5.6.29) of (5.6.27) derive the following coupled equations for X and Y

$$\frac{\partial X}{\partial t} + p\frac{\partial^2 Y}{\partial x^2} + 2qA^2 Y = 0, \tag{5.P.7}$$

$$-\frac{\partial Y}{\partial t} + p\frac{\partial^2 X}{\partial x^2} = 0. \tag{5.P.8}$$

(ii) Setting $X \simeq \cos(Kx - \Omega t), Y = \sin(Kx - \Omega t)$, find the dispersion relation between Ω and K.
(iii) Find the wavenumber region for the modulational instability and show that the maximum growth rate obtained at $K = A\sqrt{q/p}$ is $\gamma_{max} = qA^2$.

6. Kinetic Theory

In this chapter, we treat the plasma as an aggregate of many charged particles and discuss the plasma response in terms of the phase space distribution function. The chapter consists of two parts: in the first six sections, we use the Vlasov model and treat the plasma as a collisionless continuum or a continuum in the phase space. First in Sect. 6.1, we consider a small perturbation to the unperturbed state which is assumed to be stationary in time and derive the linearized Vlasov equation. Then in Sect 6.2 we consider a magnetic field free isotropic plasma and discuss the linear response of the plasma to an electrostatic perturbation. Collisionless damping of the wave due to the resonant wave-particle interaction is discussed in Sect. 6.3. In Sect. 6.4 we consider a uniform plasma in a magnetic field and derive a general expression for the electrostatic response of the plasma. Then in Sect. 6.5 we consider a spatially non-uniform plasma and present a physical mechanism of the drift wave instability. In Sect. 6.6 we briefly discuss the nonlinear effect associated with the resonant particles.

The last two sections deal with the discreteness of the plasma by considering a spatially uniform and temporally stationary unperturbed state with no magnetic field. In Sect. 6.7, we show how each charged particle acts as the source of a perturbation and estimate the thermal noise which is inherently present in any plasma. Finally in Sect. 6.8, we derive a kinetic equation that describes the Coulomb collision effect in a plasma.

6.1 Linearized Vlasov Equation

The Vlasov system of equations (4.2.4–8) is nonlinear with respect to the distribution function F_s since the electromagnetic force K_s generally depends on the distribution function. However, if the perturbed electromagnetic fields are sufficiently weak, then the perturbed part of the distribution is also small, and we can use the linear approximation for the Vlasov equation. Assuming that there is no unperturbed electric field, we write

$$E(r,t) = E^1(r,t), B(r,t) = B^0(r) + B^1(r,t) , \tag{6.1.1}$$

$$F_s(r,v,t) = F_s^0(r,v) + F_s^1(r,v,t)$$

where the superscripts 0 and 1 denote the unperturbed and perturbed parts, respectively. In the absence of perturbation, the Vlasov equation becomes

$$v \cdot \frac{\partial}{\partial \boldsymbol{r}} F_s^0(\boldsymbol{r}, \boldsymbol{v}) + \left(\boldsymbol{v} \times \boldsymbol{\Omega}_s^0\right) \cdot \frac{\partial}{\partial \boldsymbol{v}} F_s^0(\boldsymbol{r}, \boldsymbol{v}) = 0 , \tag{6.1.2}$$

where

$$\boldsymbol{\Omega}_s^0 = \frac{q_s \boldsymbol{B}^0(\boldsymbol{r})}{m_s} , \tag{6.1.3}$$

and its magnitude equals the cyclotron frequency. For the perturbation, by using the linear approximation we get

$$\frac{d}{dt} F_s^1(\boldsymbol{r}, \boldsymbol{v}, t) = \frac{q_s}{m_s} \left[\boldsymbol{E}^1(\boldsymbol{r}, t) + \boldsymbol{v} \times \boldsymbol{B}^1(\boldsymbol{r}, t)\right] \cdot \frac{\partial}{\partial \boldsymbol{v}} F_s^0(\boldsymbol{r}, \boldsymbol{v}) \tag{6.1.4}$$

where d/dt is the time derivative along the unperturbed particle orbit:

$$\frac{d}{dt} = \frac{\partial}{\partial t} + \boldsymbol{v} \cdot \frac{\partial}{\partial \boldsymbol{r}} + \left(\boldsymbol{v} \times \boldsymbol{\Omega}_s^0\right) \cdot \frac{\partial}{\partial \boldsymbol{v}} . \tag{6.1.5}$$

Note that the unperturbed particle orbit is given by

$$\frac{d\boldsymbol{r}}{dt} = \boldsymbol{v}, \quad \frac{d\boldsymbol{v}}{dt} = \boldsymbol{v} \times \boldsymbol{\Omega}_s^0 . \tag{6.1.6}$$

Equation (6.1.4) is called the *linearized Vlasov equation.* Equations (6.1.2, 4) and the electromagnetic field equations (4.2.5–8) form the basic set of equations for small perturbations in the Vlasov model.

6.2 Linear Electrostatic Response in Isotropic Plasma

In this section, we discuss the linear electrostatic response of an isotropic plasma using the linearized Vlasov equation.

For electrostatic response $\boldsymbol{B}^1 = 0$ and the electric field can be written as $\boldsymbol{E}^1 = -\nabla\phi^1$. Of the electromagnetic field equations (4.2.5–8) only the Poisson equation is needed:

$$-\varepsilon_0 \Delta\phi^1 = q_e n_e^1(\boldsymbol{r}, t) + q_i n_i^1(\boldsymbol{r}, t) \tag{6.2.1}$$

where $q_e = -e, q_i = Ze$ and

$$n_s^1(\boldsymbol{r}, t) = \int d^3\boldsymbol{v} F_s^1(\boldsymbol{r}, \boldsymbol{v}, t) . \tag{6.2.2}$$

Since we are considering a spatially uniform isotropic plasma in the absence of the magnetic field ($\boldsymbol{\Omega}_s^0 = 0$), the unperturbed distribution is a function of $v(= |\boldsymbol{v}|)$ alone,

$$F_s^0(\boldsymbol{r}, \boldsymbol{v}) = F_s^0(v) . \tag{6.2.3}$$

Note, however, that $F_s^0(v)$ is not necessarily Maxwellian due to the absence of collisions. The linearized Vlasov equation (6.1.4) becomes

$$\left(\frac{\partial}{\partial t}+\boldsymbol{v}\cdot\frac{\partial}{\partial \boldsymbol{r}}\right)F_s^1(\boldsymbol{r},\boldsymbol{v},t)=\frac{q_s}{m_s}\nabla\phi^1(\boldsymbol{r},t)\cdot\frac{\partial}{\partial \boldsymbol{v}}F_s^0(v)\,. \tag{6.2.4}$$

If $F_s^0(v)$ is given, then (6.2.1, 2, 4) form a closed set. These equations can be solved using the Fourier representation for F^1 and ϕ^1

$$\begin{pmatrix} F^1 \\ \phi^1 \end{pmatrix} \sim \begin{pmatrix} \tilde{F}^1 \\ \tilde{\phi}^1 \end{pmatrix} \exp(\mathrm{i}kx-\mathrm{i}\omega t)\,. \tag{6.2.5}$$

Although in general the wavenumber is a vector, for isotropic plasma we can choose the positive x-direction to be in the direction of $\boldsymbol{k}$, so that $\boldsymbol{k}=(k,0,0)$, $(k>0)$. We denote the x-component of the velocity by w, then (6.2.1, 4) can be written as

$$\varepsilon_0 k^2\tilde{\phi}^1 = q_{\mathrm{i}}\tilde{n}_{\mathrm{i}}^1 - e\tilde{n}_{\mathrm{e}}^1 \tag{6.2.6}$$

$$-\mathrm{i}(\omega-kw)\tilde{F}_s^1(\boldsymbol{v})=\frac{q_s}{m_s}\mathrm{i}k\tilde{\phi}^1\frac{\partial}{\partial w}F_s^0(v)\,. \tag{6.2.7}$$

Equation (6.2.2) becomes

$$\tilde{n}_s^1=\int d^3\boldsymbol{v}\,\tilde{F}_s^1(\boldsymbol{v})\,. \tag{6.2.8}$$

We first solve (6.2.7) for $\tilde{F}_s^1(\boldsymbol{v})$ and substitute the result into (6.2.6, 8). In solving (6.2.7) we encounter a divergence at $\omega = kw$. The physical meaning of this divergence will be discussed in Sect. 6.3. Here we solve this problem by assuming that the perturbation is introduced adiabatically from the infinite past $t\to-\infty$ and by adding a small positive imaginary part to ω

$$\omega\to\omega+\mathrm{i}\delta\;(\delta>0)\,. \tag{6.2.9}$$

Then $\exp(-\mathrm{i}\omega t)\to\exp(-\mathrm{i}\omega t+\delta t)$ which vanishes at $t\to-\infty$. Equation (6.2.7) can then be solved for all values of ω as

$$\tilde{F}_s^1(\boldsymbol{v})=-\frac{q_s}{m_s}\frac{k}{(\omega-kw+\mathrm{i}\delta)}\frac{\partial}{\partial w}F_s^0(v)\tilde{\phi}^1\,. \tag{6.2.10}$$

Equation (6.2.10) describes the linear response of the distribution to the electric field. Substitution of (6.2.10) into (6.2.8) yields an equation of the form

$$q_s\tilde{n}_s^1(k,\omega)=-\varepsilon_0 k^2\chi_s(k,\omega)\tilde{\phi}^1(k,\omega)\,, \tag{6.2.11}$$

where $\chi_s(k,\omega)$ represents the degree of polarization due to the linear electrostatic response of the sth species of particle and is called the *electric susceptibility* of the sth species of particle. The explicit expression for $\chi_s(k,\omega)$ is

$$\chi_s(k,\omega)=\frac{\omega_{\mathrm{p}s}^2}{k^2 n_s^0}\int d^3\boldsymbol{v}\frac{k}{\omega-kw+\mathrm{i}\delta}\frac{\partial}{\partial w}F_s^0(v)\,, \tag{6.2.12}$$

where $\omega_{\mathrm{p}s}$ is the plasma frequency ($\omega_{\mathrm{p}s}^2 = q_s^2 n_s^0/\varepsilon_0 m_s$) of the sth species of particle. Substituting (6.2.11) into (6.2.6) gives

$$\varepsilon(k,\omega)\tilde{\phi}^1(k,\omega) = 0\ , \tag{6.2.13}$$

where

$$\varepsilon(k,\omega) = \varepsilon_0\,[1 + \chi_e(k,\omega) + \chi_i(k,\omega)] \tag{6.2.14}$$

is called the *plasma dielectric function.* It represents the modification of the vacuum dielectric constant ε_0 by the plasma polarization effect $(\chi_e + \chi_i)$. A nontrivial solution $(\tilde{\phi}^1 \neq 0)$ of (6.2.13) can be found when

$$\varepsilon(k,\omega) = 0\ . \tag{6.2.15}$$

This is nothing but the *dispersion relation* for the electrostatic wave in an isotropic plasma. As we can see from (6.2.12), $\chi_s(k,\omega)$, hence $\varepsilon(k,\omega)$, is a complex function. Therefore for real k the solution of (6.2.15) for ω is in general complex. We shall dwell on this problem in the next section. Here we assume that the imaginary part of ω is sufficiently small. As can be seen from (6.2.12), such is the case for a typical value of $|w|$ either when

$$|\omega| \gg k|w| \tag{6.2.16}$$

or when

$$|\omega| \ll k|w| \tag{6.2.17}$$

The typical value of $|w|$ is the thermal speed v_{Ts}.

Let us calculate $\chi_s(k,\omega)$ for the two limiting cases (6.2.16, 17). We first carry out the velocity integral for the y-, z-components to obtain

$$\int d^3\boldsymbol{v}\frac{1}{\omega - kw + \mathrm{i}\delta}\frac{\partial}{\partial w}F_s^0(v) = \int_{-\infty}^{\infty} dw\frac{1}{\omega - kw + \mathrm{i}\delta}\frac{d}{dw}f_s^0(w)\ , \tag{6.2.18}$$

where

$$f_s^0(w) = \int_{-\infty}^{\infty} dv_y \int_{-\infty}^{\infty} dv_z F_s^0(v)\ . \tag{6.2.19}$$

First in the limit given in (6.2.16) we expand the denominator in (6.2.18) as

$$\frac{1}{\omega - kw + \mathrm{i}\delta} = \frac{1}{\omega}\left[1 + \frac{kw}{\omega} + \left(\frac{kw}{\omega}\right)^2 + \left(\frac{kw}{\omega}\right)^3 + \cdots\right] \tag{6.2.20}$$

and carry out the velocity integral term by term. Since $f_s^0(w)$ is an even function of w, $df_s^0(w)/dw$ is an odd function of w. The integral of an odd function over the entire velocity range vanishes, so that the first and the third terms inside the bracket of (6.2.20) are zero. Performing the integration by parts for the even terms under the condition that $f_s^0(w) \to 0$ at $|w| \to \infty$, we get

$$\int_{-\infty}^{\infty} dw\frac{k}{\omega - kw + \mathrm{i}\delta}\frac{d}{dw}f_s^0(w) = -\frac{k^2}{\omega^2}n_s^0\left(1 + \frac{3k^2\langle w^2\rangle}{\omega^2} + \cdots\right) \tag{6.2.21}$$

where we used the relations

$$n_s^0 = \int_{-\infty}^{\infty} dw f_s^0(w), \quad n_s^0 \langle w^2 \rangle = \int_{-\infty}^{\infty} dw w^2 f_s^0(w) . \tag{6.2.22}$$

Use of this result in (6.2.12) gives

$$\chi_s(k,\omega) = -\frac{\omega_{ps}^2}{\omega^2}\left(1 + 3\frac{k^2\langle w^2 \rangle}{\omega^2} + \cdots\right) \tag{6.2.23}$$

for $|\omega| \gg k|w|$. For a Maxwellian distribution,

$$\langle w^2 \rangle = v_{Ts}^2 = T_s/m_s . \tag{6.2.24}$$

Even if the velocity distribution is not Maxwellian, we can use (6.2.24) when we define the temperature T_s by the average particle kinetic energy in the x-direction as $\langle m_s w^2/2 \rangle = T_s/2$. Then the expansion (6.2.23) is a series expansion in powers of the square of the ratio of the thermal speed to the phase velocity.

We next consider the other limit given in (6.2.17). In this case, neglecting ω we get to the lowest order

$$\chi_s(k,\omega) \doteq \chi_s(k,0) = -\frac{\omega_{ps}^2}{k^2 n_s^0}\int_{-\infty}^{\infty} dw \frac{1}{w}\frac{df_s^0(w)}{dw} . \tag{6.2.25}$$

For a Maxwellian distribution the velocity integral gives $-n_s^0/v_{Ts}^2$ and we get

$$\chi_s(k,0) = \frac{\omega_{ps}^2}{k^2 v_{Ts}^2} = \frac{1}{k^2 \lambda_{Ds}^2} , \tag{6.2.26}$$

where $\lambda_{Ds} = v_{Ts}/\omega_{ps}$ is the Debye length of the sth species of particle. For a non-Maxwellian distribution, we can define the Debye length by the relation

$$\frac{1}{\lambda_{Ds}^2} = -\frac{\omega_{ps}^2}{n_s^0}\int_{-\infty}^{\infty} dw \frac{1}{w}\frac{df_s^0(w)}{dw} . \tag{6.2.27}$$

In most cases of interest, the right hand side is positive, and hence λ_{Ds} is real.

Using (6.2.15, 23, 26) we can derive the dispersion relations for the electron plasma wave and the ion acoustic wave. Namely, for $\omega^2 \gg k^2 v_{Te}^2 \gg k^2 v_{Ti}^2$ we use (6.2.23) for the electron susceptibility to obtain the dispersion relation (5.2.14). For the ion acoustic wave, we assume $T_e \gg T_i$ and consider the frequency region $k^2 v_{Te}^2 \gg \omega^2 \gg k^2 v_{Ti}^2$. We use (6.2.26) for the electron and (6.2.23) for the ion to obtain the dispersion relation (5.4.16). Note that in (5.4.16) we have to use (6.2.27) for the Debye length.

Finally, let us discuss the Debye screening. We consider the electron response to a static ion point charge at the origin, $q_i n_i^1(\boldsymbol{r},t) = q_i \delta[\boldsymbol{r}]$. We note that the δ-function can be represented as a superposition of the three-dimensional plane waves as

$$\begin{aligned} \delta[\boldsymbol{r}] &= \delta[x]\delta[y]\delta[z] \\ &= \int_{-\infty}^{\infty}\frac{dk_x}{2\pi}\int_{-\infty}^{\infty}\frac{dk_y}{2\pi}\int_{-\infty}^{\infty}\frac{dk_z}{2\pi}\exp[\mathrm{i}(k_x x + k_y y + k_z z)] . \end{aligned} \tag{6.2.28}$$

Since the static charge can be represented by the Fourier component at $\omega = 0$ the Fourier component of $n_i^1(\boldsymbol{r}, t)$ can be written as $\tilde{n}_i^1(\boldsymbol{k}, 0)$. From (6.2.6, 11) we can write $\tilde{\phi}^1(\boldsymbol{k}, 0)$ as

$$\begin{aligned} \tilde{\phi}^1(\boldsymbol{k},0) &= \frac{q_i \tilde{n}_i^1(\boldsymbol{k},0)}{\varepsilon_0 k^2[1+\chi_e(\boldsymbol{k},0)]} \\ &= \frac{q_i}{\varepsilon_0 k^2[1+\chi_e(\boldsymbol{k},0)]} \qquad \text{(for all } \boldsymbol{k}\text{)} \end{aligned} \tag{6.2.29}$$

from which we can calculate $\phi^1(r)$ by Fourier inversion. Since we are looking for a spherically symmetric solution, we use the spherical coordinate representation for $\boldsymbol{k} = (k, \theta, \phi)$. Noting the relations $d^3\boldsymbol{k} = -k^2 d\cos\theta d\phi$ and $\boldsymbol{k}\cdot\boldsymbol{r} = kr\cos\theta$, we get

$$\phi^1(r) = \int_0^\infty \frac{dk}{4\pi^2} \int_{-1}^1 d\cos\theta \exp[ikr\cos\theta] \frac{q_i}{\varepsilon_0[1+\chi_e(k,0)]} . \tag{6.2.30}$$

Substituting (6.2.26) into (6.2.30) and carrying out the integration, we obtain (Problem 6.1)

$$\phi^1(r) = \frac{q_i}{4\pi\varepsilon_0} \frac{1}{r} \exp\left(-\frac{r}{\lambda_{De}}\right) \tag{6.2.31}$$

where λ_{De} is defined by (6.2.27) for s = e. Equation (6.2.31) is the *Debye potential* (5.1.6).

6.3 Landau Damping

As mentioned in Sect. 6.2, the plasma dielectric function is, in general, a complex function of ω. This is due to the singular behavior of the perturbed distribution at $\omega = kw$. This singular behavior in velocity space is the characteristic of a continuum in phase space, which is different from a continuum in real space. This is the problem that we shall dwell on in this section.

6.3.1 Resonant Particles and Nonresonant Particles

We first discuss the physical meaning of $\omega = kw$. The particles satisfying this condition are moving with the phase velocity of the wave. Consider an electrostatic potential energy represented by a plane wave as shown in Fig. 6.1. In the frame moving with the phase velocity of the wave (we refer to this as the wave frame), this potential is at rest. Particles which are free from collisions can be regarded as moving with constant speed in this frame. Of course, the velocity is modulated by the potential, but the modulation is negligible if the potential is infinitely small.

We investigate the particle motion over a certain time period τ. We assume that during this time the potential can be regarded as fixed. In other words, we

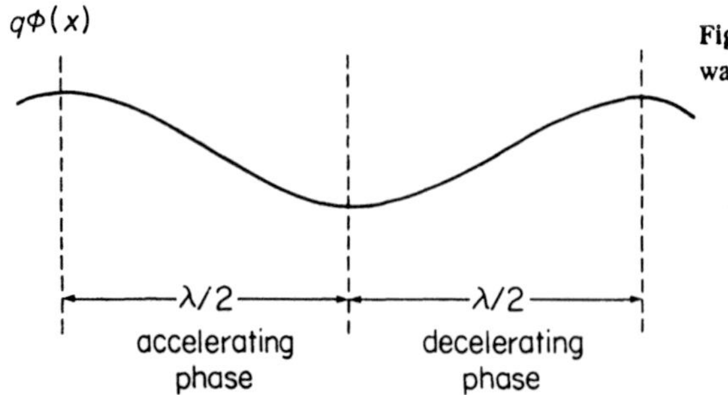

Fig. 6.1. Electrostatic potential of plane wave in the waveframe

choose τ such that it is smaller than or on the order of the lifetime of the potential. Those particles moving rapidly in the wave frame will traverse the crests and troughs of the potential during the time τ and the particle velocity undergoes a periodic modulation by the potential. When averaged over the time τ, the kinetic energy of these particles is constant. These particles will be referred to as the *nonresonant particles.*

On the other hand, those particles which are almost at rest in the wave frame will stay near the initial phase of the potential over the entire period τ. These particles will stay in the same phase of the potential and will be either accelerated or decelerated by the potential during the time τ. In other words, those particles feel the wave potential as a *dc* field. If averaged over the time τ, the kinetic energy of these particles either increases (in the acceleration phase) or decreases (in the deceleration phase) and therefore contributes to either damping or enhancement of the wave potential energy. Thus these particles contribute to the energy exchange between the wave and the particles. These particles are called the *resonant particles*. For a sufficiently weak potential, the velocity change due to the acceleration or deceleration is negligible compared to the initial velocity in the wave frame and the condition for the resonant particles (resonance condition) can be written as

$$|w - \omega/k|\tau < \lambda/2 = \pi/k \ . \tag{6.3.1}$$

We can choose an arbitrary time for τ provided that it is smaller than the lifetime of the potential. For short τ, the resonance condition of (6.3.1) can be satisfied by many particles, while for sufficiently long τ the condition $w = \omega/k$ must be met. Since the particles satisfying the condition $w = \omega/k$ satisfy the resonance condition over a long time and contribute to energy exchange with the wave, naturally these particles give rise to the wave damping or growth represented by the imaginary part of ω.

6.3.2 Effect of Resonant Particles on Electric Susceptibility

We now examine the effect of the resonant particles on the wave dispersion relation. To this end, we need an analytical continuation of the electric susceptibility in the complex ω-plane. We consider the velocity integral (6.2.18) [cf. (6.2.12)]. The integrand has a pole at

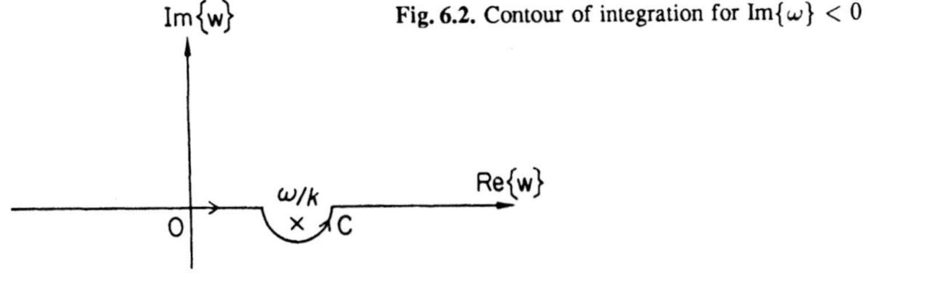

Fig. 6.2. Contour of integration for $\mathrm{Im}\{\omega\} < 0$

$$w = (\omega + \mathrm{i}\delta)/k \qquad (\delta > 0)\,. \tag{6.3.2}$$

Since $k > 0$, it is in the domain $\mathrm{Im}\{w\} > 0$. The integration path is along the real axis, so the pole is above the integration path. Therefore, for $\mathrm{Im}\{\omega\} > 0$, the integration path can be along the real axis, while for $\mathrm{Im}\{\omega\} \leq 0$, the integration path has to be modified as shown in Fig. 6.2. If we denote the correct integration path by C, then

$$\chi_s(k,\omega) = -\frac{\omega_{ps}^2}{k^2 n_s^0}\int_C dw \frac{1}{w - (\omega + \mathrm{i}\delta)/k}\frac{d}{dw}f_s^0(w) \tag{6.3.3}$$

is the analytical continuation of the electric susceptibility over the entire ω-plane.

Let us choose ω real and divide $\chi_s(k,\omega)$ into real and imaginary parts as

$$\chi_s(k,\omega) = \chi_s'(k,\omega) - \mathrm{i}\chi_s''(k,\omega)\,. \tag{6.3.4}$$

We calculate χ_s' and χ_s'' by rewriting $[w - (\omega + \mathrm{i}\delta)/k]^{-1}$ as

$$\frac{1}{w - (\omega + \mathrm{i}\delta)/k} = \frac{(w - \omega/k)}{(w - \omega/k)^2 + (\delta/k)^2} + \mathrm{i}\frac{\delta/k}{(w - \omega/k)^2 + (\delta/k)^2} \tag{6.3.5}$$

and take the limit at $\delta \to +0$. We substitute (6.3.5) into (6.3.3). The real part can be calculated as

$$\begin{aligned}&\lim_{\delta\to+0}\int_{-\infty}^{\infty} dw \frac{(w - \omega/k)}{(w - \omega/k)^2 + (\delta/k)^2}\frac{d}{dw}f_s^0(w)\\ &= P\int_{-\infty}^{\infty} dw \frac{1}{w - \omega/k}\frac{d}{dw}f_s^0(w)\end{aligned} \tag{6.3.6}$$

where P denotes the Cauchy's principal value integral. The imaginary part of (6.3.5) has a sharp peak at $w = \omega/k$ (peak value $\sim k/\delta$) and vanishes at $w \neq \omega/k$. If $df_s^0(w)/dw$ is a smooth function of w at $w = \omega/k$, we can calculate the imaginary part of (6.3.3) as

$$\begin{aligned}&\lim_{\delta\to+0}\int_{-\infty}^{\infty} dw \frac{\mathrm{i}\delta/k}{(w - \omega/k)^2 + (\delta/k)^2}\frac{d}{dw}f_s^0(w)\\ &= \left[\frac{d}{dw}f_s^0(w)\right]_{w=\omega/k}\lim_{\delta\to+0}\int_{-\infty}^{\infty} dw \frac{\mathrm{i}\delta/k}{(w - \omega/k)^2 + (\delta/k)^2}\\ &= \mathrm{i}\pi\frac{d}{dw}f_s^0(w)\bigg|_{w=\omega/k}\,.\end{aligned} \tag{6.3.7}$$

Using (6.3.6, 7) in (6.3.3), we obtain

$$\chi_s'(k,\omega) = -\frac{\omega_{ps}^2}{k^2 n_s^0} P \int_{-\infty}^{\infty} dw \frac{1}{w - \omega/k} \frac{d}{dw} f_s^0(w) \tag{6.3.8}$$

$$\chi_s''(k,\omega) = \pi \frac{\omega_{ps}^2}{k^2 n_s^0} \left[\frac{d}{dw} f_s^0(w) \right]_{w=\omega/k} \tag{6.3.9}$$

We note that only the resonant particles contribute to the imaginary part of $\chi_s(k,\omega)$.

6.3.3 Mathematical Formulation of Landau Damping

We now solve the dispersion relation (6.2.15) by assuming that $|\chi_s''| \ll |1+\chi_s'|$. We write the solution as

$$\omega = \omega_r - i\gamma \, . \tag{6.3.10}$$

If $|\chi_s''|$ is sufficiently small, we can expect

$$|\gamma| \ll |\omega_r| \, . \tag{6.3.11}$$

Therefore to lowest order we neglect γ and χ_s'' and get

$$\varepsilon'(k,\omega_r) \equiv \varepsilon_0[1 + \chi_e'(k,\omega_r) + \chi_i'(k,\omega_r)] = 0 \, . \tag{6.3.12}$$

This gives the dispersion relation for ω_r as obtained in Sect. 6.2. We next consider the analytical continuation of $\varepsilon'(k,\omega)$ and calculate it to first order in γ:

$$\begin{aligned} \varepsilon'(k,\omega_r - i\gamma) &= \varepsilon'(k,\omega_r) - i\gamma \partial\varepsilon'(k,\omega_r)/\partial\omega_r \\ &= -i\gamma \partial\varepsilon'(k,\omega_r)/\partial\omega_r \, . \end{aligned} \tag{6.3.13}$$

For $\chi''(k,\omega)$ we set $\omega = \omega_r$ since χ'' is already small,

$$\chi''(k,\omega) \doteq \chi_e''(k,\omega_r) + \chi_i''(k,\omega_r) \, .$$

Thus to first order, we have

$$\varepsilon(k,\omega) = -i\gamma \frac{\partial\varepsilon'(k,\omega_r)}{\partial\omega_r} - i\varepsilon_0[\chi_e''(k,\omega_r) + \chi_i''(k,\omega_r)] \, .$$

When this expression vanishes the imaginary part of ω is obtained from

$$\gamma = -i\varepsilon_0[\chi_e''(k,\omega_r) + \chi_i''(k,\omega_r)] \left[\frac{\partial\varepsilon'(k,\omega_r)}{\partial\omega_r} \right]^{-1} . \tag{6.3.14}$$

We can calculate γ by substituting (6.3.9) into (6.3.14). If $\gamma > 0$ the wave is damped, while if $\gamma < 0$ the wave grows. The sign of γ is determined by the sign of $\partial\varepsilon'(k,\omega_r)/\partial\omega_r$ and $[df_s^0(w)/dw]_{w=\omega_r/k}$. For the electron plasma wave and ion acoustic wave

$$\omega_r \frac{\partial\varepsilon'(k,\omega_r)}{\partial\omega_r} > 0 \tag{6.3.15}$$

(Problem 6.2). Noting that $\omega_{ps}^2/n_s^0 = q_s^2/\varepsilon_0 m_s$, the sign of γ for these waves is equal to the sign of the expression

$$-\left[e^2 \frac{w}{m_e}\frac{df_e^0(w)}{dw} + q_i^2 \frac{w}{m_i}\frac{df_i^0(w)}{dw}\right]_{w=\omega_r/k} . \tag{6.3.16}$$

If $f_s^0(w)$ is Maxwellian, then $wdf_s^0(w)/dw$ is always negative, therefore γ is always positive. This means that for Maxwellian plasma, these waves are always damped. This was first shown by Landau in 1946 and is called the *Landau damping* [6.1].

Explicit expressions for the Landau damping rates for the electron plasma wave and the ion acoustic wave can be readily calculated using (6.3.14) (Problem 6.3). For the electron plasma wave the ion contribution can be neglected and only the electron contribution needs to be calculated. At long wavelengths ($k^2\lambda_{De}^2 \ll 1$), the phase velocity of the electron plasma wave is much faster than the electron thermal speed $v_{Te} = \sqrt{T_e/m_e}$, so that the condition of (6.3.11) is satisfied. At short wavelengths, when $k\lambda_{De} \sim 1$, however, the phase velocity becomes comparable to the electron thermal speed and as a result, $\gamma \sim |\omega_r|$ and the approximation of (6.3.14) is no longer valid. For a Maxwellian distribution

$$f_s^0(w) = \frac{n_s^0}{\sqrt{2\pi}v_{Ts}} \exp\left(-\frac{w^2}{2v_{Ts}^2}\right) , \tag{6.3.17}$$

the electric susceptibility given in (6.3.3) can be written as

$$\chi_s(k,\omega) = -\frac{1}{k^2\lambda_{Ds}^2}\frac{1}{2}Z'\left(\frac{\omega}{\sqrt{2}kv_{Ts}}\right) \tag{6.3.18}$$

where the function $Z(\xi)$ is called the *plasma dispersion function* [6.2]

$$Z(\xi) = \frac{1}{\sqrt{\pi}}\int_C dt \frac{1}{t-\xi} e^{-t^2} \tag{6.3.19}$$

and its numerical values are tabulated for complex values of ξ. According to this result, the Landau damping is very strong in this case and the wave cannot propagate in a plasma at such a short wavelength.

For the case of the ion acoustic wave, both electrons and ions can contribute to the damping. For $T_e \gg T_i$ and at long wavelengths ($k\lambda_{De} \ll 1$), the phase velocity is much larger than the ion thermal speed, and the electron contribution dominates. At short wavelengths ($k\lambda_{De} \gtrsim 1$), however, the ion contribution becomes more important. At even shorter wavelengths such that $k\lambda_{Di} \sim 1$, the ion

Landau damping becomes too strong for the wave to propagate. The same is true for the case $T_e \doteq T_i$.

6.3.4 Physical Mechanism

The physical mechanism of Landau damping can be explained as follows. First, the reason that only the resonant particles contribute to the damping has already been explained. Those resonant particles which move slightly faster than the phase velocity ($w \gtrsim |(\omega_r/k|$) can stay in the resonance region given in (6.3.1) for a longer time period if they are in the deceleration phase than when they are in the acceleration phase. Therefore on the average these resonant particles tend to lose their energy to the wave. On the other hand, those resonant particles which move slightly slower than the phase velocity ($w \lesssim |\omega_r/k|$) will, on average, be accelerated by the wave and receive energy from it. On the whole, the wave either loses or gains energy depending on the relative numbers of particles at $w \gtrsim |\omega_r/k|$ and $w \lesssim |\omega_r/k|$. Therefore the sign of γ (damping rate) is determined by the sign of $wdf_s^0(w)/dw|_{w=\omega_r/k}$. If we take into account the difference of the inertia of the electron and the ion, we can understand why the sign of (6.3.16) determines the sign of γ.

6.3.5 Kinetic Instabilities

For collisionless plasma, $f_s^0(w)$ is not necessarily Maxwellian. Therefore γ is not always positive. If $\gamma < 0$, the wave energy increases in time due to the effect of the resonant particles. Such an instability is generally called the *kinetic instability*. We shall show two such examples here.

The first is the case of an electron beam with the velocity distribution $f_e^0(w)$ shown in Fig. 6.3. In this case, in the velocity region $u_1 < w < u_2$, $wdf_e^0(w)/dw$ becomes positive, so that the electron plasma wave having the phase velocity in this region becomes unstable. This instability is called the *bump-on-tail instability*. The second example is the case in which the electrons are moving relative to the ions, that is, when there is an electric current. If the electron drift velocity is much faster than the ion thermal speed, the velocity distributions $f_e^0(w)$ and $f_i^0(w)$ are as shown in Fig. 6.4. If we now consider an ion acoustic wave whose phase velocity is much faster than the ion thermal speed, the electron contri-

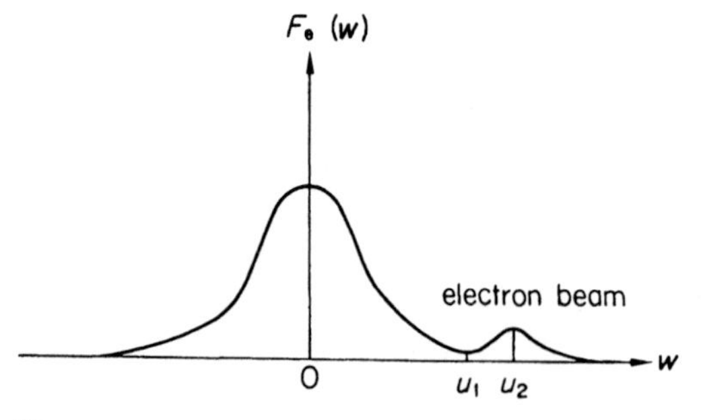

Fig. 6.3. Hypothetical electron velocity distribution function in an electron beam

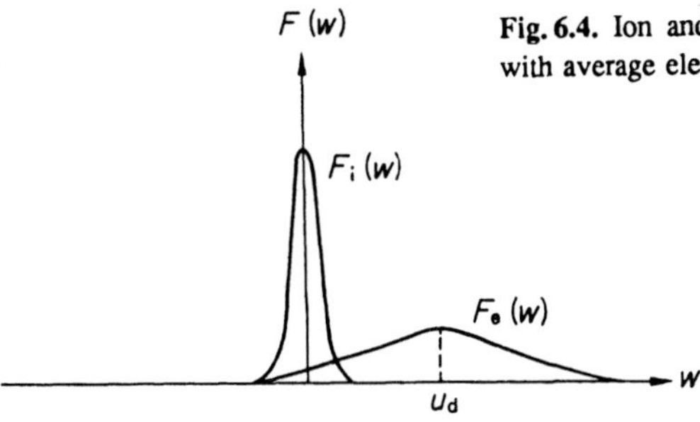

Fig. 6.4. Ion and electron velocity distribution functions with average electron drift velocity at u_d

bution becomes dominant in (6.3.16). If $\omega_r/k < u_d$, where u_d is the electron drift velocity, then $[w df_e^0(w)/dw]_{w=\omega_r/k} > 0$ and the ion acoustic wave becomes unstable. This instability is called the *current-driven ion acoustic instability*. This instability is similar to the two-beam instability discussed in Sect. 5.4, but differs in the sense that the resonant electrons play the essential role. Since the energy source of kinetic instability is limited to a small number of resonant particles, the growth rate is small compared to the case of the two-beam instability. The kinetic instability is therefore easily suppressed by a rather weak nonlinear effect as will be shown in Sect. 6.6.

6.4 Electrostatic Response in Magnetized Plasma

We consider a plasma in the presence of a uniform static magnetic field in the z-direction and discuss the electrostatic response to the electric field $\boldsymbol{E}^1 = -\nabla\phi^1$. The basic equations are (6.1.2, 4) and the Poisson equation (6.2.1).

We assume that the plasma is spatially uniform in the unperturbed state. Then (6.1.2) becomes

$$\left(\boldsymbol{v} \times \boldsymbol{\Omega}_s^0\right) \cdot \frac{\partial}{\partial \boldsymbol{v}} F_s^0(\boldsymbol{v}) = 0 , \tag{6.4.1}$$

the solution of which can be written as a function of v_z and $v_\perp$

$$F_s^0(\boldsymbol{v}) = F_s^0(v_z, v_\perp) . \tag{6.4.2}$$

In general, the unperturbed distribution is an arbitrary function of the constants of motion and which in the present case are v_z and $v_\perp$.

For the perturbed distribution, we formally integrate (6.1.4) along the unperturbed particle orbit

$$F_s^1(\boldsymbol{r}, \boldsymbol{v}, t) = \frac{q_s}{m_s} \int_{-\infty}^{t} dt' \nabla\phi^1(\boldsymbol{r}', t') \cdot \frac{\partial}{\partial \boldsymbol{v}'} F_s^0(v_z, v_\perp) \tag{6.4.3}$$

where $\boldsymbol{r}'$ and $\boldsymbol{v}'$ are the solutions of the equations of motion (6.1.6) at time t' under the condition that $\boldsymbol{r}' = \boldsymbol{r}$ and $\boldsymbol{v}' = \boldsymbol{v}$ at $t' = t$. We write the perturbed quantities in Fourier representation as

$$\begin{aligned} F_s^1(\boldsymbol{r}, \boldsymbol{v}, t) &= \tilde{F}_s^1(\boldsymbol{k}, \omega, \boldsymbol{v}) \exp[\mathrm{i}(\boldsymbol{k} \cdot \boldsymbol{r} - \omega t)] \\ \phi^1(\boldsymbol{r}', t') &= \tilde{\phi}^1(\boldsymbol{k}, \omega) \exp[\mathrm{i}(\boldsymbol{k} \cdot \boldsymbol{r}' - \omega t')] . \end{aligned} \tag{6.4.4}$$

We add a small positive imaginary part to ω which we shall not write here explicitly. Equation (6.4.3) then becomes

$$\tilde{F}_s^1 = \mathrm{i}\frac{q_s}{m_s}\tilde{\phi}^1 \int_{-\infty}^{t} dt' \boldsymbol{k} \cdot \frac{\partial}{\partial \boldsymbol{v}'} F_s^0(v_z, v_\perp) \exp[\mathrm{i}\boldsymbol{k} \cdot (\boldsymbol{r}' - \boldsymbol{r}) - \mathrm{i}\omega(t' - t)] \tag{6.4.5}$$

where we suppressed the argument of $\tilde{F}_s^1$ and $\tilde{\phi}^1$ for simplicity. In (6.4.5) $\boldsymbol{k} \cdot \partial/\partial \boldsymbol{v}'$, can be written as

$$\boldsymbol{k} \cdot \frac{\partial}{\partial \boldsymbol{v}'} = k_z \frac{\partial}{\partial v_z} + \boldsymbol{k}_\perp \cdot \boldsymbol{v}'_\perp \frac{1}{v_\perp} \frac{\partial}{\partial v_\perp} \tag{6.4.6}$$

where $\boldsymbol{k}_\perp, \boldsymbol{v}'_\perp$ are the vectors representing the perpendicular components of $\boldsymbol{k}$ and $\boldsymbol{v}'$ and $v'_z = v_z, v'_\perp = v_\perp$. Noting that

$$\boldsymbol{k}_\perp \cdot \boldsymbol{v}'_\perp = (\boldsymbol{k} \cdot \boldsymbol{v}' - \omega) + (\omega - k_z v_z)$$

and using the relations

$$\begin{aligned}(\boldsymbol{k} \cdot \boldsymbol{v}' - \omega) &\exp[\mathrm{i}\boldsymbol{k} \cdot (\boldsymbol{r}' - \boldsymbol{r}) - \mathrm{i}\omega(t' - t)] \\ &= -\mathrm{i}\frac{d}{dt'} \exp[\mathrm{i}\boldsymbol{k} \cdot (\boldsymbol{r}' - \boldsymbol{r}) - \mathrm{i}\omega(t' - t)] ,\end{aligned}$$

we can partly carry out the time integral of (6.4.5) to obtain

$$\begin{aligned}\tilde{F}_s^1 = \; & \frac{q_s}{m_s}\tilde{\phi}^1 \left\{ \frac{1}{v_\perp}\frac{\partial}{\partial v_\perp} F_s^0 + \left[k_z \frac{\partial}{\partial v_z} + (\omega - k_z v_z)\frac{1}{v_\perp}\frac{\partial}{\partial v_\perp} \right] F_s^0 \right\} \\ & \times \int_{-\infty}^{t} dt' \exp[\mathrm{i}\boldsymbol{k} \cdot (\boldsymbol{r}' - \boldsymbol{r}) - \mathrm{i}\omega(t' - t)] .\end{aligned} \tag{6.4.7}$$

To carry out the rest of the integration, we need the explicit expression for exp $(\mathrm{i}\boldsymbol{k} \cdot \boldsymbol{r}')$. The position vector $\boldsymbol{r}'$ can be written in the form of (3.2.7) as

$$\boldsymbol{r}' = \begin{pmatrix} (v_\perp/\Omega_s)\sin[\Omega_s(t' - t) + \theta_0] + x_0 \\ -(v_\perp/\Omega_s)\cos[\Omega_s(t' - t) + \theta_0] + y_0 \\ v_z t' + z_0 \end{pmatrix} . \tag{6.4.8}$$

The initial values θ_0, x_0, y_0 and z_0 are determined by the condition $\boldsymbol{r}' = \boldsymbol{r}$ at $t' = t$. Without loss of generality, we can choose the $\boldsymbol{k}$ vector to be in the xz-plane. Then from (6.4.8)

$$\begin{aligned}\exp & [\mathrm{i}\boldsymbol{k} \cdot (\boldsymbol{r}' - \boldsymbol{r})] \\ & = \exp\{\mathrm{i}k_z v_z(t' - t) + \mathrm{i}Z_s[\sin(\Omega_s[t' - t] + \theta_0) - \sin\theta_0]\}\end{aligned} \tag{6.4.9}$$

where

$$Z_s = k_\perp v_\perp / \Omega_s . \tag{6.4.10}$$

Now we use the Bessel function expansion formula

$$\exp[\pm\mathrm{i}Z\sin\varphi] = \sum_{l=-\infty}^{\infty} J_l(Z)\exp(\pm\mathrm{i}l\varphi)\,, \tag{6.4.11}$$

and (6.4.9) becomes

$$\begin{aligned}\exp[\mathrm{i}\boldsymbol{k}\cdot(\boldsymbol{r}'-\boldsymbol{r})] &= \sum_{l=-\infty}^{\infty}\sum_{l'=-\infty}^{\infty} J_l(Z_s)J_{l'}(Z_s) \\ &\times \exp[\mathrm{i}(k_z v_z + l\Omega_s)(t'-t) + \mathrm{i}(l'-l)\theta_0]\end{aligned} \tag{6.4.12}$$

Substituting this result into (6.4.7) and carrying out the time integration, we obtain

$$\begin{aligned}\tilde{F}_s^1 &= \frac{q_s}{m_s}\tilde{\phi}^1\left\{\frac{1}{v_\perp}\frac{\partial}{\partial v_\perp}F_s^0 + \left[k_z\frac{\partial}{\partial v_z} + (\omega - k_z v_z)\frac{1}{v_\perp}\frac{\partial}{\partial v_\perp}\right]F_s^0\right. \\ &\left.\times\sum_{l=-\infty}^{\infty}\sum_{l'=-\infty}^{\infty}\frac{J_l(Z_s)J_{l'}(Z_s)}{k_z v_z - \omega + l\Omega_s}\mathrm{e}^{\mathrm{i}(l-l')\theta_0}\right\}.\end{aligned} \tag{6.4.13}$$

We integrate $\tilde{F}_s^1$ with respect to $\boldsymbol{v}$ and substitute the result into the Poisson equation to obtain the dispersion relation. Noting the relations

$$\int d^3\boldsymbol{v} = \int_{-\infty}^{\infty} dv_z \int_0^{\infty} v_\perp dv_\perp \int_0^{2\pi} d\theta_0 \tag{6.4.14}$$

$$\int_0^{2\pi} d\theta_0\,\mathrm{e}^{\mathrm{i}(l-l')\theta_0} = 2\pi\delta_{l,l'} \tag{6.4.15}$$

$$\sum_{l=-\infty}^{\infty} J_l^2(z) = 1 \tag{6.4.16}$$

and rewriting

$$\frac{\omega - k_z v_z}{k_z v_z - \omega + l\Omega_s} = -1 + \frac{l\Omega_s}{k_z v_z - \omega + l\Omega_s}\,,$$

we obtain the expression

$$\tilde{n}_s^1 = \frac{q_s}{m_s}\tilde{\phi}^1\sum_{l=-\infty}^{\infty}\int d^3\boldsymbol{v}\frac{J_l^2(Z_s)}{k_z v_z - \omega + l\Omega_s}\left(\frac{l\Omega_s}{v_\perp}\frac{\partial}{\partial v_\perp} + k_z\frac{\partial}{\partial v_z}\right)F_s^0(v_z, v_\perp)\,. \tag{6.4.17}$$

Substituting (6.4.17) into (6.2.11), we finally get the dispersion relation

$$\varepsilon(\boldsymbol{k},\omega) = \varepsilon_0[1 + \chi_\mathrm{e}(\boldsymbol{k},\omega) + \chi_\mathrm{i}(\boldsymbol{k},\omega)] = 0 \tag{6.4.18}$$

where

$$\begin{aligned}\chi_s(\boldsymbol{k},\omega) &= \frac{\omega_{\mathrm{p}s}^2}{k^2 n_2^0}\sum_{l=-\infty}^{\infty}\int d^3\boldsymbol{v}\frac{J_l^2(Z_s)}{\omega - k_z v_z + l\Omega_s} \\ &\times\left\{\frac{l\Omega_s}{v_\perp}\frac{\partial}{\partial v_\perp} + k_z\frac{\partial}{\partial v_z}\right\}F_s^0(v_z, v_\perp)\,.\end{aligned} \tag{6.4.19}$$

This expression should be compared with (6.2.12).

From (6.4.18, 19) we can see two important kinetic effects (effects which never appear in the fluid model) in the dispersion relation of a magnetized plasma. One appears when the frequency ω is close to the cyclotron frequency or its overtones $l\Omega_s$. There, the effect of $k_z v_z$ plays a crucial role. In particular, those particles having the parallel velocity component

$$v_z \doteq \frac{\omega_r - l\Omega_s}{k_z} \tag{6.4.20}$$

contribute to the imaginary part of the dielectric function as in the case of the Landau damping and give rise to damping or growth of the wave. For the case of $l = 0$, condition (6.4.20) corresponds to the Landau resonance condition in the presence of a magnetic field. Since the particles can move freely only along the magnetic field line, the resonance condition is that the parallel velocity equals the parallel phase velocity ω_r/k_z. On the other hand, the case of $l \neq 0$ corresponds to a resonance accompanied by cyclotron motion. We can understand this resonance condition by writing (6.4.20) as

$$\frac{2\pi}{|\Omega_s|}\left(v_z - \frac{\omega_r}{k_z}\right) \doteq \frac{2\pi l}{k_z} \; . \tag{6.4.21}$$

The left hand side represents the distance that the particle moves relative to the wave along the magnetic field during one cyclotron period, and the right hand side is an integral multiple of the parallel wavelength $2\pi/k_z$. Under the condition given in (6.4.21), the particle comes back to the same phase of the wave after one cyclotron oscillation. A detailed calculation shows that this resonance also causes a damping of the wave if the unperturbed distribution is Maxwellian [6.3]. This damping is called the *cyclotron damping*. We note that both Landau damping and cyclotron damping take place only when the wavevector has a nonzero parallel component, i.e., $k_z \neq 0$. If $k_z = 0$, there are no resonant particles in a magnetic field, since the particles cannot move freely along the wave propagation. We have stated the effectiveness of a magnetic shear for the suppression of an instability in Sect. 5.5. This is due to the fact that as the wave propagates the parallel component of the wavevector changes and at the point where $k_\| \doteq \omega_r/v_T$ a strong Landau damping takes place.

The other kinetic effect becomes important when the particle Larmor radius becomes comparable to the perpendicular wavelength ($\sim 2\pi/k_\perp$). In this case $|Z_s|$ becomes of order of or greater than unity. For $|Z_s| \ll |$ the Bessel functions are approximately

$$J_0(Z_s) \doteq 1, \quad J_l(Z_s) \doteq 0 \qquad (l \neq 0) \; .$$

For $|Z_s| \geq 1$ they start oscillating. As a result, the dispersion relation is significantly modified. This modification is due to the fact that during the gyrating motion the particles feel different phases of the wave field or the field averaged over the Larmor radius. We illustrate this phenomenon by considering the case of very strong magnetic field, $|\Omega_s| \gg |\omega|$ and $|\Omega_s| \gg |k_z v_z|$. We ignore the terms of order $|\omega - k_z v_z|/|\Omega_s|$ in (6.4.19) and use the (6.4.16) to obtain

$$\chi_s \doteq \frac{\omega_{ps}^2}{k^2 n_s^0} \int d^3v \left\{ \frac{J_0^2(Z_s)}{\omega - k_z v_z} k_z \frac{\partial}{\partial v_z} - [1 - J_0^2(Z_s)] \frac{1}{v_\perp} \frac{\partial}{\partial v_\perp} \right\} F_s^0(v_z v_z) \, . \tag{6.4.22}$$

In the limit $Z_s \to 0$, the square of the Bessel function J_0 goes to 1 and we recover the result of the isotropic plasma with $\boldsymbol{k} = k_z \hat{\boldsymbol{z}}$. In the limit of a strong magnetic field, the perpendicular particle motion practically vanishes. As Z_s increases, $J_0^2(Z_s) \sim 1 - Z_s^2/2 + \ldots$ and the first term inside the bracket of (6.4.22) decreases while the second term increases. The decrease of the first term is due to the averaging of the electric field by the cyclotron motion while the increase of the second term is due to the appearance of the perpendicular particle motion by the cyclotron motion. These effects are known as the *finite Larmor radius effects* and are important in a high temperature plasma.

6.5 Collisionless Drift Wave Instability

The drift wave discussed in Sect. 6.5 becomes unstable by various causes [6.4]. Here we give a qualitative description of the mechanism of the drift wave instability due to the Landau resonance in a collisionless plasma in the presence of a uniform magnetic field.

We choose the magnetic field to be in the z-direction and the density gradient in the negative x-direction and assume that the wave propagates in the yz-plane (Fig. 5.6). For the electrostatic drift wave, the equi-phase surface of the electric field of the wave is perpendicular to the wavenumber vector. In Fig. 5.6 the y- and z-components of the electric field are shown. Since $\boldsymbol{E}^1 = -\nabla\phi^1$, $E_y^1 > 0$ in the phase $E_z^1 > 0$ and $E_y^1 < 0$ in the phase $E_z^1 < 0$. The presence of E_y^1 causes an $\boldsymbol{E} \times \boldsymbol{B}$ drift $v_{\mathrm{D}x}^1 = E_y^1/B_0$ in the x-direction. The direction of this $\boldsymbol{E} \times \boldsymbol{B}$ drift is also shown in Fig. 5.6. Because of the density gradient this $\boldsymbol{E} \times \boldsymbol{B}$ drift causes a density variation. In the region where the particles drift from the high density side (regions I and III in Fig. 5.6) the density increases by the $\boldsymbol{E} \times \boldsymbol{B}$ drift, while in the region where the particles drift from the low density side (region II in Fig. 5.6) the density decreases.

Let us consider the effect of the resonant particles (particles of parallel velocity $v_z = \omega/k_z$). In regions I and III of Fig. 5.6, the resonant electrons are decelerated by the parallel field E_z^1 while in region II the resonant electrons are accelerated by the parallel field E_z^1. If there is no density gradient, for a Maxwellian distribution, there are more electrons which are accelerated than those which are decelerated and therefore the wave is damped by the resonant electrons. In an inhomogeneous plasma, however, since in regions I and III the electrons drift from the high density side, while in region II the electrons drift from the low density side, the number of the resonant electrons which are accelerated may become less than those which are decelerated. Then the resonant electrons can on average contribute to increase of the wave energy. This is the physical mechanism of the collisionless drift wave instability.

Let us examine the condition for the instability. First, for a spatially uniform Maxwellian plasma, the rate of change in the number of resonant electrons due to the acceleration or deceleration by the parallel electric field E_z^1 is

$$\left[\frac{dv_z}{dt}\frac{\partial F_e^0(v_z)}{\partial v_z}\right]_{v_z=\omega/k_z} \doteq -\frac{e}{m_e}E_z^1\left[-\frac{m_e}{T_e}\frac{\omega}{k_z}\right]F_e^0(v_z=\omega/k_z)$$

where we used the equation of motion $dv_z/dt = -eE_z^1/m_e$. We evaluate as in (5.4.10)

$$E_z^1 \sim -\mathrm{i}k_z\phi^1, \quad e\phi^1/T_e \sim n_e^1/n_e^0\,, \tag{6.5.1}$$

then we can write

$$\left[\frac{dv_z}{dt}\frac{\partial F_e^0}{\partial v_z}\right]_{v_z=w/k_z} \sim -\mathrm{i}\omega\frac{n_e^1}{n_e^0}F_e^0(v_z=\omega/k_z)\,. \tag{6.5.2}$$

This contributes to the damping of the wave.

On the other hand, the rate of change in the number of resonant electrons by the $\boldsymbol{E}\times\boldsymbol{B}$ drift in an inhomogeneous plasma is

$$v_{Dx}^1\frac{\partial}{\partial x}F_e^0(v_z=\omega/k_z,x) = \frac{E_y^1}{B}\left[\frac{\partial}{\partial x}n_e^0\right]\frac{1}{n_e^0}F_e^0(v_z=\omega/k_z)\,.$$

We evaluate as $E_y^1 \sim -\mathrm{i}k_y\phi^1, e\phi^1/T_e \sim n_e^1/n_e^0$ and get

$$v_{Dx}^1\frac{\partial}{\partial x}F_e^0(v_z=\omega/k_z,x) \sim \mathrm{i}\omega_*\frac{n_e^1}{n_e^0}F_e^0(v_z=\omega/k_z) \tag{6.5.3}$$

where $\omega_* = [k_yT_e/eBn_e^0][-dn_e^0/dx] = k_yu_{ey}^0$ and is equal to the electron drift frequency (cf. Sect. 5.5). Equation (6.5.3) contributes to the growth of the wave. Comparing (6.5.2) and (6.5.3) we find that if

$$\omega < \omega_* \tag{6.5.4}$$

the growth term (6.5.3) exceeds the damping given in (6.5.2) resulting in an instability.

Now, the propagation speed of the electrostatic drift wave is determined by the propagation speed of the ion density perturbation. Electrons follow the ion density perturbation by the shielding motion along the magnetic field line. The ion density perturbation propagates by the $\boldsymbol{E}'\times\boldsymbol{B}$ drift, but because of the finite Larmor radius the electric field that the ions feel is weaker than $|E_y'|$. As a result the propagation speed of the drift wave becomes less than that calculated by the fluid model, namely, $\omega/k_y < u_{ey}^0$. Then the inequality in (6.5.4) is satisfied and the drift wave becomes unstable. Thus the drift wave instability takes place due to the combined effect of the finite ion Larmor radius effect and the $\boldsymbol{E}\times\boldsymbol{B}$ drift of the resonant electrons. So far, we have ignored the effect of the resonant ions. This is because we are assuming the region $|\omega| \gg k_zv_{Ti}$. Since the acceleration of the resonant ions by the parallel electric field occurs in the direction opposite to the resonant electrons, while the direction of the $E \times B$ drift is the same as the electrons, the Landau damping rate by the resonant ions is enhanced by the presence of the density gradient. In other words, the resonant ions can contribute to strong damping of the drift wave. Therefore, in the presence of a magnetic shear, the ion Landau damping can suppress the drift wave instability. A detailed calculation shows that the collisionless drift wave instability discussed in this section is completely suppressed by even a small magnetic shear [6.5]. Of course, drift waves can become unstable even in a sheared magnetic field by other means, such as a magnetic field curvature effect, plasma current, temperature gradient, etc.

6.6 Trapped Particle Effects and Quasilinear Effects

In Sect. 5.6, we have described typical examples of nonlinear propagation of electrostatic waves using the two-fluid model. In this model, only an average nonlinear response of the plasma can be treated. In the Vlasov model, we can treat nonlinear effects due to specific group of particles in a special velocity range, i.e., the resonant particles. These particles are nearly at rest in the wave frame. In other words, their kinetic energy in the wave frame $m[w - \omega_r/k]^2/2$ is nearly zero. Therefore, even a small amplitude fluctuating electric field can produce significant effects on the particle orbits. In this section, we illustrate such nonlinear effects by two examples. One is the case of a nonlinear wave propagating at constant phase velocity and the other is the case in which many waves of different phase velocities are present. For simplicity, we shall restrict ourselves to the case of a one-dimensional propagation of an electron plasma wave in the absence of magnetic field.

6.6.1 Bounce Frequency

Let us first consider the case of a wave propagating with a constant phase velocity. In the linear approximation, such a wave undergoes Landau damping by resonant electrons. In Sect. 6.3, we introduced a time τ which is smaller than or on the order of the lifetime of the wave potential. The lifetime can be estimated by the Landau damping rate $\tau \sim \gamma^{-1}$. Landau damping is due to the resonant particles which continue to be accelerated or decelerated during the time τ. For a sufficiently weak field, the velocity of the resonant particles can be regarded as being almost constant during the time τ and the resonance condition is given by (6.3.1). As the wave amplitude increases, the modification of the particle velocity by the wave field becomes important and, eventually, even if the initial particle velocity is equal to the wave phase velocity (that is, the particle is at rest in the wave frame), the particle starts moving due to the acceleration by the wave field and can move more than half a wavelength in time τ. Then the particles cannot stay either in an accelerating phase or in a decelerating phase of the wave, but they oscillate between acceleration and deceleration. As a result, they do not contribute to the damping of the wave. If all particles accelerate and decelerate repeatedly during the time τ, then Landau damping will practically disappear and significant nonlinear effects are expected to occur.

For a more quantitative argument, we assume that the wave electric field is constant during the time τ. The particle initially at rest in the wave frame will then move a distance on the order of $eE\tau^2/2m$ relative to the wave during the time τ. If this distance is larger than half a wavelength π/k, then all particles oscillate between acceleration and deceleration.

The condition can be written as

$$E > \frac{2m\pi}{ek\tau^2} . \tag{6.6.1}$$

In terms of the potential, we can write noting $\phi \sim E/k$,

$$e\phi > \frac{2\pi m}{k^2\tau^2} . \tag{6.6.2}$$

Setting $\tau \sim \gamma^{-1}$ we find that nonlinear effects become important when

$$\gamma < \omega_b \equiv k\sqrt{e\phi/m}$$

where we have dropped the factor $\sqrt{2\pi}$. The frequency ω_b is called the *bounce frequency* and is equal to the oscillation frequency of a particle trapped in the potential energy trough of a wave. Indeed, if we write the wave potential in the wave frame as

$$\phi = \phi_0 \cos kx$$

then the electron oscillation in the potential energy trough at $x = 0$ obeys the equation of motion

$$m\frac{d^2x}{dt^2} = e\frac{\partial\phi}{\partial x} = -ek\phi_0 \sin kx \doteq -ek^2\phi_0 x \tag{6.6.3}$$

where we assumed $|kx| \ll 1$ and used the approximation $\sin kx \sim kx$. Equation (6.6.3) describes a harmonic oscillation at frequency $k\sqrt{e\phi_0/m} = \omega_b$. We thus conclude that a significant nonlinear effect on resonant particles can be expected when the field amplitude becomes so large that the bounce frequency ω_b exceeds the Landau damping rate γ.

6.6.2 Bernstein-Greene-Kruskal (BGK) Waves

As the wave amplitude increases the Landau damping decreases and the life time of the potential becomes longer. Consider the limit $\tau \rightarrow \infty$, that is, when the wave is stationary and undamped. Such a wave must satisfy the stationary Vlasov equation in the wave frame,

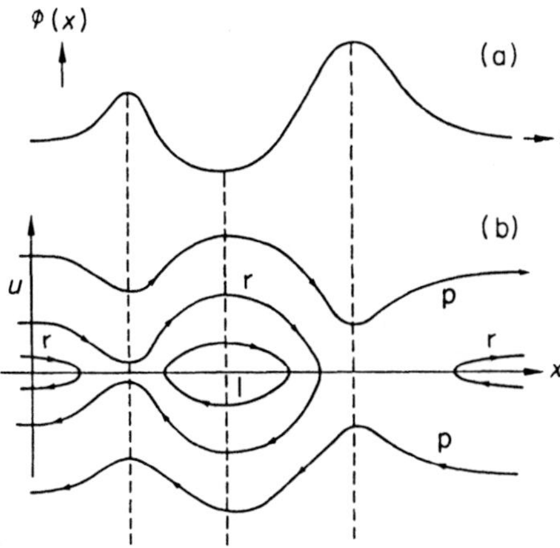

Fig. 6.5a,b. An example of BGK potential and particle orbits in phase space: (p) passing particle, (r) reflected particle and (l) localized particle

$$\frac{\partial F}{\partial t} = 0 \quad \text{or} \quad v\frac{\partial F}{\partial x} + \frac{e}{m}\frac{\partial \phi}{\partial x}\frac{\partial F}{\partial v} = 0 \tag{6.6.4}$$

and the Poisson equation (6.2.1). Suppose the solution for the potential has the form shown in Fig. 6.5a in the wave frame. In this case particle energy is conserved

$$W = mv^2/2 - e\phi = \text{const} \tag{6.6.5}$$

where v is the particle velocity in the wave frame. We can divide the particles into three classes according to their energy. The first class consists of the particles whose energy is larger than the maximum potential energy $-e\phi_{\mathrm{m}}$ and therefore pass through the potential and can move over the entire region $-\infty < x < \infty$. These particles will be called the *passing particles* or *untrapped particles*. The second class consists of those which have energy W less than $-e\phi_{\mathrm{m}}$ but can pass to either $x \to -\infty$ or $x \to +\infty$. These particles are called the *reflected particles*. The third class consists of those which are trapped in one of the potential energy troughs and are therefore localized to a finite region. These particles are called the *localized trapped particles*. The reflected particles and the localized trapped particles are collectively called the *trapped particles*. In the linear theory, there are no trapped particles. Typical phase space orbits of these three classes of particles are shown in Fig. 6.5b where p, r, l, stand for the passing particle, reflected particle, and localized particle, respectively. In a collisionless system, the particle orbit in the phase space is determined by the initial condition and particles can never move from one orbit to another. Moreover, in the stationary state, the distribution function F is constant along the particle orbit. Otherwise, the distribution function changes with time accompanied by the particle motion along the orbit. Therefore, the distribution function F that satisfies (6.6.4) is a function of the orbit alone. For the trapped particle, the orbit is uniquely determined by its energy W, while for the untrapped particle the orbit is determined by W and the direction of the particle motion (recall that W is an even function of v). Then the solution of (6.6.4) can, in general, be written in the form

$$F = \begin{cases} F_t(W) & (W < -e\phi_m) \\ F_+(W) & (W > -e\phi_m, v > 0) \\ F_-(W) & (W > -e\phi_m, v < 0) \,. \end{cases} \tag{6.6.6}$$

Substituting this distribution into the Poisson equation, we can derive an equation for the potential ϕ. Since W depends on ϕ, this equation is extremely complicated. Now, the distributions for the passing particles and the reflected particles can be determined by the boundary conditions at $x \to \pm\infty$. If the distribution of the localized trapped particles can be determined by some means, then the Poisson equation is reduced to a nonlinear eigenvalue problem for ϕ. In the linear theory $F_t = 0$ and the Poisson equation is reduced to a linear eigenvalue problem. The eigenfunction is a sinusoidal wave which contains the wavenumber k as a parameter and the eigenvalue is the frequency ω. That is, we obtain the linear dispersion relation. In the presence of trapped particles, the localized trapped particle distribution is left undetermined. In other words, because of the arbitrariness of the localized trapped particle distribution, the eigenfunction ϕ can also be arbitrary and no dispersion relation is obtained. Therefore we can have an arbitrary shape of nonlinear waves. Such waves were first pointed out by *Bernstein, Greene* and *Kruskal* and are called the BGK waves [6.6]. Note that BGK waves are stationary solutions of the Vlasov equation, but their stability is not known. It is also unknown which form of a BGK wave can be produced under given initial conditions.

6.6.3 Quasilinear Effects

We now consider the case of many waves with different phase velocities. Such a situation takes place when a warm electron beam with a velocity distribution as shown in Fig. 6.3 excites electron plasma waves (bump-on-tail instability). In this case waves of phase velocity lying in the region $u_1 < \omega/k < u_2$ grow simultaneously. When a group of waves of different phase velocities are superposed, a wave packet like that shown in Fig. 6.6 is formed. The wave packet moves with the group velocity $d\omega/dk$ while the resonant particles move with the phase

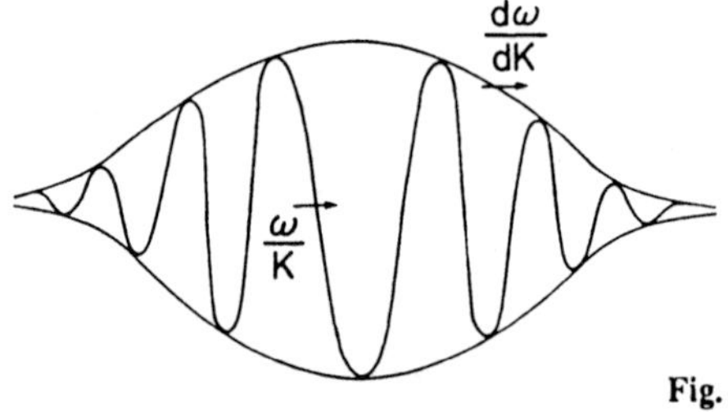

Fig. 6.6. Wave packet produced by many waves of different phase velocities

velocity. If $d\omega/dk \neq \omega/k$ then the resonant particles pass through the packet. Let L be the width of the wave packet, then the time that the resonant particles pass through the wave packet is given by [6.7]

$$\tau_c \sim L \Bigg/ \left| \frac{d\omega}{dk} - \frac{\omega}{k} \right| .$$

This is the lifetime of the potential as seen by the resonant particles. If τ_c is sufficiently long compared to the inverse of the bounce frequency ω_b^{-1}, then the trapped particle effects discussed above become the dominant nonlinear effect. On the other hand if $\tau_c < \omega_b^{-1}$, the trapped particle effect is unimportant and a new *quasilinear effect* becomes important through the resonant particles. We shall first give a qualitative description of this effect.

Let us first estimate τ_c for the case of the bump-on-tail instability of Fig. 6.3. We assume that u_1 and u_2 are close to each other and write $u_1 = \omega_{pe}/k$ and $u_2 = \omega_{pe}/(k - \Delta k) \doteq u_1[1 + \Delta k/k](\Delta k \ll k)$. We estimate L as

$$L = \Delta k^{-1} = \frac{k^{-1} u_1}{u_2 - u_1} . \tag{6.6.7}$$

Since for the electron plasma wave the group velocity is much smaller than the phase velocity, we can neglect the group velocity and estimate τ_c as

$$\tau_c = L/u_1 \doteq [k(u_2 - u_1)]^{-1} . \tag{6.6.8}$$

The condition that the trapped particle effect is negligible can then be written as

$$\omega_b \ll k(u_2 - u_1) \doteq \omega_{pe}(u_2 - u_1)/u_1 . \tag{6.6.9}$$

As the wave grows, the distribution function $F(x, v, t)$ changes. We expand F and ϕ in Fourier series as

$$F(x, v, t) = \sum_k e^{ikx} F_k(v, t) , \qquad \phi(x, t) = \sum_k e^{ikx} \phi_k(t) . \tag{6.6.10}$$

Note that $F_0(v, t)$ is the spatially averaged part of the distribution. It varies according to the Vlasov equation as

$$\frac{\partial F_0}{\partial t} = \frac{e}{m} \sum_k ik\phi_k(t) \frac{\partial}{\partial v} F_{-k}(v, t) . \tag{6.6.11}$$

To lowest order we use the linearized Vlasov equation for F_{-k}, which reads

$$\left(\frac{\partial}{\partial t} - ikv \right) F_{-k} = -\frac{e}{m} ik\phi_{-k} \frac{\partial}{\partial v} F_0(v, t) . \tag{6.6.12}$$

The solution of (6.6.12) can be written as

$$F_{-k}(v, t) = -\frac{e}{m} \int_{-\infty}^{t} dt' e^{ikv(t-t')} ik\phi_{-k}(t') \frac{\partial}{\partial v} F_0(v, t') , \tag{6.6.13}$$

where we assumed that there is no perturbation at $t \to -\infty$. We assume that

$F_0(v,t)$ depends only slowly on time and approximate $F_0(v,t')$ in (6.6.13) by $F_0(v,t)$. Then we substitute the result into (6.6.11) to obtain the diffusion equation in the form

$$\frac{\partial}{\partial t}F_0(v,t) = \frac{\partial}{\partial v}D(v,t)\frac{\partial}{\partial v}F_0(v,t) \tag{6.6.14}$$

where

$$D(v,t) = \left(\frac{e}{m}\right)^2 \sum_k k^2 \int_{-\infty}^{t} dt' \phi_k(t)\phi_{-k}(t')\mathrm{e}^{\mathrm{i}kv(t-t')} \, . \tag{6.6.15}$$

We note that $\phi_k(t)$ and $\phi_{-k}(t')$ represent the unstable electron plasma wave of frequency ω_k and $\omega_{-k} = -\omega_k$ respectively. Therefore we can write

$$\phi_k(t)\phi_{-k}(t') \sim |\phi_k(t)|^2 \exp\left[\mathrm{i}\omega_k(t'-t) + \delta t'\right]$$

where we introduced a small positive imaginary part $\delta(> 0)$ to ensure convergence at $t' \to -\infty$. The time integral in (6.6.15) then becomes

$$\frac{1}{\mathrm{i}(\omega_k - kv) + \delta} = \frac{\delta - \mathrm{i}(\omega_k - kv)}{(\omega_k - kv)^2 + \delta^2} \, . \tag{6.6.16}$$

Since ω_k and kv are odd functions of k, after summation over k the imaginary part of (6.6.16) disappears. The real part becomes a δ-function

$$\lim_{\delta \to +0} \frac{\delta}{(\omega_k - kv)^2 + \delta^2} = \pi\delta[\omega_k - kv] \, . \tag{6.6.17}$$

Using this result, we can write the diffusion constant D as follows

$$D(v) = \left(\frac{e}{m}\right)^2 \pi \sum_k k^2 |\phi_k|^2 \delta[\omega_k - kv] \, . \tag{6.6.18}$$

Equation (6.6.14, 18) are called the quasilinear equations [6.9]. They describe the following physical process. As the wave grows the diffusion constant increases. Then in the resonant velocity region, the distribution function is flattened by the diffusion. The bump in the distribution is smoothed out and disappears. Then the wave growth rate decreases and eventually the wave is stabilized when the distribution in the resonant region becomes completely flat ($\partial F_0/\partial v = 0$). The velocity space diffusion as described by the quasilinear equation is due to the average deceleration of the resonant particles moving slightly faster than the phase velocity and acceleration of those moving slightly slower than the phase velocity. Hence we can say that the quasilinear effect is a reaction to the Landau damping or growth.

In reality, the above model is too simple to describe the actual three-dimensional phenomenon, but the quasilinear effect is considered to be most important in suppressing kinetic instability.

6.7 Source and Noise Level

So far in this chapter, we have considered the plasma response in the context of the Vlasov model or the model of a collisionless continuum. In this model, the initially stationary and uniform unperturbed state remains stationary and uniform, and perturbations can be produced only by some external means. In a real plasma, perturbations are produced by fluctuations due to the discreteness of the plasma. Namely, a charged particle produces electromagnetic fields around itself which induce a plasma response. The simplest example is the Debye screening of a static point charge discussed in Sects. 5.1 and 6.2. In this section, we consider a moving charge and discuss the dynamical response of the plasma to this charge. When the velocity of the charge is equal to the phase velocity of a wave which satisfies the dispersion relation of the plasma, the wave is resonantly excited in the plasma. This is called the *Cerenkov emission* of the wave and acts as the source of the wave fluctuation. As a result of the wave emission the particle loses its energy.

For simplicity, we consider a uniform and stationary plasma in the absence of magnetic fields. We represent the charged particle by the charge density $\sigma_{es}(\boldsymbol{r},t)$ given by

$$\sigma_{es}(\boldsymbol{r},t) = q\delta[\boldsymbol{r} - \boldsymbol{r}(t)] \ . \tag{6.7.1}$$

When the particle velocity is small compared to the light speed, we can ignore the magnetic response of the plasma and use the electrostatic approximation. The linear response to $\sigma_{es}(\boldsymbol{r},t)$ is determined from the Poisson equation which can be written in the Fourier representation as (7.2.15)

$$\varepsilon(\boldsymbol{k},\omega)\phi(\boldsymbol{k},\omega) = k^{-2}\sigma_{es}(\boldsymbol{k},\omega) \tag{6.7.2}$$

where $\varepsilon(\boldsymbol{k},\omega)$ is the dielectric function given by (6.2.14), and $\sigma_{es}(\boldsymbol{k},\omega)$ and $\phi(\boldsymbol{k},\omega)$ are the Fourier transform of $\sigma_{es}(\boldsymbol{r},t)$ and the electrostatic potential induced by the source charge. Using (6.7.1), we have

$$\begin{aligned}\sigma_{es}(\boldsymbol{k},\omega) &= \int d^3\boldsymbol{r}\int_{-\infty}^{+\infty} dt \exp[\mathrm{i}(\omega t - \boldsymbol{k}\cdot\boldsymbol{r})]\sigma_{es}(\boldsymbol{r},t) \\ &= q\exp(-\boldsymbol{k}\cdot\boldsymbol{r}_0)2\pi\delta[\omega - \boldsymbol{k}\cdot\boldsymbol{v}]\end{aligned} \tag{6.7.3}$$

where we used the approximation

$$\boldsymbol{r}(t) = \boldsymbol{r}_0 + \boldsymbol{v}t \qquad (\boldsymbol{v} = \text{const}) \tag{6.7.4}$$

and the relation

$$\int_{-\infty}^{\infty} dt \exp[\mathrm{i}(\omega - \boldsymbol{k}\cdot\boldsymbol{v})t] = 2\pi\delta[\omega - \boldsymbol{k}\cdot\boldsymbol{v}] \ . \tag{6.7.5}$$

By the inverse Fourier transformation of (6.7.2) we get

$$
\begin{aligned}
\phi(\boldsymbol{r},t) &= \int \frac{d\omega}{2\pi} e^{-\mathrm{i}\omega t} \int \frac{d^3\boldsymbol{k}}{(2\pi)^3} \exp(i\boldsymbol{k}\cdot\boldsymbol{r}) \frac{\sigma_{\mathrm{es}}(\boldsymbol{k},\omega)}{k^2\varepsilon(\boldsymbol{k},\omega)} \\
&= \int \frac{d^3\boldsymbol{k}}{(2\pi)^3} \exp\{\mathrm{i}\boldsymbol{k}\cdot[\boldsymbol{r}-\boldsymbol{r}(t)]\} \frac{q}{k^2\varepsilon(\boldsymbol{k},\boldsymbol{k}\cdot\boldsymbol{v})}
\end{aligned}
\tag{6.7.6}
$$

from which the induced electric field is obtained as

$$
\tilde{\boldsymbol{E}}(\boldsymbol{r},t) = \int \frac{d^3\boldsymbol{k}}{(2\pi)^3} \frac{-\mathrm{i}\boldsymbol{k}q}{k^2\varepsilon(\boldsymbol{k},\boldsymbol{k}\cdot\boldsymbol{v})} \exp\{i\boldsymbol{k}\cdot[\boldsymbol{r}-\boldsymbol{r}(t)]\} \ . \tag{6.7.7}
$$

Knowing the induced electric field $\tilde{\boldsymbol{E}}(\boldsymbol{r},t)$, we can calculate the energy loss of the charged particle by $\tilde{\boldsymbol{E}}$ which is the field induced by charged particle itself: By using the equation of motion, we have

$$
\begin{aligned}
\frac{d}{dt}\left(\frac{1}{2}mv^2\right) &= q\boldsymbol{v}\cdot\tilde{\boldsymbol{E}}(\boldsymbol{r}(t),t) \\
&= q^2 \int \frac{d^3\boldsymbol{k}}{(2\pi)^3} \frac{-\mathrm{i}\boldsymbol{k}\cdot\boldsymbol{v}}{k^2} \frac{1}{\varepsilon(\boldsymbol{k},\boldsymbol{k}\cdot\boldsymbol{v})} \\
&= q^2 \int \frac{d^3\boldsymbol{k}}{(2\pi)^3} \frac{\boldsymbol{k}\cdot\boldsymbol{v}}{k^2} \,\mathrm{Im}\left\{\frac{1}{\varepsilon(\boldsymbol{k},\boldsymbol{k}\cdot\boldsymbol{v})}\right\}
\end{aligned}
\tag{6.7.8}
$$

where in the last line we used the fact that $\varepsilon(-\boldsymbol{k},-\boldsymbol{k}\cdot\boldsymbol{v}) = \varepsilon^*(\boldsymbol{k},\boldsymbol{k}\cdot\boldsymbol{v})$, the asterisk denoting the complex conjugate. Cerenkov emission occurs for the wave which satisfies the dispersion relation $\varepsilon(\boldsymbol{k},\boldsymbol{k}\cdot\boldsymbol{v}) = 0$. Assuming that the wave is weakly damped by the Landau damping such that (6.3.11) is satisfied, we calculate as follows:

$$
\begin{aligned}
\mathrm{Im}\left\{\frac{1}{\varepsilon(\boldsymbol{k},\omega)}\right\} &= \frac{\varepsilon_0\chi''(\boldsymbol{k},\omega)}{|\varepsilon(\boldsymbol{k},\omega)|^2} \\
&= \frac{\varepsilon_0\chi''(\boldsymbol{k},\omega)}{[\varepsilon'(\boldsymbol{k},\omega)]^2 + [\varepsilon_0\chi''(\boldsymbol{k},\omega)]^2} \\
&\doteq -\left[\frac{\partial\varepsilon'(\boldsymbol{k},\omega_{\mathrm{r}})}{\partial\omega_{\mathrm{r}}}\right]^{-1} \frac{\gamma}{[(\omega-\omega_{\mathrm{r}})^2+\gamma^2]}
\end{aligned}
\tag{6.7.9}
$$

where we used (6.3.14) and the approximation $\varepsilon'(\boldsymbol{k},\omega) \doteq [\partial\varepsilon'(\boldsymbol{k},\omega_{\mathrm{r}})/\partial\omega_{\mathrm{r}}](\omega-\omega_{\mathrm{r}})$. Since γ is small ($\gamma \ll \omega_{\mathrm{r}}$) and positive, we can use the approximation

$$
\frac{\gamma}{(\omega-\omega_{\mathrm{r}})^2+\gamma^2} \doteq \pi\delta(\omega-\omega_{\mathrm{r}}) \ . \tag{6.7.10}
$$

Then we have

$$
\mathrm{Im}\left\{\frac{1}{\varepsilon(\boldsymbol{k},\boldsymbol{k}\cdot\boldsymbol{v})}\right\} = -\pi\delta[\boldsymbol{k}\cdot\boldsymbol{v}-\omega_{\mathrm{r}}]\left[\frac{\partial\varepsilon'(\boldsymbol{k},\omega_{\mathrm{r}})}{\partial\omega_{\mathrm{r}}}\right]^{-1} \ . \tag{6.7.11}
$$

Substituting (6.7.11) into (6.7.8) yields the energy loss rate of the particle by the Cerenkov emission (or, the Cerenkov emission rate) as

$$-\frac{d}{dt}\left[\frac{1}{2}mv^2\right] = q^2 \int \frac{d^3\boldsymbol{k}}{(2\pi)^3} \frac{\pi\omega_r\delta[\boldsymbol{k}\cdot\boldsymbol{v}-\omega_r]}{k^2} \left[\frac{\partial\varepsilon'(\boldsymbol{k},\omega_r)}{\partial\omega_r}\right]^{-1} . \tag{6.7.12}$$

The resonance condition $\omega_r = \boldsymbol{k}\cdot\boldsymbol{v}$ is the same as that for the Landau damping and is often called the *Cerenkov condition.*

The above formula was derived for a single charged particle. In fact, each charged particle in the plasma can act as a source of Cerenkov emission. Cerenkov emissions by the thermal motions of the particles are called *spontaneous emission* of the wave and act as the source of thermal noise in the plasma. To calculate the spontaneous emission rate we sum up the emission rate given by (6.7.12) over all the particles whose velocity distribution is given by $F_s^0(v)$. Let $W_{\boldsymbol{k}}(t)$ be the energy of the wave mode of wavenumber $\boldsymbol{k}$ and frequency ω. We denote its rate of change due to the spontaneous emission by $[dW_{\boldsymbol{k}}/dt]_s$. Noting that $W_{\boldsymbol{k}}$ represents the energy of the two wave modes specified by $(\boldsymbol{k},\omega_r)$ and $(-\boldsymbol{k},-\omega_r)$, we have

$$-\sum_s \int d^3\boldsymbol{v} F_s^0(\boldsymbol{v}) \frac{d}{dt}\left(\frac{1}{2}mv^2\right) = \int \frac{d^3\boldsymbol{k}}{(2\pi)^3}\frac{1}{2}\left(\frac{dW_{\boldsymbol{k}}}{dt}\right)_s . \tag{6.7.13}$$

Substituting (6.7.12) into (6.7.13) we find

$$\begin{aligned}\left(\frac{dW_{\boldsymbol{k}}}{dt}\right)_s &= \sum_s q_s^2 \frac{2\pi\omega_r}{k^2}\left[\frac{\partial\varepsilon'(\boldsymbol{k},\omega_r)}{\partial\omega_r}\right]^{-1} \int d^3\boldsymbol{v} F_s^0(\boldsymbol{v})\delta[\boldsymbol{k}\cdot\boldsymbol{v}-\omega_r] \\ &= \sum_s q_s^2 \frac{2\pi}{k^2}\left[\frac{\partial\varepsilon'(\boldsymbol{k},\omega_r)}{\partial\omega_r}\right]^{-1} [w f_s^0(w)]_{w=\omega_r/k} ,\end{aligned} \tag{6.7.14}$$

where $f_s^0(w)$ is defined by (6.2.19).

In the stationary state, the spontaneous emission rate balances the Landau damping:

$$\left(\frac{dW_{\boldsymbol{k}}}{dt}\right)_s = 2\gamma W_{\boldsymbol{k}} , \tag{6.7.15}$$

where γ is given by (6.3.14). The factor 2 is needed since $W_{\boldsymbol{k}}$ is proportional to $|E_{\boldsymbol{k}}(t)|^2$. Using (6.3.9) we find

$$2\gamma = -\sum_s q_s^2 \frac{2\pi}{m_s k^2}\left[\frac{df_s^0(w)}{dw}\right]_{w=\omega_r/k} \left[\frac{\partial\varepsilon'(\boldsymbol{k},\omega_r)}{\partial\omega_r}\right]^{-1} . \tag{6.7.16}$$

Substituting (6.7.14, 16) into (6.7.15), we find

$$W_{\boldsymbol{k}} = -\frac{\sum_s q_s^2 [w f_s^0(w)]_{w=\omega_r/k}}{\sum_s (q_s^2/m_s)[df_s^0(w)/dw]_{w=\omega_r/k}} . \tag{6.7.17}$$

For the case of a Maxwell distribution of temperature $T_e = T_i = T$, we finally get

$$W_{\boldsymbol{k}} = T , \tag{6.7.18}$$

which corresponds to the equipartition of energy at thermal equilibrium. This level of fluctuation is called the *thermal level* or the *noise level.*

At thermal equilibrium in an isotropic plasma, the only electrostatic wave modes that are weakly damped are the long wavelength electron plasma waves ($k\lambda_{De} < 0.3$). Then the total number of such wave modes in volume V is

$$V \int \frac{d^3k}{(2\pi)^3}_{(k\lambda_{De}<0.3)} = \frac{1}{6\pi^2}(0.3)^3 \frac{V}{\lambda_{De}^3} \doteq \frac{5 \times 10^{-4} V}{\lambda_{De}^3} .$$

Each mode has the energy T according to (6.7.18), so that the total wave energy in an isotropic plasma at thermal equilibrium is

$$U \doteq \frac{5 \times 10^{-4} VT}{\lambda_{De}^3} .$$

On the other hand, the total energy of the plasma is $3nTV$. Taking the ratio of U to the total plasma energy we get

$$\frac{U}{3nTV} \sim \frac{10^{-4}}{n\lambda_{De}^3} ,$$

which vanishes in the ideal plasma limit ($n\lambda_{De}^3 \to \infty$). This result is consistent with the fact that the ideal plasma is a collisionless continuum and hence has no discreteness effects.

6.8 Collision Integral

Whereas the Landau damping and the Cerenkov emission result from the coupling of the individual particle motion and the collective motion, the collisional process is a result of the coupling of the individual motion of two particles. In this section, we derive the *kinetic equation* which describes the Coulomb collision of two particles in terms of the particle distribution function. Since the process arises from the discrete nature of the plasma, we use the Klimontovich distribution function (4.1.5). When we totally neglect the magnetic field (both external and induced), then the collision term arises from [(4.1.4) and (4.2.3)]

$$\left\langle \frac{q}{m} \hat{E}(\boldsymbol{r}, t) \cdot \frac{\partial}{\partial \boldsymbol{v}} \hat{F}(\boldsymbol{r}, \boldsymbol{v}, t) \right\rangle - \frac{q}{m} E(\boldsymbol{r}, t) \cdot \frac{\partial}{\partial \boldsymbol{v}} F(\boldsymbol{r}, \boldsymbol{v}, t) . \tag{6.8.1}$$

The angled bracket denotes an ensemble average over the initial coordinates of the particles, which can alternatively be regarded as a space-time average, of which the spatial average is to be taken over a scale length large compared with the effective collision radius, i.e., the Debye length. For simplicity, we choose this average to be over the entire volume V of the system. Then the average electric field vanishes $E(\boldsymbol{r}, t) = 0$ and the average distribution function becomes

$$F(\boldsymbol{r},\boldsymbol{v},t) = \frac{1}{V}\int d^3\boldsymbol{r}\sum_{j=1}^{N}\langle\delta[\boldsymbol{r}-\boldsymbol{r}_j(t)]\delta[\boldsymbol{v}-\boldsymbol{v}_j(t)]\rangle$$
$$= \frac{1}{V}\sum_{j=1}^{N}\langle\delta[\boldsymbol{v}-\boldsymbol{v}_j(t)]\rangle \equiv F^0(\boldsymbol{v},t)\ . \tag{6.8.2}$$

The time average will be discussed later.

We use the Fourier representation for the microscopic variables represented by $\hat{A}(\boldsymbol{r},t)$:

$$\hat{A}(\boldsymbol{r},t) = \sum_{\boldsymbol{k}}\exp[\mathrm{i}\boldsymbol{k}\cdot\boldsymbol{r}]A_{\boldsymbol{k}}(t) \tag{6.8.3}$$

$$A_{\boldsymbol{k}}(t) = \frac{1}{V}\int d^3\boldsymbol{r}\exp[-\mathrm{i}\boldsymbol{k}\cdot\boldsymbol{r}]\hat{A}(\boldsymbol{r},t)\ , \tag{6.8.4}$$

then $F^0(\boldsymbol{v},t) = \langle F_{\boldsymbol{k}=0}(\boldsymbol{v},t)\rangle$. The Klimontovich equation for the average distribution $F^0(\boldsymbol{v},t)$ can then be written

$$\frac{\partial F^0(\boldsymbol{v},t)}{\partial t} = -\frac{q}{m}\sum_{\boldsymbol{k}\neq 0}\left\langle \boldsymbol{E}_{\boldsymbol{k}}(t)\cdot\frac{\partial}{\partial\boldsymbol{v}}F_{-\boldsymbol{k}}(\boldsymbol{v},t)\right\rangle$$
$$\equiv \frac{\partial}{\partial\boldsymbol{v}}\cdot\boldsymbol{J}(\boldsymbol{v},t)\ , \tag{6.8.5}$$

where

$$\boldsymbol{J}(\boldsymbol{v},t) = -\frac{q}{m}\sum_{\boldsymbol{k}=0}\langle\boldsymbol{E}_{\boldsymbol{k}}(t)F_{-\boldsymbol{k}}(\boldsymbol{v},t)\rangle\ . \tag{6.8.6}$$

The right hand side of (6.8.5) corresponds to the collision term (6.8.1).

The equation for $F_{-\boldsymbol{k}}(\boldsymbol{v},t)$ is written as

$$\left(\frac{\partial}{\partial t}-\mathrm{i}\boldsymbol{k}\cdot\boldsymbol{v}\right)F_{-\boldsymbol{k}}(\boldsymbol{v},t)+\frac{q}{m}\boldsymbol{E}_{-\boldsymbol{k}}(t)\cdot\frac{\partial}{\partial\boldsymbol{v}}F_0(\boldsymbol{v},t)$$
$$= -\frac{q}{m}\sum_{\boldsymbol{k}'\neq\boldsymbol{k}}\boldsymbol{E}_{-\boldsymbol{k}'}(t)\cdot\frac{\partial}{\partial\boldsymbol{v}}F_{\boldsymbol{k}'-\boldsymbol{k}}(\boldsymbol{v},t)\ . \tag{6.8.7}$$

If the right hand side is neglected, this equation is in the same form as the linearized Vlasov equation (6.6.12). The only difference is that the solution of (6.8.7) must satisfy the initial condition

$$F_{-\boldsymbol{k}}(\boldsymbol{v},t=0) = \frac{1}{V}\sum_{j=1}^{N}\exp(\mathrm{i}\boldsymbol{k}\cdot\boldsymbol{r}_j)\delta[\boldsymbol{v}-\boldsymbol{v}_j]\ , \tag{6.8.8}$$

where $\boldsymbol{r}_j$ and $\boldsymbol{v}_j$ are the position and the velocity of the jth particle at $t = 0$. As we shall show later, the right hand side of (6.8.7) is indeed negligible under certain conditions, so that we can ignore it for the moment and solve (6.8.7) subject to the initial condition of (6.8.8) to obtain

$$
\begin{aligned}
F_{-k}(v,t) &= \exp(\mathrm{i}k\cdot vt)F_{-k}(v,t=0) \\
&\quad -\frac{q}{m}\int_0^t dt'\exp[\mathrm{i}k\cdot v(t-t')]E_{-k}(t')\cdot\frac{\partial}{\partial v}F_0(v,t')\,. \qquad (6.8.9)
\end{aligned}
$$

Substitution of (6.8.9) into (6.8.6) yields the following two terms:

$$
J(v,t) = J_1(v,t) + J_2(v,t)\,, \qquad (6.8.10)
$$

where

$$
\begin{aligned}
J_1(v,t) &= -\frac{q}{m}\sum_{k\neq 0}\exp(\mathrm{i}k\cdot vt)\,\langle E_k(t)F_{-k}(v,t=0)\rangle \\
&\doteq -\frac{q}{m}\sum_{k\neq 0}\frac{1}{V}\sum_{j=1}^{N}\langle\exp[\mathrm{i}k\cdot r_j(t)]E_k(t)\delta[v-v_j]\rangle\,, \qquad (6.8.11)
\end{aligned}
$$

$$
\begin{aligned}
J_2(v,t) &= \left(\frac{q}{m}\right)^2\sum_{k\neq 0}\int_0^t dt'\exp[\mathrm{i}k\cdot v(t-t')] \\
&\quad\times\left\langle E_k(t)E_{-k}(t')\cdot\frac{\partial}{\partial v}F_0(v,t')\right\rangle \\
&\doteq \left(\frac{q}{m}\right)^2\sum_{k\neq 0}\int_0^t dt'\exp[\mathrm{i}k\cdot v(t-t')] \\
&\quad\times\langle E_k(t)E_{-k}(t')\rangle\cdot\frac{\partial}{\partial v}F^0(v,t') \qquad (6.8.12)
\end{aligned}
$$

where in (6.8.11) we used the "straight-orbit" approximation

$$
r_j(t)\doteq r_j+v_jt \qquad (6.8.13)
$$

and in (6.8.12) we used the "decoupling" approximation

$$
\begin{aligned}
&\left\langle E_k(t)E_{-k}(t')\cdot\frac{\partial}{\partial v}F_0(v,t')\right\rangle \\
&\quad\doteq\langle E_k(t)E_{-k}(t')\rangle\cdot\frac{\partial}{\partial v}\left\langle F^0(v,t')\right\rangle\,. \qquad (6.8.14)
\end{aligned}
$$

The Fourier component $E_k(t)$ can be calculated from (6.7.7) which is the electric field produced by a single charge at $\{r,v\}=\{r(t),v\}$. Noting the relation

$$
\sum_{k\neq 0}=V\int\frac{d^3k}{(2\pi)^3} \qquad (6.8.15)
$$

we find

$$
\begin{aligned}
E_k(t) &= \sum_{s=\mathrm{e,i}}\frac{q_s}{V}\int d^3r'\int d^3v'\hat{F}_s(r',v',t)\,\frac{(-\mathrm{i}k)\exp(-\mathrm{i}k\cdot r')}{k^2\varepsilon(k,k\cdot v')} \\
&= \frac{1}{V}\sum_s q_s\sum_{i=1}^{N_s}\int d^3v'\delta[v'-v_i]\,\frac{(-\mathrm{i}k)\exp[-\mathrm{i}k\cdot r_i'(t)]}{k^2\varepsilon(k,k\cdot v')}\,. \qquad (6.8.16)
\end{aligned}
$$

Substitution of (6.8.16) into (6.8.11) then yields

$$J_1(v,t) = \frac{q}{m}\sum_s q_s \frac{1}{V^2}\sum_{k\neq 0}\int d^3v' \frac{\mathrm{i}k}{k^2\varepsilon(k, k\cdot v')}$$
$$\times \sum_{j=1}^{N}\sum_{i=1}^{N_s}\langle \exp\{\mathrm{i}k\cdot[r_j(t)-r_i(t)]\}\delta[v-v_j]\delta[v'-v_i]\rangle \,. \tag{6.8.17}$$

At this point, we bring in the *time* average. It should be noted that the time t has to be sufficiently short compared to the mean free time in order for the straight-orbit approximation of (6.8.13) to be valid. Yet the mean free time in a near ideal plasma is extremely long, so that we can choose t to be much longer than the time $(k\cdot v)^{-1}$ which is typically on the order of the plasma oscillation period (note that if $k = \lambda_D^{-1}$ and $v = v_T$ then $kv = \omega_p$). Then after averaging over such a time scale, the exponential factor in (6.8.17) will vanish unless s is the same as the particle under consideration, i.e., $q_s = q$, and $i = j$. Under these conditions

$$\langle \exp\{\mathrm{i}k\cdot[r_j(t)-r_i(t)]\}\delta[v-v_j]\delta[v'-v_i]\rangle$$
$$\doteq \delta_{ij}\,\langle\delta[v-v_j]\rangle\,\delta[v-v'] \,. \tag{6.8.18}$$

Such an approximation is called the *random phase approximation*. Equation (6.8.17) can then be reduced to the form

$$J_1(v,t) = \frac{q^2}{m}\frac{1}{V^2}\sum_{k\neq 0}\frac{\mathrm{i}k}{k^2\varepsilon(k,k\cdot v)}\sum_{j=1}^{N}\langle\delta[v-v_j]\rangle$$
$$= \frac{q^2}{m}\frac{1}{V}\sum_{k\neq 0}\frac{\mathrm{i}k}{k^2\varepsilon(k,k\cdot v)}F^0(v,t) \,, \tag{6.8.19}$$

where in the last line we ignored the time dependence of $v_j(t)$, or equivalently $F^0\ (v,\ t=0)\ \doteq\ F^0\ (v,\ t)$. In carrying out the k-summation in (6.8.19), we note that $\varepsilon(-k,-\omega) = \varepsilon^*(k,\omega)$ from which we have

$$\sum_{k\neq 0}\frac{\mathrm{i}k}{k^2\varepsilon(k,k\cdot v)} = \sum_{k\neq 0}\frac{k}{k^2|\varepsilon(k,k\cdot v)|^2}\,\mathrm{Im}\,\{\varepsilon(k,k\cdot v)\} \,. \tag{6.8.20}$$

Now from (6.2.12, 14) we have

$$\mathrm{Im}\{\varepsilon(k,k\cdot v)\} = \varepsilon_0\sum_{s=e,i}\mathrm{Im}\{\chi_s(k,k\cdot v)\}$$
$$= \sum_s\frac{q_s^2}{k^2 m_s}\int d^3v'\,\mathrm{Im}\frac{1}{k\cdot(v-v')+\mathrm{i}\delta}k\cdot\frac{\partial}{\partial v'}F_s^0(v')$$
$$\doteq -\sum_s\frac{q_s^2}{k^2 m_s}\int d^3v'\pi\delta[k\cdot(v-v')]k\cdot\frac{\partial}{\partial v}F_s^0(v',t) \,, \tag{6.8.21}$$

where again we used the approximation $F_s^0(\boldsymbol{v}') \doteq F_s^0(\boldsymbol{v}', t)$. Use of (6.8.21) in (6.8.20) and then in (6.8.19) finally yields the expression

$$J_1(\boldsymbol{v},t) = -\frac{q^2}{m}\sum_s \frac{q_s^2}{m_s}\int\frac{d^3\boldsymbol{k}}{(2\pi)^3}\int d^3\boldsymbol{v}'\frac{\pi\delta[\boldsymbol{k}\cdot(\boldsymbol{v}-\boldsymbol{v}')]}{|\varepsilon(\boldsymbol{k},\boldsymbol{k}\cdot\boldsymbol{v})|^2}\frac{\boldsymbol{k}\boldsymbol{k}}{k^4}\cdot\frac{\partial}{\partial\boldsymbol{v}'} \times F_s^0(\boldsymbol{v}',t)F^0(\boldsymbol{v},t)\,. \qquad (6.8.22)$$

In calculating $J_2(\boldsymbol{v},t)$, we first calculate the auto-correlation function $\langle \boldsymbol{E}_{\boldsymbol{k}}(t)\boldsymbol{E}_{-\boldsymbol{k}}(t')\rangle$ as follows: using (6.8.16) we have

$$\begin{aligned}\langle \boldsymbol{E}_{\boldsymbol{k}}(t)\boldsymbol{E}_{-\boldsymbol{k}}(t')\rangle &= \frac{1}{V^2}\sum_s\sum_{s'}q_sq_{s'}\sum_{j=1}^{N_s}\sum_{i=1}^{N_{s'}}\int d^3\boldsymbol{v}'\int d^3\boldsymbol{v}'' \\ &\times\frac{\boldsymbol{k}\boldsymbol{k}}{k^4}\frac{\langle\exp\{\mathrm{i}\boldsymbol{k}\cdot[\boldsymbol{r}_i(t')-\boldsymbol{r}_j(t)]\}\delta[\boldsymbol{v}'-\boldsymbol{v}_j]\delta[\boldsymbol{v}''-\boldsymbol{v}_i]\rangle}{\varepsilon(\boldsymbol{k},\boldsymbol{k}\cdot\boldsymbol{v}')\varepsilon(-\boldsymbol{k},-\boldsymbol{k}\cdot\boldsymbol{v}'')} \\ &= \frac{1}{V^2}\sum_s q_s^2\sum_{j=1}^{N_s}\int d^3\boldsymbol{v}'\,\langle\delta[\boldsymbol{v}'-\boldsymbol{v}_j]\rangle\frac{\boldsymbol{k}\boldsymbol{k}}{k^4}\frac{\exp[-\mathrm{i}\boldsymbol{k}\cdot\boldsymbol{v}'(t-t')]}{|\varepsilon(\boldsymbol{k},\boldsymbol{k}\cdot\boldsymbol{v}')|^2} \\ &= \frac{1}{V}\sum_s q_s^2\int d^3\boldsymbol{v}'\frac{\boldsymbol{k}\boldsymbol{k}}{k^4}\frac{F_s^0(\boldsymbol{v}',t)}{|\varepsilon(\boldsymbol{k},\boldsymbol{k}\cdot\boldsymbol{v}')|^2}\exp[-\mathrm{i}\boldsymbol{k}\cdot\boldsymbol{v}'(t-t')]\,,\end{aligned} \qquad (6.8.23)$$

where again we used the random phase approximation (6.8.18) in the form

$$\begin{aligned}&\langle\exp\{\mathrm{i}\boldsymbol{k}\cdot[\boldsymbol{r}_i(t')-\boldsymbol{r}_j(t)]\}\delta[\boldsymbol{v}'-\boldsymbol{v}_j]\delta[\boldsymbol{v}''-\boldsymbol{v}_i]\rangle \\ &\quad\doteq \exp[-\mathrm{i}\boldsymbol{k}\cdot\boldsymbol{v}'(t-t')]\,\langle\delta[\boldsymbol{v}'-\boldsymbol{v}_j]\rangle\,\delta_{ij}\delta[\boldsymbol{v}'-\boldsymbol{v}'']\,.\end{aligned}$$

If we substitute (6.8.23) into (6.8.12) and neglect the time dependence of $F^0(\boldsymbol{v},t')$ as before, we have a time integral of the form

$$\begin{aligned}\int_0^t dt'\exp[\mathrm{i}\boldsymbol{k}\cdot(\boldsymbol{v}-\boldsymbol{v}')(t-t')] &= \frac{1-\exp[\mathrm{i}\boldsymbol{k}\cdot(\boldsymbol{v}-\boldsymbol{v}')t]}{\mathrm{i}\boldsymbol{k}\cdot(\boldsymbol{v}'-\boldsymbol{v})} \\ &= \frac{\sin[\boldsymbol{k}\cdot(\boldsymbol{v}'-\boldsymbol{v})t]}{\boldsymbol{k}\cdot(\boldsymbol{v}'-\boldsymbol{v})} + (\text{odd function of } \boldsymbol{k})\,.\end{aligned}$$

We can approximate the first term by $\pi\delta[\boldsymbol{k}\cdot(\boldsymbol{v}-\boldsymbol{v}')]$ for large t, while the odd function part vanishes after summation over $\boldsymbol{k}$, so that for $J_2(\boldsymbol{v},t)$

$$J_2(\boldsymbol{v},t) = \frac{q^2}{m^2}\sum_s q_s^2\int\frac{d^3\boldsymbol{k}}{(2\pi)^3}\int d^3\boldsymbol{v}'\frac{\pi\delta[\boldsymbol{k}\cdot[\boldsymbol{v}-\boldsymbol{v}']}{|\varepsilon(\boldsymbol{k},\boldsymbol{k}\cdot\boldsymbol{v})|^2}\frac{\boldsymbol{k}\boldsymbol{k}}{k^4} \times\frac{\partial}{\partial\boldsymbol{v}}F^0(\boldsymbol{v},t)F_s^0(\boldsymbol{v}',t)\,. \qquad (6.8.24)$$

Substituting (6.8.22, 24) into (6.8.10), we obtain

$$J(\boldsymbol{v},t) = \frac{q^2}{m}\sum_s q_s^2\int\frac{d^3\boldsymbol{k}}{(2\pi)^3}\int d^3\boldsymbol{v}'\frac{\pi\delta[\boldsymbol{k}\cdot(\boldsymbol{v}-\boldsymbol{v}')]}{|\varepsilon(\boldsymbol{k},\boldsymbol{k}\cdot\boldsymbol{v})|^2}\frac{\boldsymbol{k}\boldsymbol{k}}{k^4} \times\left[F_s^0(\boldsymbol{v}',t)\frac{1}{m}\frac{\partial}{\partial\boldsymbol{v}}F^0(\boldsymbol{v},t)-F^0(\boldsymbol{v},t)\frac{1}{m_s}\frac{\partial}{\partial\boldsymbol{v}'}F_s^0(\boldsymbol{v}',t)\right] \qquad (6.8.25)$$

Our final form of the kinetic equation for the sth species of particle colliding with the s'th species of particle can be obtained by substitution of (6.8.25) into (6.8.5) as follows:

$$\frac{\partial}{\partial t} F_s^0(\boldsymbol{v}, t) = \int d^3 \boldsymbol{v}' \frac{\partial}{\partial \boldsymbol{v}} \cdot \overleftrightarrow{Q}_{ss'} \cdot \left(\frac{1}{m_s} \frac{\partial}{\partial \boldsymbol{v}} - \frac{1}{m_{s'}} \frac{\partial}{\partial \boldsymbol{v}'} \right) F_s^0(\boldsymbol{v}, t) F_{s'}^0(\boldsymbol{v}', t) , \tag{6.8.26}$$

where

$$\overleftrightarrow{Q}_{ss'} = \frac{q_s^2 q_{s'}^2}{m_s} \int \frac{d^3 \boldsymbol{k}}{(2\pi)^3} \frac{\pi \delta[\boldsymbol{k} \cdot (\boldsymbol{v} - \boldsymbol{v}')]}{|\varepsilon(\boldsymbol{k}, \boldsymbol{k} \cdot \boldsymbol{v})|^2} \frac{\boldsymbol{k}\boldsymbol{k}}{k^4} . \tag{6.8.27}$$

Equation (6.8.26) is called the *Balescu-Lenard equation* [6.8].

A few remarks are in order. First, (6.8.26) is in the form of the *Fokker-Planck equation* for the probability distribution function $P(\boldsymbol{v}, t)$:

$$\frac{\partial}{\partial t} P(\boldsymbol{v}, t) = \frac{\partial}{\partial \boldsymbol{v}} \cdot \left[-\left\langle \frac{\Delta \boldsymbol{v}}{\Delta t} \right\rangle P(\boldsymbol{v}, t) + \frac{1}{2} \frac{\partial}{\partial \boldsymbol{v}} \cdot \left\langle \frac{\Delta \boldsymbol{v}^2}{\Delta t} \right\rangle P(\boldsymbol{v}, t) \right] , \tag{6.8.28}$$

where $\Delta \boldsymbol{v}$ is the velocity change in a short time scale Δt. This equation can be derived for a general class of Markov random processes with small momentum transfer, i.e., when $\langle \Delta \boldsymbol{v}^n / \Delta t \rangle$ are negligible for $n \geq 3$ [6.10]. Indeed, it is possible to show by solving the equation of motion that in the present case [6.11]

$$\left\langle \frac{\Delta \boldsymbol{v}^2}{\Delta t} \right\rangle = \int d^3 \boldsymbol{v}' \frac{1}{m_s} \overleftrightarrow{Q}_{ss'} F_{s'}^0(\boldsymbol{v}', t) \tag{6.8.29}$$

$$\left\langle \frac{\Delta \boldsymbol{v}}{\Delta t} \right\rangle - \frac{1}{2} \frac{\partial}{\partial \boldsymbol{v}} \cdot \left\langle \frac{\Delta \boldsymbol{v}^2}{\Delta t} \right\rangle = \int d^3 \boldsymbol{v}' \frac{1}{m_{s'}} \overleftrightarrow{Q}_{ss'} \cdot \frac{\partial}{\partial \boldsymbol{v}'} F_{s'}^0(\boldsymbol{v}', t) .$$

The Fokker-Planck equation consists of a diffusion term in velocity space [the second term in (6.8.28)] and the dynamical friction term [the first term in (6.8.28)]. The diffusion term represents the cumulative small angle scattering effect, while the dynamical friction term represents the slowing down process accompanied by the angular scattering.

Second, the kinetic equation (6.8.26) includes the collective shielding effect on the colliding particles through the factor $|\varepsilon(\boldsymbol{k}, \boldsymbol{k} \cdot \boldsymbol{v})|^{-2}$. As seen from (6.8.27), without this shielding factor, the k-integral shows a logarithmic divergence at both small and large values of k. Divergence at large values of k is due to the inadequacy of the present calculation for a very short distance encounter. It can either be remedied by including the quantum effect due to finite de Broglie wavelength of the particle or by retaining the right hand side of (6.8.7) under certain approximations. On the other hand, the divergence at small values of k, which results from the long range character of the Coulomb interaction, is removed in (6.8.27) by the collective shielding factor $|\varepsilon(\boldsymbol{k}, \boldsymbol{k} \cdot \boldsymbol{v})|^{-2}$. Indeed, at $k \to 0$ or $\boldsymbol{k} \cdot \boldsymbol{v} \to 0$, we have the Debye shielding factor

$$\frac{1}{|\varepsilon(\boldsymbol{k}, \boldsymbol{k} \cdot \boldsymbol{v})|^2} \doteq \frac{1}{[1 + (k\lambda_{\mathrm{D}})^{-2}]^2} \doteq k^4 \lambda_{\mathrm{D}}^4$$

which removes the divergence.

Third, because of the complexity of (6.8.27), we often use a simpler form of the kinetic equation. The simplest form is the one-dimensional *Lenard-Bernstein model* [6.12]:

$$\frac{\partial F^0}{\partial t} = \nu \frac{\partial}{\partial v}\left(vF^0 + \frac{T}{m}\frac{\partial F^0}{\partial v}\right) \tag{6.8.30}$$

whose solution can be written as [6.13]

$$F^0(v,t) = \int dv_0 P(v_0 \rightarrow v, t) F^0(v_0, t=0) \tag{6.8.31}$$

$$P(v_0 \rightarrow v, t) = \sqrt{\frac{m}{2\pi T[1-\exp(-2\nu t)]}} \exp\left\{-\frac{m}{2T}\frac{[v - v_0 \exp(-\nu t)]^2}{1-\exp(-2\nu t)}\right\} \tag{6.8.32}$$

where ν is the collision frequency and T is the final temperature. Obviously, the long time solution ($t \gg 1$) of (6.8.30) is a Maxwellian distribution independent of the initial distribution.

Finally, we need to comment on the neglect of the right hand side of (6.8.7). As is seen from (6.8.9, 16), the right hand side of (6.8.7) contains a factor of the form

$$\langle \exp\{\mathrm{i}\boldsymbol{k}' \cdot \boldsymbol{r}_j(t) - \mathrm{i}(\boldsymbol{k}' - \boldsymbol{k}) \cdot \boldsymbol{r}_i(t)\}\rangle \ .$$

For $j \neq i$, this average vanishes by the random phase approximation as before. For $j = i$, it becomes $\langle \exp[-\mathrm{i}\boldsymbol{k} \cdot \boldsymbol{r}_i(t)]\rangle$ which again vanishes after the time average. Therefore, the contribution of the right hand side of (6.8.7) is negligible, provided that the random phase approximation is applicable. The validity of this approximation depends on the Coulomb interaction intensity. For very short distance collisions, the Coulomb force becomes strong, and the assumption of random phase becomes invalid. Therefore the present derivation of the kinetic equation is valid only for relatively long distance collisions, or small angle scattering processes. For a more complete discussion of the validity of this approximation we refer the reader to the original articles [6.8].

PROBLEMS

6.1. Carrying out the integration in (6.2.30), derive the Debye potential (6.2.31).

6.2. Using the dispersion relations (5.2.14) for the electron plasma wave and (5.4.16) for the ion acoustic wave, show that

$$\frac{\partial \varepsilon'(k,\omega_\mathrm{r})}{\partial \omega_\mathrm{r}} \doteq \varepsilon_0 \frac{2}{\omega_\mathrm{pe}} \quad \text{(electron plasma wave)}$$

$$\doteq \varepsilon_0 \frac{2}{\omega_\mathrm{pi}} \left(\frac{1 + k^2\lambda_\mathrm{De}^2}{k^2\lambda_\mathrm{De}^2}\right)^{3/2} \quad \text{(ion acoustic wave with } T_\mathrm{i} = 0) \,. \tag{6.8.33}$$

6.3. Using (6.3.9) in (6.3.14) and the results of the preceding problem, show that the Landau damping rates for the electron plasma wave and the ion acoustic wave for the Maxwellian distribution functions (6.3.17) (s = e, i) are given by

$$\begin{aligned}\gamma &\doteq \sqrt{\frac{\pi}{8}}\omega_{\mathrm{pe}}\frac{1}{k^3\lambda_{\mathrm{De}}^3}\exp\left[-\frac{1}{2k^2\lambda_{\mathrm{De}}^2}-\frac{3}{2}\right] \quad \text{(electron plasma wave)}\\ &\doteq \sqrt{\frac{\pi}{8}}\frac{kc_s}{(1+k^2\lambda_{\mathrm{De}}^2)^2}\left\{\sqrt{\frac{m_{\mathrm{e}}}{m_{\mathrm{i}}}}+\left(\frac{T_{\mathrm{e}}}{T_{\mathrm{i}}}\right)^{3/2}\right.\\ &\quad\left.\times\exp\left[-\frac{1}{2(1+k^2\lambda_{\mathrm{De}}^2)}\frac{T_{\mathrm{e}}}{T_{\mathrm{i}}}-\frac{3}{2}\right]\right\} \quad \text{(ion acoustic wave)}\end{aligned}$$

6.4. Suppose that the electrons are drifting against the ions with average speed u_{d} in the x-direction and that the velocity distributions of the electrons and ions are Maxwellian with $T_{\mathrm{e}} = 20T_{\mathrm{i}}$. Find the condition for the current driven ion accoustic instability for the hydrogen plasma in the limit $k^2\lambda_{\mathrm{D}}^2 \ll 1$.

6.5. Using the relations for the Bessel functions

$$\int_0^{\infty} dv_{\perp}v_{\perp}J_n^2\left(\frac{k_{\perp}v_{\perp}}{\Omega_{\mathrm{e}}}\right)\exp\left[-\frac{v_{\perp}^2}{2v_{T\mathrm{e}}^2}\right] = v_{T\mathrm{e}}^2 I_n(b)\,\mathrm{e}^{-b}\,,$$

where I_n is the modified Bessel function and $b = k_{\perp}^2 v_{T\mathrm{e}}^2/\Omega_{\mathrm{e}}^2$ show that the electrostatic electron cyclotron wave (*Bernstein mode*) propagating perpendicular to the magnetic field satisfies the following dispersion relation:

$$1 = \frac{\omega_{\mathrm{pe}}^2}{k_{\perp}^2 v_{T\mathrm{e}}^2}\sum_{l=-\infty}^{\infty}\mathrm{e}^{-b}I_l(b)\frac{l\Omega_{\mathrm{e}}}{\omega - l\Omega_{\mathrm{e}}}\,.$$

Assume that the electron velocity distribution is Maxwellian.

6.6. Using the equation of motion, derive (6.8.29) without the shielding factor $|\varepsilon(\boldsymbol{k},\boldsymbol{k}\cdot\boldsymbol{v})|^{-2}$.

6.7. Suppose the collision model (6.8.30) is applicable to a Fourier component of the averaged distribution function $F_{-k}(v,t)$ given by (6.8.9). Show that the effective collision frequency becomes $\nu(kv_T/\nu)^{2/3}$.

7. General Theory of Linear Waves

Plasma sustains various types of waves. Waves in a plasma often act as the carriers of energy and momentum and can be considered to be elementary excitations in a plasma. Waves in a plasma can generally be represented by electromagnetic fluctuations and/or density perturbations. The propagation characteristics of the wave are determined by the average properties of the plasma. In many cases of interest, waves can be treated by linear approximations. If in the absence of wave fluctuation a plasma is spatially uniform and temporally stationary, then the waves propagate as plane waves and the frequency and wavenumber satisfy the linear dispersion relation determined by the properties of the unperturbed state.

In this chapter, we first describe the wave source and the linear response to the source in Sect. 7.1 and then in Sect. 7.2 we derive the general dielectric tensor for the case of spatially uniform and temporally stationary plasma. A general expression for the linear dispersion relation is derived. The result is applied to the cold plasma model in Sect. 7.3 and finally the wave energy and momentum equations are derived in Sect. 7.4.

7.1 Source and Response

Waves are excited by some kind of source. The wave source can be either a fluctuation produced by some external means or a spontaneous fluctuation inside the plasma. In either case, the wave source in a plasma can be considered to have an electromagnetic origin. For the case of external excitation, the source can be represented by a fluctuating current or charge on a plasma boundary, while for the case of spontaneous excitation, the source can be represented by a fluctuating current or charge due to particle motion in a plasma. In either case, the wave source is represented by a source current or charge denoted by $\boldsymbol{J}_s(\boldsymbol{r},t), \sigma_{es}(\boldsymbol{r},t)$. Since they satisfy the charge conservation relation

$$\frac{\partial}{\partial t}\sigma_{es}(\boldsymbol{r},t) + \nabla\cdot\boldsymbol{J}_s(\boldsymbol{r},t) = 0\,, \tag{7.1.1}$$

we only have to consider the source current $\boldsymbol{J}_s(\boldsymbol{r},t)$.

By Ampere's law the source current produces an electromagnetic field denoted by $\boldsymbol{E}'(\boldsymbol{r},t)$ and $\boldsymbol{B}'(\boldsymbol{r},t)$. They satisfy the Faraday induction law

$$\nabla\times\boldsymbol{E}'(\boldsymbol{r},t) = -\frac{\partial}{\partial t}\boldsymbol{B}'(\boldsymbol{r},t)\,. \tag{7.1.2}$$

Substituting for $\boldsymbol{B}'$ using (7.1.2), we can write the Ampere law as

$$\nabla \times [\nabla \times \boldsymbol{E}'(\boldsymbol{r},t)] + \frac{1}{c^2}\frac{\partial^2}{\partial t^2}\boldsymbol{E}'(\boldsymbol{r},t) = -\mu_0 \frac{\partial}{\partial t}\boldsymbol{J}'(\boldsymbol{r},t) , \tag{7.1.3}$$

where $\boldsymbol{J}'(\boldsymbol{r},t)$ is the fluctuating current density. It consists of the source current and the current induced in the plasma by the fluctuating electric field. This induced current density, denoted by $\boldsymbol{J}_\mathrm{i}(\boldsymbol{r},t)$, is due to the particle orbit modification by the electric field $\boldsymbol{E}'$. We have

$$\boldsymbol{J}'(\boldsymbol{r},t) = \boldsymbol{J}_\mathrm{s}(\boldsymbol{r},t) + \boldsymbol{J}_\mathrm{i}(\boldsymbol{r},t) . \tag{7.1.4}$$

In order to derive the expression for $\boldsymbol{J}_\mathrm{i}$, we need to solve the basic equations for the plasma, depending on the description of the plasma, the MHD equations, two-fluid equations, Vlasov equation, etc. We shall not derive the expression for $\boldsymbol{J}_\mathrm{i}$ here, but restrict ourselves to a general phenomenological argument. In general, we can write $\boldsymbol{J}_\mathrm{i}(\boldsymbol{r},t)$ as a linear response to $\boldsymbol{E}'$ in the form

$$\boldsymbol{J}_\mathrm{i}(\boldsymbol{r},t) = \int d^3\boldsymbol{r}' \int_{-\infty}^{t} dt' \, \overleftrightarrow{\sigma}(\boldsymbol{r},\boldsymbol{r}';t,t') \cdot \boldsymbol{E}'(\boldsymbol{r}',t') . \tag{7.1.5}$$

We note here that the space integration is over the entire volume while the time integration is from $-\infty$ to t. This equation expresses the fact that the induced current is a linear superposition of the plasma response to the electric field $\boldsymbol{E}'$ at some point $\boldsymbol{r}'$ in the plasma at some time in the past t'. The tensor $\overleftrightarrow{\sigma}$ is determined by the average properties of the plasma and is to be calculated from the basic equations. If we assume that $\overleftrightarrow{\sigma}$ is known then for given source current (7.1.3–5) form a closed set.

In the special case of electrostatic response, the electric field is expressed in terms of the scalar potential $\phi'(\boldsymbol{r},t')$, where $\boldsymbol{E}' = -\nabla\phi'$. Then the source can also be represented by a scalar fluctuation. Since electrostatic fluctuation is represented by charge oscillation, we can use $\sigma_\mathrm{es}(\boldsymbol{r},t)$ instead of $\boldsymbol{J}_\mathrm{s}(\boldsymbol{r},t)$. In place of (7.1.3), we can use the Poisson equation

$$\varepsilon_0 \Delta\phi'(\boldsymbol{r},t) = -\sigma_\mathrm{e}'(\boldsymbol{r},t) . \tag{7.1.6}$$

Again we can write $\sigma_\mathrm{e}'(\boldsymbol{r},t)$ as the sum of the source charge and the induced charge $\sigma_\mathrm{ei}(\boldsymbol{r},t)$;

$$\sigma_\mathrm{e}'(\boldsymbol{r},t) = \sigma_\mathrm{es}(\boldsymbol{r},t) + \sigma_\mathrm{ei}(\boldsymbol{r},t) . \tag{7.1.7}$$

The induced charge can be expressed as the linear response to the potential $\phi'(\boldsymbol{r},t)$ as

$$\sigma_\mathrm{ei}(\boldsymbol{r},t) = \int d^3\boldsymbol{r}' \int_{-\infty}^{t} dt' Q(\boldsymbol{r},\boldsymbol{r}';t,t')\phi'(\boldsymbol{r}',t') , \tag{7.1.8}$$

where Q is determined by the average properties of the plasma.

7.2 Dielectric Tensor and the Dispersion Relation in a Uniform Plasma

If the plasma in the absence of the fluctuation is spatially uniform and temporally stationary, the linear response coefficients $\overleftrightarrow{\sigma}$ and Q become functions of $(\boldsymbol{r}-\boldsymbol{r}')$ and $(t-t')$ alone, since these coefficients must be invariant against the different choice of the space time origin. Equation (7.1.5) then becomes

$$\boldsymbol{J}_{\rm i}(\boldsymbol{r},t)=\int d^3\boldsymbol{r}'\int_{-\infty}^{t}dt'\,\overleftrightarrow{\sigma}(\boldsymbol{r}-\boldsymbol{r}',t-t')\cdot\boldsymbol{E}'(\boldsymbol{r}',t')\,. \tag{7.2.1}$$

Since the right hand side has the form of a convolution, this equation can be reduced to an algebraic equation by the Fourier-Laplace transformation. We denote the Fourier transform of a function $g(\boldsymbol{r},t)$ by

$$g(\boldsymbol{k},\omega)=\int d^3\boldsymbol{r}\int_{-\infty}^{\infty}dt\exp[\mathrm{i}(\omega t-\boldsymbol{k}\cdot\boldsymbol{r})]g(\boldsymbol{r},t)\,. \tag{7.2.2}$$

By Fourier inversion we have

$$g(\boldsymbol{r},t)=\int\frac{d^3\boldsymbol{k}}{(2\pi)^3}\int\frac{d\omega}{2\pi}\exp[\mathrm{i}(\boldsymbol{k}\cdot\boldsymbol{r}-\omega t)]g(\boldsymbol{k},\omega)\,. \tag{7.2.3}$$

Note that $g(\boldsymbol{r},t)$ stands for a vector component of $\boldsymbol{E}'$, $\boldsymbol{J}_{\rm i}$ and $\boldsymbol{J}_{\rm s}$ and we assume that they satisfy appropriate boundary conditions so that the space integral of g converges and that it vanishes at $t\to-\infty$. To ensure the convergence of (7.2.2) at $t\to+\infty$, we introduce a small imaginary part in ω

$$\omega\to\omega+\mathrm{i}\delta\qquad(\delta>0)\,. \tag{7.2.4}$$

The ω-integral in (7.2.3) is taken for $\mathrm{Im}\{\omega\}>\delta$ and $\mathrm{Re}\{\omega\}$ from $-\infty$ to ∞. Applying the above Fourier transformation to (7.2.1) we obtain

$$\boldsymbol{J}_{\rm i}(\boldsymbol{k},\omega)=\overleftrightarrow{\sigma}(\boldsymbol{k},\omega)\cdot\boldsymbol{E}'(\boldsymbol{k},\omega)\,, \tag{7.2.5}$$

where $\overleftrightarrow{\sigma}(\boldsymbol{k},\omega)$ is the Fourier transform in space and the Laplace transform in time of $\overleftrightarrow{\sigma}(\boldsymbol{r},t)$:

$$\overleftrightarrow{\sigma}(\boldsymbol{k},\omega)=\int d^3\boldsymbol{r}\int_{0}^{\infty}dt\exp[\mathrm{i}(\omega t-\boldsymbol{k}\cdot\boldsymbol{r})]\,\overleftrightarrow{\sigma}(\boldsymbol{r},t)\,. \tag{7.2.6}$$

This relation represents a generalized Ohm's law and we refer to $\overleftrightarrow{\sigma}(\boldsymbol{k},\omega)$ as the *conductivity tensor*. Applying the Fourier transform to (7.1.3) we get

$$\begin{aligned}\boldsymbol{k}\times[\boldsymbol{k}\times\boldsymbol{E}'(\boldsymbol{k},\omega)]+\frac{\omega^2}{c^2}\boldsymbol{E}'(\boldsymbol{k},\omega)&=-\mathrm{i}\omega\mu_0\boldsymbol{J}(\boldsymbol{k},\omega)\\&=-\mathrm{i}\omega\mu_0\boldsymbol{J}_{\rm s}(\boldsymbol{k},\omega)-\mathrm{i}\omega\mu_0\,\overleftrightarrow{\sigma}(\boldsymbol{k},\omega)\cdot\boldsymbol{E}'(\boldsymbol{k},\omega)\,.\end{aligned} \tag{7.2.7}$$

We use the vectorial relation (3.3.3) to obtain

$$\overleftrightarrow{D}(\boldsymbol{k},\omega)\cdot\boldsymbol{E}'(\boldsymbol{k},\omega)=\boldsymbol{S}(\boldsymbol{k},\omega)\,, \tag{7.2.8}$$

where the right hand side denotes the contribution of the source current

$$S(\boldsymbol{k},\omega) = \frac{1}{\mathrm{i}\omega} \boldsymbol{J}_s(\boldsymbol{k},\omega) \tag{7.2.9}$$

while $\overleftrightarrow{D}(\boldsymbol{k},\omega)$ is a tensor determined by the average properties of the plasma:

$$\overleftrightarrow{D}(\boldsymbol{k},\omega) = \overleftrightarrow{\varepsilon}(\boldsymbol{k},\omega) - \varepsilon_0 N^2 \left[\overleftrightarrow{I} - \frac{\boldsymbol{k}\boldsymbol{k}}{k^2}\right] , \tag{7.2.10}$$

where $\overleftrightarrow{I}$ is the unit tensor, N is the index of refraction

$$N^2 = k^2c^2/\omega^2 \tag{7.2.11}$$

and $\overleftrightarrow{\varepsilon}(\boldsymbol{k},\omega)$ is the *dielectric tensor* given by

$$\overleftrightarrow{\varepsilon}(\boldsymbol{k},\omega) = \varepsilon_0 \overleftrightarrow{I} - \frac{1}{\mathrm{i}\omega} \overleftrightarrow{\sigma}(\boldsymbol{k},\omega) . \tag{7.2.12}$$

In (7.2.10) the term $\varepsilon_0 N^2(\overleftrightarrow{I} - \boldsymbol{k}\boldsymbol{k}/k^2)$ is the contribution of $\nabla \times \boldsymbol{B}'$ and $\overleftrightarrow{\varepsilon}(\boldsymbol{k},\omega)$ is that of the current. The first term in (7.2.12) is the contribution of the displacement current $c^{-2}\partial^2\boldsymbol{E}'/\partial t^2$ and the second term is that of the induced plasma current. We can see from this expression that the plasma effect appears only through $\overleftrightarrow{\sigma}(\boldsymbol{k},\omega)$. We therefore write this term as

$$-\frac{1}{\mathrm{i}\omega} \overleftrightarrow{\sigma}(\boldsymbol{k},\omega) = \varepsilon_0 \overleftrightarrow{\chi}(\boldsymbol{k},\omega) , \tag{7.2.13}$$

and call $\overleftrightarrow{\chi}(\boldsymbol{k},\omega)$ the *electric susceptibility tensor*.

The dielectric tensor can then be written as

$$\overleftrightarrow{\varepsilon}(\boldsymbol{k},\omega) = \varepsilon_0[\overleftrightarrow{I} + \overleftrightarrow{\chi}(\boldsymbol{k},\omega)] . \tag{7.2.14}$$

We note that $\overleftrightarrow{\chi}$ can be divided into the electron contribution and the ion contribution. We note that the conductivity tensor, susceptibility tensor and dielectric tensor not only characterize the wave propagation in the plasma, but reflect the basic physical properties of the plasma.

In the special case of electrostatic response, we note that Q is a function of $(\boldsymbol{r} - \boldsymbol{r}')$ and $(t - t')$ and apply the Fourier transform to (7.1.6, 8) to obtain the relation

$$\varepsilon^\ell(\boldsymbol{k},\omega)\phi'(\boldsymbol{k},\omega) = \frac{1}{k^2}\sigma_{\mathrm{es}}(\boldsymbol{k},\omega) , \tag{7.2.15}$$

where $\varepsilon^\ell(\boldsymbol{k},\omega)$ is called the *longitudinal dielectric function* given by

$$\varepsilon^\ell(\boldsymbol{k},\omega) = \varepsilon_0 \left[1 + \chi^\ell(\boldsymbol{k},\omega)\right] , \tag{7.2.16}$$

where

$$\chi^\ell(\boldsymbol{k},\omega) = -\frac{1}{k^2\varepsilon_0} Q(\boldsymbol{k},\omega) \tag{7.2.17}$$

with

$$Q(\boldsymbol{k},\omega) = \int d^3\boldsymbol{r} \int_0^\infty dt \exp[\mathrm{i}(\omega t - \boldsymbol{k}\cdot\boldsymbol{r})] Q(\boldsymbol{r},t) . \tag{7.2.18}$$

The quantity $\chi^\ell(\boldsymbol{k},\omega)$ is called the *longitudinal electric susceptibility* and is given by (6.2.12) for the case of the Vlasov model.

The relations between ε^ℓ and χ^ℓ to $\overleftrightarrow{\varepsilon}$ and $\overleftrightarrow{\chi}$ can be derived by taking the divergence of (7.2.9): Noting (7.1.1) we have

$$\mathrm{i}\boldsymbol{k}\cdot\boldsymbol{S}(\boldsymbol{k},\omega) = \sigma_{\mathrm{es}}(\boldsymbol{k},\omega) . \tag{7.2.19}$$

From (7.2.10) we have

$$\mathrm{i}\boldsymbol{k}\cdot \overleftrightarrow{D}(\boldsymbol{k},\omega)\cdot\boldsymbol{E}'(\boldsymbol{k},\omega) = \boldsymbol{k}\cdot\overleftrightarrow{\varepsilon}\cdot\boldsymbol{k}\phi'(\boldsymbol{k},\omega) . \tag{7.2.20}$$

Using these results in (7.2.8) and comparing the result with (7.2.15) we find the relations

$$\varepsilon^\ell(\boldsymbol{k},\omega) = \boldsymbol{k}\cdot\overleftrightarrow{\varepsilon}(\boldsymbol{k},\omega)\cdot\boldsymbol{k}/k^2 , \tag{7.2.21}$$

$$\chi^\ell(\boldsymbol{k},\omega) = \boldsymbol{k}\cdot\overleftrightarrow{\chi}(\boldsymbol{k},\omega)\cdot\boldsymbol{k}/k^2 . \tag{7.2.22}$$

We shall now derive the dispersion relation of the wave. We first formally solve (7.2.8) as

$$\boldsymbol{E}'(\boldsymbol{k},\omega) = \overleftrightarrow{D}^{-1}(\boldsymbol{k},\omega)\cdot\boldsymbol{S}(\boldsymbol{k},\omega) , \tag{7.2.23}$$

where $\overleftrightarrow{D}^{-1}$ is the inverse matrix of $\overleftrightarrow{D}$. The inverse Laplace transformation of (7.2.23) gives

$$\boldsymbol{E}'(\boldsymbol{k},t) = \int_{-\infty+\mathrm{i}\delta'}^{\infty+\mathrm{i}\delta'} \frac{1}{2\pi}\mathrm{e}^{-\mathrm{i}\omega t}\,\overleftrightarrow{D}^{-1}(\boldsymbol{k},\omega)\cdot\boldsymbol{S}(\boldsymbol{k},\omega)d\omega , \tag{7.2.24}$$

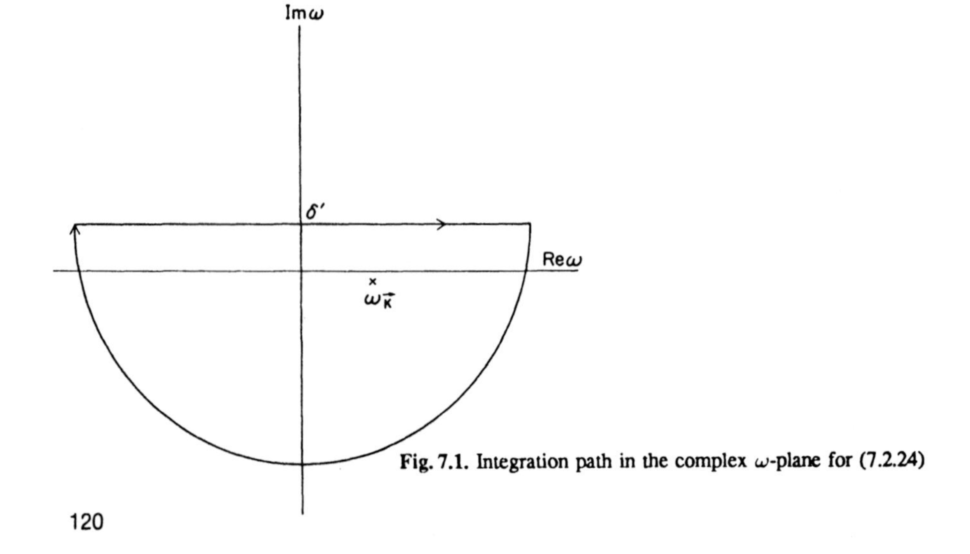

Fig. 7.1. Integration path in the complex ω-plane for (7.2.24)

where $\delta' > \delta > 0$. To carry out the integration we first analytically continue the integrand to the region $\mathrm{Im}\{\omega\} < \delta$. An example of analytical continuation was shown in Sect. 6.3. Then we close the integration path by adding a half circle in the plane $\mathrm{Im}\{\omega\} < 0$, as shown in Fig. 7.1. Because of the convergence factor $\exp(-\mathrm{i}\omega t)$, the added section of the path makes no contribution to the integral. The integral can be evaluated by calculating the residues of the poles inside the integration path. Suppose that there is a pole inside the integration path at

$$\omega = \omega_{\boldsymbol{k}} \equiv \omega_{\mathrm{r}} - \mathrm{i}\gamma \, . \tag{7.2.25}$$

Its contribution to $\boldsymbol{E}'(\boldsymbol{r}, t)$ takes the form

$$\boldsymbol{E}'(\boldsymbol{k}, t) \sim \exp[-\mathrm{i}\omega_{\mathrm{r}} t - \gamma t]$$

therefore if $|\gamma|$ is sufficiently small it describes a wave.

In actual calculations there are many singularities inside the integration path. Among these, we are specifically interested in the solution of the equation

$$\det \overleftrightarrow{D}(\boldsymbol{k}, \omega) = 0 \tag{7.2.26}$$

which comes from the singularity of $\overleftrightarrow{D}^{-1}$. The relation (7.2.26) is called the *dispersion relation* of the plasma. It is determined by the averaged properties of the plasma and is independent of the method of exciting the wave. On the other hand, the singularity of $S(\boldsymbol{k}, \omega)$ depends on the source.

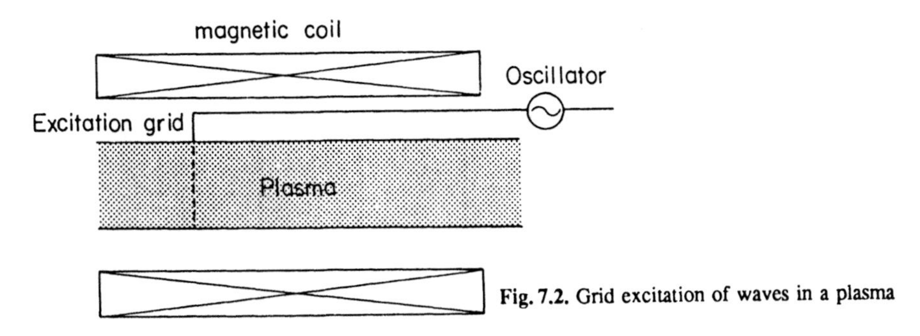

Fig. 7.2. Grid excitation of waves in a plasma

In actual experiments, the singularity of $S(\boldsymbol{k}, \omega)$ can have an important effect on the behavior of $\boldsymbol{E}(\boldsymbol{k}, t)$. For instance, when a grid is inserted inside the plasma flow with an applied oscillating voltage at frequency ω as shown in Fig. 7.2, the grid excites the wave that satisfies the dispersion relation, but at the same time, because of the periodic modulation of the plasma flow by the grid a wave-like density perturbation propagates downstream [7.1–3]. This is called the *ballistic mode* and is an oscillation which is unique to the grid excitation. For the case of an electrostatic wave the dispersion relation becomes

$$\varepsilon^{\ell}(\boldsymbol{k}, \omega) = 0 \, . \tag{7.2.27}$$

As seen from (7.2.8, 15), the wave that satisfies the dispersion relation corresponds to a finite response to an infinitesimal source:

$$\overleftrightarrow{D}(\boldsymbol{k},\omega)\cdot\boldsymbol{E}'(\boldsymbol{k},\omega)=0\,, \tag{7.2.28}$$

$$\varepsilon^{\ell}(\boldsymbol{k},\omega)\phi'(\boldsymbol{k},\omega)=0\,. \tag{7.2.29}$$

In this sense, the dispersion relation determines the eigenmodes in the plasma.

As shown in Chap. 6, the solution of the dispersion relation is in general complex. In other words, the tensor $\overleftrightarrow{D}(\boldsymbol{k},\omega)$ is not necessarily hermitian. We divide $\overleftrightarrow{D}$ into the hermitian part $\overleftrightarrow{D}^{\mathrm{H}}$ and the antihermitian part $\overleftrightarrow{D}^{\mathrm{AH}}$ as

$$\overleftrightarrow{D}(\boldsymbol{k},\omega)=\overleftrightarrow{D}^{\mathrm{H}}(\boldsymbol{k},\omega)+\mathrm{i}\,\overleftrightarrow{D}^{\mathrm{AH}}(\boldsymbol{k},\omega)\,, \tag{7.2.30}$$

where

$$\begin{aligned}\overleftrightarrow{D}^{\mathrm{H}}(\boldsymbol{k},\omega)&=[\overleftrightarrow{D}(\boldsymbol{k},\omega)+\overleftrightarrow{D}^{*}(\boldsymbol{k},\omega)]/2\\ \overleftrightarrow{D}^{\mathrm{AH}}(\boldsymbol{k},\omega)&=[\overleftrightarrow{D}(\boldsymbol{k},\omega)-\overleftrightarrow{D}^{*}(\boldsymbol{k},\omega)]/2\mathrm{i}\,,\end{aligned} \tag{7.2.31}$$

the star denoting the hermitian conjugate. For the case $|\overleftrightarrow{D}^{\mathrm{H}}|\gg|\overleftrightarrow{D}^{\mathrm{AH}}|$, we can derive the equations to calculate the real and imaginary parts of the frequency in the same way as in Sect. 6.3:

$$\det\overleftrightarrow{D}^{\mathrm{H}}(\boldsymbol{k},\omega_{\mathrm{r}})=0\,, \tag{7.2.32}$$

$$\gamma=\frac{\boldsymbol{E}^{*}\cdot\overleftrightarrow{D}^{\mathrm{AH}}(\boldsymbol{k},\omega_{\mathrm{r}})\cdot\boldsymbol{E}}{(\partial/\partial\omega_{\mathrm{r}})\boldsymbol{E}\cdot\overleftrightarrow{D}^{\mathrm{H}}(\boldsymbol{k},\omega_{\mathrm{r}})\cdot\boldsymbol{E}}\,. \tag{7.2.33}$$

For the case of the electrostatic wave, we write

$$\varepsilon^{\ell}(\boldsymbol{k},\omega)=\varepsilon^{\ell}_{\mathrm{r}}(\boldsymbol{k},\omega)+\mathrm{i}\varepsilon^{\ell}_{\mathrm{i}}(\boldsymbol{k},\omega)\,, \tag{7.2.34}$$

and the approximate dispersion relation becomes

$$\varepsilon^{\ell}_{\mathrm{r}}(\boldsymbol{k},\omega_{\mathrm{r}})=0 \tag{7.2.35}$$

$$\gamma=\frac{\varepsilon^{\ell}_{\mathrm{i}}(\boldsymbol{k},\omega_{\mathrm{r}})}{(\partial/\partial\omega_{\mathrm{r}})\varepsilon^{\ell}_{\mathrm{r}}(\boldsymbol{k},\omega_{\mathrm{r}})} \tag{7.2.36}$$

in agreement with (6.3.12, 14).

7.3 Cold Plasma Model

A variety of waves can be excited in a plasma in a magnetic field. Here we consider the simplest model called the *cold plasma model* which is useful to study the waves whose phase velocity is much faster than the thermal speed of the particles [7.4]. In particular, we consider a uniform and stationary plasma in a uniform magnetic field described by the two-fluid model with no pressure term. In this model the plasma motion consists only of the $\boldsymbol{E} \times \boldsymbol{B}$ drift and the polarization drift.

We consider the geometry where the magnetic field is in the z-direction and the wave propagates in the xz-plane. The linearized equations of motion for the electron and the ion are

$$\frac{\partial \boldsymbol{u}_s}{\partial t} = \frac{q_s}{m_s}(\boldsymbol{E}' + \boldsymbol{u}_s \times \boldsymbol{B}_0) . \tag{7.3.1}$$

We use the Fourier representation and write

$$\begin{pmatrix} \boldsymbol{u} \\ \boldsymbol{E}' \end{pmatrix} \sim \begin{pmatrix} \tilde{\boldsymbol{u}} \\ \tilde{\boldsymbol{E}} \end{pmatrix} \exp[\mathrm{i}(\boldsymbol{k} \cdot \boldsymbol{r} - \omega t)] . \tag{7.3.2}$$

Equation (7.3.1) can then be solved as

$$\tilde{u}_x = \frac{1}{\omega^2 - \Omega^2} \frac{q}{m}(\mathrm{i}\omega \tilde{E}_x - \Omega \tilde{E}_y) \tag{7.3.3}$$

$$\tilde{u}_y = \frac{1}{\omega^2 - \Omega^2} \frac{q}{m}(\mathrm{i}\omega \tilde{E}_y + \Omega \tilde{E}_x) \tag{7.3.4}$$

$$\tilde{u}_z = -\frac{1}{\mathrm{i}\omega} \frac{q}{m} \tilde{E}_z , \tag{7.3.5}$$

where $\Omega = qB_0/m$ and for simplicity we suppressed the suffix s denoting the species of the particle. In addition to this solution, we could add the terms describing the cyclotron motion, but such a solution is independent of the electric field and contributes to the source current. Since we are primarily interested in the induced current, we shall neglect the cyclotron motion. The second terms inside the brackets of (7.3.3, 4) are the $\boldsymbol{E} \times \boldsymbol{B}$ drift terms and the other terms are for the polarization drift. We note that at sufficiently strong magnetic fields ($|\Omega| \gg |\omega|$) in (7.3.3, 4) the $\boldsymbol{E} \times \boldsymbol{B}$ drift is larger than the polarization drift while at sufficiently high frequency ($|\omega| \gg |\Omega|$) the polarization drift dominates. The induced current can be calculated using (7.3.3–5) as

$$\boldsymbol{J}_\mathrm{i} = \sum_s n_{0s} q_s \tilde{\boldsymbol{u}}_s \equiv \overleftrightarrow{\sigma} (\boldsymbol{k}, \omega) \cdot \tilde{\boldsymbol{E}} , \tag{7.3.6}$$

where the conductivity tensor is given by

$$\overleftrightarrow{\sigma} (\boldsymbol{k}, \omega) = \sum_s \varepsilon_0 \omega_{ps}^2 \begin{pmatrix} [\mathrm{i}\omega/(\omega^2 - \Omega_s^2)], & -[\Omega_s/(\omega^2 - \Omega_s^2)], & 0 \\ [\Omega_s/(\omega^2 - \Omega_s^2)], & [\mathrm{i}\omega/(\omega^2 - \Omega_s^2)], & 0 \\ 0, & 0, & -(1/\mathrm{i}\omega) \end{pmatrix} . \tag{7.3.7}$$

Note that the diagonal elements consist of the polarization drift, while the off-diagonal elements consist of the $\boldsymbol{E} \times \boldsymbol{B}$ drift. In the limit of strong magnetic field, the electron $\boldsymbol{E} \times \boldsymbol{B}$ drift and the ion $\boldsymbol{E} \times \boldsymbol{B}$ drift cancel each other and the off-diagonal elements vanish. The dielectric tensor can be calculated by substituting (7.3.7) into (7.2.12) as

$$\overleftrightarrow{\varepsilon}(\boldsymbol{k},\omega) = \varepsilon_0 \begin{pmatrix} \chi_1, & -\mathrm{i}\chi_2, & 0 \\ \mathrm{i}\chi_2, & \chi_1, & 0 \\ 0, & 0, & \chi_3 \end{pmatrix} \tag{7.3.8}$$

where

$$\begin{aligned} \chi_1 &= 1 - \sum_s \frac{\omega_{ps}^2}{\omega^2 - \Omega_s^2}, \quad \chi_2 = \sum_s \frac{\Omega_s}{\omega} \frac{\omega_{ps}^2}{\omega^2 - \Omega_s^2}, \\ \chi_3 &= 1 - \sum_s \frac{\omega_{ps}^2}{\omega^2}. \end{aligned} \tag{7.3.9}$$

The polarization drift contributes to χ_1 and χ_3 while the $\boldsymbol{E} \times \boldsymbol{B}$ drift to χ_2. The displacement current contributes 1 to χ_1 and χ_3. Substituting (7.3.8) into (7.2.10) and then into (7.2.26), we can derive the dispersion relation as

$$\frac{1}{\varepsilon_0} \det \overleftrightarrow{D}(\boldsymbol{k},\omega) = \begin{vmatrix} \chi_1 - N^2\cos^2\theta, & -\mathrm{i}\chi_2, & N^2\cos\theta\sin\theta \\ \mathrm{i}\chi_2, & \chi_1 - N^2, & 0 \\ N^2\cos\theta\sin\theta, & 0, & \chi_3 - N^2\sin^2\theta \end{vmatrix} = 0 \tag{7.3.10}$$

where we used the representation

$$\boldsymbol{k} = (k\sin\theta, 0, k\cos\theta).$$

Equation (7.2.28) can then be written as

$$\begin{pmatrix} \chi_1 - N^2\cos^2\theta, & -\mathrm{i}\chi_2, & N^2\cos\theta\sin\theta \\ \mathrm{i}\chi_2, & \chi_1 - N^2, & 0 \\ N^2\cos\theta\sin\theta, & 0, & \chi_3 - N^2\sin^2\theta \end{pmatrix} \begin{pmatrix} \tilde{E}_x \\ \tilde{E}_y \\ \tilde{E}_z \end{pmatrix} = 0. \tag{7.3.11}$$

We now consider the solutions of the dispersion relation (7.3.10) for some special cases.

7.3.1 Unmagnetized Plasma

First we consider the case of zero magnetic field ($\Omega = 0$). In this case, $\chi_1 = \chi_3$ and $\chi_2 = 0$. There are two solutions,

$$\chi_1 = \chi_3 = N^2, \quad \tilde{E}_x = \tilde{E}_z = 0, \quad \tilde{E}_y \neq 0 \tag{7.3.12}$$

and

$$\chi_3 = 0, \quad \tilde{E}_y = 0, \quad \tilde{E}_x, \tilde{E}_z \neq 0 \tag{7.3.13}$$

Equation (7.3.12) is a transverse electromagnetic wave ($\boldsymbol{k} \cdot \boldsymbol{E} = 0$) with the dispersion relation

$$\omega^2 = k^2c^2 + \sum_s \omega_{ps}^2. \tag{7.3.14}$$

In the absence of a plasma, $\omega_{ps} = 0$ and we recover the dispersion relation for an electromagnetic wave in vacuum. The presence of a plasma ($\omega_{ps} \neq 0$) limits the frequency range of the electromagnetic wave propagation to the region

$$\omega^2 \geq \sum_s \omega_{ps}^2 \doteq \omega_{pe}^2 = \frac{n_0 e^2}{m_e \varepsilon_0} . \tag{7.3.15}$$

Below this frequency, the wavenumber k becomes imaginary and the wave cannot propagate. This is due to the shielding of the displacement current by the plasma current (conduction current) and is called the *cut-off* of the electromagnetic wave. For given frequency, the plasma density at which the cut-off takes place is called the *critical density* or the *cut-off density* and is given by

$$n_c = \frac{m_e \varepsilon_0}{e^2} \omega^2 . \tag{7.3.16}$$

Since the vacuum wavelength is inversely proportional to the frequency, the critical density is higher, the shorter the wavelength.

The solution (7.3.13) gives the electron plasma wave (Sect. 5.2) in the cold plasma model $\omega^2 = \sum_s \omega_{ps}^2 \doteq \omega_{pe}^2$. Since the thermal effect is ignored in the cold plasma model, there is no solution for the ion acoustic wave.

7.3.2 Parallel Propagation

We next consider the case of the wave propagation parallel to the magnetic field, i.e., $\theta = 0$. In this case, the dispersion relation can be reduced to

$$(\chi_1 - N^2)^2 - \chi_2^2 = 0, \qquad \tilde{E}_z = 0, \qquad \tilde{E}_x, \tilde{E}_y \neq 0 \tag{7.3.17}$$

$$\chi_3 = 0, \qquad \tilde{E}_x = \tilde{E}_y = 0, \qquad \tilde{E}_z \neq 0 . \tag{7.3.18}$$

The solution (7.3.17) describes transverse waves ($\boldsymbol{k} \cdot \tilde{\boldsymbol{E}} = 0$) while (7.3.18) describes a longitudinal wave ($\boldsymbol{k} \times \tilde{\boldsymbol{E}} = 0$). Obviously, (7.3.18) gives the electron plasma wave propagating along the magnetic field. Since the fluid motion along the magnetic field is unaffected by the magnetic field, the longitudinal wave propagating along the magnetic field is the same as that in the absence of the magnetic field.

Consider next the transverse waves. The solution of (7.3.17) can be written as

$$\chi_1 - N^2 = \pm \chi_2 \tag{7.3.19}$$

or more explicitly

$$1 - \sum_s \frac{\omega_{ps}^2}{\omega^2 - \Omega_s^2} - \frac{k^2 c^2}{\omega^2} = \pm \sum_s \frac{\Omega_s}{\omega} \frac{\omega_{ps}^2}{\omega^2 - \Omega_s^2} . \tag{7.3.20}$$

On the left hand side, the first term stands for the displacement current, the second term for the polarization drift current, the third term is $\nabla \times \boldsymbol{B}'$ and the right hand side is the $\boldsymbol{E} \times \boldsymbol{B}$ drift current. Obviously, these transverse waves are the electromagnetic waves in a magnetized plasma. Note that in a magnetic field the electromagnetic wave can propagate even above the critical density because the plasma current cannot flow freely across the magnetic field and hence, cannot shield the displacement current. We also note that because of the $\boldsymbol{E}\times\boldsymbol{B}$ drift, the plasma current, and hence the electromagnetic wave fields, rotate around the magnetic field. As a result the transverse waves propagate as circularly polarized waves. At sufficiently high frequencies, $\omega^2 \gg \Omega_s^2$, the polarization drift dominates over the $\boldsymbol{E}\times\boldsymbol{B}$ drift, and the dispersion relation becomes the same as that in the absence of the magnetic field (7.3.14).

In the frequency range $\Omega_e^2 \gg \omega^2 \gg \Omega_i^2$, the $\boldsymbol{E}\times\boldsymbol{B}$ drift dominates for electrons, while the polarization drift dominates for ions to yield

$$\frac{k^2c^2+\omega_{pi}^2}{\omega^2} = 1 - \frac{\omega_{pe}^2}{(\omega \pm \Omega_e)\omega} . \tag{7.3.21}$$

When the wave fields rotate in the same direction as the electron cyclotron motion, a resonance occurs at $\omega = |\Omega_e|$. This mode is called the *right hand polarized wave*. Such a wave, called the *whistler wave*, propagates in the earth's magnetic field and is excited by thunder near the equator. In particular, for the case where $k^2c^2 \gg \omega^2, \omega_{pi}^2$, the dispersion relation (7.3.21) is reduced to

$$\omega = \frac{|\Omega_e|k^2c^2}{k^2c^2+\omega_{pe}^2} . \tag{7.3.22}$$

At long wavelengths, $k^2c^2 \ll \omega_{pe}^2$, ω is proportional to k^2 and at short wavelengths, $k^2c^2 \gg \omega_{pe}^2$, ω approaches the electron cyclotron frequency $|\Omega_e|$. There, however, since the phase velocity becomes small, we can no longer ignore the kinetic effects.

At sufficiently low frequencies, $\omega^2 \ll \Omega_i^2$, the $\boldsymbol{E}\times\boldsymbol{B}$ drift current vanishes and the dominant term is the ion polarization drift. Thus we get

$$\omega^2 = \frac{k^2c^2}{1+\omega_{pi}^2/\Omega_i^2} . \tag{7.3.23}$$

This mode is called the *Alfven wave*. For $\omega_{pi}^2 \gg \Omega_i^2$

$$\omega^2 = k^2v_A^2, v_A^2 = \frac{B_0^2}{\mu_0 n_0 m_i} , \tag{7.3.24}$$

where v_A is called the Alfven speed.

7.3.3 Perpendicular Propagation

We now consider the wave propagation perpendicular to the magnetic field, i.e., $\theta = \pi/2$. In this case, the dispersion relation (7.3.10) is reduced to

$$\chi_1(\chi_1 - N^2) = \chi_2^2, \quad \tilde{E}_z = 0, \quad \tilde{E}_x, \tilde{E}_y \neq 0 , \tag{7.3.25}$$

$$\chi_3 = N^2, \quad \tilde{E}_x = \tilde{E}_y = 0, \quad \tilde{E}_z \neq 0 . \tag{7.3.26}$$

The solution of (7.3.26) is a transverse wave, while that of (7.3.25) is a mixed longitudinal and transverse mode. The dispersion relation (7.3.26) is the same as (7.3.12). Since in this *ordinary mode* the electric field is along the magnetic field, the induced current is unaffected by the magnetic field. On the other hand, the dispersion relation (7.3.25) describes an *extraordinary mode*. For this mode the electric field components $\tilde{E}_x$ and $\tilde{E}_y$ satisfy the relation

$$\chi_1 \tilde{E}_x - \mathrm{i}\chi_2 \tilde{E}_y = 0 \, ,$$

$$\mathrm{i}\chi_2 \tilde{E}_x + (\chi_1 - N^2)\tilde{E}_y = 0 \, . \tag{7.3.27}$$

For the case $\chi_2 \doteq 0$, $|\tilde{E}_y| \gg |\tilde{E}_x|$ and the wave becomes almost transverse. Such is the case either when $\omega^2 \gg \omega_{\mathrm{pe}}^2, \Omega_{\mathrm{e}}^2$ or when $\omega^2 \ll \Omega_{\mathrm{i}}^2$. The high frequency case is reduced to the vacuum electromagnetic wave, $\omega^2 = k^2c^2$, while the low frequency case to the Alfven wave.

On the other hand, when $\chi_1 \doteq 0$ and $N^2 \gg 1$, the wave becomes nearly longitudinal $|\tilde{E}_x| \gg |\tilde{E}_y|$. From $\chi_1 \doteq 0$ we have

$$1 \doteq \sum_s \frac{\omega_{\mathrm{p}s}^2}{\omega^2 - \Omega_s^2} \, . \tag{7.3.28}$$

Physically, it describes the balance of the displacement current with the polarization drift current in the x-direction.

$$\frac{1}{c^2}\frac{\partial E_x'}{\partial t} = -\mu_0 J_x' \, .$$

Two different modes can be obtained from (7.3.27). One is at $\omega^2 > \omega_{\mathrm{pe}}^2, \Omega_{\mathrm{e}}^2$;

$$\omega^2 \doteq \omega_{\mathrm{pe}}^2 + \Omega_{\mathrm{e}}^2 \, . \tag{7.3.29}$$

This mode is called the *upper hybrid mode*. The other is at $\Omega_{\mathrm{e}}^2 \gg \omega^2 \gg \Omega_{\mathrm{i}}^2$ and is given by

$$\omega^2 = \frac{\omega_{\mathrm{pi}}^2}{1 + \omega_{\mathrm{pe}}^2/\Omega_{\mathrm{e}}^2} \, . \tag{7.3.30}$$

This mode is called the *lower hybrid mode* and is used for the current drive in a tokamak plasma (Sect. 12.3). In this mode, the electrons are frozen in the magnetic field while the ions are only slightly affected by the magnetic field. At low density, $\omega_{\mathrm{pe}}^2 \ll \Omega_{\mathrm{e}}^2$; $\omega^2 \doteq \omega_{\mathrm{pi}}^2$ corresponding to the ion plasma oscillation, while at high density, $\omega_{\mathrm{pe}}^2 \gg \Omega_{\mathrm{e}}^2$, $\omega^2 = |\Omega_{\mathrm{e}}\Omega_{\mathrm{i}}|$ in which the electron polarization drift balances the ion polarization drift and the displacement current is negligible. We note that in the limit as $N^2 \to \infty$, $|\tilde{E}_x|$ becomes resonantly large. This limit is thus called the *upper hybrid resonance* (for high frequency modes) and the *lower hybrid resonance* (for low frequency modes). In this case, however, the thermal or kinetic effects have to be taken into account and we have to use the dispersion relation (6.4.18).

7.3.4 Oblique Propagation

Let us finally discuss the case of slightly oblique propagation near the lower hybrid resonance. We consider the case when $N^2 \gg 1$ and set the coefficient of N^2 in (7.3.10) equal to zero to obtain

$$\chi_3/\chi_1 = -\tan^2\theta \,. \tag{7.3.31}$$

Considering the frequency range $\Omega_e^2 \gg \omega^2 \gg \Omega_i^2$ and approximating χ_1 and χ_3 as

$$\begin{aligned} \chi_1 &\doteq 1 + \omega_{pe}^2/\Omega_e^2 - \omega_{pi}^2/\omega^2 \\ \chi_3 &\doteq -\omega_{pe}^2/\omega^2 \,, \end{aligned} \tag{7.3.32}$$

we get

$$\omega^2 = \frac{\omega_{pi}^2}{1 + \omega_{pe}^2/\Omega_e^2}\left(1 + \frac{m_i}{m_e}\cot^2\theta\right) \,. \tag{7.3.33}$$

This dispersion relation corresponds to the lower hybrid mode propagating obliquely to the magnetic field. We note that even for a small angle of order $\cot^2\theta \sim m_e/m_i$, the dispersion relation is significantly modified from the case of the perpendicular propagation.

7.4 Wave Energy and Momentum

Consider a wave represented by a wavenumber $\boldsymbol{k}$ and a real frequency ω and denote the wave electric field $\boldsymbol{E}(\boldsymbol{r},t)$ and the source current $\boldsymbol{J}_s(\boldsymbol{r},t)$ which excites this wave as

$$\begin{aligned} \boldsymbol{E}(\boldsymbol{r},t) &= \tilde{\boldsymbol{E}}(\boldsymbol{r},t)\exp[\mathrm{i}(\boldsymbol{k}\cdot\boldsymbol{r} - \omega t)] + \text{c.c.} \\ \boldsymbol{J}_s(\boldsymbol{r},t) &= \tilde{\boldsymbol{J}}_s(\boldsymbol{r},t)\exp[\mathrm{i}(\boldsymbol{k}\cdot\boldsymbol{r} - \omega t)] + \text{c.c.}\,, \end{aligned} \tag{7.4.1}$$

where c.c. stands for the complex conjugate, and the amplitudes $\tilde{\boldsymbol{E}}(\boldsymbol{r},t)$, $\tilde{\boldsymbol{J}}_s(\boldsymbol{r},t)$ are assumed to be slowly varying functions of space and time, due to the wave excitation by a source and by damping from, for example, wave-particle interaction. We shall assume that the unperturbed plasma is spatially uniform and temporally stationary.

7.4.1 Wave Energy Equation

The source current supplies energy to the wave by inverse Joule heating. Per unit time and per unit volume it is given by

$$-\langle \boldsymbol{E}(\boldsymbol{r},t)\cdot\boldsymbol{J}_s(\boldsymbol{r},t)\rangle = -\left\langle \tilde{\boldsymbol{E}}^*\cdot\tilde{\boldsymbol{J}}_s\right\rangle - \left\langle \tilde{\boldsymbol{E}}\cdot\tilde{\boldsymbol{J}}_s^*\right\rangle \,, \tag{7.4.2}$$

where the angle bracket denotes the average over one oscillation period. The minus sign comes from the fact that it is the work done by the source on the wave.

We use the Fourier representation for $\tilde{\boldsymbol{E}}$ and $\tilde{\boldsymbol{J}}_s$ as

$$\left\{ \begin{array}{c} \tilde{\boldsymbol{E}}(\boldsymbol{r},t) \\ \tilde{\boldsymbol{J}}_s(\boldsymbol{r},t) \end{array} \right\} = \int \frac{d\alpha}{2\pi} \int \frac{d^3\boldsymbol{q}}{(2\pi)^3} \exp\left[\mathrm{i}(\boldsymbol{q}\cdot\boldsymbol{r} - \alpha t)\right] \left\{ \begin{array}{c} \tilde{\boldsymbol{E}}(\boldsymbol{q},\alpha) \\ \tilde{\boldsymbol{J}}_s(\boldsymbol{q},\alpha) \end{array} \right\} . \quad (7.4.3)$$

Then from (7.2.8, 9) we have

$$\tilde{\boldsymbol{J}}_s(\boldsymbol{q},\alpha) = \mathrm{i}(\omega+\alpha) \overleftrightarrow{D}(\boldsymbol{k}+\boldsymbol{q}, \omega+\alpha) \cdot \tilde{\boldsymbol{E}}(\boldsymbol{q},\alpha) . \quad (7.4.4)$$

Since $\tilde{\boldsymbol{E}}$ and $\tilde{\boldsymbol{J}}_s$ are slowly varying in space and time, we can naturally assume that $|\boldsymbol{q}|$ and $|\alpha|$ are sufficiently small and retain only first order terms:

$$\begin{aligned} &(\omega+\alpha) \overleftrightarrow{D}(\boldsymbol{k}+\boldsymbol{q}, \omega+\alpha) \\ &\quad \doteq \omega \overleftrightarrow{D}(\boldsymbol{k},\omega) + \alpha \frac{\partial}{\partial\omega} \overleftrightarrow{D}(\boldsymbol{k},\omega) + \boldsymbol{q}\cdot\frac{\partial}{\partial\boldsymbol{k}} \overleftrightarrow{D}(\boldsymbol{k},\omega) , \end{aligned} \quad (7.4.5)$$

from which we get

$$\begin{aligned} &\tilde{\boldsymbol{J}}_s(\boldsymbol{r},t) \doteq \mathrm{i}\omega \overleftrightarrow{D}(\boldsymbol{k},\omega) \cdot \tilde{\boldsymbol{E}}(\boldsymbol{r},t) \\ &\quad - \frac{\partial}{\partial t}\frac{\partial}{\partial\omega}[\omega \overleftrightarrow{D}(\boldsymbol{k},\omega)] \cdot \tilde{\boldsymbol{E}}(\boldsymbol{r},t) + \frac{\partial}{\partial\boldsymbol{r}} \cdot \frac{\partial}{\partial\boldsymbol{k}}[\omega \overleftrightarrow{D}(\boldsymbol{k},\omega)] \cdot \tilde{\boldsymbol{E}}(\boldsymbol{r},t) , \end{aligned} \quad (7.4.6)$$

where we used the relations

$$\left\{ \begin{array}{c} \alpha \\ \boldsymbol{q} \end{array} \right\} \exp\left[\mathrm{i}(\boldsymbol{q}\cdot\boldsymbol{r} - \alpha t)\right] = \left\{ \begin{array}{c} \mathrm{i}(\partial/\partial t) \\ -\mathrm{i}(\partial/\partial\boldsymbol{r}) \end{array} \right\} \exp\left[\mathrm{i}(\boldsymbol{q}\cdot\boldsymbol{r} - \alpha t)\right] . \quad (7.4.7)$$

A similar equation can be derived for $\tilde{\boldsymbol{J}}_s^*(\boldsymbol{r},t)$. Using these relations in (7.4.2), we obtain

$$\begin{aligned} &-\langle \boldsymbol{E}\cdot\boldsymbol{J}_s \rangle \\ &\quad = -\mathrm{i}\omega[\tilde{\boldsymbol{E}}^*\cdot \overleftrightarrow{D}(\boldsymbol{k},\omega) \cdot \tilde{\boldsymbol{E}} - \tilde{\boldsymbol{E}}\cdot \overleftrightarrow{D}^*(\boldsymbol{k},\omega) \cdot \tilde{\boldsymbol{E}}^*] \\ &\qquad + \tilde{\boldsymbol{E}}^* \cdot \left(\frac{\partial}{\partial t}\frac{\partial}{\partial\omega}\right)[\omega \overleftrightarrow{D}(\boldsymbol{k},\omega)] \cdot \tilde{\boldsymbol{E}} + \tilde{\boldsymbol{E}} \cdot \left(\frac{\partial}{\partial t}\frac{\partial}{\partial\omega}\right)[\omega \overleftrightarrow{D}^*(\boldsymbol{k},\omega)] \cdot \tilde{\boldsymbol{E}}^* \\ &\qquad - \tilde{\boldsymbol{E}}^* \cdot \left(\frac{\partial}{\partial\boldsymbol{r}}\cdot\frac{\partial}{\partial\boldsymbol{k}}\right)[\omega \overleftrightarrow{D}(\boldsymbol{k},\omega)] \cdot \tilde{\boldsymbol{E}} - \tilde{\boldsymbol{E}} \cdot \left(\frac{\partial}{\partial\boldsymbol{r}}\cdot\frac{\partial}{\partial\boldsymbol{k}}\right) \\ &\qquad \times [\omega \overleftrightarrow{D}^*(\boldsymbol{k},\omega)] \cdot \tilde{\boldsymbol{E}}^* , \end{aligned} \quad (7.4.8)$$

where $\partial/\partial\omega$ and $\partial/\partial\boldsymbol{k}$ operate only on $\omega \overleftrightarrow{D}$ and $\omega \overleftrightarrow{D}^*$. We divide $\overleftrightarrow{D}(\boldsymbol{k},\omega)$ into the hermitian part $\overleftrightarrow{D}^{\mathrm{H}}$ and the antihermitian part $\overleftrightarrow{D}^{\mathrm{AH}}$ as in (7.2.30). Then

$$D^{\mathrm{H}}_{\alpha\beta} = D^{\mathrm{H}*}_{\beta\alpha}, \qquad D^{\mathrm{AH}}_{\alpha\beta} = D^{\mathrm{AH}*}_{\beta\alpha} \quad (7.4.9)$$

where the suffixes α, β denote the components. As in Sect. 7.2, we assume $|\overleftrightarrow{D}^{\mathrm{H}}| \gg |\overleftrightarrow{D}^{\mathrm{AH}}|$. The terms on the right hand side of (7.4.8) can then be written

$$-\mathrm{i}\omega\{\tilde{\boldsymbol{E}}^*\cdot \overleftrightarrow{D} \cdot \tilde{\boldsymbol{E}} - \tilde{\boldsymbol{E}}\cdot \overleftrightarrow{D}^* \cdot \tilde{\boldsymbol{E}}^*\} = 2\omega \tilde{\boldsymbol{E}}^*\cdot \overleftrightarrow{D}^{\mathrm{AH}} \cdot \tilde{\boldsymbol{E}} , \quad (7.4.10)$$

$$\tilde{\boldsymbol{E}}^{*}\cdot\left(\frac{\partial}{\partial t}\frac{\partial}{\partial\omega}\right)[\omega\,\overleftrightarrow{D}]\cdot\tilde{\boldsymbol{E}}+\tilde{\boldsymbol{E}}\cdot\left(\frac{\partial}{\partial t}\frac{\partial}{\partial\omega}\right)[\omega\,\overleftrightarrow{D}^{*}]\cdot\tilde{\boldsymbol{E}}^{*}$$

$$\doteq\sum_{\alpha,\beta}\left[\tilde{E}_{\alpha}^{*}\frac{\partial}{\partial\omega}(\omega D_{\alpha,\beta}^{\mathrm{H}})\frac{\partial\tilde{E}_{\beta}}{\partial t}+\tilde{E}_{\beta}\frac{\partial}{\partial\omega}(\omega D_{\beta,\alpha}^{\mathrm{H}*})\frac{\partial\tilde{E}_{\alpha}^{*}}{\partial t}\right]$$

$$=\frac{\partial}{\partial t}\left\{\tilde{\boldsymbol{E}}^{*}\cdot\frac{\partial}{\partial\omega}[\omega\,\overleftrightarrow{D}^{\mathrm{H}}(\boldsymbol{k},\omega)]\cdot\tilde{\boldsymbol{E}}\right\}, \tag{7.4.11}$$

$$-\tilde{\boldsymbol{E}}^{*}\cdot\left(\frac{\partial}{\partial\boldsymbol{r}}\cdot\frac{\partial}{\partial\boldsymbol{k}}\right)[\omega\,\overleftrightarrow{D}]\cdot\tilde{\boldsymbol{E}}-\tilde{\boldsymbol{E}}\cdot\left(\frac{\partial}{\partial\boldsymbol{r}}\cdot\frac{\partial}{\partial\boldsymbol{k}}\right)[\omega\,\overleftrightarrow{D}^{*}]\cdot\tilde{\boldsymbol{E}}^{*}$$

$$=-\frac{\partial}{\partial\boldsymbol{r}}\cdot\frac{\partial}{\partial\boldsymbol{k}}\left\{\tilde{\boldsymbol{E}}^{*}\cdot[\omega\,\overleftrightarrow{D}^{\mathrm{H}}(\boldsymbol{k},\omega)]\cdot\tilde{\boldsymbol{E}}\right\}, \tag{7.4.12}$$

where in (7.4.11, 12) the antihermitian part is ignored. Combining (7.4.10–12), we obtain an equation of the form

$$\frac{\partial}{\partial t}W(\boldsymbol{r},t)+\nabla\cdot\boldsymbol{\Sigma}(\boldsymbol{r},t)=-\langle\boldsymbol{E}(\boldsymbol{r},t)\cdot\boldsymbol{J}_{s}(\boldsymbol{r},t)\rangle-2\omega\tilde{\boldsymbol{E}}^{*}\cdot\overleftrightarrow{D}^{\mathrm{AH}}\cdot\tilde{\boldsymbol{E}}\,, \tag{7.4.13}$$

where

$$W(\boldsymbol{r},t)=\tilde{\boldsymbol{E}}^{*}\cdot\frac{\partial}{\partial\omega}[\omega\,\overleftrightarrow{D}^{\mathrm{H}}(\boldsymbol{k},\omega)]\cdot\tilde{\boldsymbol{E}}, \tag{7.4.14}$$

$$\boldsymbol{\Sigma}(\boldsymbol{r},t)=-\omega\frac{\partial}{\partial\boldsymbol{k}}[\tilde{\boldsymbol{E}}^{*}\cdot\overleftrightarrow{D}^{\mathrm{H}}(\boldsymbol{k},\omega)\cdot\tilde{\boldsymbol{E}}] \tag{7.4.15}$$

where $\partial/\partial\omega$ and $\partial/\partial\boldsymbol{k}$ operate only on $\omega\,\overleftrightarrow{D}^{\mathrm{H}}(\boldsymbol{k},\omega)$. Equation (7.4.13) is called the wave energy equation, where $W(\boldsymbol{r},t)$ is the wave energy and $\boldsymbol{\Sigma}(\boldsymbol{r},t)$ the wave energy flux. The first term on the right hand side is the energy supply by the source to the wave and the second term stands for the wave energy dissipation, as we shall show shortly.

So far, we have not made any assumption regarding the relation between ω and $\boldsymbol{k}$, except that $\tilde{\boldsymbol{E}}(\boldsymbol{r},t)$ etc. are slowly varying in space and time. We now assume the linear dispersion relation (7.2.32) in the hermitian approximation and choose $\tilde{\boldsymbol{E}}$ and $\tilde{\boldsymbol{E}}^{*}$ to be the right and left eigenvectors of $\overleftrightarrow{D}^{\mathrm{H}}$:

$$\overleftrightarrow{D}^{\mathrm{H}}(\boldsymbol{k},\omega)\cdot\tilde{\boldsymbol{E}}(\boldsymbol{r},t)=\tilde{\boldsymbol{E}}^{*}(\boldsymbol{r},t)\cdot\overleftrightarrow{D}^{\mathrm{H}}(\boldsymbol{k},\omega)=0\,. \tag{7.4.16}$$

As mentioned before, the amplitude of these eigenvectors can change with $\boldsymbol{r}$ and t due to the source excitation and the damping, although their directions are unchanged in a uniform and stationary medium. Then from (7.4.14, 16), the wave energy becomes

$$W(\boldsymbol{r},t)=\omega\tilde{\boldsymbol{E}}^{*}(\boldsymbol{r},t)\cdot\left\{\frac{\partial}{\partial\omega}\,\overleftrightarrow{D}^{\mathrm{H}}(\boldsymbol{k},\omega)\right\}\cdot\tilde{\boldsymbol{E}}(\boldsymbol{r},t)\,. \tag{7.4.17}$$

We now slightly shift the wavenumber from $\boldsymbol{k}$ to $\boldsymbol{k}+d\boldsymbol{k}$. The solution of (7.2.32) and (7.4.16) then changes as $\omega \to \omega + d\omega$ and $\tilde{\boldsymbol{E}} \to \tilde{\boldsymbol{E}} + d\tilde{\boldsymbol{E}}$. Using the relations

$$
\begin{aligned}
0 = \frac{d}{d\boldsymbol{k}}\left\{\tilde{\boldsymbol{E}}^{*}\cdot \overleftrightarrow{D}^{\mathrm{H}}\cdot \tilde{\boldsymbol{E}}\right\} &= \frac{\partial}{\partial \boldsymbol{k}}\left\{\tilde{\boldsymbol{E}}^{*}\cdot \overleftrightarrow{D}^{\mathrm{H}}\cdot \tilde{\boldsymbol{E}}\right\} + \frac{d\omega}{d\boldsymbol{k}}\frac{\partial}{\partial\omega}\left\{\tilde{\boldsymbol{E}}^{*}\cdot \overleftrightarrow{D}^{\mathrm{H}}\cdot \tilde{\boldsymbol{E}}\right\} , \\
\frac{\partial}{\partial k_\alpha}\left\{\tilde{\boldsymbol{E}}^{*}\cdot \overleftrightarrow{D}^{\mathrm{H}}\cdot \tilde{\boldsymbol{E}}\right\} &= \tilde{\boldsymbol{E}}^{*}\cdot \frac{\partial \overleftrightarrow{D}^{\mathrm{H}}}{\partial k_\alpha}\cdot \tilde{\boldsymbol{E}} , \\
\frac{\partial}{\partial \omega}\left\{\tilde{\boldsymbol{E}}^{*}\cdot \overleftrightarrow{D}^{\mathrm{H}}\cdot \tilde{\boldsymbol{E}}\right\} &= \tilde{\boldsymbol{E}}^{*}\cdot \frac{\partial}{\partial \omega}\overleftrightarrow{D}^{\mathrm{H}}\cdot \tilde{\boldsymbol{E}} ,
\end{aligned}
$$

we find

$$
\boldsymbol{\Sigma}(\boldsymbol{r},t) = -\omega\frac{\partial}{\partial \boldsymbol{k}}\left\{\tilde{\boldsymbol{E}}^{*}\cdot \overleftrightarrow{D}^{\mathrm{H}}\cdot \tilde{\boldsymbol{E}}\right\} = \frac{d\omega}{d\boldsymbol{k}}W(\boldsymbol{r},t) , \tag{7.4.18}
$$

which implies that the wave energy is transported with the group velocity of the wave.

The second term on the right hand side of (7.4.13) can be calculated as follows. Suppose the eigenmode equation for the complex eigenfrequency, $\omega - \mathrm{i}\gamma$, be given by

$$
\overleftrightarrow{D}(\omega - \mathrm{i}\gamma)\cdot(\tilde{\boldsymbol{E}} + \Delta\tilde{\boldsymbol{E}}) = (\tilde{\boldsymbol{E}}^{*} + \Delta\tilde{\boldsymbol{E}}\,')\cdot \overleftrightarrow{D}(\omega - \mathrm{i}\gamma) = 0 .
$$

We assume that $|\overleftrightarrow{D}^{\mathrm{AH}}|, \gamma, |\Delta\tilde{\boldsymbol{E}}|, |\Delta\tilde{\boldsymbol{E}}'|$ are all sufficiently small and linearize the equation

$$
(\tilde{\boldsymbol{E}}^{*} + \Delta\tilde{\boldsymbol{E}}\,')\cdot \overleftrightarrow{D}(\omega - \mathrm{i}\gamma)\cdot(\tilde{\boldsymbol{E}} + \Delta\tilde{\boldsymbol{E}}) = 0
$$

with respect to these small quantities. By using (7.4.16), we obtain

$$
-\mathrm{i}\gamma\tilde{\boldsymbol{E}}^{*}\cdot\frac{\partial}{\partial\omega}\overleftrightarrow{D}^{\mathrm{H}}(\boldsymbol{k},\omega)\cdot\tilde{\boldsymbol{E}} + \mathrm{i}\tilde{\boldsymbol{E}}^{*}\cdot\overleftrightarrow{D}^{\mathrm{AH}}(\boldsymbol{k},\omega)\cdot\tilde{\boldsymbol{E}} = 0
$$

or

$$
\gamma = \tilde{\boldsymbol{E}}^{*}\cdot\overleftrightarrow{D}^{\mathrm{AH}}(\boldsymbol{k},\omega)\cdot\tilde{\boldsymbol{E}} \Big/ \left[\tilde{\boldsymbol{E}}^{*}\cdot\frac{\partial}{\partial\omega}\overleftrightarrow{D}^{\mathrm{H}}(\boldsymbol{k},\omega)\cdot\tilde{\boldsymbol{E}}\right] , \tag{7.4.19}
$$

from which we obtain, noting (7.4.17)

$$
-2\omega\tilde{\boldsymbol{E}}^{*}\cdot\overleftrightarrow{D}^{\mathrm{AH}}(\boldsymbol{k},\omega)\cdot\tilde{\boldsymbol{E}} = -2\gamma W(\boldsymbol{r},t) \tag{7.4.20}
$$

which is the rate of damping of the wave energy. The factor of 2 comes from the fact that the wave energy is proportional to square of the wave amplitude $\tilde{\boldsymbol{E}}(\boldsymbol{r},t)$.

Using (7.4.18, 20), we can write the wave energy equation (7.4.13) as:

$$
\frac{\partial}{\partial t}W(\boldsymbol{r},t) + \nabla\cdot\left\{\frac{d\omega}{d\boldsymbol{k}}W(\boldsymbol{r},t)\right\} = -2\gamma W(\boldsymbol{r},t) - \langle \boldsymbol{E}(\boldsymbol{r},t)\cdot\boldsymbol{J}_s(\boldsymbol{r},t)\rangle . \tag{7.4.21}
$$

7.4.2 Wave Momentum

The wave momentum equation can be derived in essentially the same way as the wave energy equation. The momentum supply to the wave from the source charge $\sigma_{es}(\boldsymbol{r},t)$ and the source current $\boldsymbol{J}_s(\boldsymbol{r},t)$ per unit time per unit volume is

$$\frac{d\boldsymbol{p}}{dt} = -\langle \sigma_{es}(\boldsymbol{r},t)\boldsymbol{E}(\boldsymbol{r},t)\rangle - \langle \boldsymbol{J}_s(\boldsymbol{r},t) \times \boldsymbol{B}(\boldsymbol{r},t)\rangle \ . \tag{7.4.22}$$

In addition to (7.4.1) we use the representation

$$\left\{ \begin{array}{c} \sigma_{es}(\boldsymbol{r},t) \\ \boldsymbol{B}(\boldsymbol{r},t) \end{array} \right\} = \left\{ \begin{array}{c} \tilde{\sigma}_{es}(\boldsymbol{r},t) \\ \tilde{\boldsymbol{B}}(\boldsymbol{r},t) \end{array} \right\} \exp[\mathrm{i}(\boldsymbol{k}\cdot\boldsymbol{r} - \omega t)] \tag{7.4.23}$$

where $\tilde{\sigma}_{es}$ and $\tilde{\boldsymbol{B}}$ are assumed to be slowly varying in space and time. From the charge continuity equation (7.1.1) and the Faraday induction law (7.1.2), we can write in the lowest order approximation

$$\begin{aligned} \tilde{\sigma}_{es}(\boldsymbol{r},t) &= \frac{(\boldsymbol{k}+\boldsymbol{q})}{\omega+\alpha}\cdot\tilde{\boldsymbol{J}}_s(\boldsymbol{r},t) \doteq \frac{\boldsymbol{k}}{\omega}\cdot\tilde{\boldsymbol{J}}_s(\boldsymbol{r},t) \\ \tilde{\boldsymbol{B}}(\boldsymbol{r},t) &= \frac{1}{\omega+\alpha}(\boldsymbol{k}+\boldsymbol{q})\times\tilde{\boldsymbol{E}}(\boldsymbol{r},t) \doteq \frac{\boldsymbol{k}}{\omega}\times\tilde{\boldsymbol{E}}(\boldsymbol{r},t) \ . \end{aligned} \tag{7.4.24}$$

Then (7.4.22) becomes

$$\begin{aligned} \frac{d\boldsymbol{p}(\boldsymbol{r},t)}{dt} &\doteq -\frac{1}{\omega}\Big\{\big\langle \boldsymbol{k}\cdot\tilde{\boldsymbol{J}}_s\tilde{\boldsymbol{E}}^*\big\rangle + \big\langle \boldsymbol{k}\cdot\tilde{\boldsymbol{J}}_s^*\tilde{\boldsymbol{E}}\big\rangle \\ &\quad + \big\langle\tilde{\boldsymbol{J}}_s\times(\boldsymbol{k}\times\tilde{\boldsymbol{E}}^*)\big\rangle + \big\langle\tilde{\boldsymbol{J}}_s^*\times(\boldsymbol{k}\times\tilde{\boldsymbol{E}})\big\rangle\Big\} \\ &= -\frac{\boldsymbol{k}}{\omega}\langle \boldsymbol{J}_s(\boldsymbol{r},t)\cdot\boldsymbol{E}(\boldsymbol{r},t)\rangle \ , \end{aligned} \tag{7.4.25}$$

where we used the vectorial relation (3.3.3). The right hand side of (7.4.25) is $\boldsymbol{k}/\omega$ times the rate of wave energy supply by the source current, which has already been calculated. Therefore, by defining the wave momentum $\boldsymbol{p}$ and the wave momentum flux $\overleftrightarrow{\Pi}$ by the relations

$$\boldsymbol{p}(\boldsymbol{r},t) = \frac{\boldsymbol{k}}{\omega}W(\boldsymbol{r},t) \tag{7.4.26}$$

$$\overleftrightarrow{\Pi}(\boldsymbol{r},t) = \frac{\boldsymbol{k}}{\omega}\boldsymbol{\Sigma}(\boldsymbol{r},t) = \frac{d\omega}{d\boldsymbol{k}}\boldsymbol{p}(\boldsymbol{r},t) \tag{7.4.27}$$

we obtain the wave momentum equation in the form

$$\frac{\partial\boldsymbol{p}}{\partial t} + \nabla\cdot\overleftrightarrow{\Pi} = -2\gamma\boldsymbol{p} - \langle\sigma_{es}\boldsymbol{E} + \boldsymbol{J}_s\times\boldsymbol{B}\rangle \ , \tag{7.4.28}$$

where we used (7.4.21, 22, 25).

7.4.3 Action

Noting (7.4.26), we introduce the action $N(\boldsymbol{k},\boldsymbol{r},t)$ given by the relation

$$N(\boldsymbol{k},\boldsymbol{r},t)=\frac{\partial}{\partial\omega}[\tilde{\boldsymbol{E}}^{*}(\boldsymbol{r},t)\cdot\overleftrightarrow{D}^{\mathrm{H}}(\boldsymbol{k},\omega)\cdot\tilde{\boldsymbol{E}}(\boldsymbol{r},t)]\,, \tag{7.4.29}$$

from which we have

$$W(\boldsymbol{r},t)=\omega N(\boldsymbol{k},\boldsymbol{r},t),\qquad \boldsymbol{p}(\boldsymbol{r},t)=\boldsymbol{k}N(\boldsymbol{k},\boldsymbol{r},t)\,. \tag{7.4.30}$$

The action obeys the wave kinetic equation

$$\frac{\partial}{\partial t}N(\boldsymbol{k},\boldsymbol{r},t)+\nabla\cdot[\boldsymbol{v}_{\mathrm{g}}N(\boldsymbol{k},\boldsymbol{r},t)]=-2\gamma N(\boldsymbol{k},\boldsymbol{r},t)+S(\boldsymbol{k},\boldsymbol{r},t) \tag{7.4.31}$$

where $\boldsymbol{v}_{\mathrm{g}}(=\partial\omega/\partial\boldsymbol{k})$ is the group velocity of the wave and $S(\boldsymbol{k},\boldsymbol{r},t)$ is the source given by

$$S(\boldsymbol{k},\boldsymbol{r},t)=-\frac{\langle \boldsymbol{J}_{\mathrm{s}}(\boldsymbol{r},t)\cdot\boldsymbol{E}(\boldsymbol{r},t)\rangle}{\omega}\,. \tag{7.4.32}$$

We can interpret $N(\boldsymbol{k},\boldsymbol{r},t)$ as the "plasmon number" of wavenumber $\boldsymbol{k}$, since $N/\hbar$ corresponds to the plasmon number in the quantum description of a wave.

7.4.4 Non-uniformity Effect

When the unperturbed plasma is spatially non-uniform, we have to take into account the variation of the propagation direction of the wave, that is, the refraction effect. The eigenmode in a non-uniform plasma can no longer be specified by the wavenumber vector since it varies in space. However, as far as the spatial non-uniformity scale length is long compared to the characteristic wavelength, we can, to lowest order, use the geometrical optic approximation. This will be explained in Sect. 12.1. Then we can derive the equation for the action $N(\boldsymbol{k},\boldsymbol{r},t)$ in the form [7.5],

$$\begin{aligned}\frac{\partial}{\partial t}N(\boldsymbol{k},\boldsymbol{r},t)+\frac{\partial\omega}{\partial\boldsymbol{k}}\cdot\frac{\partial}{\partial\boldsymbol{r}}N(\boldsymbol{k},\boldsymbol{r},t)-\frac{\partial\omega}{\partial\boldsymbol{r}}\cdot\frac{\partial}{\partial\boldsymbol{k}}N(\boldsymbol{k},\boldsymbol{r},t)\\ =-2\gamma N(\boldsymbol{k},\boldsymbol{r},t)+S(\boldsymbol{k},\boldsymbol{r},t)\,,\end{aligned} \tag{7.4.33}$$

where the dependence of ω on $\boldsymbol{k}$ and $\boldsymbol{r}$ is determined from the condition that for given ω, $\boldsymbol{k}$ is given by the local dispersion relation (12.1.3), or

$$\boldsymbol{k}=\boldsymbol{k}(\omega,\boldsymbol{r})\,. \tag{7.4.34}$$

The third term on the left hand side of (7.4.33) represents the refraction effect. Since the mathematical derivation of (7.4.33) is somewhat involved, we shall not present it here. Instead, we give a physical argument based on the geometrical optics equations (12.1.6, 7). Using these equations, the left hand side of (7.4.33) is written as

$$\frac{d}{dt}N(\boldsymbol{k},\boldsymbol{r},t) = \left(\frac{\partial}{\partial t} + \frac{d\boldsymbol{r}}{dt}\cdot\frac{\partial}{\partial \boldsymbol{r}} + \frac{d\boldsymbol{k}}{dt}\cdot\frac{\partial}{\partial \boldsymbol{k}}\right) N(\boldsymbol{k},\boldsymbol{r},t) , \tag{7.4.35}$$

which is the rate of change of the plasmon number by the motion of the plasmon. In the absence of the damping and source, the equation reduces to the same form as the Vlasov equation. Namely, (7.4.33) describes conservation of the plasmon number (or conservation of the action) in the absence of the damping and source.

PROBLEMS

7.1. Suppose that in a uniform and stationary plasma a source charge given by

$$\sigma_{es}(\boldsymbol{r},t) = \sigma_{e0}\cos(kx)\delta[y]\delta[z]\delta[t]$$

(σ_{e0} = const) is introduced. Discuss the linear response to this source charge.

7.2. Show that the linear electrostatic response to the source charge produced by an external electric field $\boldsymbol{E}_{ext}(\boldsymbol{r},t)$ is given by

$$\boldsymbol{E}(\boldsymbol{k},\omega) = \frac{\varepsilon_0}{\varepsilon^{\ell}(\boldsymbol{k},\omega)}\frac{\boldsymbol{k}\boldsymbol{k}}{k^2}\cdot\boldsymbol{E}_{ext}(\boldsymbol{k},\omega) .$$

Assume that the plasma is uniform and stationary in the absence of the external field.

7.3. Consider a plasma in the slab model. Discuss the electrostatic response at frequency ω when the source charge is given by

$$\sigma_{es}(x,t) = \sigma_0\frac{d}{dx}\delta[x]\cos\omega t \qquad (\varrho_0 = \text{const}) .$$

Note that since $\varepsilon(-k,-\omega) = \varepsilon^*(k,\omega)$ the dielectric functions in the regions $k > 0$ and $k < 0$ are in different branches and are defined in different Riemann spaces.

7.4. Consider the electromagnetic wave propagating parallel to the magnetic field in the cold plasma model. Estimate the ratios, $\tilde{E}_x/\tilde{E}_y$, $\tilde{B}_x/\tilde{B}_y$ and $\tilde{E}_x/\tilde{B}_y$.

7.5. Discuss the dispersion relation for the left hand polarized wave propagating parallel to the magnetic field in the frequency range, $\omega \lesssim \Omega_i$ by using the cold plasma model. Such a wave is called an *electromagnetic ion cyclotron wave*.

7.6. Calculate the magnetic field produced by the current due to the $\boldsymbol{E}\times\boldsymbol{B}$ drift for the upper hybrid and lower hybrid wave propagating perpendicular to the magnetic field. Estimate the electromagnetic correction to these waves.

7.7. Draw the plasma motion ($\boldsymbol{u}_e$ and $\boldsymbol{u}_i$) in the xy-plane associated with the lower hybrid mode for the case $\omega_{pe}^2 \gg \omega_{ce}^2$.

7.8. Discuss the obliquely propagating Alfven waves. There are two branches, one propagating with the Alfven speed which is called the *fast mode* and another propagating with a slower speed which is called the *slow mode*.

7.9. Show that when the ion contribution is negligible in (7.3.28), we have the solutions

$$\begin{aligned}\omega^2 &\doteq \Omega_e^2 \cos^2\theta &\quad (\omega_{pe}^2 \gg \Omega_e^2)\\ &\doteq \omega_{pe}^2 \cos^2\theta &\quad (\omega_{pe}^2 \ll \Omega_e^2)\,.\end{aligned}$$

The latter (the case of strong magnetic field) is called the *Gould-Trivelpiece mode.*

7.10. Show that the Whistler wave propagating obliquely to the magnetic field in high density plasma ($\omega_{pe}^2 \gg \Omega_e^2$) satisfies the dispersion relation given by

$$\omega = \frac{k^2c^2\omega_{ce}\cos\theta}{k^2c^2 + \omega_{pe}^2}\,.$$

8. Parametric Excitation and Mode Coupling

We have described simple examples of nonlinear wave propagation in Sects.5.6 and 6.6. There we were mainly concerned with a single nonlinear wave mode. In many cases of interest, nonlinear coupling of different linear wave modes can also be important. Such couplings typically occur as a result of periodic modulation of a plasma parameter characterizing the dispersion relation of the wave. The modulation is due to the presence of another wave, therefore it can be considered as a *mode coupling process*. Such a mode coupling often causes an unstable beat wave, that is, a beat between the original wave and the modulation. This instability is called the *parametric instability* [8.1.2]. In this chapter, we first discuss some general properties of parametric instabilities by using model equations and then specifically discuss the case of the excitation of an electron plasma wave and an ion acoustic wave in an isotropic plasma. We further discuss the nonlinear wave-particle interaction for the case of the electron plasma wave and finally discuss the Langmuir wave turbulence with particular attention to the condensation and collapse of the wave.

8.1 Mathieu Equation Model

The simplest example of parametric excitation is the amplification of a pendulum oscillation by a periodic modulation of the string length. The equation describing the pendulum oscillation can be written as

$$\frac{d^2X(t)}{dt^2} + \Omega^2 X(t) = 0 \qquad (8.1.1)$$

where $X(t)$ is the displacement from the equilibrium position and Ω is the frequency which depends on the string length. We denote its periodic modulation by

$$\Omega^2 = \Omega_0^2(1 + 2\varepsilon \cos \omega_0 t) , \qquad (8.1.2)$$

where Ω_0 is the natural frequency of the pendulum, ε stands for the strength of the modulation and ω_0 the modulation frequency. Substitution of (8.1.2) into (8.1.1) gives the Mathieu equation whose periodic solutions are given by the Mathieu functions. For small $|\varepsilon|$ ($|\varepsilon| \ll 1$), the solution of the Mathieu equation becomes unstable when [8.3]

$$\omega_0 \simeq 2\Omega_0/n \quad (n = 1,2,3,\dots) \,. \tag{8.1.3}$$

Physically, when a natural oscillation at frequency Ω_0 couples to the modulation, a response is produced at frequencies $\Omega_0 \pm n\omega_0$. When one of these frequencies is close to $-\Omega_0$, it resonantly couples to another natural oscillation and the natural oscillations resonantly absorb the energy of the modulation. The *resonance condition* $\Omega_0 - n\omega_0 \simeq -\Omega_0$ gives the condition (8.1.3). If the natural oscillation is a damped oscillation, instability occurs when the growth rate exceeds the damping. Therefore there is a clear *threshold* for the instability in the modulation amplitude in addition to the resonance condition. These features are characteristic of parametric excitation. To demonstrate these features more clearly, let us assume $|\varepsilon| \ll 1$ and solve the Mathieu equation by a perturbation method. We use the Fourier representation

$$X(t) = \int \frac{d\omega}{2\pi} \mathrm{e}^{-\mathrm{i}\omega t} \tilde{X}(\omega) \,, \tag{8.1.4}$$

then the Mathieu equation can be written as

$$D(\omega)\tilde{X}(\omega) = -\varepsilon\Omega_0^2[\tilde{X}(\omega - \omega_0) + \tilde{X}(\omega + \omega_0)] \tag{8.1.5}$$

where

$$D(\omega) = \omega^2 - \Omega_0^2 \tag{8.1.6}$$

stands for the linear contribution, the vanishing of which gives the linear dispersion relation. The right hand side of (8.1.5) stands for the mode coupling to the responses at frequencies shifted by $\pm\omega_0$ due to the modulation. We can write similar equations for $X(\omega \pm \omega_0)$, obtaining

$$D(\omega \pm \omega_0)\tilde{X}(\omega \pm \omega_0) = -\varepsilon\Omega_0^2[\tilde{X}(\omega) + \tilde{X}(\omega \pm 2\omega_0)] \,. \tag{8.1.7}$$

The equation now contains the responses at frequencies $\omega \pm 2\omega_0$ in addition to that at the original frequency ω, and therefore we must also solve for $X(\omega \pm 2\omega_0)$. If we continue this procedure, we obtain a hierarchy of equations which is never closed because of the subsequent appearance of new responses at frequencies shifted by integral multiples of ω_0 from the original frequency.

Note that because $|\varepsilon| \ll 1$, (8.1.5) implies that unless ω is close to either Ω_0 or $-\Omega_0$, the response $\tilde{X}(\omega)$ becomes very small (of order ε). The same consideration can be applied to (8.1.7), where the response $\tilde{X}(\omega \pm \omega_0)$ is very small unless $\omega \pm \omega_0$ is close to either Ω_0 or $-\Omega_0$. The obvious approximation which we can immediately think of is to retain only those responses which have frequencies near $\pm\Omega_0$ and neglect all others. This approximation leads to our first example, the case of $n = 1$ in (8.1.3).

8.1.1 The Case of $\omega_0 \simeq 2\Omega_0$

Let ω be close to $+\Omega_0$ in (8.1.5). Then $\omega - \omega_0 \simeq -\Omega_0$ which yields a large response, but $\omega + \omega_0 \simeq 3\Omega_0$ is off-resonant from the natural frequency. We therefore keep only $X(\omega - \omega_0)$ and neglect $X(\omega + \omega_0)$ in (8.1.5). Similarly, in

(8.1.7) we keep only $X(\omega)$ but neglect $X(\omega \pm 2\omega_0)$. In this way, we obtain a closed set of equations for $X(\omega)$ and $X(\omega - \omega_0)$. A nontrivial solution can be obtained by setting the determinant of the coefficients equal to zero, i.e.,

$$D(\omega)D(\omega - \omega_0) = \varepsilon^2 \Omega_0^4 , \tag{8.1.8}$$

which gives the nonlinear dispersion relation to determine ω. Now we recall that we are considering the case $\omega \simeq \Omega_0$ and $\omega - \omega_0 \simeq -\Omega_0$, so that we can make the so-called *resonance approximation* which amounts to

$$\begin{aligned} D(\omega) &= (\omega - \Omega_0)(\omega + \Omega_0) \simeq 2\Omega_0(\omega - \Omega_0) \\ D(\omega - \omega_0) &= (\omega - \omega_0 - \Omega_0)(\omega - \omega_0 + \Omega_0) \simeq -2\Omega_0(\omega - \Omega_0 - \Delta) , \end{aligned} \tag{8.1.9}$$

where we introduced the *frequency mismatch* Δ defined by

$$\Delta = \omega_0 - 2\Omega_0 . \tag{8.1.10}$$

Use of the approximation (8.1.9) reduces (8.1.8) to a quadratic form which can immediately be solved as

$$\omega = \Omega_0 + \frac{1}{2}\left(\Delta \pm \sqrt{\Delta^2 - \varepsilon^2\Omega_0^2}\right) , \tag{8.1.11}$$

In the limit of small $|\varepsilon|$, this equation yields two real solutions $\omega \simeq \Omega_0$ and $\omega \simeq \omega_0 - \Omega_0$, the former being the natural oscillation and the latter the driven oscillation. The modulation yields a coupling of these two oscillations and produces an instability when

$$\varepsilon^2 > \Delta^2/\Omega_0^2 . \tag{8.1.12}$$

For the given mismatch Δ, the right hand side gives the *threshold* intensity for the modulation to cause instability. The threshold becomes zero at exact matching, $\Delta = 0$. The growing solutions have real frequency given by

$$\omega_r = \Omega_0 + \frac{\Delta}{2} = \frac{\omega_0}{2} \tag{8.1.13}$$

which is independent of the original natural frequency indicating *frequency locking*. The growth rate above threshold is given by

$$\gamma = \frac{1}{2}\sqrt{\varepsilon^2\Omega_0^2 - \Delta^2} \leq \frac{|\varepsilon|\Omega_0}{2} , \tag{8.1.14}$$

where the right hand expression gives the *maximum growth rate*, $\gamma_{max} = |\varepsilon|\Omega_0/2$ which is obtained at exact matching $\Delta = 0$.

8.1.2 The Case of $\omega_0 \simeq \Omega_0$

The foregoing approximation of neglecting all the off-resonant responses is no longer useful for the case where $n > 1$. In these cases, we have to retain some nonresonant responses which connect the two resonant responses at $\pm\Omega_0$. Here we consider the case $n = 2$, i.e., $\omega_0 \simeq \Omega_0$; then neither of the functions $X(\omega \pm \omega_0)$ in (8.1.5) becomes resonant. Of these, the $X(\omega - \omega_0)$ mode couples directly to $X(\omega - 2\omega_0) \simeq X(-\Omega_0)$ which is another resonant mode, while the $X(\omega + \omega_0)$ mode couples back to the original resonant mode $X(\omega)$. Although both couplings have to be retained for a precise quantitative argument, here we shall keep only the one which couples to $X(\omega - 2\omega_0)$ and neglect $X(\omega + \omega_0)$. The latter simply yields a frequency shift which dose not contribute to instability (Problem 8.1).

To make the equations symmetric, we choose ω to be close to zero. Then $X(\omega)$ is the mode which connects the two resonant responses $X(\omega \pm \omega_0 \simeq \pm\Omega_0)$ and our approximation amounts to neglecting $X(\omega \pm 2\omega_0)$ in (8.1.7), retaining all terms in (8.1.5). The equations are closed among $X(\omega \pm \omega_0)$ and $X(\omega)$ and we get the dispersion relation in the form

$$1 = \frac{\varepsilon^2 \Omega_0^4}{D(\omega)} \left[\frac{1}{D(\omega + \omega_0)} + \frac{1}{D(\omega - \omega_0)} \right] . \tag{8.1.15}$$

As before, we make the resonance approximation for $D(\omega \pm \omega_0 \simeq \pm\Omega_0)$. Introducing the frequency mismatch δ by

$$\delta = \omega_0 - \Omega_0 , \tag{8.1.16}$$

we have

$$D(\omega \pm \Omega_0) \simeq \pm 2\Omega_0(\omega \pm \delta) . \tag{8.1.17}$$

If we further approximate $D(\omega)$ by $D(0) = -\Omega_0^2$, (8.1.15) again becomes quadratic in ω and can be solved as

$$\omega^2 = \delta(\varepsilon^2 \Omega_0 + \delta) . \tag{8.1.18}$$

This equation has a growing solution when

$$-\varepsilon^2 \Omega_0 < \delta < 0 , \tag{8.1.19}$$

i.e., when the modulation frequency is slightly less than the natural frequency. The threshold modulation intensity for a given frequency mismatch is given by

$$\varepsilon^2 = -\delta / \Omega_0 . \tag{8.1.20}$$

There, however, the growth rate γ also vanishes. For a given ε^2, the *maximum growth rate* is obtained at

$$\delta = -\varepsilon^2 \Omega_0 / 2$$

with

$$\gamma_{\max} = \varepsilon^2 \Omega_0 / 2 . \tag{8.1.21}$$

Note that it is smaller than $\gamma_{\max}$ for the case of $n = 1$ by a factor of ε. This is because we had to invoke a nonresonant response in order to produce the coupling.

An interesting feature of the present instability is that the nonresonant growing mode has zero frequency ($\omega_r = 0$). Therefore, the instability is called the *purely growing mode instability*. It also implies that the accompanying resonant responses $X(\omega \pm \omega_0)$ have frequency exactly equal to the modulation frequency $\pm\omega_0$, again indicating *frequency locking*. We note that all three modes, $X(\omega)$ and $X(\omega \pm \omega_0)$, grow with the same growth rate γ, but their amplitudes are different, that is, $|X(\omega)| \sim \varepsilon|X(\omega \pm \omega_0)|$ as seen from (8.1.5) with $D(\omega) \doteq -\Omega_0^2$.

8.1.3 Effect of Damping

When the natural oscillation is a damped mode, we can use instead of (8.1.1),

$$\frac{d^2}{dt^2}X(t) + 2\Gamma\frac{dX(t)}{dt} + (\Omega^2 + \Gamma^2)X(t) = 0 \,, \tag{8.1.22}$$

which in the absence of modulation yields the two damped natural oscillations $X(t) \simeq \exp(\pm \mathrm{i}\Omega_0 t - \Gamma t)$. Now (8.1.22) can be reduced to (8.1.1) by the transformation

$$\tilde{X}(t) \equiv X(t)\mathrm{e}^{\Gamma t} \,. \tag{8.1.23}$$

The analyses given above can then be used for $\tilde{X}(t)$. The only modification is to reduce the growth rate from γ to $(\gamma - \Gamma)$. As a result, we now get a *finite threshold* for the instability independent of the frequency mismatch, since we need a growth rate γ for $\tilde{X}(t)$ to be greater than Γ. The *minimum threshold* is obtained by setting the maximum growth rate $\gamma_{\max}$ equal to Γ which gives

$$\varepsilon^2 > \frac{4\Gamma^2}{\Omega_0^2} \qquad \text{for} \quad \omega_0 \simeq 2\Omega_0 \tag{8.1.24}$$

$$\varepsilon^2 > \frac{2\Gamma}{\Omega_0} \qquad \text{for} \quad \omega_0 \simeq \Omega_0 \,. \tag{8.1.25}$$

The former is obtained at exact matching $\Delta = 0$, while the latter at $\delta = -\varepsilon^2\Omega_0/2$.

8.2 Coupled-Mode Parametric Excitation

The foregoing example shows that the parametric instability occurs as a result of the coupling of natural oscillations at different frequencies. In the above example, it is assumed that all the excited oscillations have the same natural frequency $|\Omega_0|$. An obvious extension of this analysis is to consider a coupling of oscillations or wave modes which have different natural frequencies, the difference being due either to different wavenumbers or to different branches [8.4]. A common problem in plasma physics is when the modulator, which we call the *pump*, produces a coupling of high frequency waves with a low frequency wave. This is the situation considered in this section.

We first derive a simple dispersion relation which is a generalization of (8.1.15) based on a model set of equations. Then we consider two special cases: that of a spatially uniform pump and that in which the process describes a stimulated scattering of the pump wave. Some terminologies which are often used in the literature to classify various instabilities are briefly summarized.

8.2.1 Dispersion Relation

We consider two branches of normal modes, a low frequency mode represented by $X_L(\boldsymbol{k},\omega)$ and a high frequency mode represented by $X_H(\boldsymbol{k},\omega)$. In the absence of the pump they satisfy the linear dispersion relations

$$D_L(\boldsymbol{k},\omega) \equiv \omega^2 - \omega_L^2(\boldsymbol{k}) = 0 \qquad \text{for} \quad X_L \tag{8.2.1}$$

$$D_H(\boldsymbol{k},\omega) \equiv \omega^2 - \omega_H^2(\boldsymbol{k}) = 0 \qquad \text{for} \quad X_H\,, \tag{8.2.2}$$

where for simplicity we neglected the damping.

The pump, or modulator, which we denote by

$$Z(r,t) = 2Z_0 \cos(\boldsymbol{k}_0 \cdot \boldsymbol{r} - \omega_0 t)\,, \qquad Z_0 = \text{const} \tag{8.2.3}$$

produces a coupling of the two modes. In analogy to the Mathieu equation model, we assume the following form for the coupling:

$$D_L(\boldsymbol{k},\omega)X_L(\boldsymbol{k},\omega) = Z_0[\lambda_+ X_H(\boldsymbol{k}+\boldsymbol{k}_0,\omega+\omega_0) + \lambda_- X_H(\boldsymbol{k}-\boldsymbol{k}_0,\omega-\omega_0)] \tag{8.2.4}$$

$$D_H(\boldsymbol{k}\pm\boldsymbol{k}_0,\omega\pm\omega_0)X_H(\boldsymbol{k}\pm\boldsymbol{k}_0,\omega\pm\omega_0) = Z_0\mu_\pm X_L(\boldsymbol{k},\omega)\,, \tag{8.2.5}$$

where the coupling coefficients $\lambda_\pm$ and $\mu_\pm$ are constants. Solving (8.2.5) for X_H and substituting the result into (8.2.4) yields the nonlinear dispersion relation

$$1 = \frac{Z_0^2}{D_L(\boldsymbol{k},\omega)}\left[\frac{\lambda_+\mu_+}{D_H(\boldsymbol{k}+\boldsymbol{k}_0,\omega+\omega_0)} + \frac{\lambda_-\mu_-}{D_H(\boldsymbol{k}-\boldsymbol{k}_0,\omega-\omega_0)}\right]\,, \tag{8.2.6}$$

which resembles (8.1.15).

As before, we shall restrict ourselves to the case of weak pumping, i.e., $Z_0^2 \ll 1$. Then (8.2.6) can be satisfied only when one of the dispersion functions D_L or D_H, becomes nearly equal to zero. This is not sufficient to cause instability, however. Instability occurs when at least two of the zeroes of the dispersion functions merge, as in the case of the two-stream instability discussed in Sect.5.3. There are two such situations:

(a) $D_L(\boldsymbol{k},\omega) \simeq 0$ and $D_H(\boldsymbol{k}+\boldsymbol{k}_0,\omega+\omega_0) \simeq 0$ or $D_H(\boldsymbol{k}-\boldsymbol{k}_0,\omega-\omega_0) \simeq 0$

(b) $D_H(\boldsymbol{k}+\boldsymbol{k}_0,\omega+\omega_0) \simeq 0$ and $D_H(\boldsymbol{k}-\boldsymbol{k}_0,\omega-\omega_0) \simeq 0$.

From the analogy to the Mathieu equation model, case (a) corresponds to *resonant coupling*, i.e., $n = 1$, and case (b) to *nonresonant coupling*, i.e., $n = 2$. The resonant coupling occurs when the following resonance condition is satisfied:

$$\omega_0(\boldsymbol{k}_0) \simeq \omega_{\rm H}(\boldsymbol{k}_0 \pm \boldsymbol{k}) + \omega_{\rm L}(\mp \boldsymbol{k}) \,. \tag{8.2.7}$$

In either case, $|\omega|$ is assumed to be much less than ω_0 and hence $|\omega \pm \omega_0| \simeq \omega_0 \simeq \omega_{\rm H}$. We can then make the resonance approximation (8.1.9) for $D_{\rm H}$ to obtain

$$\begin{aligned} D_{\rm H}(\boldsymbol{k} \pm \boldsymbol{k}_0, \omega \pm \omega_0) &\simeq \pm 2\omega_0[\omega \pm \omega_0 \mp \omega_{\rm H}(\boldsymbol{k} \pm \boldsymbol{k}_0)] \\ &= \pm 2\omega_0[(\omega - \alpha) \pm \delta] \,, \end{aligned} \tag{8.2.8}$$

where we introduced two parameters

$$\begin{aligned} \alpha &= [\omega_{\rm H}(\boldsymbol{k} + \boldsymbol{k}_0) - \omega_{\rm H}(\boldsymbol{k} - \boldsymbol{k}_0)]/2 \\ \delta &= \omega_0 - [\omega_{\rm H}(\boldsymbol{k} + \boldsymbol{k}_0) + \omega_{\rm H}(\boldsymbol{k} - \boldsymbol{k}_0)]/2 \,. \end{aligned} \tag{8.2.9}$$

Here δ is the mismatch of the pump frequency from the average of the two natural frequencies of the high frequency modes, analogous to the mismatch introduced by (8.1.16), and α is their frequency difference which arises from the finiteness of the pump wavenumber $\boldsymbol{k}_0$. Note that we are choosing $\omega_{\rm H}$ to be positive. Substitution of (8.2.1, 8) into (8.2.6) yields a biquadratic equation for ω.

In many cases of interest, particularly when the growth rate assumes its maximum value, the coupling coefficients $\lambda_+\mu_+$ and $\lambda_-\mu_-$ in (8.2.6) become real and identical [8.5]. For simplicity, we restrict ourselves to such situations and introduce a dimensionless small parameter ε by the relation

$$Z_0^2\lambda_+\mu_+ = Z_0^2\lambda_-\mu_- \equiv \varepsilon^2\omega_0^2\omega_{\rm L}^2(\boldsymbol{k}) \,. \tag{8.2.10}$$

The dispersion relation is then reduced to the simple form

$$(\omega^2 - \omega_{\rm L}^2)[(\omega - \alpha)^2 - \delta^2] = -\varepsilon^2\omega_0\omega_{\rm L}^2\delta \tag{8.2.11}$$

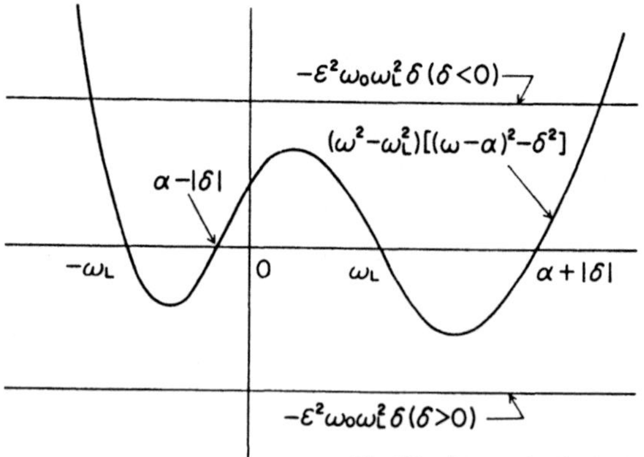

Fig. 8.1. Curves for both sides of (8.2.11) versus ω for the case $\alpha > 0$

Figure 8.1 shows curves for both sides of this equation for the case $\alpha > 0$. It can be clearly seen from this figure that there are four complex solutions or two growing solutions for the case $\delta > 0$ and two complex or one growing solution for the case $\delta < 0$. Setting

$$\omega = \omega_r + i\gamma$$

and taking the imaginary part of (8.2.11), we find that these growing solutions arise in the following frequency ranges: (Problem 8.2)

$$\delta > 0: \quad \omega_r \geq \alpha \quad \text{(Mode I)}$$

$$0 \geq \omega_r \quad \text{(Mode II)} \tag{8.2.12}$$

$$\delta < 0: \quad \alpha \geq \omega_r \geq 0\,. \quad \text{(Mode III)} \tag{8.2.13}$$

8.2.2 Uniform or Dipole Pump: $\boldsymbol{k}_0 = 0$

In this case, α vanishes and δ becomes identical to (8.1.16) with $\omega_H(\boldsymbol{k}) = \Omega_0$. The dispersion relation (8.2.11) becomes quadratic in ω^2 and can hence be solved.

Let us first consider the very weak pump case $\varepsilon^2 \ll 1$. We have two types of solutions similar to those obtained in Sect. 8.1.

(a) Resonant, for $\delta \simeq \omega_L(\boldsymbol{k}) > 0$ with the growth rate

$$\gamma \simeq \sqrt{\gamma_{max}^2 - \Delta^2/4}\,, \tag{8.2.14}$$

where $\Delta(= \delta - \omega_L(\boldsymbol{k}))$ is the frequency mismatch and γ_{max} is the maximum growth rate given by

$$\gamma_{max} = |\varepsilon|\sqrt{\omega_0\omega_L}/2\,, \tag{8.2.15}$$

which can be compared with (8.1.14). The excited modes have frequencies close to $\omega_H(\boldsymbol{k})$ and $\omega_L(\boldsymbol{k})$, having the sum frequency exactly equal to the pump frequency ω_0. This type of instability is called the *resonant decay instability*. Note that in this special case of $\boldsymbol{k}_0 = 0$, Modes I and II are degenerate, i.e., $\omega_r \simeq \pm\omega_L(\boldsymbol{k})$.

(b) Nonresonant, for the case $\delta < 0$ (Mode III) with the same growth characteristics as those derived in Sect. 8.1, provided that the growth rate is much smaller than ω_L. This nonresonant type of instability is called the *oscillating two-stream instability* (OTSI).

In both cases we can derive the *minimum threshold* for instability, if we introduce the phenomenological damping rates, Γ_L and Γ_H, by the relations

$$D_{H(L)}(\boldsymbol{k},\omega) = [\omega - \omega_{H(L)} + i\Gamma_{H(L)}][\omega + \omega_{H(L)} + i\Gamma_{H(L)}]\,. \tag{8.2.16}$$

The results can be written as

$$\gamma_{max} > \sqrt{\Gamma_H\Gamma_L} \quad \text{(resonant)} \tag{8.2.17}$$

$$\gamma_{max} > \Gamma_H \quad \text{(nonresonant)}\,, \tag{8.2.18}$$

which should be compared with (8.1.24, 25).

A new type of solution arises when the pump intensity becomes sufficiently large to satisfy the inequality

$$\varepsilon^2 \gg \omega_{\rm L}/\omega_0 \,. \tag{8.2.19}$$

In this case, the growth rate becomes greater than $\omega_{\rm L}$ and therefore the growth characteristics substantially deviate from those obtained in Sect. 8.1. Maximization of the growth rate with respect to δ then yields (Problem 8.3),

$$\omega = \begin{cases} 1/2(\varepsilon^2\omega_0\omega_{\rm L}^2/4)^{1/3}(\pm\sqrt{5}+{\rm i}\sqrt{3}) & \text{for} \quad \delta > 0 \\ {\rm i}(\varepsilon^2\omega_0\omega_{\rm L}^2/2)^{1/3} & \text{for} \quad \delta < 0 \,. \end{cases} \tag{8.2.20}$$

We note the cube root dependence of the maximum growth rate on the pumping power ε^2 in both cases. These modes are sometimes called the *quasi-reactive modes*.

8.2.3 Stimulated Scattering

We next consider the situation where the pump wave satisfies the same linear dispersion relation as the excited high frequency wave, i.e.,

$$\omega_0 = \omega_{\rm H}(\boldsymbol{k}_0) \,. \tag{8.2.21}$$

This corresponds to the solution where the pump wave is coherently scattered by the low frequency wave. In this case, the mismatch (8.2.9) becomes

$$\delta = \omega_{\rm H}(\boldsymbol{k}_0) - [\omega_{\rm H}(\boldsymbol{k}_0+\boldsymbol{k}) + \omega_{\rm H}(\boldsymbol{k}_0-\boldsymbol{k})]/2 \,, \tag{8.2.22}$$

where we used the relation $\omega_{\rm H}(\boldsymbol{k}) = \omega_{\rm H}(-\boldsymbol{k})$ since we are choosing $\omega_{\rm H}$ to be positive. In the particular case where the group dispersion of the high frequency wave, i.e., $d^2\omega_{\rm H}/dk^2$, calculated along the line connecting the three wavenumber vectors $\boldsymbol{k}_0$ and $\boldsymbol{k}_0 \pm \boldsymbol{k}$ has a fixed sign, the mismatch δ also has a fixed sign:

$$\begin{aligned} \delta > 0 \quad \text{if} \quad & \frac{d^2\omega_{\rm H}}{dk^2} < 0 \\ \delta < 0 \quad \text{if} \quad & \frac{d^2\omega_{\rm H}}{dk^2} > 0 \,. \end{aligned} \tag{8.2.23}$$

Combining this relation with (8.2.12, 13), we find that in this case we can have only Modes I and II or Mode III, depending on the linear dispersion characteristics of the high frequency mode. This situation is quite similar to the modulational instability which can be obtained from the nonlinear Schrödinger equation discussed in Sect. 5.6 [8.6]. The maximum growth rate and the minimum threshold in the presence of damping are obtained under the resonance condition (8.2.7)

$$\omega_{\rm H}(\boldsymbol{k}_0) = \omega_{\rm H}(\boldsymbol{k}_0-\boldsymbol{k}) + \omega_{\rm L}(\boldsymbol{k}) \,. \tag{8.2.24}$$

This is also called the *resonant decay instability*. In an isotropic plasma the condition is satisfied when $|\boldsymbol{k}_0 - \boldsymbol{k}| \simeq \boldsymbol{k}_0$, with the maximum growth rate under the *backscattering* condition

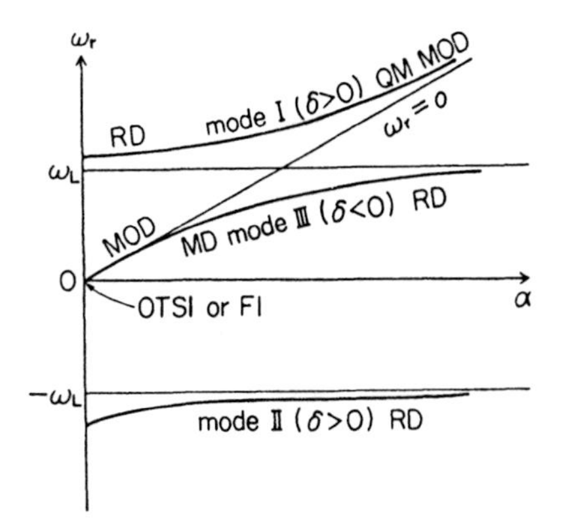

RD : resonant decay instability
QM : quasi-mode instability
MOD: modulational instability
MD : modified decay instability
OTSI: oscillating two-stream instability
FI : filamentation instability

Fig. 8.2. Real frequency of an excited low frequency oscillation as a function of α for weak pumping with the various types of possible parametric instabilities

$$\boldsymbol{k}_0 \simeq \boldsymbol{k} - \boldsymbol{k}_0 \qquad \boldsymbol{k} \simeq 2\boldsymbol{k}_0 \tag{8.2.25}$$

Figure 8.2 shows an example of the real frequency of the excited low frequency oscillation as a function of α for a relatively weak pump case. The term *modified decay instability* is used for the nonresonant region of Mode III instability. When α vanishes due to a particular vectorial relation of $\boldsymbol{k}_0$ and $\boldsymbol{k}$, a purely growing mode can be excited when $\delta < 0$. This corresponds to the *filamentation instability* in which the pump wave is scattered in the forward direction by a low frequency wave propagating nearly perpendicularly to the pump wave splitting the pump wave into many small filaments. Finally, the *modulational instability* occurs when $k \gg k_0$, although in some literature this terminology is used to describe a wider class of instabilities, including OTSI and modified decay.

8.3 Coupled Electron Plasma Wave and Ion Acoustic Wave Parametric Excitation

In this section, we consider two examples of coupled mode parametric excitation for the case where the high frequency wave is an electron plasma wave and the low frequency wave is an ion acoustic wave. In the first example an electromagnetic wave acts as a pump and in the other example we examine stimulated scattering. Throughout this section, we consider a spatially uniform plasma with no magnetic field and with $T_e \gg T_i$.

8.3.1 Electromagnetic Pump

The linear dispersion relations for the electromagnetic wave, the electron plasma wave, and the ion acoustic wave are respectively given by [(7.3.14), (5.2.14) and (5.4.13)]

$$\omega^2 \equiv \omega_0^2 = k_0^2c^2 + \omega_{pe}^2 \qquad \text{(electromagnetic wave)} \tag{8.3.1}$$

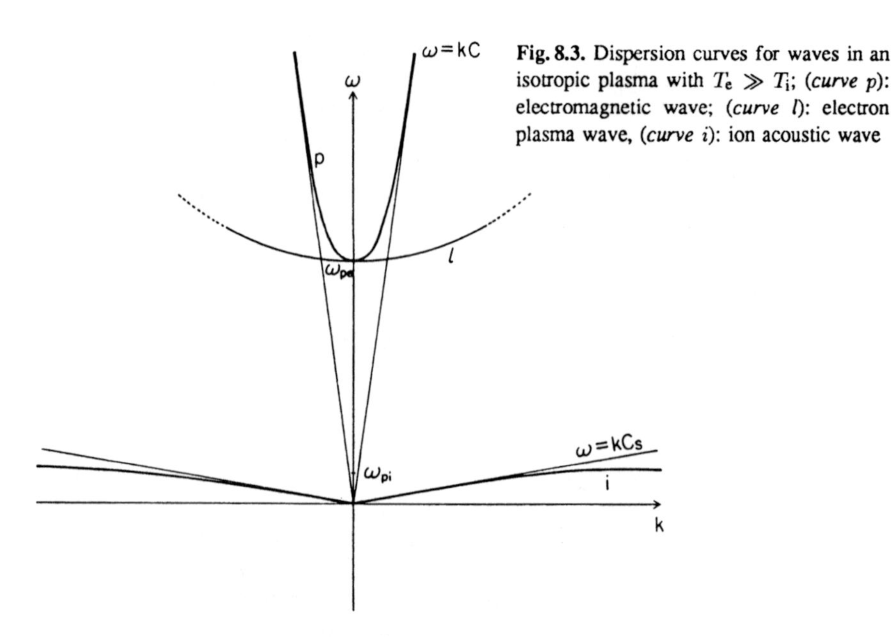

Fig. 8.3. Dispersion curves for waves in an isotropic plasma with $T_e \gg T_i$; (*curve p*): electromagnetic wave; (*curve l*): electron plasma wave, (*curve i*): ion acoustic wave

$$\omega^2 \equiv \omega_H^2 = \omega_{pe}^2(1 + 3k^2\lambda_D^2) \quad \text{(electron plasma wave, } k\lambda_D < 1) \tag{8.3.2}$$

$$\omega^2 \equiv \omega_L^2 = k^2 c_s^2 \quad \text{(ion acoustic wave, } c_s = \sqrt{T_e/m_i}) \tag{8.3.3}$$

They are shown in Fig. 8.3. The condition for the resonant three mode coupling

$$\begin{aligned} \boldsymbol{k}_0 &= \boldsymbol{k}_H + \boldsymbol{k}_L \\ \omega_0 &= \omega_H(\boldsymbol{k}_H) + \omega_L(\boldsymbol{k}_L) , \end{aligned} \tag{8.3.4}$$

where $\boldsymbol{k}_0, \boldsymbol{k}_H, \boldsymbol{k}_L$ are the wavenumbers of the electromagnetic wave, the electron plasma wave and the ion acoustic wave, respectively, can be satisfied only when $\boldsymbol{k}_0 \simeq 0, \boldsymbol{k}_L \simeq -\boldsymbol{k}_H \equiv \boldsymbol{k}$ and $\omega_0 \simeq \omega_{pe}$, as seen from Fig. 8.3. We therefore represent the electromagnetic pump wave by a spatially uniform oscillating electric field (*dipole pump*)

$$\boldsymbol{E} = 2\boldsymbol{E}_0 \cos(\omega_0 t) . \tag{8.3.5}$$

The linear propagation of the long wavelength electron plasma wave can be described by the equation

$$D_H(\boldsymbol{k},\omega)\tilde{E}(\boldsymbol{k},\omega) \equiv [\omega^2 - \omega_H^2(\boldsymbol{k})]\tilde{E}(\boldsymbol{k},\omega) = 0 , \tag{8.3.6}$$

where $\tilde{E}(\boldsymbol{k},\omega)$ is the Fourier representation of the electric field component parallel to the wave propagation $\boldsymbol{k}$. This equation can be derived from the Ampere equation

$$\frac{1}{c^2}\frac{\partial^2 \boldsymbol{E}}{\partial t^2} + \mu_0 \frac{\partial \boldsymbol{J}}{\partial t} = 0$$

or in the Fourier representation

$$\frac{\omega^2}{c^2}\tilde{E}(\boldsymbol{k},\omega) + \mu_0 \mathrm{i}\omega \tilde{J}(\boldsymbol{k},\omega) = 0 , \tag{8.3.7}$$

with

$$\tilde{J}(\boldsymbol{k},\omega) = -en_0\tilde{u}_e(\boldsymbol{k},\omega) \tag{8.3.8}$$

where $\tilde{u}_e(\boldsymbol{k},\omega)$ is the Fourier representation for the electron fluid velocity component along $\boldsymbol{k}$. At the long wavelength limit, we can evaluate $\tilde{u}_e$ by the cold plasma model

$$\tilde{u}_e(\boldsymbol{k},\omega) = \frac{e}{\mathrm{i}\omega m_e}\tilde{E}(\boldsymbol{k},\omega) . \tag{8.3.9}$$

Substituting (8.3.9) into (8.3.8) and then into (8.3.7), we get

$$[\omega^2 - \omega_{pe}^2]\tilde{E}(\boldsymbol{k},\omega) = 0 ; \qquad \omega_{pe}^2 = \frac{n_0 e^2}{m_e \varepsilon_0} , \tag{8.3.10}$$

where we used the relation $\varepsilon_0\mu_0 = c^{-2}$. Allowance for a finite wavenumber correction yields (8.3.6).

The presence of an electromagnetic wave represented by (8.3.5) produces an electron fluid oscillation (polarization drift) which can be written in the Fourier representation

$$\boldsymbol{u}_0 = \frac{e}{\mathrm{i}\omega_0 m_e}\boldsymbol{E}_0 . \tag{8.3.11}$$

Coupling of this oscillation with the oscillation (8.3.9) produces a ponderomotive potential force on the electron fluid. In the Fourier representation, the ponderomotive potential can be written as

$$\begin{aligned} \tilde{\Phi}_p(\boldsymbol{k},\omega) &= \frac{m_e}{2}(\tilde{u}_e u_0)(\boldsymbol{k},\omega) \\ &= \frac{m_e}{2}[\tilde{u}_e(\boldsymbol{k},\omega-\omega_0)u_0 + \tilde{u}_e^*(\boldsymbol{k},\omega+\omega_0)u_0^*] \\ &\simeq \frac{e^2 E_0}{2\omega_0^2 m_e}[\tilde{E}(\boldsymbol{k},\omega-\omega_0) + \tilde{E}(\boldsymbol{k},\omega+\omega_0)] \end{aligned} \tag{8.3.12}$$

where in the last formula we used the approximation $(\omega \pm \omega_0) \simeq \pm\omega_0$ and u_0, E_0 are the components of $\boldsymbol{u}_0, \boldsymbol{E}_0$ parallel to $\boldsymbol{k}, u_0 = \boldsymbol{k}\cdot\boldsymbol{u}_0/k, E_0 = \boldsymbol{k}\cdot\boldsymbol{E}_0/k$. The ponderomotive force causes an electron density perturbation $\delta n_e(\boldsymbol{k},\omega)$, as shown in Sect. 5.5. This density perturbation in turn modifies the plasma current $\boldsymbol{J}$, or the plasma frequency, as

$$\omega_{pe}^2 \to \omega_{pe}^2\left(1 + \frac{\delta n_e}{n_0}\right) . \tag{8.3.13}$$

This modifies the electron plasma wave equation (8.3.10) or (8.3.6). Choosing ω to be low frequency, we have

$$D_H(\boldsymbol{k},\omega \pm \omega_0)\tilde{E}(\boldsymbol{k},\omega \pm \omega_0) = \omega_{pe}^2 \frac{\delta n_e(\boldsymbol{k},\omega)}{n_0} E_0 \tag{8.3.14}$$

We next consider the equation for the density perturbation. In the absence of a pump, the linear wave equation is that for the ion acoustic wave which reads

$$D_L(\boldsymbol{k},\omega)\delta n_e(\boldsymbol{k},\omega) \equiv [\omega^2 - \omega_L^2(\boldsymbol{k})]\delta n_e(\boldsymbol{k},\omega) = 0 \, . \tag{8.3.15}$$

In order to derive this equation, we use (6.2.11) for the electron and the ion:

$$-e\delta n_e(\boldsymbol{k},\omega) = -\varepsilon_0 k^2 \chi_e(\boldsymbol{k},\omega)\tilde{\phi}(\boldsymbol{k},\omega) \tag{8.3.16}$$

$$e\delta n_i(\boldsymbol{k},\omega) = -\varepsilon_0 k^2 \chi_i(\boldsymbol{k},\omega)\tilde{\phi}(\boldsymbol{k},\omega) \, . \tag{8.3.17}$$

Using the local charge neutrality $\delta n_e = \delta n_i$, we eliminate $\tilde{\phi}(\boldsymbol{k},\omega)$ to obtain

$$[\chi_e(\boldsymbol{k},\omega) + \chi_i(\boldsymbol{k},\omega)]\delta n_e(\boldsymbol{k},\omega) = 0 \, . \tag{8.3.18}$$

We use the low frequency approximation (6.2.26) for χ_e and the high frequency approximation (6.2.23) for χ_i:

$$\chi_e(\boldsymbol{k},\omega) \simeq \frac{1}{k^2\lambda_D^2} \tag{8.3.19}$$

$$\chi_i(\boldsymbol{k},\omega) \simeq -\frac{\omega_{pi}^2}{\omega^2}, \qquad \omega_{pi}^2 = \frac{n_0 e^2}{m_i \varepsilon_0} \, . \tag{8.3.20}$$

Use of these approximations in (8.3.18) immediately yields (8.3.15) when we substitute $\omega_L^2(k) = k^2 c_s^2 = k^2 \lambda_D^2 \omega_{pi}^2$.

Now, as mentioned, the pump wave coupled to the electron plasma wave produces a ponderomotive force on the electron. Then on the right hand side of (8.3.16) we have to add the ponderomotive potential (8.3.12) to $-e\tilde{\phi}(k,\omega)$ to obtain

$$\delta n_e(\boldsymbol{k},\omega) = \frac{\varepsilon_0}{e^2} k^2 \chi_e(\boldsymbol{k},\omega) \left[e\tilde{\phi}(\boldsymbol{k},\omega) - \tilde{\Phi}_p(\boldsymbol{k},\omega)\right] \, . \tag{8.3.21}$$

Again, due to charge neutrality we can eliminate $\tilde{\phi}(\boldsymbol{k},\omega)$ from (8.3.17, 21). Noting $\chi_e + \chi_i \simeq D_L/k^2\lambda_D^2\omega^2$, we get

$$\begin{aligned} D_L(\boldsymbol{k},\omega)\delta n_e(\boldsymbol{k},\omega) &= -\omega^2 k^2 \lambda_D^2 \chi_i \frac{\varepsilon_0 k^2}{e^2} \chi_e \tilde{\Phi}_p(\boldsymbol{k},\omega) \\ &= \frac{\varepsilon_0 k^2}{2m_e}\left(\frac{\omega_{pi}}{\omega_0}\right)^2 \boldsymbol{E}_0 \cdot \{\tilde{\boldsymbol{E}}(\boldsymbol{k},\omega - \omega_0) + \tilde{\boldsymbol{E}}(\boldsymbol{k},\omega + \omega_0)\} \end{aligned} \tag{8.3.22}$$

where in the last line we used (8.3.19, 20). Equations (8.3.14, 22) have the same form as (8.2.4, 5), so we can use the results obtained in Sect. 8.2, for the dipole pump case.

First, in this case the coupling coefficient ε^2 in (8.2.10) is given by

$$\varepsilon^2 = \frac{|\boldsymbol{E}_0|^2}{\omega_0^2\omega_L^2(k)} \frac{\varepsilon_0 k^2}{2m_e}\left(\frac{\omega_{pi}}{\omega_0}\right)^2 \frac{\omega_{pe}^2}{n_0} \simeq \frac{1}{2}\left(\frac{eE_0}{m_e\lambda_D\omega_0^2}\right)^2 , \tag{8.3.23}$$

where we used the approximation $\omega_0 \simeq \omega_{pe}$. We note that $|eE_0/m_e\omega_0^2|$ is the electron excursion length due to the pump electric field. Thus the parameter ε is the ratio of the electron excursion length to the Debye length. If we note that $|eE_0/m_e\omega_0|$ is the electron polarization drift velocity $|u_0|$, and $\lambda_D\omega_0 \simeq \lambda_D\omega_{pe}$ is the electron thermal speed $v_{Te} = \sqrt{T_e/m_e}$, we find $\varepsilon = |u_0|/v_{Te}$. Another way of looking at this coefficient is to note that $\omega_0^4 m_e^2 \lambda_D^2/e^2 \simeq \omega_{pe}^2 m_e T_e/e^2 = n_0 T_e/\varepsilon_0$. We then find

$$\varepsilon^2 = \frac{\varepsilon_0 E_0^2}{2}\frac{1}{n_0 T_e}, \tag{8.3.24}$$

which is the ratio of the pump field energy density to the thermal energy density of the plasma. The maximum growth rate and threshold for the instability in the presence of linear damping (Landau damping) of the waves for both the resonant decay instability and the oscillating two-stream instability can be readily derived by using the general formulas (8.2.15, 17). We leave the details as an exercise for the reader.

8.3.2 Stimulated Scattering

Since the electron plasma wave has a positive group dispersion $d^2\omega_H/dk^2 > 0$, the stimulated scattering instability is the Mode III instability discussed in Sect. 8.2. Namely, the electron plasma wave decays into another electron plasma wave with a frequency downshifted by the ion acoustic frequency. The physical mechanism of the nonlinear coupling is the same as the case of the electromagnetic pump. Namely, the density perturbation $\delta n_e(k,\omega)$ produces a coupling of two high frequency electron plasma waves and the ponderomotive force due to coupling of these high frequency wave fields excites the density perturbation. The only difference is the wavenumber matching condition, that is, we now have

$$\boldsymbol{k}_H \simeq -\boldsymbol{k}_0 \qquad \text{and} \qquad \boldsymbol{k}_L \equiv \boldsymbol{k} \simeq 2\boldsymbol{k}_0 \quad \text{(backscattering)}\,. \tag{8.3.25}$$

The electron plasma wave equation corresponding to (8.3.14) then becomes

$$D_H(-\boldsymbol{k}_H,\omega-\omega_0)\tilde{E}(-\boldsymbol{k}_H,\omega-\omega_0) = \omega_{pe}^2\frac{\delta n_e(\boldsymbol{k},\omega)}{n_0}\tilde{E}(-\boldsymbol{k}_0,-\omega_0) \tag{8.3.26}$$

and the ion acoustic wave equation corresponding to (8.3.22) is

$$D_L(\boldsymbol{k}_L,\omega)\delta n_e(k_L,\omega) = \frac{\varepsilon_0 k^2}{2m_e}\left(\frac{\omega_{pi}}{\omega_{pe}}\right)^2 \tilde{E}(\boldsymbol{k}_0,\omega_0)\cdot\tilde{E}(-\boldsymbol{k}_H,\omega-\omega_0)\,. \tag{8.3.27}$$

The coupling coefficient ε^2 is the same as in (8.3.23) or (8.3.24) except that E_0^2 is replace by $|\tilde{E}(\boldsymbol{k}_0,\omega_0)|^2$.

We remark that the same physical mechanisms as discussed above apply to stimulated scattering of an electromagnetic wave. When the scatterer is an electron plasma wave, it is called *stimulated Raman scattering*, while when an ion acoustic wave acts as the scatterer it is called *stimulated Brillouin scattering*.

When the electromagnetic wave is scattered in the forward direction by a static density perturbation, it corresponds to the filamentation instability. In all these cases, the density perturbation causing a coupling of two high frequency waves and the ponderomotive force of these waves exciting the density perturbation are the dominant nonlinear coupling mechanisms and the coupled mode equations can be derived in essentially the same way as above [8.5].

8.4 Nonlinear Wave-Particle Interaction – Nonlinear Landau Damping of an Electron Plasma Wave

In the foregoing example we have considered the case where either of the excited waves is a weakly damped wave. This restriction is not necessary, however. One of the excited waves can be a highly damped mode since one of the modes can be a driven mode, driven by the coupling of another mode and the pump wave. In this case the energy absorbed by the driven mode is immediately absorbed by the plasma particles. Therefore the process can be regarded as a nonlinear wave-particle interaction. In particular, in stimulated scattering the scatterer can be a highly damped mode or the plasma particles that satisfy a resonant condition. In this case we refer to the process as *nonlinear Landau damping* of a high frequency wave. Here we consider the nonlinear Landau damping of the electron plasma wave. This is important when $T_e \simeq T_i$, since then the ion acoustic wave is highly damped by the resonant ions. This process is sometimes called the *induced scattering of the electron plasma wave on ions* and is often the dominant nonlinear mode coupling process in the Langmuir wave turbulence produced, for example, by an electron beam [8.7].

The basic equation for the electron plasma wave is the same as (8.3.26). For a low frequency wave, we can no longer use the approximation given in (8.3.20) since $\chi_i(\boldsymbol{k},\omega)$ has a large imaginary part due to the resonant ions. Eliminating $\tilde{\phi}(\boldsymbol{k},\omega)$ from (8.3.17, 21) and solving for $\delta n_e = \delta n_i$ we find

$$\delta n_e(\boldsymbol{k},\omega) = -\frac{\chi_i(\boldsymbol{k},\omega)\varepsilon_0}{\varepsilon(\boldsymbol{k},\omega)e^2}k^2\chi_e(\boldsymbol{k},\omega)\tilde{\Phi}_p(\boldsymbol{k},\omega)\,, \tag{8.4.1}$$

where

$$\varepsilon(\boldsymbol{k},\omega) = \chi_e(\boldsymbol{k},\omega) + \chi_i(\boldsymbol{k},\omega)\,. \tag{8.4.2}$$

Substituting (8.3.12) into (8.4.1) and retaining only the term proportional to $\tilde{E}(\boldsymbol{k},\omega-\omega_0)$ we obtain

$$\delta n_e(\boldsymbol{k},\omega) = -\frac{\chi_i(\boldsymbol{k},\omega)}{\varepsilon(\boldsymbol{k},\omega)}\frac{\varepsilon_0}{2m_e\omega_0^2\lambda_D^2}\tilde{E}(\boldsymbol{k}_0,\omega_0)\cdot\tilde{E}(-\boldsymbol{k}_H,\omega-\omega_0)\,, \tag{8.4.3}$$

where we used (8.3.19). Substitution of (8.4.3) into (8.3.26) yields the following nonlinear dispersion relation:

$$\left[D_H(-\boldsymbol{k}_H,\omega-\omega_0) + \frac{\varepsilon_0\omega_{pe}^2}{2m_e n_0\omega_0^2\lambda_D^2}|\tilde{E}(\boldsymbol{k}_0,\omega_0)|^2\frac{\chi_i(\boldsymbol{k},\omega)}{\varepsilon(\boldsymbol{k},\omega)}\right] = 0\,, \tag{8.4.4}$$

or using the polarization drift velocity $u_0 = e\tilde{E}(\boldsymbol{k},\omega_0)/m_e\omega_0$,

$$(\omega-\omega_0)^2 - \omega_H^2(\boldsymbol{k}_H) + \frac{\omega_{pe}^2}{2}\frac{u_0^2}{v_{Te}^2}\frac{\chi_i(\boldsymbol{k},\omega)}{\varepsilon(\boldsymbol{k},\omega)} = 0 \, . \tag{8.4.5}$$

We set

$$\omega = \omega_r - i\gamma \tag{8.4.6}$$

and equate the imaginary part of (8.4.5) to zero

$$\gamma = -\frac{\omega_0}{4}\frac{u_0^2}{v_{Te}^2}\frac{1}{|\varepsilon(\boldsymbol{k},\omega)|^2}\frac{1}{k^2\lambda_D^2}\mathrm{Im}\{\chi_i(\boldsymbol{k},\omega)\} \, . \tag{8.4.7}$$

Since for a Maxwellian distribution of ions $\mathrm{Im}\{\chi_i(\boldsymbol{k},\omega)\}$ is positive, γ is negative, implying that the electron plasma wave of wavenumber $\boldsymbol{k}_H$ is unstable. This in turn implies a damping of the pump electron plasma wave. In other words, this process describes a nonlinear Landau damping of the pump electron plasma wave by exciting another electron plasma wave of frequency downshifted by ω_r. Note that $\mathrm{Im}\{\chi_i(\boldsymbol{k},\omega)\}$ comes from the resonant ions which satisfy the resonance condition $\omega = \boldsymbol{k}\cdot\boldsymbol{v}$ where v is the velocity of the resonant ion. Therefore the resonance condition for the nonlinear Landau damping can be written as

$$\omega_0(\boldsymbol{k}_0) - \omega_H(\boldsymbol{k}_H) = \boldsymbol{k}\cdot\boldsymbol{v} = (\boldsymbol{k}_0 - \boldsymbol{k}_H)\cdot\boldsymbol{v} \, . \tag{8.4.8}$$

We can look at this resonance condition as

$$\omega_0 - \boldsymbol{k}_0\cdot\boldsymbol{v} = \omega_H - \boldsymbol{k}_H\cdot\boldsymbol{v} \, , \tag{8.4.9}$$

which implies that the nonlinear wave-particle interaction takes place between the two waves which have the same Doppler shifted frequency as seen from the particle. Physically, the current induced by the incident wave of wavenumber $\boldsymbol{k}_0$ and frequency ω_0 acts as a source for the scattered wave of the same frequency as seen from the particle that carries the induced current.

As a final remark, we note that whereas the resonant decay instability is limited to a local region where the resonance condition is satisfied in a spatially nonuniform plasma, the nonlinear wave-particle interaction is not limited since the resonance condition (8.4.9) is less sensitive to the plasma density.

8.5 Condensation and Collapse of Langmuir Waves

Electron plasma waves or Langmuir waves can be excited by various instabilities, such as the bump-on-tail instability, the parametric instability by electromagnetic pump, etc. When these waves are excited well above the thermal level, nonlinear mode coupling takes place and the excited wave energy is carried to a region outside the unstable region. In the nonlinear mode coupling of the electron plasma waves the wave energy is transferred from a high frequency mode to a low frequency mode, either as in the case of the resonant decay instability via an ion acoustic wave or as in the case of the nonlinear Landau damping by the ions. In other words, the electron plasma wave energy is carried toward the long wavelength region via nonlinear mode coupling process. However, linear Landau damping of the electron plasma wave rapidly decreases as the wavenumber decreases. As a result, the wave energy tends to pile up in the longest wavelength region where no Landau damping takes place. This phenomenon is called the *condensation* of Langmuir wave to $k \to 0$ [8.8]. In the long wavelength region, the electron plasma wave can be described by the fluid model. The nonlinear wave equation in the fluid model has been derived in Sect. 5.6 for the case of one-dimensional propagation (5.6.17). There we have noted the analogy to the quantum-mechanical Schrödinger equation. The nonlinear term acts to localize the wave by trapping in the "potential well" represented by the density perturbation which is negative since it is produced by the ponderomotive force. On the other hand, the second term in (5.6.17) acts as a diffraction term and tends to spread the wave packet. The balance of the nonlinear localization effect and the diffraction or delocalization effect forms an envelope soliton.

This discussion is valid for the one-dimensional propagation. A markedly different situation takes place for the case of a three-dimensional propagation. We denote the electron plasma wave by an oscillating potential as

$$\phi(\boldsymbol{r},t) = \Psi(\boldsymbol{r},t)\mathrm{e}^{-\mathrm{i}\omega_{\mathrm{pe}}t} + \Psi^*(\boldsymbol{r},t)\mathrm{e}^{\mathrm{i}\omega_{\mathrm{pe}}t} , \tag{8.5.1}$$

where $\Psi(\boldsymbol{r},t)$ is a slowly varying function of $\boldsymbol{r}$ and t. It satisfies the three-dimentional version of (5.6.17);

$$\mathrm{i}\frac{\partial}{\partial t}\Psi(\boldsymbol{r},t) + \frac{3}{2}\frac{v_{T\mathrm{e}}^2}{\omega_{\mathrm{pe}}}\nabla^2\Psi(\boldsymbol{r},t) - \frac{\omega_{\mathrm{pe}}}{2}\frac{\delta n_{\mathrm{e}}(\boldsymbol{r},t)}{n_0}\Psi(\boldsymbol{r},t) = 0 . \tag{8.5.2}$$

Similarly, for $\Psi^*(\boldsymbol{r},t)$ we have

$$-\mathrm{i}\frac{\partial}{\partial t}\Psi^*(\boldsymbol{r},t) + \frac{3}{2}\frac{v_{T\mathrm{e}}^2}{\omega_{\mathrm{pe}}}\nabla^2\Psi^*(\boldsymbol{r},t) - \frac{\omega_{\mathrm{pe}}}{2}\frac{\delta n_{\mathrm{e}}(\boldsymbol{r},t)}{n_0}\Psi^*(\boldsymbol{r},t) = 0 . \tag{8.5.3}$$

We multiply (8.5.2) by Ψ^* and (8.5.3) by Ψ and integrate over space, with the boundary condition that $\nabla\Psi$ vanish at $|r| \to \infty$. Equations (8.5.2) and (8.5.3), give, respectively,

$$\begin{aligned}
\mathrm{i}\int d^3\boldsymbol{r}\Psi^*\frac{\partial\Psi}{\partial t} - \frac{3}{2}\frac{v_{T\mathrm{e}}^2}{\omega_{\mathrm{pe}}}\int d^3\boldsymbol{r}|\nabla\Psi|^2 - \frac{\omega_{\mathrm{pe}}}{2}\frac{\delta n_{\mathrm{e}}}{n_0}\int d^3\boldsymbol{r}|\Psi|^2 &= 0 \\
-\mathrm{i}\int d^3\boldsymbol{r}\Psi\frac{\partial\Psi^*}{\partial t} - \frac{3}{2}\frac{v_{T\mathrm{e}}^2}{\omega_{\mathrm{pe}}}\int d^3\boldsymbol{r}|\nabla\Psi|^2 - \frac{\omega_{\mathrm{pe}}}{2}\frac{\delta n_{\mathrm{e}}}{n_0}\int d^3\boldsymbol{r}|\Psi|^2 &= 0 .
\end{aligned}$$

Taking the difference of these two equations yields the following conservation law

$$\int d^3\boldsymbol{r}|\Psi|^2 = \mathrm{const} . \tag{8.5.4}$$

In the long wavelength limit, the frequencies of the electron plasma waves are degenerate, i.e., $\omega_k \simeq \omega_{pe}$. Therefore, the ponderomotive force due to two electron plasma wave fields is almost static, $\Phi_p(k,\omega) \simeq \Phi_p(k,0)$. The density perturbation can be evaluated by the static approximation (5.6.24) to get (5.6.25)

$$\frac{\delta n_e(\boldsymbol{r},t)}{n_0} = -e^2|\Psi(\boldsymbol{r},t)|^2/(T_e+T_i)\,. \tag{8.5.5}$$

Suppose $|\Psi(r,t)|^2$ has a spherically symmetric profile of amplitude $A^2(t)$ and radius $a(t)$. Then from the conservation law of (8.5.4) we have

$$A^2(t)a^3(t) = \text{const}\,. \tag{8.5.6}$$

The nonlinear localization term is proportional to $|\Psi|^2\Psi$ or $A^3(t)$ while the diffraction term is proportional to $A(t)a^{-2}(t)$ which is equal to $A^{7/3}(t)$ according to (8.5.6). As the localization proceeds, the nonlinear term increases as A^3 while the diffraction term grows as $A^{7/3}$. In other words the nonlinear localization is further enhanced and the diffraction term can never catch up to the nonlinear term. Localization proceeds indefinitely until the long wavelength approximation used above breaks down. This nonlinear localization of a three-dimensional electron plasma wave is called the *Langmuir wave collapse* [8.9]. It proceeds until $a(t)$ becomes so small that the Landau damping starts playing an important role. Within the frame work of the fluid theory, the process can be analyzed by the following set of coupled equations:

$$\mathrm{i}\frac{\partial\Psi(\boldsymbol{r},t)}{\partial t} + \frac{3}{2}\frac{v_{Te}^2}{\omega_{pe}}\nabla^2\Psi(\boldsymbol{r},t) - \frac{\omega_{pe}}{2}\frac{\delta n_e(\boldsymbol{r},t)}{n_0}\Psi(\boldsymbol{r},t) = 0$$

$$\left\{\frac{\partial^2}{\partial t^2} - c_s^2\nabla^2\right\}\delta n_e(\boldsymbol{r},t) = \frac{\varepsilon_0}{2m_i}\nabla^2|\nabla\Psi(\boldsymbol{r},t)|^2$$

where the first equation is the same as (8.5.2) while the second equation follows from (8.3.22) and describes the ion acoustic wave equation driven by the ponderomotive force. These set of equations are known as the Zakharov equations. For a more complete analysis, the effects of the Landau damping have to be included, which can be studied only by a numerical analysis. According to numerical simulation, high energy electrons are produced by the Landau damping of the long wavelength waves and these electrons enhance the Landau damping and eventually stop the collapse [8.10–12].

PROBLEMS

8.1. Discuss the solution of the Mathieu equation model with $\omega_0 = \Omega_0$ by retaining $(\omega+\omega_0)$ as well as $(\omega-\omega_0)$ in (8.1.5), but neglecting $(\omega+2\omega_0)$ in the equation for $(\omega+\omega_0)$.

8.2. Setting $\omega = \omega_r + \mathrm{i}\gamma$ in (8.2.11), derive the following relation for the case $\gamma \neq 0$.

$$\omega_r(\omega_r-\alpha)\left[4\gamma^2 + \frac{(\omega_r^2-\gamma^2-\omega_L^2)^2}{\omega_r^2}\right] = \varepsilon^2\omega_0\omega_L^2\delta$$

which proves (8.2.12, 13).

8.3. Consider the case where ω_L^2 is negligible compared to ω^2 in (8.2.11) with $\alpha = 0$ (dipole pump).

i. Show that if $\delta < 0$ a purely growing solution is possible with the maximum growth rate given by (8.2.20) for the case $\delta < 0$ at

$$\delta = -(\varepsilon_0 \omega_0 \omega_L^2/2)^{1/3}.$$

ii. Show that if $\delta > 0$ the growth rate γ is given by

$$\gamma^2 = -\frac{\delta^2}{4} + \frac{\sqrt{\varepsilon_0 \omega_0 \omega_L^2 \delta}}{2}$$

which assumes the maximum value at $\delta = (\varepsilon_0 \omega_0 \omega_L^2/4)^{1/3}$. Derive (8.2.20) for the case $\delta > 0$.

iii. Show that the neglect of ω_L^2 is justifiable under the conditions (8.2.19, 20).

8.4. Consider the parametric excitation of an electron plasma wave and an ion acoustic wave by a dipole electromagnetic pump and derive the threshold intensity and the maximum growth rate above threshold.

8.5. Show that the stimulated scattering of an electron plasma wave has a maximum growth rate under the backscattering condition (8.3.25).

8.6. Let $\Delta N_0, \Delta N_H, \Delta N_L$ be the change of the actions of the pump wave, high frequency excited wave and low frequency excited wave, respectively, due to their parametric coupling. Using the wave energy and momentum conservation relations,

$$\omega_0 \Delta N_0 = \omega_H \Delta N_H + \omega_L \Delta N_L$$
$$\boldsymbol{k}_0 \Delta N_0 = \boldsymbol{k}_H \Delta N_H + \boldsymbol{k}_L \Delta N_L$$

show that the pump energy is distributed to the excited waves by their frequency ratio. This relation is a special case of the Manley-Rowe relation [8.13] for the mode coupling.

8.7. By using the Manley-Rowe relation, show that in the stimulated scattering the wave energy is always transported to the lower frequency side.

Part II

Applications to Fusion Plasmas

9. Plasma Confinement for Fusion

Up to now, the principal efforts of fusion research have been devoted to better confinement of high temperature plasmas. In this chapter, we briefly outline these efforts. First in Sect. 9.1, we present the principle of thermonuclear fusion, with particular attention paid to the basic requirements imposed on a plasma used for the core of the thermonuclear fusion reactor. Two different approaches to achieving the confiniment of such core plasmas – magnetic confinement and inertial confinement – are briefly described in the remaining two sections.

9.1 Principle of Thermonuclear Fusion

9.1.1 D-D Reaction

The primary nuclear reaction that has been considered for a long-lasting energy source is the so-called D-D reaction which takes place when two nuclei of deuterium collide with each other. Deuterium can be found in sea water, its abundancy being about 0.0148% that of hydrogen, and as a fuel resource, this amount can be regarded as almost inexhaustible.

The D-D reaction consists of the following two nuclear reactions:

$$D + D \rightarrow {}^3He + n + 3.27\ \mathrm{MeV} \tag{9.1.1}$$

$$D + D \rightarrow T + H + 4.03\ \mathrm{MeV} \tag{9.1.2}$$

In reaction (9.1.1), an isotope of helium (^{3}He) and a neutron (n) are produced by the collision of two deuteriums (D), while in reaction (9.1.2), a tritium (T) and a proton (H) are produced. The numbers on the right hand sides denote the kinetic energy released by the reactions, which can be calculated as follows: If we denote the mass defect of each particle in the unit of MeV (10^6eV), we have D: 13.1359 MeV, He: 14.9313 MeV, and n: 8.0714 MeV, so that the energy released by reaction (9.1.1) is $2 \times 13.1359 - (14.9313 + 8.0714) = 3.2691 = 3.27$ MeV. For reaction (9.1.2), we can use for T: 14.9500 MeV and for H: 7.289 MeV. The partition of the released energy among the reaction products can be estimated from energy and momentum conservation. We first note that the kinetic energy of the deuteriums before collision is very small compared to the energy released by the reaction. We can therefore ignore the initial kinetic energy and treat the deuteriums as being at rest. Denoting the mass and the speed of the helium and neutron by $m_{\mathrm{He}}, m_{\mathrm{n}}, v_{\mathrm{He}}, v_{\mathrm{n}}$, respectively, we have for reaction (9.1.1)

$$\frac{1}{2}m_{\mathrm{He}}v_{\mathrm{He}}^2 + \frac{1}{2}m_{\mathrm{n}}v_{\mathrm{n}}^2 = 3.27 \text{ MeV}$$

$$m_{\mathrm{He}}v_{\mathrm{He}} = m_{\mathrm{n}}v_{\mathrm{n}}$$

where in the second formula we assumed that He and n fly out in the opposite directions. From these relations, we find

$$E_{\mathrm{He}} \equiv \frac{1}{2}m_{\mathrm{He}}v_{\mathrm{He}}^2 = \frac{3.27}{1 + m_{\mathrm{He}}/m_{\mathrm{n}}} = 0.82 \text{ MeV}$$

$$E_{\mathrm{n}} \equiv \frac{1}{2}m_{\mathrm{n}}v_{\mathrm{n}}^2 = \frac{3.27}{1 + m_{\mathrm{n}}/m_{\mathrm{He}}} = 2.45 \text{ MeV} \tag{9.1.3}$$

We can see from (9.1.3) that the lighter particle acquires more energy than the heavier particle.

In the above estimation, we have ignored the initial kinetic energy of the deuterium. However, since the nuclei are positively charged, they must have enough energy to overcome the Coulomb repulsion between them, in order for them to be able to combine. The required energy can be estimated as follows. Let the charges of the two nuclei be $Z_1 e$ and $Z_2 e$ and let the collision radius be R_0, then the Coulomb potential energy (*Coulomb barrier*) to be overcome is

$$E_{\mathrm{Coul}} = \frac{Z_1 Z_2 e^2}{4\pi\varepsilon_0 R_0} \tag{9.1.4}$$

For deuterium $Z_1 = Z_2 = 1$, we find $E_{\mathrm{Coul}} = 1.43 \times 10^{-9}/R_0$ eV, where we used $e = 1.6 \times 10^{-19}$ C, $\varepsilon_0 = (36\pi)^{-1} \times 10^{-9}$F/m. The collision radius R_0 can be estimated by the radius of the nucleus, $R_0 = 5 \times 10^{-15}$ m. Then we get $E_{\mathrm{Coul}} = 2.86 \times 10^5$ eV = 286 keV. In practice, because of quantum mechanical tunneling, the reaction can take place at lower energy. In any case, this energy is much smaller than that released by the reaction, that is, on the order of a few MeV, justifying the neglect of the initial kinetic energy in the estimation (9.1.3).

We can imagine a simple method to induce the nuclear reaction which would be to accelerate a deuterium beam to an energy exceeding the Coulomb barrier and allow it to bombard a solid deuterium target. This method does not work, however, since most of the beam energy is used simply for ionization and heating of the target material. The probability of the nuclear reaction to occur is therefore too small to get a net energy gain. In order to avoid the loss of the initial kinetic energy to ionization and heating, the target deuterium must be heated. That is, rather than impinging a fast deuterium beam on a solid target, we simply produce a deuterium gas at sufficiently high temperature such that a sufficient amount of collisions overcoming the Coulomb barrier can take place among the deuterium nuclei constituting the gas. In reality, because of the large amount of energy released by the reaction, only a small fraction of the deuterium nuclei must have energy as large as the Coulomb barrier energy in order to acquire a net energy gain. In a system at thermal equilibrium, the energy of the particles is partitioned among them according to the Maxwell-Boltzmann distribution. From this distribution we can evaluate the gas temperature needed in order to have a sufficient fraction of deuterium atoms fast enough to overcome the barrier, as we shall show later in this section. In any case, at such high temperatures the fuel is in a state of a fully ionized plasma. This method of inducing nuclear reactions by thermally heating the reactants is called thermonuclear fusion and is currently considered as the most promising method of generating a fusion reactor.

9.1.2 D-T Reaction

Current nuclear fusion research is focused on the D-T thermonuclear fusion reaction

$$\mathrm{D} + \mathrm{T} \rightarrow {}^4\mathrm{He}(3.5\ \mathrm{MeV}) + \mathrm{n}(14.1\ \mathrm{MeV}) \tag{9.1.5}$$

Reaction (9.1.5) can occur in a high temperature deuterium-tritium plasma. Most of the energy released by the reaction is converted to the kinetic energy of the neutron. Since the neutron is not confined by a magnetic field, it hits the vessel wall immediately after reaction. The neutron kinetic energy is converted to heat in the *blanket* of a fusion reactor shown in Fig. 9.1. The heat is taken away from the blanket by a coolant and is used to run an electric generator. If we add ^{6}Li inside the blanket, then tritium can be produced by reaction

$$\mathrm{n} + {}^6\mathrm{Li} \rightarrow {}^4\mathrm{He}(2.1\ \mathrm{MeV}) + \mathrm{T}(2.7\ \mathrm{MeV}) \tag{9.1.6}$$

and then used as the fuel. Another reaction product is the alpha particle ^{4}He carrying 3.5 MeV which can be confined inside the vacuum vessel by electromagnetic fields, and can be used to heat the fuel plasma (*alpha-particle heating*). If the alpha-particle heating works correctly, the nuclear reaction persists without additional heating of the fuel. All we need is to continue to supply the cold

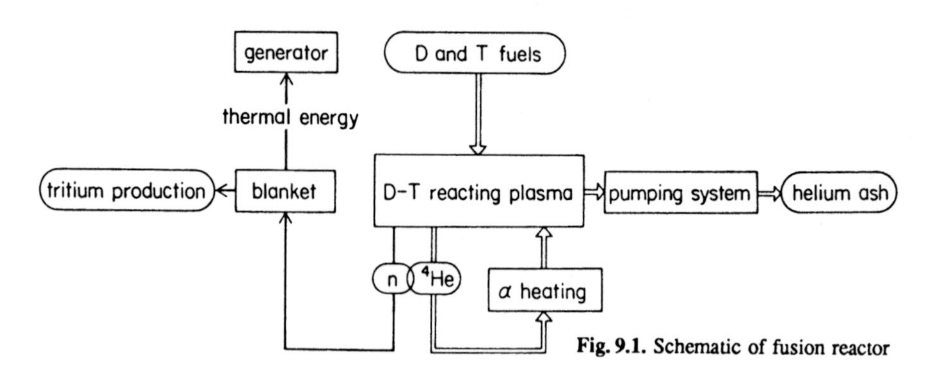

Fig. 9.1. Schematic of fusion reactor

deuterium-tritium fuel. However, those alpha particles which are cooled after giving up their energy to heat the fuel plasma have to be pumped out.

One problem of the D-T reaction is that the tritium is radioactive and requires careful handling. Another problem is the high energy neutron product. With an energy of 14.1 MeV, its irradiation of the wall and structural materials can result in their induced radioactivity. Therefore, designing the composition of the building materials is nontrivial. In reality, some amount of radioactivity is unavoidable and must be taken into consideration when designing a reactor.

9.1.3 Ignition Condition

Let us estimate the minimum temperature needed to induce the self-sustaining alpha-particle heating [9.1]. This is simply a matter of calculating the total power produced by the reactions and subtracting power loss.

The number of reactions per unit time and unit volume is $R_{12} = n_1 n_2 \sigma v$, were n_1 and n_2 are the number densities of the reacting particles 1 and 2, σ is the reaction cross section at a relative particle speed v. If we assume that both types of particles are at thermal equilibrium at the same temperature, then they have a Maxwell velocity distribution and we can get an average value for the rate R_{12} by integrating it over this distribution:

$$\langle R_{12} \rangle = n_1 n_2 \langle \sigma v \rangle = \frac{n_1 n_2 \int \sigma v f_M(\boldsymbol{v}) d^3\boldsymbol{v}}{\int f_M(\boldsymbol{v}) d^3\boldsymbol{v}} , \tag{9.1.7}$$

where $f_M(\boldsymbol{v})$ is the Maxwellian distribution at temperature T and reduced mass $m_1 m_2/(m_1+m_2)$. By multiplying $\langle R_{12} \rangle$ by the energy released in a single reaction event W, we get the reaction power density P_r:

$$P_r = \langle R_{12} \rangle W \tag{9.1.8}$$

In case of the D-T reaction, the energy released into the neutron is effectively lost to the walls, thus, the only available remaining energy is that of the alpha-particles, i.e., $W = 3.5$ MeV.

We next consider the energy loss from the confined plasma. When the fuel plasma is well confined, the principal energy loss arises from Bremsstrahlung due to the perturbation of electron orbits by ions. The power density of this loss is given by [9.1]

$$P_b = 3.34 \times 10^{-15} n_e \sum_s (n_s Z_s^2) T_e^{1/2} (\text{keV/cm}^3\text{s}) \tag{9.1.9}$$

where the temperature T_e is measured in units of keV, $\sum_s$ stands for the sum over all ion species s, and Z_s is the charge number of the sth ion species.

If P_r exceeds P_b then we can gain energy from the fusion reactor. Figure 9.2 shows the temperature dependence of P_r and P_b for the D-T reaction where $n_D = n_T = n_e/2 = 5 \times 10^{20}\ \text{m}^{-3} (Z = 1)$. The temperature at which $P_r = P_b$ is

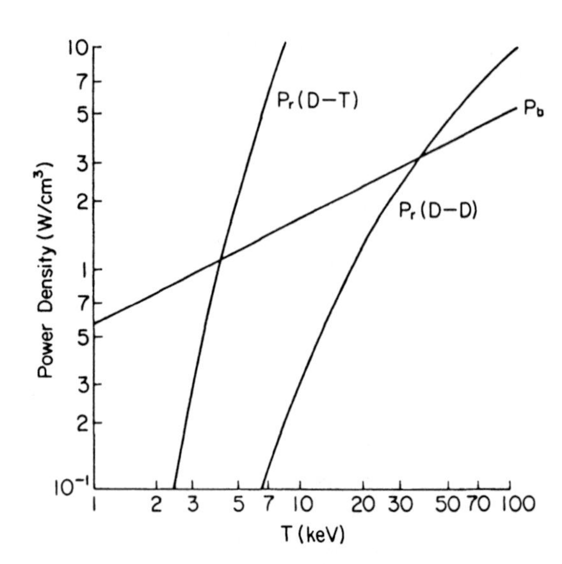

Fig. 9.2. Reaction and Bremsstrahlung power densities P_r and P_b versus temperature for $n_D = n_T = n_e/2 = 5 \times 10^{20} \mathrm{m}^{-3}$ and ionic charge $Z = 1$

called the *ignition temperature*. For the D-T reaction, it's at about 4 keV. We have also shown the reaction power density P_r for the D-D reaction where we now use 8.3 MeV for W, the sum of the energy released to charged particles by the D-D reactions (9.1.1, 2) and by the secondary D-T reaction that occurs with the tritium produced by (9.1.2). As seen in the figure, the ignition temperature for the D-D reaction is about 40 keV. Thus, the ignition condition for the D-T reaction is much more feasible than that for the D-D reaction.

9.1.4 Lawson Criterion

In the above argument, we have tacitly assumed that the fuel plasma is well confined. In reality, the confinement of such a high temperature plasma is extremely difficult and the plasma energy leaks out by various means, such as heat conduction, particle diffusion, radiation emission, etc. Thus we are forced to extract the excess energy within the confinement time of the plasma energy. Suppose a plasma of temperature T and density $n_e = n_D + n_T$ is maintained during a time τ and the fusion reaction persists during that time. If we denote the plasma volume by V, the energy released by the nuclear reaction during the time τ is given by

$$E_{fus} = \tau P_{fus} V \,, \qquad (9.1.10)$$

where P_{fus} is the fusion power density including the kinetic energy released into the neutron as well as to the alpha particle. We assume that some fraction of E_{fus} given by ηE_{fus} ($\eta < 1$) is available to heat the plasma. The energy lost during the time τ is given by

$$E_{lost} = (\tau P_b + 3n_e T)V \,, \qquad (9.1.11)$$

where the term $3n_e TV$ is the total kinetic energy of the plasma particles,

$(3/2)n_eTV + (3/2)(n_D + n_T)TV$. We assume that a fraction of $\eta' < 1$ of E_{lost} can still be collected and used to heat the plasma. Then the condition for heating the plasma is given by

$$\eta E_{fus} > (1 - \eta')E_{lost} . \quad (9.1.12)$$

This condition is called the *Lawson criterion*. Choosing $\eta = \eta' = 1/3$, we can rewrite (9.1.12) as

$$\frac{P_{fus}/3n_e^2T}{P_b/3n_e^2T + 1/n_e\tau} > 2 . \quad (9.1.13)$$

For $n_D = n_T = n_e/2$, we have $P_{fus} \propto n_e^2$, $P_b \propto n_e^2$ as seen from (9.1.8, 9). Therefore only dependence on the density comes from the term $1/n_e\tau$ in (9.1.13). The values of $n_e\tau$ and T satisfying (9.1.13) are shown in the *Lawson diagram* in Fig. 9.3. From the figure, we see that the Lawson criterion for the D-T reaction requires $n_e\tau > 5 \times 10^{19}$ sm^{-3} which occurs at $T = 20$ keV. To satisfy the Lawson criterion we can choose a lower temperature but then require higher value of $n_e\tau$. In any case, we must make a trade-off between the temperature and the product $n_e\tau$. The ratio E_{fus}/E_{lost} is called the *Q-value*

$$Q = \frac{P_{fus}}{P_b + 3n_eT/\tau} , \quad (9.1.14)$$

which gives a measure of the "quality" of the fuel plasma.

The problems for developing a fusion reactor can thus be divided into i) how to produce and confine a high temperature plasma (*plasma confinement*) and ii) how to utilize the energy released by the fusion reaction (*reactor technology*). The aim of plasma confinement is to achieve an $n\tau$ value which satisfies the Lawson criterion for the plasma above the ignition temperature. For the D-T reaction, the requirements are $T = 10$ keV and $n\tau > 10^{20}$ m^{-3} s. There are two

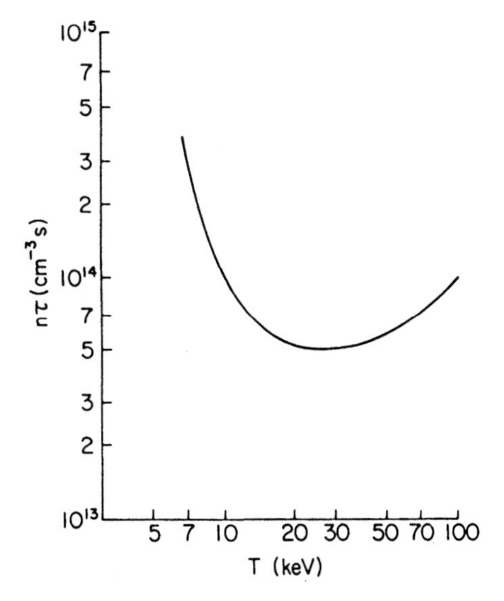

Fig. 9.3. Lawson diagram for the D-T reaction

fundamentally different approaches to achieve these conditions. One is to confine a relatively low density plasma ($n \sim 10^{20}\ \mathrm{m}^{-3}$) for a time on the order of $\tau \sim 1$ s by using a magnetic field (*magnetic confinement*). The upper bound of the plasma density is limited by the technological limit of the magnetic field strength. The magnetic pressure $B_0^2/2\mu_0$ has to be greater than the plasma pressure nT, and since $T = 10$ keV, n has to be less than $10^{20} - 10^{21}\ \mathrm{m}^{-3}$ for $B_0 \leq 10\ T$. The other is to compress the plasma to an ultra-high density on the order of $n \sim 10^{31}\ \mathrm{m}^{-3}$ and let the Lawson criterion be satisfied within the time that the plasma blows apart (*inertial confinement*). In the following two sections, we shall briefly outline the principles of these two approaches.

9.2 Magnetic Confinement and Heating of Plasmas

Various magnetic field configurations have been proposed and tested to confine a plasma. They can be roughly divided into two classes. i) *open-ended confinement* by a straight arrangement of the magnetic coils and ii) *toroidal confinement* by toroidally arranged magnetic coils.

9.2.1 Open-End Confinement

The open-ended confinement scheme makes use of the magnetic mirror traps (Sect.3.2) [9.2–4]. A simple magnetic mirror configuration is unable to confine a plasma, however, because of unfavorable magnetic field curvature. When a small perturbation of the plasma-vacuum boundary occurs along the magnetic field line, the curvature drift of the particles produces a charge separation which yields an electric field as shown in Fig.9.4. The resultant $\boldsymbol{E} \times \boldsymbol{B}$ drift occurs in the direction that enhances the perturbation of the boundary. This instability is called the *flute instability* or the *interchange instability* (Sect.10.3). Physically, this is due to the fact that in a simple mirror the magnetic pressure is higher in the plasma region than in the vacuum region. Thus in the direction perpendicular to the magnetic field, the magnetic field pressure tends to expel the plasma. This instability can be suppressed, however, by introducing additional magnetic

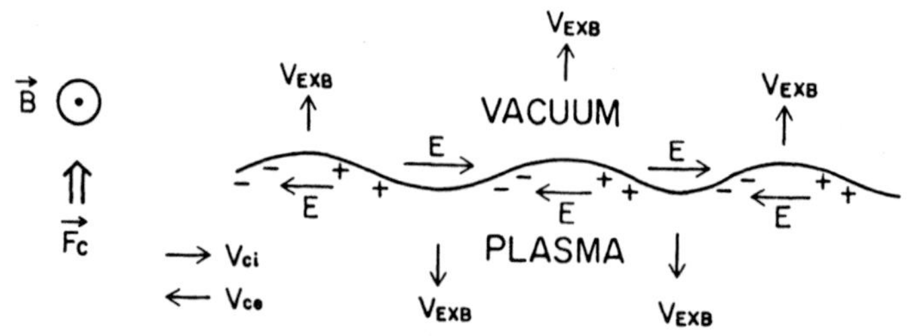

Fig. 9.4. Mechanism of flute instability; v_{ci} and v_{ce} are the curvature drift velocity of an ion and an electron in the presence of a curvature force $\boldsymbol{F}_c$

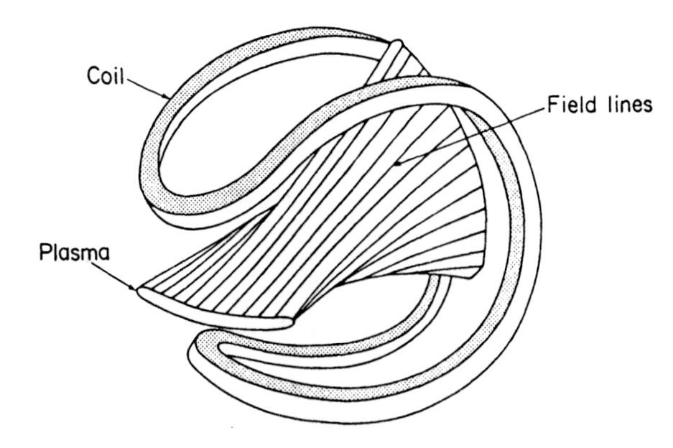

Fig. 9.5. Minimum B magnetic field configuration in a baseball seam coil

coils such that the magnetic pressure becomes stronger in the vacuum region than in the plasma region. An analysis shows that a current flowing along a coil in the shape of a baseball seam can produce a magnetic field configuration with a minimum magnetic field strength at the center (*minimum B configuration*, Fig. 9.5).

Another serious problem of the open-ended confinement is *the end loss* of particles. As shown in Sect. 3.2, the magnetic mirror can trap only those particles whose pitch angles are greater than θ_c given by (3.2.16). The particles with small pitch angles escape through the mirror throat. As a result, the velocity distribution of the particles inside the mirror significantly deviates from Maxwellian (loss-cone distribution, Fig. 9.6). The plasma becomes unstable against waves with ion cyclotron range frequency (ICRF) and the plasma particles diffuse rapidly to the region of small pitch angles in the velocity space and then escape from the end. To prevent this end loss of the particles, a device was invented to introduce an electrostatic potential in both ends of the mirror field. The idea is to create such an electrostatic potential by using high density, high temperature plasmas on the ends of the central cell region where the main plasma is confined (*plug cells*). Between the plug cells and central cells are minimum B configuration fields (*anchor cells*) to suppress the flute instability. This configuration is called a *tandem mirror* and is illustrated in Fig. 9.7 and Fig. 14.3 [9.2–4].

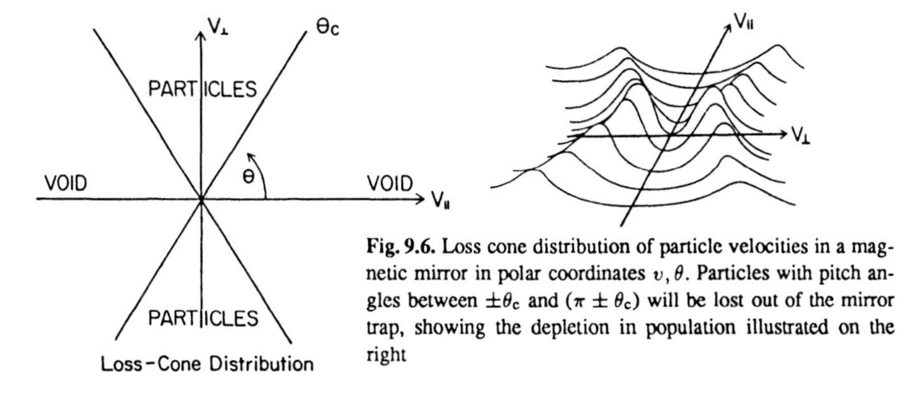

Fig. 9.6. Loss cone distribution of particle velocities in a magnetic mirror in polar coordinates v, θ. Particles with pitch angles between $\pm\theta_c$ and $(\pi \pm \theta_c)$ will be lost out of the mirror trap, showing the depletion in population illustrated on the right

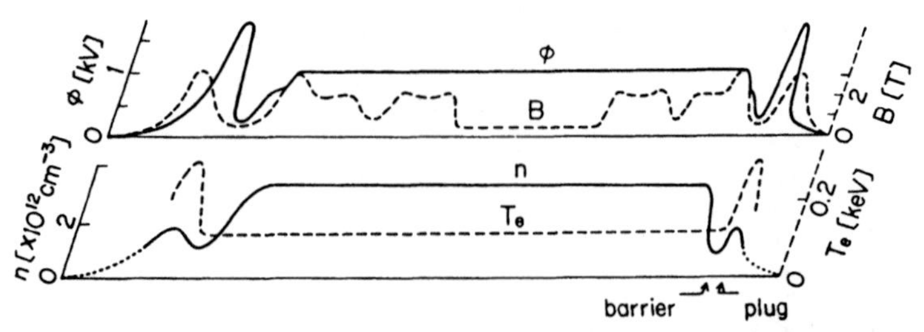

Fig. 9.7. Distribution of plasma parameters in a tandem mirror configuration: Φ – electrostatic potential; B – magnetic field; n – particle density; T_e – electron temperature

Other types of open-ended confinement devices such as the theta pinch, field reversal configuration (FRC), RF end plugging, among others, have been developed.

9.2.2 Toroidal Confinement

In toroidal confinement, the magnetic coils are arranged such that they produce a toroidal field. However, a simple torus consisting of closed magnetic field lines is unable to confine a plasma (Sect. 10.2). The curvature drift and the gradient B drift of the particles produce the charge separation shown in Fig. 9.8a. The electric field due to the charge separation drives an outgoing plasma flow ($\boldsymbol{E} \times \boldsymbol{B}$ drift) across the toroidal field (Fig. 9.8b). The process is the same as the flute instability discussed above and can be suppressed by introducing a poloidal magnetic field inside the cross section of the torus (Fig. 9.9a). The net magnetic field line (composed of toroidal and poloidal components) is helically twisted. If we plot the cross points of a given field line on a poloidal cross section for seven turns around the torus, they move as shown in Fig. 9.9b. The average rotation angle on the plane divided by 2π is called the *rotational transform*. The particles moving along the field lines cancel or short-circuit the charge separation by the Debye shielding, so that the electric field does not grow, whereby the flute instability is suppressed.

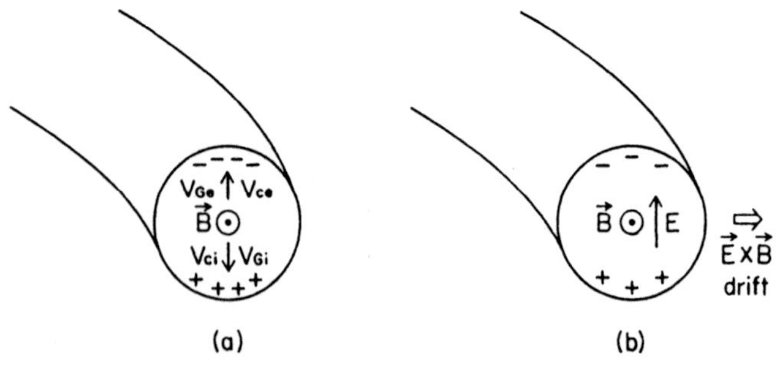

Fig. 9.8. (a) charge separation produced by curvature drift v_c and the gradient B drift v_G in a simple torus; (b) electric field due to the charge separation and direction of $\boldsymbol{E} \times \boldsymbol{B}$ drift

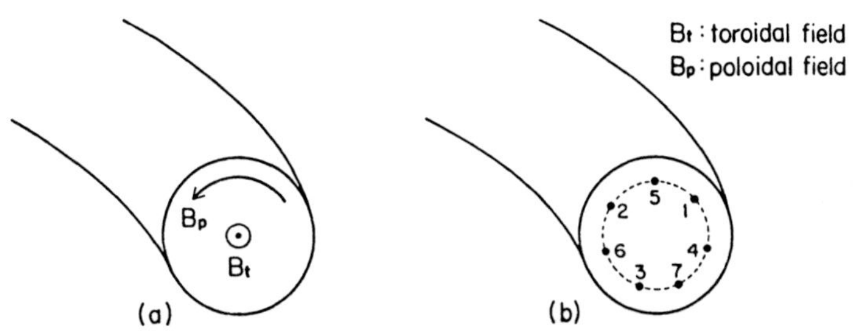

Fig. 9.9. (a) superposition of poloidal B_p and toroidal B_t magnetic fields; (b) points of intersection between magnetic field lines and a poloidal plane (labels show numbers of rotation around the torus along the magnetic field line)

9.2.3 Tokamak

The most advanced toroidal confinement system is the tokamak [9.5, 6], in which the poloidal field component of the helical magnetic field, or the rotational transform, is produced by a plasma current flowing along the toroidal field (*toroidal current*, Fig. 9.10, Refs. 9.4, 5). The toroidal current is driven in pulses by magnetic induction using the electric transformer method (Fig. 9.10). Initially the toroidal current acts as a discharge current and heats the plasma by *ohmic heating*, and is therefore called the *ohmic discharge*. Ohmic heating can heat a plasma with a density $n \sim 10^{19}$ m^{-3} up to about 1–2 keV. In order to heat the plasma up to the ignition temperature of $\sim$ 10 keV, however, additional heating power must be applied. One method is *neutral beam injection* (NBI) across the magnetic field.

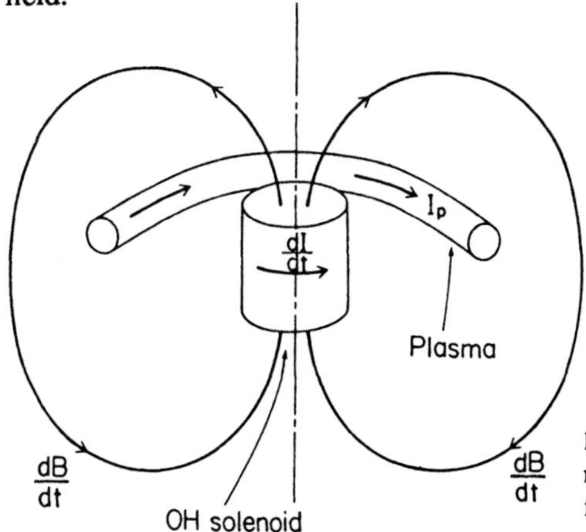

Fig. 9.10. Toroidal current I_p driven by magnetic induction $dB/dt \neq 0$ using electric transformer method

In NBI a proton or charged deuterium beam is passed through a neutralizing electron cloud. Although the neutralization efficiency is relatively low ($\sim$ 10%) for high energy beams ($E > 100$ keV), once the neutral beam is injected inside the plasma, the ionization and heating take place by collisional relaxation. The effectiveness of NBI heating has already been established [9.7, 8].

Another heating method is to apply high power microwaves to the plasma. These microwave techniques are also well established methods [9.9], but in order to understand them the physics of the wave-plasma interaction must be known. This is the primary topic of Chap. 6 and Chap. 12. Whereas NBI is also useful for fuel injection, microwave heating can be used for current control, whence the stability control (Sect. 12.3).

A great advantage of the tokamak confinement scheme is the axisymmetry of the configuration. Apart from small ripple fields due to the discrete coil arrangement, the system can be assumed to be uniform in the toroidal direction. Due to this axisymmetry, the plasma confinement theory based on the plasma physics has been well established. It is also able to confine energetic particles such as those produced by additional heating when toroidal plasma current is sufficiently large. A possible disadvantage is the pulsed operation as the toroidal current is produced by magnetic induction. Every time the current is terminated, the plasma hits the wall due to the lack of magnetohydrodynamic (MHD) equilibrium (Sect. 10.2) and wall loading (heat and electromagnetic stress) becomes severe. In order to reduce this difficulty, a method has been invented to noninductively drive a steady current by microwaves (*current drive*). The usefulness of current drive has already been established for relatively low density plasmas.

A more essential difficulty arises from the presence of the toroidal current itself. Current driven instabilities often destroy the plasma confinement (*disruptive instabilities*). This type of instability is unavoidable in some parameter regime as long as the confining poloidal field is produced by the plasma current. In order to control the plasma stability, elaborate feedback control is needed. Thus the entire coil configuration becomes quite complex.

9.2.4 Reversed Field Pinch and Spheromak

There are several other types of toroidal axisymmetric plasma confinement devices. The plasma current in tokamaks is limited by the MHD kink instability (Sect. 10.3) called the Kruskal-Shafranov limit $q(a) > 1$, where $q(a)$ is a safety factor which measures the stability of the plasma current (Sect. 10.2). *Reversed field pinch* (RFP) can overcome this limitation keeping $q(a) < 1$. In this configuration a ratio of the plasma pressure to the magnetic pressure $\beta \sim 10\%$, equivalent to stable MHD conditions, is expected theoretically and this was already confirmed experimentally [9.10, 11]. Reversed Field Pinch is produced by applying a high magnetic shear (Sect. 5.5 and Sect. 10.3) near the edge region of the plasma which suppresses the MHD instability. This high shear configuration reverses the toroidal magnetic field in the edge region (Fig. 9.11). Recently, it was suggested that the RFP configuration emerges from a turbulent state as a self-organization mechanism. The resultant force-free equilibrium is called a *Taylor state*. Strictly speaking, the Taylor state confines only $\beta = 0$ plasmas; however, this contradicts the experimental result of $\beta \sim 10\%$. The theoretical explanation to solve this discrepancy is a topic of current study.

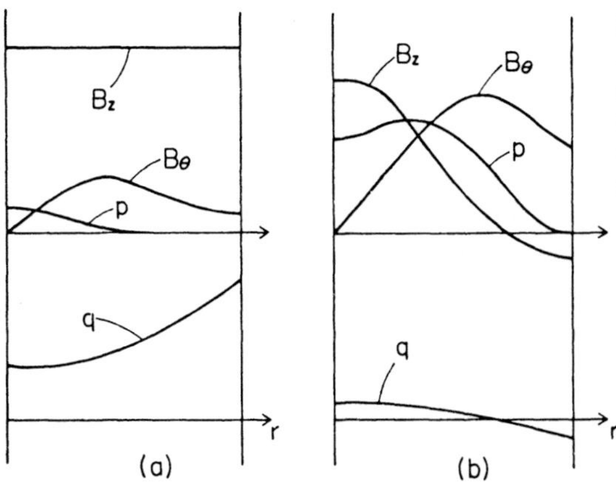

Fig. 9.11. (a) Toroidal B_z and poloidal B_θ magnetic fields, pressure P, and safety factor q as a function of radius, calculated assuming a cylindrically symmetric plasma in a tokamak; (b) reversed field pinch

The advantage of not having a system restricted by the Kruskal-Shafranov limitation is that ignition is possible by ohmic heating, if the energy confinement is good. In all experiments thus far, the confinement time has scaled with the plasma current but the absolute value is lower than that in tokamaks of similar size. The mechanism of this anomalous transport is also a current theoretical problem in research on RFP.

Spheromaks are advanced toroidal plasma confinement devices [9.12]. The magnetic field configuration is characterized by extremely low aspect ratios $R/a \sim 1$ and no external toroidal fields (Fig. 9.12). The inside region has a

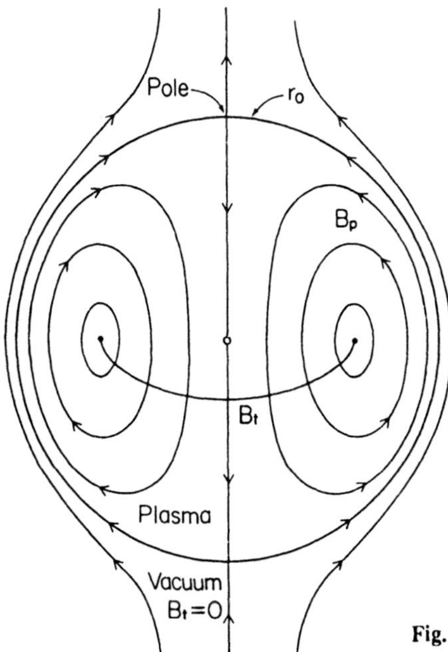

Fig. 9.12. Magnetic fields B_p and B_t in a spheromak configuration; r_0 is a separatrix, and beyond r_0, $B_t = 0$

safety factor less than unity, similar to RFP; however, $q(a) = 0$ at the plasma boundary. The spheromak configuration was experimentally produced without toroidal field coils. This means that spheromaks are topologically identical to open confinement systems. From the point of view of reactor design, poloidal coils linked with toroidal coils, as usually seen in tokamaks and RFP devices, are undesireable for maintenance and accessibility of a reactor. Thus, spheromaks may resolve this difficulty. Spheromak experiments are currently in progress and plasmas of $T_e = 100 - 150$ eV have been stably confined. Further investigations into plasma stability and transport are also under way.

9.2.5 Helical System (Stellarator/Heliotron)

As discussed in Sect. 9.2.2, to confine a plasma in a closed system with a toroidal magnetic field, it is necessary to provide a helical twist to the magnetic field lines, called a rotational transform, as they pass around the major axis of the torus. By adjustment of the transform, nested, closed magnetic surfaces can be formed. In tokamaks, rotational transform is obtained by inducing a toroidal plasma current, which produces the required poloidal magnetic field. A stellarator can be defined as a toroidal device in which the rotational transform is produced by the helical fields generated by coils outside of the plasma [9.13–16]. Stellarators can therefore confine a plasma in steady-state operation without driven plasma currents. In a currentless stellarator, Ampere's law requires that the line integral of the poloidal component of the magnetic field on a contour s encircling the minor axis on each magnetic flux surface vanishes:

$$\oint \boldsymbol{B} \cdot d\boldsymbol{s} = 0 \tag{9.2.1}$$

The poloidal field must therefore change sign and magnitude along the contour. As a consequence, the field lines in a stellarator do not wrap monotonically around the torus minor cross section as they do in a tokamak, but progress instead back and forth, as is illustrated in Fig. 9.13. The magnetic field structure

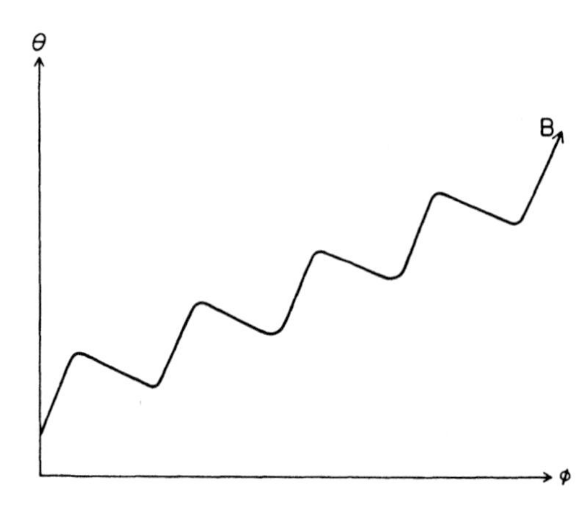

Fig. 9.13. Behavior of a magnetic field line in a stellarator on a flux surface; θ is a poloidal angle and ϕ is a toroidal angle

of a stellarator is thus a series of M fundamental units, called the field periods, each of which incrementally rotates the field lines in the poloidal direction.

In stellarator/heliotron devices, the helical magnetic field necessary for closed flux surfaces is generated by windings which are themselves helical, covering the plasma in both toroidal and poloidal directions. Thus continuous-coil devices can be separated into two categories, depending upon whether helical windings are also used to generate a net toroidal magnetic field.

In a classical stellarator, 2 l helical windings carrying currents in alternating directions produce the helical components of a stellarator field with l-fold poloidal symmetry, but not a net toroidal field. Therefore, additional toroidal field coils are required. In all classical stellarators built thus far, the toroidal field coils were located outside of the helical winding, but in principle, this is not necessary. One of the advantages of the classical stellarator is the flexibility afforded by the possibility of independent variation of the helical and toroidal field components, which makes it possible to vary the flux surface volume and transform profile. This feature allows a wide range of configurations in a single experimental device. A serious disadvantage of the classical stellarator configuration is that the helical windings and toroidal field coils are intertwined. Thus, disassembly and maintenance are difficult, and experimental access for heating and diagnostics is limited.

Some of the difficulties of the classical stellarator are alleviated in the heliotron/torsatron configuration. Here, a field with l-fold poloidal symmetry is generated by l helical windings, all carrying current in the same direction (Fig. 9.14). The helical coil set thus generates both toroidal and poloidal field components, and, in principle, no other coils are needed. There is a constraining relationship between the number of field periods M, the poloidal multipolarity number l and the coil aspect ratio, a_c/R. For example, in Fig. 9.14, we can see 13 field periods and two coils at each poloidal cross section, which correspond to $M = 13$ and $l = 2$. When M is increased for a fixed a_c/R, the helical component of the magnetic field decreases and the toroidal component

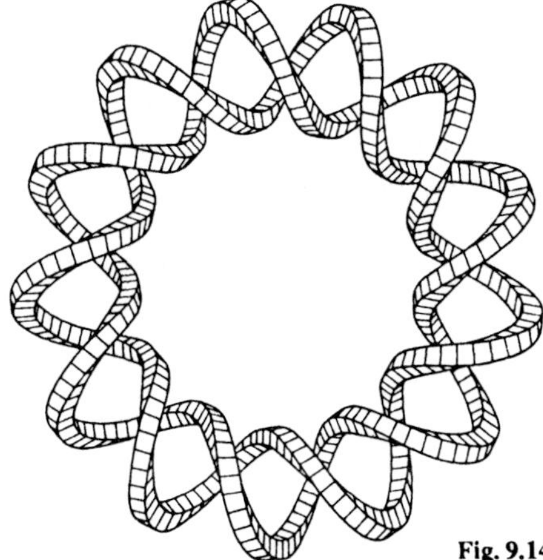

Fig. 9.14. An example of a heliotron/torsatron coil

increases (the outermost magnetic surface expands); when M is decreased, the opposite situation is obtained (the outermost magnetic surface shrinks). Usually $(a_c/R) \times (M/l) = 1.3 - 1.4$ is chosen to allow reasonable volume utilization.

There are other types of stellarator. Recently spatial axis stellarators, where a magnetic axis has a spiral line, have been intensively studied. An example belonging to this categary is the helliac.

9.3 Inertial Confinement and Compression of Plasmas

In inertial confinement fusion [9.17, 18], we need to achieve the Lawson criterion $n\tau > 10^{20}\mathrm{m}^{-3}\mathrm{s}$ before the fuel flies apart. The fuel particles fly with a speed on the order of the speed of sound c_s which is about $10^6\mathrm{ms}^{-1}$. For a spherical fuel pellet of radius r, the confinement time τ is then estimated as $\tau = r/c_s$. The Lawson criterion then reads

$$nr \gtrsim 10^{26}\mathrm{m}^{-2}\,, \tag{9.3.1}$$

or when using the mass density ϱ for a D-T mixture with specific gravity 2.5 g/mole,

$$\varrho r \geq 0.4\ \mathrm{kg\ m}^{-2}\,. \tag{9.3.2}$$

The energy required to heat the fuel to the ignition temperature is

$$U = \frac{4}{3}\pi r^3 nT \sim \left(\frac{n}{n_s}\right) r^3 \times 1.6 \times 10^8 \mathrm{MJ}\,, \tag{9.3.3}$$

where $n_s = 4.5 \times 10^{28}\ \mathrm{m}^{-3}$ is the solid deuterium density. Obviously the required energy becomes enormously large unless a very small fuel pellet is used, say $r < 10^{-2}$ m. Thus, from the relation $\tau \sim r/c_s < 10^{-8}$ s the energy which has to be focused onto this tiny fuel pellet must also be delivered in the form of a very short pulse, that is, very high power density is required. The first important problem in inertial confinement fusion, therefore, is to develop an *energy driver* which can deliver such high power densities. The presently conceivable energy driver is an intense laser or high power particle beam. If we take into account the efficiency of the energy driver (driver eficiency) and the fact that not all of the energy delivered can be used to heat the fuel, the energy required to power the driver can become very large.

Let E be the nominal driver output energy, and the desired fusion output energy is κ times E. After taking into account the efficiency η introduced in (9.1.12) we have

$$\tau P_{\mathrm{fus}} \frac{4}{3}\pi r^3 \eta = \kappa E \tag{9.3.4}$$

Suppose δE of the driver energy can be used to heat the fuel plasma ($\delta < 1$) then

$$\delta E = \frac{4}{3}\pi r^3 n(T_e + T_i) . \tag{9.3.5}$$

Using the relation $\tau = r/c_s$ and $T_e = T_i = 10$ keV, we eliminate r and obtain

$$E = \left(\frac{n_s}{n}\right)^2 \frac{\kappa^3}{\eta^3 \delta^4} \times 1.6\,\text{MJ} \tag{9.3.6}$$

by noting (9.1.8). Typical values for laser beam drivers are $E < 10$ MJ, $\kappa \simeq 10 - 10^2$, and $\eta \sim 1/3$. We therefore have to reduce the factor $(n_s/n)^2\delta^{-4}$ to the lowest possible value. In other words, we have to compress the fuel to an ultra-high density. This problem depends largely on the pellet configuration or design. The second important problem in inertial confinement fusion, therefore, is to design a high gain pellet, which can only be done after we understand the physics at high density plasmas. Let us discuss this problem in more detail in the case of an intense laser beam driver.

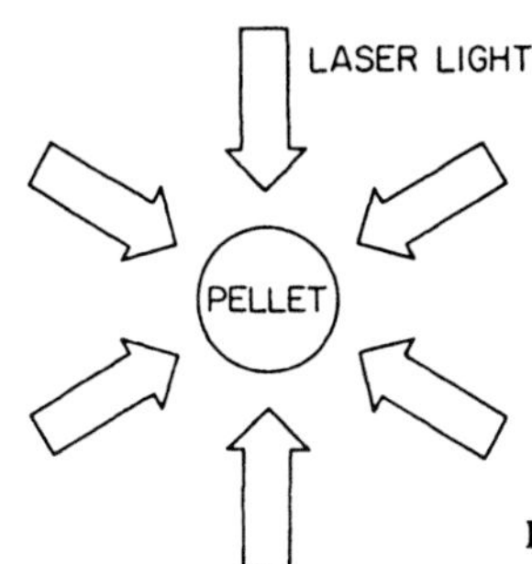

Fig. 9.15. Target pellet and laser beams

The simplest configuration is to shine laser light on a solid spherical target from all directions (Fig. 9.15). The pellet can be compressed by the light pressure itself, which is equal to I/c for laser power density I and speed of light c. The highest power densities that can be achieved today are about 10^{21} W/m^2, so that light pressure alone is insufficient to compress the target of 10 keV beyond the solid density.

As shown in Sect. 7.3, the laser light cannot propagate inside a high density plasma because of shielding by the plasma current. For neodymium glass lasers which are the most widely used, the critical density above which the light cannot propagate is about $0.025n_s$, and for CO_2 laser it is about $0.0025n_s$. Thus the laser light incident on a solid target first ionizes the solid on surface and produces a low density expanding plasma. This plasma then absorbs the light with an efficiency that is dependent on the light wavelength. The absorption efficiency is high if we use a short wavelength laser since its cut-off or critical density is high. It is known both theoretically and experimentally that nearly 100% absorption is possible if a wavelength on the order of 0.3μm, which corresponds to the third harmonic of the neodymium glass laser, is used. Once the laser energy is absorbed by the surface plasma, it continues to expand and in reaction to this expansion, the core high-density plasma is compressed. The ablation pressure is now estimated as I/c_s, where c_s is the sound velocity, and is c/c_s times greater than the light pressure. During the compression phase the core plasma has to be kept at a temperature low enough to keep the core pressure low (*adiabatic compression*). By appropriate pulse shaping we can concentrate the shock waves at the center and compress the central core plasma to a very high density on the order of $n \sim 10^3 n_s$.

This is the basic scenario for laser fusion. There are several problems. First, the laser absorption by the plasma has to be high. As mentioned earlier, the absorption is almost perfect if we use a short wavelength laser. In many cases, the laser-plasma interaction on the surface produces energetic electrons and since these hot electrons have long mean free paths, they can penetrate into the core region and preheat the core plasma. Because core preheating inhibits the compression, we have to shield the hot electrons by some means. If the compression becomes nonuniform, then an instability (Rayleigh–Taylor instability) develops and the compression efficiency is strongly reduced [9.17, 18].

Methods have been developed to avoid these problems of core preheating and nonuniform compression by indirect illumination of the target. Namely, a double shell is used for the pellet and the laser is directed onto the outer shell. This shell has a surface coating composed of heavy metals which convert it to X-ray blackbody radiation. This radiation then acts as the energy driver to implode the inner shell of the target. High compression efficiency in this method has been experimentally confirmed [9.19, 20].

Use of such complicated compression techniques naturally reduces the factor δ. In order to solve this problem, a technique was invented to reduce the required energy: the fuel is separated into two parts, the main fuel and the igniter. The external driver energy is used only to heat the small igniter plasma to the ignition temperature. The main fuel is then heated by the alpha particles produced by the nuclear reaction inside the igniter. The condition for the igniter to trap the thermonuclear α-particles to heat the main fuel requires the density-radius product $\rho R \geq 3\,\mathrm{kg/m^3}$.

The driver efficiency of the neodymium glass laser is less than 1 %. Therefore the development of a high efficiency high power laser is needed. Particle beams have relatively high driver efficiencies and can deliver a very large energy (100 MJ or more). The principal problem of the particle beam as a driver is that because they are composed of charged particles, focusing and transport are difficult. However, inertial confinement fusion is generally considered to have an advantage because it does not require a high vacuum inside the reactor vessel. This simplifies the reactor construction in that there are no restrictions on the type of wall coatings that can be used for neutron radiation shielding; even liquids could be applied.

10. Ideal Magnetohydrodynamics

The simplest model to describe the dynamics of plasmas immersed in a magnetic field is the one-fluid magnetohydrodynamics (MHD), which treats the plasma composed of many charged particles with locally neutral charge as a continuous single fluid [10.1]. This theory does not provide information on the velocity distribution and neglects the physics relating to wave-particle interactions, as does the two-fluid theory as well. It does have the advantage that the macroscopic dynamics of the magnetized plasma can be analyzed in realistic three-dimensional geometries. From this point of view the one-fluid MHD is often more useful than the two-fluid theory.

One-fluid MHD including resistivity, or resistive MHD, has already been derived in Chap. 4. One-fluid MHD neglecting resistivity is called the *ideal MHD*. In this chapter, we discuss the basic properties of ideal MHD plasmas. First in Sect. 10.1, we derive the basic equations from the resistive MHD equations derived in Sect. 4.5. Then the equilibrium properties are explained in Sect. 10.2. Several types of instabilities are described based on the energy principle in the following three sections. Finally, MHD waves, or Alfven waves, are presented in the last section.

10.1 Basic Equations

Here the ideal MHD equations are deduced from the resistive MHD equations and their properties are discussed [10.1–3].

When $\boldsymbol{u} \times \boldsymbol{B}$ is greater than $\eta \boldsymbol{J}$ or

$$R_{\mathrm{m}} = \frac{uB}{\eta J} = \frac{\mu_0 u L}{\eta} \gg 1 , \tag{10.1.1}$$

where the resistivity η is given by (3.5.11), Ohm's law becomes

$$\boldsymbol{E} + \boldsymbol{u} \times \boldsymbol{B} = 0 . \tag{10.1.2}$$

Here L is the characteristic length of the system and R_{m} is called the *magnetic Reynolds number* in analogy to the Reynolds number $R = uL/\mu_{\mathrm{n}}$ (μ_{n} is the viscosity) in neutral fluid dynamics. Usually $u = v_{\mathrm{A}}$, the Alfven velocity given by (7.3.24). The magnetic Reynolds number becomes

$$S \equiv \frac{\mu_0 v_A L}{\eta} . \tag{10.1.3}$$

In resistive MHD, Ohm's law is used to determine $\boldsymbol{J}$ from $\boldsymbol{E}$, $\boldsymbol{u}$ and $\boldsymbol{B}$. However, when $\eta \boldsymbol{J}$ is small, another equation to determine the current density $\boldsymbol{J}$ is required. We neglect the displacement current in (4.5.16), which is valid for $u^2/c^2 \ll 1$. Then we have

$$\mu_0 \boldsymbol{J} = \nabla \times \boldsymbol{B} . \tag{10.1.4}$$

By taking the divergence of (10.1.4), we obtain $\nabla \cdot \boldsymbol{J} = 0$. This contradicts the charge continuity equation (4.5.14) since in general, charge oscillation can take place in plasma dynamics. In ideal MHD, we drop the condition of local charge neutrality equation. The term $\sigma_e E$ in the equation of motion in (4.5.4) is assumed to be negligibly small. In the case that we need to calculate the charge density σ_e, we use the Poisson equation (4.5.17) as $\sigma_e = \varepsilon_0 \nabla \cdot \boldsymbol{E}$. Then the ideal MHD equations are

$$\frac{\partial \varrho}{\partial t} + \nabla \cdot (\varrho \boldsymbol{u}) = 0 , \tag{10.1.5}$$

$$\varrho \frac{d\boldsymbol{u}}{dt} = -\nabla P + \boldsymbol{J} \times \boldsymbol{B} , \tag{10.1.6}$$

$$\frac{\partial P}{\partial t} + (\boldsymbol{u} \cdot \nabla) P + \gamma P \nabla \cdot \boldsymbol{u} = 0 , \tag{10.1.7}$$

$$\frac{\partial \boldsymbol{B}}{\partial t} = -\nabla \times \boldsymbol{E} , \tag{10.1.8}$$

$$\mu_0 \boldsymbol{J} = \nabla \times \boldsymbol{B} , \tag{10.1.9}$$

$$\boldsymbol{E} = -\boldsymbol{u} \times \boldsymbol{B} . \tag{10.1.10}$$

Here the variables are mass density, pressure, velocity, magnetic field, electric field and current density (ϱ, P, $\boldsymbol{u}$, $\boldsymbol{B}$, $\boldsymbol{E}$, $\boldsymbol{J}$) of which the time evolution is determined only by (ϱ, P, $\boldsymbol{u}$, $\boldsymbol{B}$). In (10.1.7) γ denotes the specific heat ratio. The electric field $\boldsymbol{E}$ and the current density $\boldsymbol{J}$ are determined by (10.1.9, 10). In ideal MHD, $\nabla \cdot \boldsymbol{B} = 0$ is imposed as the initial condition, which is understood from the divergence of (10.1.8).

Though ideal MHD does not include kinetic effects, it is the simplest and the most useful theory to describe plasmas confined by a magnetic field. Various realistic geometries of magnetic configuration have been analyzed by using the approximation of ideal MHD.

Sometimes the assumption of incompressibility, or $\nabla \cdot \boldsymbol{u} = 0$, is imposed. Then the incompressible ideal MHD equations are written in terms of (ϱ, P, $\boldsymbol{u}, \boldsymbol{B}$) as

$$\varrho \frac{d\boldsymbol{u}}{dt} = -\nabla P + \frac{1}{\mu_0} (\nabla \times \boldsymbol{B}) \times \boldsymbol{B} , \tag{10.1.11}$$

$$\frac{d\boldsymbol{B}}{dt} = (\boldsymbol{B} \cdot \nabla)\boldsymbol{u} \,, \tag{10.1.12}$$

$$\frac{dP}{dt} = 0 \,, \tag{10.1.13}$$

$$\nabla \cdot \boldsymbol{u} = 0 \,. \tag{10.1.14}$$

We can prove that the equations for ideal MHD satisfy the conservation of energy law by considering the example of magnetized plasmas confined in a metal wall chamber with conductivity $\sigma \to \infty$. The left hand side of the equation of motion multiplied by $\boldsymbol{u}$ is written as

$$\begin{aligned} &\varrho \boldsymbol{u} \cdot \left[\frac{\partial \boldsymbol{u}}{\partial t} + (\boldsymbol{u} \cdot \nabla)\boldsymbol{u} \right] \\ &\quad = \frac{\partial}{\partial t}\left(\frac{1}{2}\varrho u^2\right) + \frac{u^2}{2}\nabla \cdot (\varrho \boldsymbol{u}) + \frac{1}{2}\varrho(\boldsymbol{u} \cdot \nabla)u^2 \\ &\quad = \frac{\partial}{\partial t}\left(\frac{1}{2}\varrho u^2\right) + \nabla \cdot \left(\frac{1}{2}\varrho u^2 \boldsymbol{u}\right) \,. \end{aligned} \tag{10.1.15}$$

In the second equality the continuity equation was used. The pressure equation can alternatively be written as

$$\boldsymbol{u} \cdot \nabla P = \frac{1}{\gamma - 1}\frac{\partial P}{\partial t} + \frac{\gamma}{\gamma - 1}\nabla \cdot (P\boldsymbol{u}) \,. \tag{10.1.16}$$

The second term of the right hand side of the equation of motion multiplied by $\boldsymbol{u}$ is also written as

$$\begin{aligned} \boldsymbol{u} \cdot [(\nabla \times \boldsymbol{B}) \times \boldsymbol{B}] &= -(\boldsymbol{u} \times \boldsymbol{B}) \cdot (\nabla \times \boldsymbol{B}) \\ &= \boldsymbol{E} \cdot (\nabla \times \boldsymbol{B}) = \boldsymbol{B} \cdot (\nabla \times \boldsymbol{E}) - \nabla \cdot (\boldsymbol{E} \times \boldsymbol{B}) \\ &= -\frac{\partial}{\partial t}\frac{B^2}{2} - \nabla \cdot (\boldsymbol{E} \times \boldsymbol{B}) \,. \end{aligned} \tag{10.1.17}$$

When (10.1.15–17) are summed up,

$$\frac{\partial}{\partial t}\left(\frac{1}{2}\varrho u^2 + \frac{P}{\gamma - 1} + \frac{B^2}{2\mu_0}\right) + \nabla \cdot \left(\frac{1}{2}\varrho u^2 \boldsymbol{u} + \frac{\gamma}{\gamma - 1}P\boldsymbol{u} + \frac{\boldsymbol{E} \times \boldsymbol{B}}{\mu_0}\right) = 0 \tag{10.1.18}$$

is obtained. It is noted that the momentum is also shown in the conservation form. The stress tensor is given by

$$\overleftrightarrow{T} = \varrho \boldsymbol{u}\boldsymbol{u} + \left(P + \frac{B^2}{2\mu_0}\right)\overleftrightarrow{I} - \frac{\boldsymbol{B}\boldsymbol{B}}{\mu_0} \,. \tag{10.1.19}$$

By noting the equality $\boldsymbol{B} \times (\nabla \times \boldsymbol{B}) = \nabla B^2/2 - \nabla \cdot (\boldsymbol{B}\boldsymbol{B})$, the equation of motion is expressed by

$$\frac{\partial}{\partial t}(\varrho \boldsymbol{u}) + \nabla \cdot \overleftrightarrow{T} = 0 \,. \tag{10.1.20}$$

Here $\boldsymbol{BB}$ in (10.1.19) denotes a dyadic form and $\overleftrightarrow{T}$ denotes a tensor.

We are considering magnetized plasmas extending to a perfectly conducting wall and, therefore, the boundary conditions at the wall are that the normal components of $\boldsymbol{B}$ and $\boldsymbol{u}$ vanish,

$$\boldsymbol{B} \cdot \boldsymbol{n} = 0 \, , \qquad \boldsymbol{u} \cdot \boldsymbol{n} = 0 \tag{10.1.21}$$

where $\boldsymbol{n}$ denotes the unit normal vector on the wall. By using these boundary conditions, when the conservation of energy law (10.1.18) is integrated over the plasma region bounded by the wall, we find

$$W = \iiint \left(\frac{1}{2} \varrho u^2 + \frac{P}{\gamma - 1} + \frac{B^2}{2\mu_0} \right) dV = \text{const} \, . \tag{10.1.22}$$

The first term denotes the kinetic energy of the plasma, the second term corresponds to the internal energy and the third term is the magnetic energy. The sum of the second and the third terms is the potential energy. Equation (10.1.22) shows that the total energy is conserved for isolated plasmas inside a conducting wall.

The expression for energy conservation in a plasma expanding in a vacuum is more complicated than for the case of a conducting wall boundary, since the plasma boundary is now allowed to move. To show this, first note that for a quantity Z defined by $Z(t) = \iiint Z(\boldsymbol{r}, t) dV$, the total time derivative in a volume whose boundary is moving with velocity, $\boldsymbol{u}$, is given by

$$\frac{d}{dt} Z(t) = \iiint \frac{\partial Z}{\partial t} dV + \iint Z \boldsymbol{n} \cdot \boldsymbol{u} \, dS \, . \tag{10.1.23}$$

At the plasma-vacuum interface there are boundary conditions to be met within the ideal MHD. Since the plasma surface $r = R(\theta, z, t)$ is by definition a constant-pressure or a flux surface (Sect. 10.2), it follows that

$$\hat{\boldsymbol{n}} \cdot \hat{\boldsymbol{B}}|_R = \boldsymbol{n} \cdot \boldsymbol{B}|_R = 0 \, . \tag{10.1.24}$$

Here quantities with a hat denote vacuum variables. In general, the plasma surface can move and $\boldsymbol{n} \cdot \boldsymbol{u}|_R$ is arbitrary. It is possible to have jumps in the pressure and the tangential magnetic field across the surface. Integration of the equation of motion in the neighborhood of the surface requires

$$\left. \left[P + \frac{B^2}{2\mu_0} \right] \right|_R = 0 \tag{10.1.25}$$

where $[Q] \equiv \hat{Q} - Q$. If we apply Ohm's law (10.1.10) and the plasma-vacuum boundary conditions given in (10.1.24), it follows that

$$\frac{dW}{dt} = - \iint \left(P + \frac{B^2}{2\mu_0} \right) \boldsymbol{n} \cdot \boldsymbol{u} \, dS \, . \tag{10.1.26}$$

Here W is the plasma energy given by (10.1.22). Note that if the plasma surface is moving, the boundary term is, in general, non-zero.

Now consider the vacuum region. Here the total energy is given by

$$\hat{W} = \iiint \frac{\hat{B}^2}{2\mu_0}\, dV \tag{10.1.27}$$

and from this it follows that

$$\frac{d\hat{W}}{dt} = \iiint \frac{1}{\mu_0}\hat{\boldsymbol{B}} \cdot \frac{\partial \hat{\boldsymbol{B}}}{\partial t} dV - \iint \frac{\hat{B}^2}{2\mu_0} \boldsymbol{n} \cdot \boldsymbol{u}\, dS\,. \tag{10.1.28}$$

This equation is simplified by substituting in (10.1.8,10). After some algebra, where we again apply the boundary condition (10.1.24), one obtains

$$\frac{d\hat{W}}{dt} = \iint \frac{\hat{B}^2}{2\mu_0} \boldsymbol{n} \cdot \boldsymbol{u}\, dS\,. \tag{10.1.29}$$

The final energy conservation relation is obtained by adding (10.1.26) and (10.1.29),

$$\frac{d}{dt}(W + \hat{W}) = 0\,. \tag{10.1.30}$$

Here the boundary terms cancel by virtue of (10.1.25) which is the pressure balance condition. Equation (10.1.30) implies that if an ideal MHD plasma is isolated from a conducting wall by a vacuum region, the combined energy of the plasma-vacuum system is conserved. The fact that only the total energy is conserved indicates that, in general, energy will flow from the plasma to the vacuum or vice versa as the plasma moves. From (10.1.8,10), we obtain

$$\frac{\partial \boldsymbol{B}}{\partial t} = \nabla \times (\boldsymbol{u} \times \boldsymbol{B})\,. \tag{10.1.31}$$

This equation describes the coupling between the magnetic field and the fluid motion under the ideal MHD. To examine the relationship we integrate over an arbitrary region inside the plasma and apply the Stokes theorem to obtain

$$\frac{\partial}{\partial t} \iint \boldsymbol{B} \cdot d\boldsymbol{S} - \oint (\boldsymbol{u} \times \boldsymbol{B}) \cdot d\boldsymbol{s} = 0\,, \tag{10.1.32}$$

where the line integral is extended over the periphery of surface S. This expression can be rewritten as

$$\frac{\partial \Phi}{\partial t} + \oint \boldsymbol{B} \cdot (\boldsymbol{u} \times d\boldsymbol{s}) = 0\,. \tag{10.1.33}$$

The first term represents the rate of change of the flux through the fixed surface S, while the second is the additional increment of flux swept out per unit time by the periphery moving with the local fluid velocity $\boldsymbol{u}$ (Fig. 10.1). Then the left hand side of (10.1.33) yields the total rate of change of the flux through a surface fixed to and moving with the plasma. Thus (10.1.33) expresses the constancy of the flux through any surface in a perfectly conducting fluid

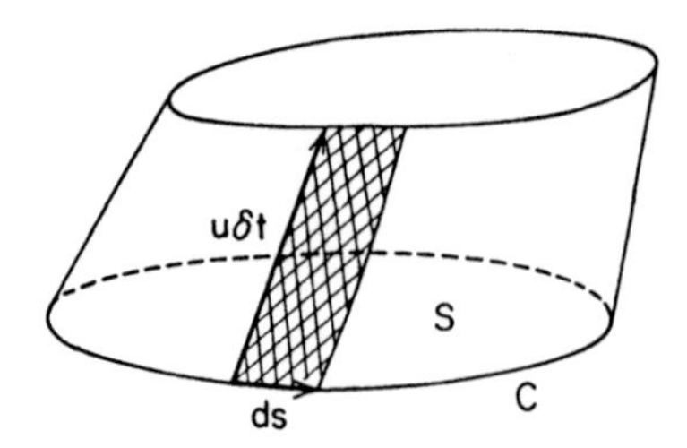

Fig. 10.1. Displacement of a surface S with a boundary C during δt. The displacement velocity is u and ds is a line element

$$\frac{d\Phi}{dt} = 0\,. \tag{10.1.34}$$

The concept of the magnetic line of force is an abstraction. In general, no identity can be attached to these lines. However, (10.1.34) means that the local fluid velocity $\boldsymbol{u}(\boldsymbol{r}, t)$ is the velocity of the local section of a field line. This introduces a concept which is essential in the ideal MHD, namely that the field line is 'frozen-in' the fluid.

10.2 MHD Equilibrium

In the study of magnetically confined plasmas, the most important requirement is the existence of an MHD equilibrium suitable to confinement [10.4–9]. This is defined by a stationary state without plasma flow, i.e., $\partial/\partial t = 0$ and $\boldsymbol{u}_0 = 0$. Then the ideal MHD equations give

$$\mu_0 \nabla P = (\nabla \times \boldsymbol{B}) \times \boldsymbol{B}\,, \tag{10.2.1}$$

$$\nabla \cdot \boldsymbol{B} = 0\,. \tag{10.2.2}$$

When the plasma is confined in a symmetric geometry such as axisymmetry or helical symmetry, the existence of nested flux surfaces has been theoretically proven. From (10.2.1), we have $\boldsymbol{B} \cdot \nabla P = 0$. This means that the magnetic field lines stay on the isobaric surface (i.e., P = const at the surface). The magnetic field satisfying (10.2.2) is described by $\boldsymbol{B} = \nabla \Psi \times \nabla \theta$, where Ψ and θ are two scalar functions. When P is a function of Ψ only, $\boldsymbol{B} \cdot \nabla P = 0$ is satisfied automatically. The surfaces Ψ = const are called the flux surfaces. When these surfaces are nested closed surfaces, the configuration can be used for plasma confinement. The function θ is related to the angular dependence inside the Ψ = const surface. There is no positive proof that nested flux surfaces can exist in a three-dimensional system without any symmetry.

Here we consider axisymmetric plasmas to explain the properties of the MHD equilibrium. The cylindrical coordinates (r, φ, z) are employed. The axisymmetry imposes $\partial/\partial\varphi = 0$. The flux function Ψ is introduced by

$$rB_r = -\frac{\partial \Psi}{\partial z}, \qquad rB_z = \frac{\partial \Psi}{\partial r}. \tag{10.2.3}$$

This corresponds to the case when the scalar function θ is equal to the angle variable φ. The φ component of the magnetic field is $B_\varphi = B_\varphi(r, z)$. These relations automatically satisfy $\nabla \cdot \boldsymbol{B} = 0$ in the cylindrical coordinates. By substituting these into the right hand side of (10.2.1), the relations

$$\begin{aligned}
\mu_0 \frac{\partial P}{\partial r} + \frac{B_\varphi}{r}\frac{\partial}{\partial r}(rB_\varphi) + \frac{1}{r^2}\frac{\partial \Psi}{\partial r}\Delta^*\Psi &= 0 \\
\frac{\partial \Psi}{\partial r}\frac{\partial}{\partial z}(rB_\varphi) - \frac{\partial \Psi}{\partial z}\frac{\partial}{\partial r}(rB_\varphi) &= 0 \\
\mu_0 \frac{\partial P}{\partial z} + B_\varphi \frac{\partial B_\varphi}{\partial z} + \frac{1}{r^2}\frac{\partial \Psi}{\partial z}\Delta^*\Psi &= 0
\end{aligned} \tag{10.2.4}$$

are obtained, where

$$\Delta^*\Psi = \frac{\partial^2 \Psi}{\partial r^2} - \frac{1}{r}\frac{\partial \Psi}{\partial r} + \frac{\partial^2 \Psi}{\partial z^2}. \tag{10.2.5}$$

The second equation of (10.2.4) means that rB_φ is a function of Ψ only, i.e., $rB_\varphi = f(\Psi)$, where $f(\Psi)$ is an arbitrary function of Ψ. When this result is substituted into the first and the third equations of (10.2.4), we have

$$\begin{aligned}
\mu_0 \frac{\partial P}{\partial r} + \frac{\partial \Psi}{\partial r}\left(\frac{1}{r^2}\Delta^*\Psi + \frac{1}{r^2} f \frac{\partial f}{\partial \Psi}\right) = 0 \\
\mu_0 \frac{\partial P}{\partial z} + \frac{\partial \Psi}{\partial z}\left(\frac{1}{r^2}\Delta^*\Psi + \frac{1}{r^2} f \frac{\partial f}{\partial \Psi}\right) = 0 .
\end{aligned} \tag{10.2.6}$$

We multiply the first equation of (10.2.6) by $\partial\Psi/\partial z$ and the second equation by $\partial\Psi/\partial r$. Taking their difference gives

$$\frac{\partial P}{\partial r}\frac{\partial \Psi}{\partial z} - \frac{\partial P}{\partial z}\frac{\partial \Psi}{\partial r} = 0 . \tag{10.2.7}$$

This means that P is also a function of Ψ only, $P = g(\Psi)$, where g is an arbitrary function of Ψ. By using (10.2.7) in (10.2.6), we obtain

$$\Delta^*\Psi + f\frac{\partial f}{\partial \Psi} + r^2\frac{\partial g}{\partial \Psi} = 0 . \tag{10.2.8}$$

This equation is called the *Grad-Shafranov* equation. Usually, first $f(\Psi)$ and $g(\Psi)$ are defined, and then $\Psi(r, z)$ is solved as a boundary-value problem. Solutions of (10.2.8) showing nested flux surfaces correspond to MHD equilibrium.

Note that a simple torus having a magnetic field $\boldsymbol{B} = (0, B_\varphi(r, z), 0)$ does not have MHD equilibrium. From (10.2.1, 2), we obtain for the simple torus

$$\nabla \times [(\nabla \times \boldsymbol{B}) \times \boldsymbol{B}] = \nabla \times (\boldsymbol{B} \cdot \nabla)\boldsymbol{B} = -\frac{1}{r}\frac{\partial B_\varphi^2}{\partial z}\hat{\varphi} = 0 . \tag{10.2.9}$$

This means that B_φ should be independent of z for MHD equilibrium. However, to confine a doughnut-like toroidal plasma B_φ must depend on z. As discussed in Chap. 9 this proves that the simple torus cannot sustain MHD equilibrium; an additional poloidal magnetic field is necessary.

Next we will apply the MHD equilibrium equation (10.2.1, 2) to a cylindrical plasma model. Let the plasma have a length $2\pi R$ with a periodic boundary condition. This is the most simple model for a toroidal plasma assuming negligibly small toroidal curvature. For a cylindrical plasma the magnetic field is assumed to be $\boldsymbol{B} = (0, B_\theta(r), B_z(r))$. Then the cylindrical MHD equilibrium equation (10.2.1) gives

$$\frac{d}{dr}\left[P(r) + \frac{B_\theta^2 + B_z^2}{2\mu_0}\right] + \frac{B_\theta^2}{r\mu_0} = 0 \,. \tag{10.2.10}$$

There are three types of cylindrical equilibrium models. One is the *theta pinch* model. The plasma is confined by the $J_\theta B_z$ force produced by the 'theta' component of the plasma current. In this case, from (10.2.10)

$$P + \frac{B_z^2}{2\mu_0} = \text{const} \tag{10.2.11}$$

is required for the MHD equilibrium. An important parameter β indicating the ratio between the plasma pressure and the magnetic field pressure is defined by

$$\beta(r) = \frac{P(r)}{(P + B_z^2/2\mu_0)|_{r=a}} \,. \tag{10.2.12}$$

Note that $0 < \beta < 1$ and $\beta(0) = 2\mu_0 P(0)/[B_z(a)]^2$, where a is the plasma radius corresponding to $P(a) = 0$. The typical behavior of B_z and P in a theta pinch configuration are shown in Fig. 10.2. Another cylindrical equilibrium is the *Z pinch* model which confines a plasma by $J_z B_\theta$ force produced by the 'z' component of the plasma current. An example of the Z pinch equilibrium is given by

$$\begin{aligned}
B_\theta &= \frac{\mu_0 I}{\pi}\frac{r}{r^2 + a^2} \,, \\
J_z &= \frac{2I}{\pi}\frac{a^2}{(r^2 + a^2)^2} \,, \\
P &= \frac{\mu_0 I^2}{2\pi^2}\frac{a^2}{(r^2 + a^2)^2} \,,
\end{aligned} \tag{10.2.13}$$

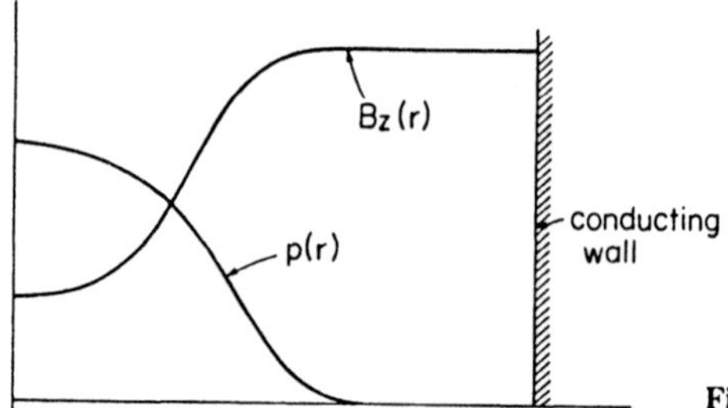

Fig. 10.2. MHD equilibrium of a theta pinch

also called the *Benette pinch*. These B_θ and P satisfy the equilibrium equation

$$\frac{d}{dr}\left(P+\frac{B_\theta^2}{2\mu_0}\right)+\frac{B_\theta^2}{r\mu_0}=0\,, \tag{10.2.14}$$

and the configuration shown by (10.2.13) satisfies

$$\frac{\mu_0 I^2}{4\pi^2}=2\int P(r)r dr\,, \tag{10.2.15}$$

where I is the total current.

The last cylindrical equilibrium model is the *screw pinch* satisfying (10.2.10). Tokamak and RFP (reversed field pinch) are typical examples of configurations belonging to this category. The configurations are characterized by the safety factor profile $q(r)$ which is related to the rotation angle of the magnetic field line. When the magnetic field line goes one turn around a torus along the field line itself, the change in angle $\Delta\theta$ is given by

$$\Delta\theta=\int_0^{2\pi R}\frac{d\theta}{dz}dz\,. \tag{10.2.16}$$

For the screw pinch, the equations for the magnetic field line are written,

$$\frac{dr}{dz}=\frac{B_r}{B_z}=0\,,\qquad \frac{d\theta}{dz}=\frac{B_\theta}{rB_z}\,. \tag{10.2.17}$$

The angle $\Delta\theta$ defines the rotational transform angle, $\iota(r)=2\pi RB_\theta/rB_z$. Then the safety factor is defined by the relation

$$q(r)=\frac{2\pi}{\iota(r)}=\frac{rB_z(r)}{RB_\theta(r)}\,. \tag{10.2.18}$$

In tokamaks, typically $q(0)<1$ and $q(a)\sim 3$, which corresponds to $|B_\theta|\ll|B_z|$ (Fig. 9.11a). In contrast, in RFP $q(0)\sim 0.1$ and $q(a)<0$, which corresponds to $|B_\theta|\sim|B_z|$. The beta value is in the range of $\beta\sim 0.1$. The sign of B_z changes in the neighborhood of the edge region which induces this 'reversed field' pinch (Fig. 9.11b).

Another example of a solution to the Grad-Shafranov equation (10.2.8) in cylindrical coordinates is

$$\begin{aligned}\Psi(r,z)&=-\frac{3}{4}r^2B_0\left[1-\frac{r^2+z^2}{a^2}\right]\quad\text{for } r^2+z^2\le a^2\\ \Psi(r,z)&=\frac{1}{2}r^2B_0\left[1-\frac{a^3}{(r^2+z^2)^{3/2}}\right]\quad\text{for } r^2+z^2>a^2\,.\end{aligned} \tag{10.2.19}$$

The derivatives of Ψ with respect to r and z are continuous at the plasma boundary of $r^2+z^2=a^2$. The surfaces of constant flux are shown in Fig. 10.3. This solution is similar to a vortex ring and this particular example is called Hill's spherical vortex. If we let $f=0$ and assume that the pressure is proportional to Ψ, then

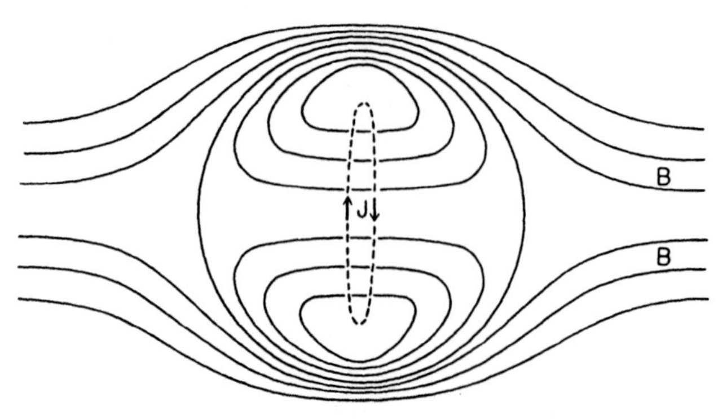

Fig. 10.3. Magnetic surfaces and plasma current in an MHD equilibrium corresponding to Hill's spherical vortex

$$P = -\frac{15}{2a^2} B_0 \Psi \quad \text{for } r^2 + z^2 \leq a^2$$

$$P = 0 \qquad \text{for } r^2 + z^2 > a^2 . \tag{10.2.20}$$

This MHD equilibrium configuration is called the *field reversal configuration* (FRC).

We now consider a different aspect of the equilibrium problem of toroidal plasma which reduces to finding the dependence of the position and shape of the boundary on the mean pressure inside the toroidal plasma. We write the system of the equilibrium equations in toroidal coordinates $(\varrho, \theta, \varphi)$ (Fig. 10.4):

$$\frac{\partial P}{\partial \varrho} = J_\theta B_\varphi - J_\varphi B_\theta , \tag{10.2.21}$$

$$B_\varrho \frac{\partial P}{\partial \varrho} + \frac{B_\theta}{\varrho} \frac{\partial P}{\partial \theta} = 0 , \tag{10.2.22}$$

$$J_\varrho \frac{\partial P}{\partial \varrho} + \frac{J_\theta}{\varrho} \frac{\partial P}{\partial \theta} = 0 , \tag{10.2.23}$$

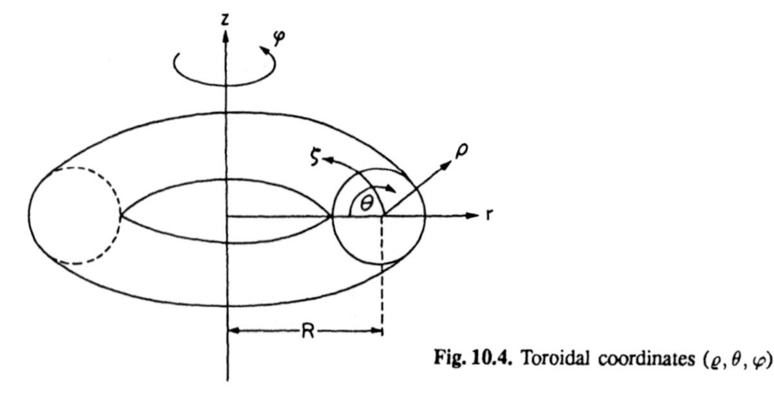

Fig. 10.4. Toroidal coordinates $(\varrho, \theta, \varphi)$

$$\mu_0 J_\varrho = \frac{1}{[1-(\varrho/R)\cos\theta]} \frac{1}{\varrho}\frac{\partial}{\partial\theta}[1-(\varrho/R)\cos\theta]B_\varphi \,, \tag{10.2.24}$$

$$\mu_0 J_\theta = \frac{1}{[1-(\varrho/R)\cos\theta]} \frac{\partial}{\partial\varrho}[1-(\varrho/R)\cos\theta]B_\varphi \,, \tag{10.2.25}$$

$$\mu_0 J_\varphi = \frac{1}{\varrho}\frac{\partial}{\partial\varrho}(\varrho B_\theta) - \frac{1}{\varrho}\frac{\partial B_\varrho}{\partial\theta} \,, \tag{10.2.26}$$

$$\frac{\partial}{\partial\varrho}[(\varrho-(\varrho^2/R)\cos\theta)B_\varrho] + \frac{\partial}{\partial\theta}[(1-(\varrho/R)\cos\theta)B_\theta] = 0 \,. \tag{10.2.27}$$

Equations (10.2.22, 23) correspond to $\boldsymbol{B}\cdot\nabla P = 0$ and $\boldsymbol{J}\cdot\nabla P = 0$, respectively, and (10.2.27) is $\nabla\cdot\boldsymbol{B} = 0$ where we assumed an axisymmetric toroidal plasma. We then assume that we can separate the angular dependence of $P, \boldsymbol{B}$ and $\boldsymbol{J}$:

$$\begin{aligned} P &= P_0(\varrho) + P_1(\varrho,\theta) \\ \boldsymbol{B} &= \boldsymbol{B}_0(\varrho) + \boldsymbol{B}_1(\varrho,\theta) \\ \boldsymbol{J} &= \boldsymbol{J}_0(\varrho) + \boldsymbol{J}_1(\varrho,\theta) \end{aligned} \tag{10.2.28}$$

where

$$\begin{aligned} \boldsymbol{B}_0(\varrho) &= B_{\theta 0}(\varrho)\hat{\mathbf{e}}_\theta + B_{\varphi 0}(\varrho)\hat{\mathbf{e}}_\varphi \\ \boldsymbol{J}_0(\varrho) &= J_{\theta 0}(\varrho)\hat{\mathbf{e}}_\theta + J_{\varphi 0}(\varrho)\hat{\mathbf{e}}_\varphi \,, \end{aligned} \tag{10.2.29}$$

which are the cylindrically symmetric solutions of the equilibrium equation in the zeroth order approximation of $1/R \to 0$. The pressure profile is characterized by

$$2\mu_0\,[\langle P\rangle - P_0(a)] + \left\langle B_{\varphi 0}^2\right\rangle = B_{\varphi 0}^2(a) + B_{\theta 0}^2(a) \,, \tag{10.2.30}$$

which is obtained from (10.2.10) by multiplying by ϱ^2 and integrating over the radius a, where

$$\begin{aligned} \langle P_0\rangle &= \frac{1}{\pi a^2}\int_0^a 2\pi P_0(\varrho)\varrho d\varrho \\ \left\langle B_{\varphi 0}^2\right\rangle &= \frac{1}{\pi a^2}\int_0^a 2\pi B_{\varphi 0}^2(\varrho)\varrho d\varrho \,. \end{aligned} \tag{10.2.31}$$

The corrections associated with the toroidal curvature are expressed as

$$\begin{aligned} P_1(\varrho,\theta) &= P_1(\varrho)\cos\theta \\ \boldsymbol{B}_1(\varrho,\theta) &= (\boldsymbol{B}_1(\varrho) - B_{1\varrho}\hat{\mathbf{e}}_\varrho)\cos\theta + B_{1\varrho}\hat{\mathbf{e}}_\varrho\sin\theta \\ \boldsymbol{J}_1(\varrho,\theta) &= (\boldsymbol{J}_1(\varrho) - J_{1\varrho}\hat{\mathbf{e}}_\varrho)\cos\theta + J_{1\varrho}\hat{\mathbf{e}}_\varrho\sin\theta \,. \end{aligned} \tag{10.2.32}$$

In the first order approximation in the expansion with respect to the curvature, the cross sections of the magnetic surfaces remain circles but with displaced centers. The magnetic surface with a radius ϱ' is related to ϱ and θ in the following way,

$$\varrho = \varrho' + \xi(\varrho',\theta) = \varrho' + \xi_1(\varrho)\cos\theta \,. \tag{10.2.33}$$

By definition, the plasma pressure on this surface has a relation $P(\varrho' + \xi, \theta) = P_0(\varrho')$. Expanding the left side of this equation,

$$P(\varrho' + \xi, \theta) = P_0(\varrho') + P_1(\varrho, \theta) + \xi(\varrho, \theta)\frac{dP_0}{d\varrho} + \dots , \tag{10.2.34}$$

we can find the relation between the displacement ξ and the pressure correction P_1 in the linear approximation,

$$P_1(\varrho, \theta) = -\xi(\varrho, \theta)\frac{dP_0}{d\varrho} = -\xi_1(\varrho)\frac{dP_0}{d\varrho}\cos\theta \, . \tag{10.2.35}$$

Linearizing (10.2.21–27), we obtain equations for the corrections proportional to $\cos\theta$ or $\sin\theta$,

$$\frac{dP_1}{d\varrho} = J_{\theta 0}B_{\varphi 1} - J_{\varphi 0}B_{\theta 1} + J_{\theta 1}B_{\varphi 0} - J_{\varphi 1}B_{\theta 0} \, , \tag{10.2.36}$$

$$B_{\varrho 1} = -\frac{\xi_1}{\varrho}B_{\theta 0} \, , \tag{10.2.37}$$

$$J_{\varrho 1} = -\frac{\xi_1}{\varrho}J_{\theta 0} \, , \tag{10.2.38}$$

$$B_{\varphi 1} = \frac{\varrho}{R}B_{\varphi 0} + \mu_0\xi_1 J_{\theta 0} \, , \tag{10.2.39}$$

$$J_{\theta 1} = \frac{\varrho}{R}J_{\theta 0} - \frac{d}{d\varrho}(\xi_1 J_{\theta 0}) \, , \tag{10.2.40}$$

$$\mu_0 J_{\varphi 1} = \frac{1}{\varrho}\frac{d}{d\varrho}(\varrho B_{\theta 1}) - \frac{1}{\varrho}B_{\varrho 1} \, , \tag{10.2.41}$$

$$B_{\theta 1} = \frac{\varrho}{R}B_{\theta 0} - \frac{d}{d\varrho}(\xi_1 B_{\theta 0}) \, . \tag{10.2.42}$$

We note that $B_{\theta 1}, B_{\varphi 1}$ and $J_{\theta 1}$ are expressed by using ξ_1 and the zeroth order quantities. Substituting these first order quantities into the right hand side of (10.2.36) we can find

$$J_{\varphi 1} = \left[\frac{\varrho}{R}(2J_{\theta 0}B_{\varphi 0} - J_{\varphi 0}B_{\theta 0}) - \xi_1 B_{\theta 0}\frac{dJ_{\varphi 0}}{d\varrho}\right] \Big/ B_{\theta 0} \, . \tag{10.2.43}$$

Now, substituting $J_{\varphi 1}, B_{\theta 1}$ and $B_{\varrho 1}$ into (10.2.41), we obtain the following differential equation for ξ_1,

$$\frac{d}{d\varrho}\left(\varrho B_{\theta 0}^2\frac{d\xi_1}{d\varrho}\right) - \frac{\varrho}{R}\left(B_{\theta 0}^2 - 2\mu_0\varrho\frac{dP_0}{d\varrho}\right) = 0 \, . \tag{10.2.44}$$

Integrating this equation, we have

$$\frac{d\xi_1}{d\varrho} = \frac{\varrho}{R}\frac{G}{D}, \tag{10.2.45}$$

where

$$D = [B_{\theta 0}(\varrho)]^2$$

and

$$G = 2\mu_0\left[\langle P_0\rangle_\varrho - P_0(\varrho)\right] + \frac{1}{2}\left\langle B_{\theta 0}^2\right\rangle_\varrho,$$

and by assuming $\xi_1(a) = 0$, the expression

$$\xi_1 = \frac{1}{R}\int_a^\varrho \varrho'\frac{G}{D}d\varrho' \tag{10.2.46}$$

shows the relative displacement of the magnetic surface of radius ϱ with respect to the plasma of radius a ($\varrho > a$). Here

$$\langle P_0\rangle_\varrho = \frac{1}{\pi\varrho^2}\int_0^\varrho 2\pi P_0(\varrho')\varrho'\,d\varrho' \tag{10.2.47}$$

and $\left\langle B_{\theta 0}^2\right\rangle_\varrho$ has a similar expression. Then we have, noting $P_0(\varrho > a) = 0$ and $B_{\theta 0}^2(\varrho > a) = B_{\theta 0}^2(a)a^2/\varrho^2$,

$$\begin{aligned}\langle P_0\rangle_\varrho &= \langle P_0\rangle_a \frac{a^2}{\varrho^2}\\ \left\langle B_{\theta 0}^2\right\rangle_\varrho &= \left\langle B_{\theta 0}^2\right\rangle_a \frac{a^2}{\varrho^2}\left[1 + \frac{2}{l_i}\ln\left(\frac{\varrho}{a}\right)\right].\end{aligned} \tag{10.2.48}$$

where $l_i = \left\langle B_{\theta 0}^2\right\rangle_a / B_{\theta 0}^2(a)$ is the internal inductance. As a result, we obtain

$$\xi_1(\varrho) = \frac{\varrho^2}{2R}\left[\ln\frac{\varrho}{a} + \left(1 - \frac{a^2}{\varrho^2}\right)\left(\frac{2\mu_0\langle P\rangle_a}{B_{\theta 0}^2(a)} + \frac{l_i - 1}{2}\right)\right]. \tag{10.2.49}$$

If the plasma is located inside an ideally conducting chamber of radius $\varrho = b$, the expression for $\xi_1(b)$ determines the magnitude of the displacement of the center of the cross section of the plasma with respect to the center of the chamber. Expression (10.2.49) shows that $\xi_1(b)$ is proportional to the poloidal beta $\beta_p = 2\mu_0\langle P\rangle_a / B_{\theta 0}^2(a)$, which means that there is a limit on beta determined by the shift of the plasma column. This limit is called the *equilibrium beta limit* and usually corresponds to the beta value with $\xi_1(b) \sim a/2$. In actual experiments the conducting walls are usually made of copper which has a high but not infinite conductivity. Hence, the flux can only remain compressed about a resistive diffusion time of magnetic field in the conducting walls, which is often shorter than the experimental times of interest. We can choose the magnitude and sign of the vertical magnetic field generated by the external coils such that it produces an inward $\boldsymbol{J} \times \boldsymbol{B}_v$ force to reach MHD equilibrium. When the vertical magnetic field is imposed, the displacement of the plasma column becomes

$$\frac{\xi_1(b)}{b} = \frac{b}{2R}\left[\left(\beta_p + \frac{l_i - 1}{2}\right)\left(1 - \frac{a^2}{b^2}\right) + \ln\frac{b}{a}\right] - \frac{B_v}{B_{\theta 0}(b)}, \tag{10.2.50}$$

where $B_{\theta 0}(b) = \mu_0 I_0 / 2\pi b$ and I_0 is the total plasma current. The internal inductance depends on the current profile and is typically $l_i > 1/2$.

The evolution of MHD equilibrium is studied by using Ohm's law along the magnetic field,

$$\boldsymbol{E} \cdot \boldsymbol{B} = \eta \boldsymbol{J} \cdot \boldsymbol{B} . \tag{10.2.51}$$

By using the definition of the toroidal flux Φ_t and the poloidal flux Φ_p, the magnetic fields along θ and z in cylindrical coordinates are given by

$$B_\theta = -\frac{\partial \Phi_p}{\partial r}, \qquad B_z = \frac{1}{2\pi r}\frac{\partial \Phi_t}{\partial r} . \tag{10.2.52}$$

The electric fields are found from

$$E_\theta = -\frac{1}{2\pi r}\frac{\partial \Phi_t}{\partial t}, \qquad E_z = -\frac{\partial \Phi_p}{\partial t}, \tag{10.2.53}$$

where we have assumed $E_r = 0$. Thus (10.2.51) becomes

$$\frac{\partial \Phi_p}{\partial t} + \frac{B_\theta}{2\pi r B_z}\frac{\partial \Phi_t}{\partial t} = \frac{\eta}{\mu_0}\frac{B^2}{B_z^2}\frac{1}{r}\frac{\partial}{\partial r} r \frac{\partial \Phi_p}{\partial r} - \eta \frac{B_\theta}{B_z^2}\frac{\partial P}{\partial r} . \tag{10.2.54}$$

Here the equilibrium relation $J_\theta = (\partial P/\partial r + J_z B_\theta)/B_z$ was used. The plasma evolution according to (10.2.54) can be viewed as a series of quasistatic equilibria. Now we consider plasma heating on a time scale assumed to be short compared to the magnetic diffusion time $\tau_s = \mu_0 a^2/\eta$. In this situation the electrical properties of the plasma during the heating process are the same as those of a perfect conductor and the right hand side of (10.2.54) is negligible. Then the frozenness of the safety factor $q = -(d\Phi_t/d\Phi_p)/2\pi R$, is assured by noting that (10.2.54) reduces to $\partial \Phi_p/\partial t + (1/2\pi R q)\partial \Phi_t/\partial t = 0$. This sequence of MHD equilibrium evolution is called the *flux conserving torus* (FCT). In the numerical study of tokamak MHD equilibria by using the Grad-Shafranov equation, the FCT concept is useful to obtain a high beta equilibrium under a constant safety factor profile.

10.3 Interchange Instability

First consider two separate thin flux tubes in a plasma along the magnetic field line as shown in Fig. 10.5. If we interchange these tubes without disturbing the rest of the system no changes in the energy should occur. If the interchange does lower the potential energy of the system then it must be unstable [10.10–15]. The energy of this instability, that is, the change in magnetic energy due to interchange of the flux tubes is

$$\delta W_m = \delta \left(\int \frac{B^2}{2\mu_0} A dl \right) = \delta \left(\frac{\Phi^2}{2\mu_0} \int \frac{dl}{A} \right) , \tag{10.3.1}$$

where the integral is taken along the length of the flux tube. Here $\Phi = BA$ =const and A is the cross-sectional area. By adopting the notation of Fig. 10.5,

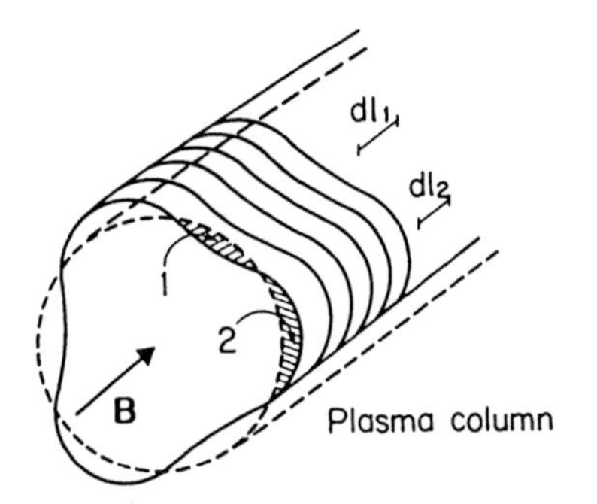

Fig. 10.5. Interchange displacement of flux tube 1 with flux tube 2. *Dashed lines* show a cylindrical plasma column; dl_1 and dl_2 are line elements along the magnetic field for the each flux tube

$$\begin{aligned}\delta W_{\mathrm{m}} &= \frac{1}{2\mu_0}\left[\Phi_2^2\left(\int\frac{dl_1}{A_1}-\int\frac{dl_2}{A_2}\right)+\Phi_1^2\left(\int\frac{dl_2}{A_2}-\int\frac{dl_1}{A_1}\right)\right]\\ &= \frac{1}{2\mu_0}\left(\Phi_2^2-\Phi_1^2\right)\left(\int\frac{dl_1}{A_1}-\int\frac{dl_2}{A_2}\right)\ .\end{aligned} \tag{10.3.2}$$

In the case of an incompressible fluid the specific heat ratio γ goes to infinity. We shall later show [in (10.3.20)] that the larger the value of γ, the more stable the system. Therefore, an incompressible fluid is at least as stable as a compressible one in the same configuration. The condition of incompressibility implies that the two tubes have equal volume. In this case $A_1 dl_1 = A_2 dl_2$ and (10.3.2) can be written as

$$\delta W_{\mathrm{m}} = \frac{1}{2\mu_0}(\Phi_2^2-\Phi_1^2)\int\left[1-\left(\frac{dl_2}{dl_1}\right)^2\right]\frac{dl_1}{A_1}\ , \tag{10.3.3}$$

If a fluid is completely penetrated by an external field in a mirror geometry with $dl_2 > dl_1$, then $\Phi_2 < \Phi_1$ (since $B_2 < B_1$) and there is no interchange instability for an incompressible fluid.

To calculate the change in the potential energy when the pressure is changed we first take two infinitesimally short tube sections of lengths dl_1 and dl_2 and perform integration along the tubes afterward. The change in energy is then

$$\begin{aligned}\delta W_{\mathrm{pot}} &= \frac{1}{\gamma-1}\delta(PV) = \frac{1}{\gamma-1}\delta\left(\frac{PV^\gamma}{V^{\gamma-1}}\right)\\ &= \frac{1}{\gamma-1}\left[(PV^\gamma)_2\left(\frac{1}{V_1^{\gamma-1}}-\frac{1}{V_2^{\gamma-1}}\right)+(PV^\gamma)_1\left(\frac{1}{V_2^{\gamma-1}}-\frac{1}{V_1^{\gamma-1}}\right)\right]\\ &= \frac{1}{\gamma-1}[(PV^\gamma)_2-(PV^\gamma)_1]\left(\frac{1}{V_1^{\gamma-1}}-\frac{1}{V_2^{\gamma-1}}\right)\ ,\end{aligned} \tag{10.3.4}$$

where use has been made of the adiabatic equation of state. Since

$$\frac{1}{V_2^{\gamma-1}}-\frac{1}{V_1^{\gamma-1}} = \delta\left(\frac{1}{V^{\gamma-1}}\right) = \frac{1-\gamma}{V^\gamma}\delta V\ , \tag{10.3.5}$$

it follows that

$$\delta W_{\mathrm{pot}} = \frac{\delta(PV^\gamma)}{V^\gamma}\delta V\ . \tag{10.3.6}$$

Let us interchange two flux tubes which contain equal flux. If $\Phi_2 = \Phi_1$, $\delta W_m = 0$, and the only energy change results from δW_{pot}. The system is then certainly unstable if

$$\delta(PV^\gamma)\delta V = (\delta PV^\gamma + \gamma PV^{\gamma-1}\delta V)\delta V < 0 . \tag{10.3.7}$$

Near the fluid boundary $P \to 0$ and the first term dominates over the second term. If we let $\delta P = P_2 - P_1 < 0$ instability develops if

$$\delta V > 0 \tag{10.3.8}$$

or, after integration along the length of the tube, if

$$\delta\left(\int A dl\right) = \Phi\delta\left(\int \frac{dl}{B}\right) > 0 . \tag{10.3.9}$$

Considering a mirror configuration where the fluid is penetrated by the magnetic field (i.e., the field is approximately the same as it is in the absence of the fluid), the interchange of an inner flux tube with an outer flux tube results in an increase of dl/B since dl is large in the outer region where B is small. Therefore, (10.3.9) is satisfied, hence the system is unstable.

This can be generalized by the statement that geometries in which the magnetic field lines curve toward the fluid along the entire fluid boundary are interchange-unstable.

One can suppress this instability. For instance, a combination of a mirror geometry with quadrupole stabilizing coils prevents interchange instabilities by introducing an "outwardly increasing" magnetic field at the boundary with a cusp-like curvature (Fig.10.6). The magnetic field increasing toward the boundary is called the *magnetic well*.

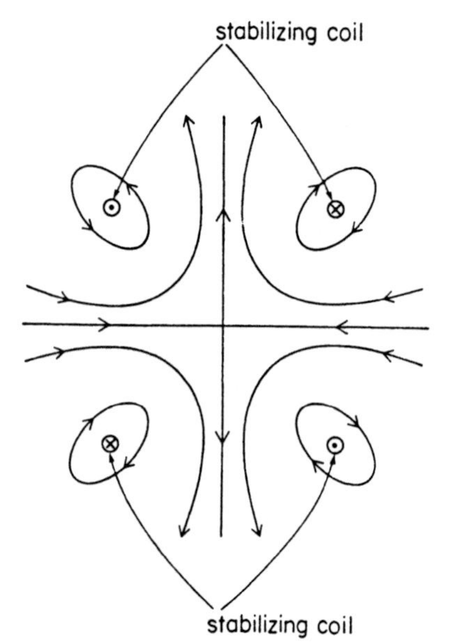

Fig. 10.6. A cusp field which stabilizes interchange instability

The MHD linear stability is analyzed more systematically based on the energy principle. First, a small perturbation with displacement vector $\boldsymbol{\xi}$ is added to the MHD equilibrium equations. Then the pressure and magnetic field are described as

$$\begin{aligned} P &= P_0 + P_1 + P_2 + \ldots \\ \boldsymbol{B} &= \boldsymbol{B}_0 + \boldsymbol{B}_1 + \boldsymbol{B}_2 + \ldots \end{aligned} \tag{10.3.10}$$

where $P_1, \boldsymbol{B}_1$ are of first order in $\boldsymbol{\xi}$ and $P_2, \boldsymbol{B}_2$ are of second order in $\boldsymbol{\xi}$. Equation (10.3.10) is substituted into the potential energy

$$W = \int \left(\frac{B_0^2}{2\mu_0} + \frac{\boldsymbol{B}_0 \cdot \boldsymbol{B}_1}{\mu_0} + \frac{B_1^2}{2\mu_0} + \frac{\boldsymbol{B}_0 \cdot \boldsymbol{B}_2}{\mu_0} + \frac{P_0}{\gamma - 1} + \frac{P_1}{\gamma - 1} + \frac{P_2}{\gamma - 1} \right) dV . \tag{10.3.11}$$

Here $\boldsymbol{B}_1, P_1, \boldsymbol{B}_2$ and P_2 are given through (10.1.7, 8, 10) as

$$\boldsymbol{B}_1 = \int \nabla \times \left(\frac{\partial \boldsymbol{\xi}}{\partial t} \times \boldsymbol{B}_0 \right) dt = \nabla \times (\boldsymbol{\xi} \times \boldsymbol{B}_0) , \tag{10.3.12}$$

$$P_1 = -\int \left(\gamma P_0 \nabla \cdot \frac{\partial \boldsymbol{\xi}}{\partial t} + \frac{\partial \boldsymbol{\xi}}{\partial t} \cdot \nabla P_0 \right) dt = -(\gamma P_0 \nabla \cdot \boldsymbol{\xi} + \boldsymbol{\xi} \cdot \nabla P_0) , \tag{10.3.13}$$

$$\boldsymbol{B}_2 = \int \nabla \times \left(\frac{\partial \boldsymbol{\xi}}{\partial t} \times \boldsymbol{B}_1 \right) dt = \frac{1}{2} \nabla \times (\boldsymbol{\xi} \times \boldsymbol{B}_1) , \tag{10.3.14}$$

$$\begin{aligned} P_2 &= -\int \left(\gamma P_1 \nabla \cdot \frac{\partial \boldsymbol{\xi}}{\partial t} + \frac{\partial \boldsymbol{\xi}}{\partial t} \cdot \nabla P_1 \right) dt \\ &= \frac{1}{2} [\gamma (\gamma P_0 \nabla \cdot \boldsymbol{\xi} + \boldsymbol{\xi} \cdot \nabla P_0) \nabla \cdot \boldsymbol{\xi} + \boldsymbol{\xi} \cdot \nabla (\gamma P_0 \nabla \cdot \boldsymbol{\xi} + \boldsymbol{\xi} \cdot \nabla P_0)] , \end{aligned} \tag{10.3.15}$$

where $\boldsymbol{\xi}(0) = 0$, $\partial \boldsymbol{\xi} / \partial t |_{t=0} \neq 0$ and exponential time dependence is assumed for $\boldsymbol{\xi}, P_1$ and $\boldsymbol{B}_1$. The second term in (10.3.11) is written as

$$\begin{aligned} \int \frac{\boldsymbol{B}_0 \cdot \boldsymbol{B}_1}{\mu_0} dV &= \frac{1}{\mu_0} \int (\nabla \times \boldsymbol{B}_0) \times (\boldsymbol{\xi} \times \boldsymbol{B}_0) dV \\ &= \int \boldsymbol{J}_0 \cdot (\boldsymbol{\xi} \times \boldsymbol{B}_0) dV = -\int \boldsymbol{\xi} \cdot (\boldsymbol{J}_0 \times \boldsymbol{B}_0) dV \end{aligned} \tag{10.3.16}$$

where $\boldsymbol{\xi} \cdot \boldsymbol{n}|_s = 0$ is imposed at the conducting boundary. The normal unit vector on the boundary is denoted by $\boldsymbol{n}$. It is noted that $\boldsymbol{B}_0$ is tangential to the conducting boundary from the ideal MHD boundary condition. The sixth term of (10.3.11) becomes

$$-\int \frac{1}{\gamma - 1} (\gamma P_0 \nabla \cdot \boldsymbol{\xi} + \boldsymbol{\xi} \cdot \nabla P_0) dV = -\int \frac{\gamma P_0}{\gamma - 1} \boldsymbol{\xi} \cdot \boldsymbol{n} \, dS + \int \boldsymbol{\xi} \cdot \nabla P_0 dV . \tag{10.3.17}$$

The first term of the right hand side vanishes by the condition $\boldsymbol{\xi} \cdot \boldsymbol{n}|_s = 0$ on the conducting boundary. Thus the sum of the second term and the sixth term of (10.3.11) becomes zero from the MHD equilibrium $\nabla P_0 = \boldsymbol{J}_0 \times \boldsymbol{B}_0$. The remaining terms are written as

$$\int \frac{B_0 \cdot B_2}{\mu_0} dV = \frac{1}{2\mu_0} \int B_0 \cdot \nabla \times (\xi \times B_1) dV = \frac{1}{2} \int \xi \cdot (B_1 \times J_0) dV \,, \tag{10.3.18}$$

by the same calculation as (10.3.17) and

$$\begin{aligned} \int \frac{P_2}{\gamma - 1} dV &= -\frac{1}{2} \frac{1}{(\gamma - 1)} \int (\gamma P_0 \nabla \cdot \xi + \xi \cdot \nabla P_0) \xi \cdot n \, dS \\ &\quad + \frac{1}{2} \int (\gamma P_0 \nabla \cdot \xi + \xi \cdot \nabla P_0) \nabla \cdot \xi \, dV \,. \end{aligned} \tag{10.3.19}$$

Thus the second order term of the potential energy (10.3.11) designated by δW is obtained as

$$\delta W = \frac{1}{2} \int \left[\frac{B_1^2}{\mu_0} - \xi \cdot (J_0 \times B_1) + (\gamma P_0 \nabla \cdot \xi + \xi \cdot \nabla P_0) \nabla \cdot \xi \right] dV \,. \tag{10.3.20}$$

MHD stability depends on the sign of δW by analogy to particle motion in a potential field. For a given perturbation, if $\delta W < 0$, the MHD equilibrium is unstable. If $\delta W > 0$ for any ξ which satisfies the boundary condition, then the equilibrium is stable. This criterion is called the *energy principle* for linear stability analyses.

As an example of the stability analysis based on the energy principle, we discuss the interchange mode in a cylindrical plasma. The perturbation is Fourier analyzed along θ and z directions with mode number m and wave number k: $\xi(r, \theta, z) = \sum_{m,k} \xi_{mk}(r) \exp(\mathrm{i}m\theta + \mathrm{i}kz)$. The potential energy can be checked for each Fourier component with $\{m, k\}$. By noting that a cylindrical MHD equilibrium has $P_0(r)$ and $B_0 = (0, B_\theta^0(r), B_z^0(r))$, the variables are changed from $(\xi_r, \xi_\theta, \xi_z)$ to (ξ, η, ζ) by

$$\begin{aligned} \xi &= \xi_r \\ \eta &= \nabla \cdot \xi - \frac{1}{r} \frac{d}{dr}(r\xi_r) = \frac{\mathrm{i}m}{r} \xi_\theta + \mathrm{i}k\xi_z \\ \zeta &= \mathrm{i}(\xi \times B_0)_r = \mathrm{i}\xi_\theta B_z^0 - \mathrm{i}\xi_z B_\theta^0 \,. \end{aligned} \tag{10.3.21}$$

The magnetic perturbation B_1 is expressed by

$$\begin{aligned} B_{1r} &= \mathrm{i}F\xi \\ B_{1\theta} &= k\zeta - \frac{d}{dr}(\xi B_\theta^0) \,, \\ B_{1z} &= -\frac{m}{r}\zeta - \frac{1}{r}\frac{d}{dr}(\xi r B_z^0) \,, \end{aligned} \tag{10.3.22}$$

where $F \equiv mB_\theta^0/r + kB_z^0$.

Then the energy integral (10.3.20) is written explicitly for the $\{m, k\}$ mode by performing integration over θ and z,

$$\begin{aligned}
\delta W_{mk} &= \frac{2\pi L}{\mu_0}\int_0^a \left[F^2|\xi|^2 + \left| k\zeta - \frac{d}{dr}(\xi B_\theta^0)\right|^2 + \left|\frac{m}{r}\zeta + \frac{1}{r}\frac{d}{dr}(\xi r B_z^0)\right|^2 \right. \\
&+ \mu_0 \mathrm{Re}\left\{ J_\theta^0 \xi \left(\frac{m}{r}\zeta^* + \frac{1}{r}\frac{d}{dr}(\xi^* r B_z^0)\right) + J_z^0 \xi \left(k\zeta^* - \frac{d}{dr}(\xi^* B_\theta^0)\right) \right. \\
&+ \left. J_z^0 \xi (B_\theta^0 \eta^* + k\zeta^*) - J_\theta^0 \xi \left(B_z^0 \eta^* - \frac{m}{r}\zeta^*\right)\right\} \\
&+ \gamma \mu_0 P_0 \left| \eta + \frac{1}{r}\frac{d}{dr}(r\xi)\right|^2 \\
&+ \left. \mu_0 \frac{dP_0}{dr} \mathrm{Re}\left\{ \xi \left[\eta^* + \frac{1}{r}\frac{d}{dr}(r\xi^*)\right]\right\}\right] r\, dr , \qquad (10.3.23)
\end{aligned}$$

where a and L denote the radius and length of the cylindrical plasma. By using the equilibrium condition, the variable η disappears except in the eighth term and minimization of δW_{mk} with respect to η reduces to $\eta = -(1/r)d(r\xi)/dr$. Then (10.3.23) can be reduced to the form

$$\delta W_{mk} = \frac{2\pi L}{\mu_0}\int_0^a \left(A|\zeta|^2 + 2\mathrm{Re}\{B\zeta^*\} + C\right) r dr$$

where

$$\begin{aligned}
A &= k^2 + \frac{m^2}{r^2} , \\
B &= -k\frac{d}{dr}(\xi B_\theta^0) + \frac{m}{r^2}\frac{d}{dr}(\xi r B_z^0) - \frac{m}{r}\xi\frac{dB_z^0}{dr} + k\xi\left(\frac{B_\theta^0}{r} + \frac{dB_\theta^0}{dr}\right) \\
&= \left(-kB_\theta^0 + \frac{m}{r}B_z^0\right)\frac{d\xi}{dr} + \left(kB_\theta^0 + \frac{m}{r}B_z^0\right)\frac{\xi}{r} , \\
C &= \left(kB_z^0 + \frac{m}{r}B_z^0\right)^2 |\xi|^2 + |\frac{d}{dr}(\xi B_\theta^0)|^2 + |\frac{1}{r}\frac{d}{dr}(\xi r B_z^0)|^2 \\
&+ \mathrm{Re}\left\{ -\xi\frac{dB_z^0}{dr}\left[\frac{1}{r}\frac{d}{dr}(\xi^* r B_z^0) + \frac{1}{r}\frac{d}{dr}(\xi^* r)B_z^0\right] \right. \\
&+ \left. \frac{\xi}{r}\frac{d}{dr}(rB_\theta^0)\left[-\frac{d}{dr}(\xi^* B_\theta^0) - \frac{1}{r}\frac{d}{dr}(r\xi^* B_\theta^0)\right]\right\} .
\end{aligned}$$

Minimization with respect to ζ then gives

$$\zeta = -\frac{B}{A} = \frac{r}{k^2r^2 + m^2}\left[(krB_\theta^0 - mB_z^0)\frac{d\xi}{dr} - (krB_\theta^0 + mB_z^0)\frac{\xi}{r}\right] . \qquad (10.3.24)$$

After this minimization, we get

$$\begin{aligned}
\delta W_{mk} &= \frac{2\pi L}{\mu_0}\int_0^a \left(-\frac{|B|^2}{A} + C\right) r\, dr \\
&= \frac{2\pi L}{\mu_0}\int_0^a \left[\tilde{a}\left|\frac{\xi}{r}\right|^2 + \tilde{b}\left|\frac{d\xi}{dr}\right|^2 + 2\tilde{c}\mathrm{Re}\left\{\frac{d\xi^*}{dr}\frac{\xi}{r}\right\}\right] r\, dr ,
\end{aligned}$$

where

$$\tilde{a} = -\frac{1}{m^2+k^2r^2}(krB_\theta^0 + mB_z^0)^2 + (krB_z^0 + mB_\theta^0)^2$$
$$-2B_\theta^0 \frac{d}{dr}(rB_\theta^0) + (B_\theta^0)^2 + (B_z^0)^2$$
$$= \frac{(mB_\theta^0 - krB_z^0)^2}{m^2+k^2r^2} + (krB_z^0 + mB_\theta^0)^2 - 2B_\theta^0 \frac{d}{dr}(rB_\theta^0) ,$$
$$\tilde{b} = -\frac{1}{m^2+k^2r^2}(-krB_\theta^0 + mB_z^0)^2 + (B_\theta^0)^2 + (B_z^0)^2 = \frac{(mB_\theta^0 + krB_z^0)^2}{m^2+k^2r^2}$$
$$\tilde{c} = -\frac{1}{m^2+k^2r^2}[m^2(B_z^0)^2 - k^2r^2(B_\theta^0)^2] + (B_z^0)^2 - (B_\theta^0)^2$$
$$= \frac{1}{m^2+k^2r^2}(krB_z^0 + mB_\theta^0)(krB_z^0 - mB_\theta^0) ,$$

from which we have

$$\delta W_{mk} = \frac{2\pi L}{\mu_0}\int_0^a \Lambda\left(\xi, \frac{d\xi}{dr}\right) r\, dr , \tag{10.3.25}$$

where

$$\Lambda\left(\xi, \frac{d\xi}{dr}\right) = \frac{1}{k^2r^2+m^2}\left[(krB_z^0 + mB_\theta^0)\frac{d\xi}{dr} + (krB_z^0 - mB_\theta^0)\frac{\xi}{r}\right]^2$$
$$+\left[(krB_z^0 + mB_\theta^0)^2 - 2B_\theta^0\frac{d}{dr}(rB_\theta^0)\right]\frac{\xi^2}{r^2} . \tag{10.3.26}$$

The minimization with respect to η means that $\nabla \cdot \boldsymbol{\xi} = 0$. This means that the perturbation to minimize the potential energy is incompressible although the plasma itself is not incompressible.

From (10.3.26), if a configuration has

$$B_\theta^0 \frac{d}{dr}(rB_\theta^0) < 0 \quad \text{or} \quad B_\theta^0 J_z^0 < 0 \tag{10.3.27}$$

for any r, then $\delta W_{mk} > 0$ or it is magnetohydrodynamically stable. The potential energy of the cylindrical plasma (10.3.25) is expressed in another form by using the relation

$$2\int \tilde{c}\mathrm{Re}\left\{\frac{\xi}{r}\frac{d\xi^*}{dr}\right\} r\, dr = \int \tilde{c}\left[\xi\frac{d\xi^*}{dr} + \xi^*\frac{d\xi}{dr}\right] dr$$
$$= -\int \frac{d\tilde{c}}{dr}|\xi|^2 dr$$

and using the equilibrium condition $dP_0/dr = J_\theta^0 B_z^0 - J_z^0 B_\theta^0$, as

$$\delta W_{mk} = \frac{2\pi L}{\mu_0}\int_0^a \left[f\left|\frac{d\xi}{dr}\right|^2 + g|\xi|^2\right] dr , \tag{10.3.28}$$

where

$$f = \frac{r(krB_z^0 + mB_\theta^0)^2}{k^2r^2+m^2} , \tag{10.3.29}$$

$$g = \frac{2k^2r^2}{k^2r^2+m^2}\mu_0\frac{dP}{dr} + \frac{1}{r}(krB_z^0 + mB_\theta^0)^2\frac{k^2r^2+m^2-1}{k^2r^2+m^2} + \frac{2k^2r}{(k^2r^2+m^2)^2}[k^2r^2(B_z^0)^2 - m^2(B_\theta^0)^2] \,. \tag{10.3.30}$$

The radial perturbation ξ which minimizes δW_{mk} can be obtained by solving the Euler-Lagrange equation

$$\frac{d}{dr}\left(f\frac{d\xi}{dr}\right) - g\xi = 0 \,. \tag{10.3.31}$$

When $krB_z^0+mB_\theta^0 = 0$ at the position of $r = r_s$ inside the plasma column, the Euler-Lagrange equation (10.3.31) has a regular singular point at r_s since f has a zero of second order. By noting that the parallel wavenumber of the perturbation is $k_\| = (m/r)B_\theta^0/B_0 + kB_z^0/B_0$, $f = 0$ is equivalent to $k_\| = 0$. By using the safety factor $q = (r/R)B_z^0/B_\theta^0$, the parallel wavenumber is equal to $(B_\theta^0/rB_0)(m - nq)$, where the wavenumber in the x direction is $k = -n/R$ for a cylindrical plasma with the periodic length $2\pi R$ and $B_0^2 = (B_\theta^0)^2 + (B_z^0)^2$. Therefore, the singular point appears for an equilibrium configuration with a spatial dependence of q or a sheared magnetic field of $dq/dr \neq 0$.

We now assume that $\xi(r)$ is continuous and bounded in $|r - r_s| < \varepsilon$ and $\xi(r) = 0$ in $|r - r_s| > \varepsilon$, where ε is a very small length. For the perturbation, (10.3.28) becomes

$$\delta W_{mk} \doteq \frac{2\pi L}{\mu_0}\frac{r_s^3 k_\|'^2 B_0^2}{m^2 + k^2 r_s^2}\varepsilon \int_{-1}^{1}\left(x^2\left|\frac{d\xi}{dx}\right|^2 + D|\xi|^2\right) dx \,, \tag{10.3.32}$$

by expanding f and g around $r = r_s$ as $f \doteq k_\|'^2 B_0^2 r_s^3 (r - r_s)^2/(m^2 + k^2 r_s^2)$ and $g \doteq 2k^2 r_s^2 \mu_0 (dP_0/dr)/(m^2 + k^2 r_s^2) \doteq D k_\|'^2 B_0^2 r_s^3/(m^2 + k^2 r_s^2)$, where

$$k_\|' = \frac{kB_z^0}{B_0}\frac{d}{dr}\ln\left(\frac{B_\theta^0}{rB_z^0}\right) = k\frac{B_z^0}{B_0}\frac{\iota'}{\iota} \,, \tag{10.3.33}$$

$$D = \frac{2\mu_0 P_0'}{r_s (B_z^0)^2}\left(\frac{\iota'}{\iota}\right)^2 \,. \tag{10.3.34}$$

Here the rotational transform ι is equal to the inverse of the safety factor. The normalized length is $x = (r - r_s)/\varepsilon$ and the prime denotes a derivative with respect to r. For the integral of (10.3.32), we use the inequality

$$0 \leq \int_{-1}^{1}\left|x\frac{d\xi}{dx} + \frac{\xi}{2}\right|^2 dx = \int_{-1}^{1}\left[x^2\left|\frac{d\xi}{dx}\right|^2 + \frac{|\xi|^2}{4} - \frac{|\xi|^2}{2}\right] dx \,, \tag{10.3.35}$$

where we used the boundary condition $\xi(1) = \xi(-1) = 0$. By using (10.3.35), $\delta W_{mk} > 0$ requires $D + 1/4 > 0$. This criterion is written as

$$\frac{2\mu_0}{(B_z^0)^2}\frac{dP_0}{dr} + \frac{r_s}{4}\left(\frac{\iota'}{\iota}\right)^2 \geq 0 \tag{10.3.36}$$

for stability against the localized mode near the singular point. The inequality (10.3.36) is called the *Suydam criterion.* The pressure gradient is normally negative in a magnetically confined plasma and the first term of (10.3.36) has a negative contribution or destabilizing effect. The second term is positive and is stabilizing due to the magnetic shear effect. Magnetic shear generally has a stabilizing effect on interchange instability.

10.4 Rayleigh-Taylor Instability

As another example of a pressure-driven mode, we now consider the magnetohydrodynamic version of the Rayleigh-Taylor instability [10.1, 3, 10]. We consider a magnetohydrodynamic fluid supported by a magnetic field without shear against an effective gravitational field. To treat this case we have to include the effective gravitational force density in the equation of motion

$$\boldsymbol{F}_g = -\varrho\nabla\phi_g \tag{10.4.1}$$

where ϕ_g is the gravitational potential. In the presence of a displacement $\boldsymbol{\xi}$, the first order gravitational force density is written as

$$\boldsymbol{F}_{g1} = -\varrho_1\nabla\phi_g = \nabla\cdot(\varrho_0\boldsymbol{\xi})\nabla\phi_g\ , \tag{10.4.2}$$

where the equation of continuity was used. For a uniform gravitational field acting in the x-direction $\boldsymbol{F}_g = -\varrho g\hat{\boldsymbol{e}}_x(g > 0)$, the equilibrium equation reads

$$\frac{\partial}{\partial x}\left(P + \frac{B^2}{2\mu_0}\right) + \varrho g = 0 \tag{10.4.3}$$

and the first order equation of motion becomes

$$\begin{aligned}-\varrho_0\omega^2\boldsymbol{\xi} \quad &= \quad -\nabla P_1 + \frac{1}{\mu_0}[(\nabla\times\boldsymbol{B}_1)\times\boldsymbol{B}_0 \\ &\qquad + (\nabla\times\boldsymbol{B}_0)\times\boldsymbol{B}_1] + g\nabla\cdot(\varrho_0\boldsymbol{\xi})\hat{\boldsymbol{e}}_x\ .\end{aligned} \tag{10.4.4}$$

We are going to investigate the stability of an incompressible fluid by considering strong magnetic fields. We take $\boldsymbol{B} = B_0\hat{\boldsymbol{e}}_z$, assume that equilibrium quantities depend on x only, and consider only perturbations which do not vary along the magnetic field (Fig. 10.7),

$$(\hat{\boldsymbol{e}}_z\cdot\nabla)\boldsymbol{\xi} = 0\ . \tag{10.4.5}$$

For perturbed magnetic fields one may write

$$\boldsymbol{B}_1 = \nabla\times(\boldsymbol{\xi}\times\boldsymbol{B}_0) = -\xi_x\frac{dB_0}{dx}\hat{\boldsymbol{e}}_z \tag{10.4.6}$$

$$(\nabla\times\boldsymbol{B}_1)\times\boldsymbol{B}_0 + (\nabla\times\boldsymbol{B}_0)\times\boldsymbol{B}_1 = -\nabla(\boldsymbol{B}_0\cdot\boldsymbol{B}_1)\ , \tag{10.4.7}$$

by noting that B_0 is independent of z. Equation (10.4.4) now becomes

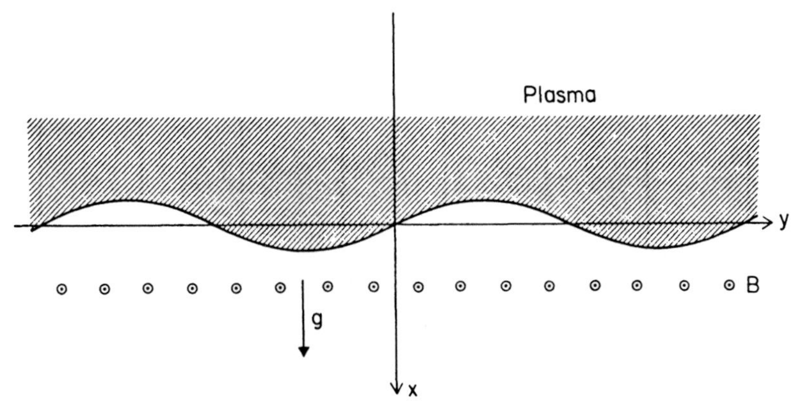

Fig. 10.7. Deformation of a plasma surface at equilibrium by a gravitational force g

$$\varrho_0\omega^2\boldsymbol{\xi} = \nabla\left(P_1 + \frac{\boldsymbol{B}_0\cdot\boldsymbol{B}_1}{\mu_0}\right) - g\boldsymbol{\xi}\cdot\nabla\varrho_0\hat{e}_x\,, \tag{10.4.8}$$

where use has been made of the incompressibility condition $\nabla\cdot\boldsymbol{\xi} = 0$.

The translational symmetry of the system in the y-direction permits solutions of the form $\mathrm{e}^{iky}(k > 0)$ and the corresponding x and y component equations of (10.4.8) become

$$\varrho_0\omega^2\xi_x = \frac{\partial}{\partial x}\left(P_1 + \frac{1}{\mu_0}\boldsymbol{B}_0\cdot\boldsymbol{B}_1\right) - g\xi_x\frac{d\varrho_0}{dx} \tag{10.4.9}$$

$$\varrho_0\omega^2\xi_y = \mathrm{i}k\left(P_1 + \frac{1}{\mu_0}\boldsymbol{B}_0\cdot\boldsymbol{B}_1\right)\,. \tag{10.4.10}$$

Dividing (10.4.10) by $\mathrm{i}k$ we get an expression for the sum of the first order fluid and magnetic pressures which we can then substitute into (10.4.9) to get

$$\varrho_0\omega^2\xi_x = \frac{\omega^2}{\mathrm{i}k}\frac{\partial}{\partial x}(\varrho_0\xi_y) - g\frac{d\varrho_0}{dx}\xi_x\,. \tag{10.4.11}$$

The displacement ξ_y can be eliminated by differentiating (10.4.11) with respect to y and using $\nabla\cdot\boldsymbol{\xi} = 0$ to obtain

$$\mathrm{i}k\varrho_0\omega^2\xi_x = -\frac{\omega^2}{\mathrm{i}k}\frac{\partial}{\partial x}\left(\varrho_0\frac{\partial\xi_x}{\partial x}\right) - \mathrm{i}kg\frac{d\varrho_0}{dx}\xi_x\,. \tag{10.4.12}$$

We thus arrive at the differential equation which describes the variation of ξ_x in the x-direction:

$$\frac{\partial}{\partial x}\left(\varrho_0\omega^2\frac{\partial\xi_x}{\partial x}\right) - k^2\left(\varrho_0\omega^2 + g\frac{d\varrho_0}{dx}\right)\xi_x = 0\,. \tag{10.4.13}$$

Assume that the fluid is of uniform density ϱ_0 in the region $x < 0$ and is separated from the vacuum by a thin current carrying layer at $x = 0$. Note that this assumption does not exclude the presence of a magnetic field in the fluid interior. For $x < 0$, $d\varrho_0/dx = 0$ and we get

$$\frac{\partial^2 \xi_x}{\partial x^2} - k^2 \xi_x = 0 . \qquad (10.4.14)$$

The only solution of (10.4.14) which does not diverge for $x \to -\infty$ is

$$\xi_x = \xi_x^0 e^{iky} e^{kx} \qquad (10.4.15)$$

Since $d\varrho_0/dx = \varrho_0 \delta(x)$, one may integrate (10.4.13) across the boundary from $x = -\varepsilon$ to $x = \varepsilon(\varepsilon \to +0)$, to obtain $\varrho_0 \omega^2 k \xi_x + k^2 g \varrho_0 \xi_x = 0$, where we used (10.4.15). The dispersion relation for the Rayleigh-Taylor instability is then

$$\omega^2 = -kg(< 0) . \qquad (10.4.16)$$

The solution is unstable for every wavenumber. In fact the instability grows faster at larger k. This is also called the ***Kruskal-Schwarzchild instability***. Its significance lies not so much in the unstable behavior of a fluid supported against a gravitational field but in the instability exhibited by systems supported by a magnetic field against inertial forces in the magnetic confinement system. Clearly a fluid accelerated by a pusher in a typical inertial confinement configuration, with the acceleration vector pointing toward the fluid, can be viewed from the moving system as being subject to the equivalent of a gravitational field and must be unstable (Fig. 9.15).

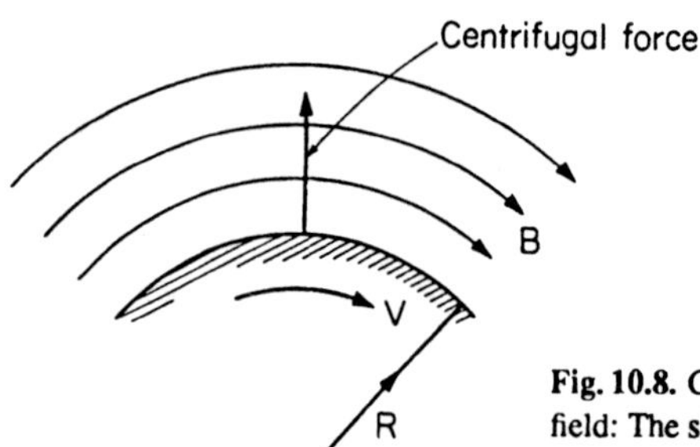

Fig. 10.8. Centrifugal force produced by a curved magnetic field: The shaded region shows a plasma and V is a velocity along the magnetic field line

A fluid streaming along slightly inward-curving field lines (Fig. 10.8) is another example of such an unstable system usually occurring in magnetically confined plasmas, where now the centrifugal force plays the role of the gravitational force.

When the magnetic lines of force that confine the plasma have finite radius of curvature, the radius vector R_c of the curvature is given by

$$\frac{\boldsymbol{R}_c}{R_c^2} = -\left(\frac{\boldsymbol{B}}{B} \cdot \nabla\right) \frac{\boldsymbol{B}}{B} . \qquad (10.4.17)$$

By noting Problem 3.3, we may apply the result of the Rayleigh-Taylor instability analysis to the present case by the substitution,

$$F_g \to \frac{W_\perp + 2W_\parallel}{M} \frac{R_c}{R_c^2} , \qquad (10.4.18)$$

where $W_\perp$ and $W_\parallel$ are the energies of the particle motions perpendicular and parallel to the magnetic field.

From the dispersion relation (10.4.16), we then find that a system in which the magnetic field confines the plasma in a convex shape is unstable because $\boldsymbol{R}_c$ is directed from the plasma to vacuum (Fig. 10.8).

10.5 Current Driven Instability (I)

The stability of a current carrying plasma in a strong longitudinal magnetic field strongly depends on whether the highly conductive plasma extends to the surface of the metal chamber (*fixed boundary*) or whether there is a vacuum region with zero conductivity between the boundary and the chamber (*free boundary*) [10.1, 10, 13, 16, 17].

Here we examine the *kink instability* in a cylindrical plasma of periodic length $L = 2\pi R$ as a model for a toroidal plasma with minor radius a and major radius R. The radius of the conducting wall is denoted by b. In such a cylindrical plasma, it is sufficient to examine perturbations of the form $\boldsymbol{\xi}(r, \theta, z) = \boldsymbol{\xi}(r) \exp[\mathrm{i}(m\theta - kz) + \gamma t]$, where the wavenumber k, by virtue of the finite length of the cylinder, takes on the discrete values $k = 2\pi n/L$ $(n = 0, 1, 2 \ldots)$.

The problem of the stability of plasmas with free boundaries can be solved using the energy principle. The expression for the potential energy δW, appearing as a function of the plasma displacement $\boldsymbol{\xi}$, can be extended into the vacuum region where $\boldsymbol{\xi}$ is determined from the expression for the perturbation of the magnetic field $\boldsymbol{B}_1 = \nabla \times (\boldsymbol{\xi} \times \boldsymbol{B}_0)$. The radial component ξ of the displacement vector $\boldsymbol{\xi}$ is associated with the radial component of the magnetic field B_{1r} by the relation

$$B_{1r} = \mathrm{i}\frac{B_\theta^0}{r}(m - nq)\xi . \qquad (10.5.1)$$

It can be seen that the displacement function ξ can have a strong singularity in the vacuum region at $m = nq$. The presence of singularities in the vacuum region is the defining characteristic of kink instabilities. First we consider the expression for δW in (10.3.28) by replacing a with b. This is not suitable for investigating the stability of plasmas with free boundaries, since it was obtained by partial integration of (10.3.25) over the region containing the singularity. We must therefore separate (10.3.25) to isolate this region by integrating the terms with $\mathrm{Re}\{\xi d\xi^*/dr\}$ by parts. Then δW can be represented as a sum of three terms

$$\delta W = \delta W_{\mathrm{pot}} + \delta W_b + \delta W_{\mathrm{v}} , \qquad (10.5.2)$$

where δW_{pot} is the same as (10.3.25) for the plasma and δW_b and δW_{v} the energy of the plasma-vacuum interface and that of the vacuum region are given by

$$\delta W_b = \frac{2\pi L}{\mu_0} \frac{k^2a^2(B_z^0)^2 - m^2(B_\theta^a)^2}{m^2 + k^2a^2} |\xi_a|^2 \qquad (10.5.3)$$

and

$$\delta W_v = \frac{2\pi L}{\mu_0} \int_a^b \left(\frac{1}{m^2 + k^2 r^2} \left| \frac{d\eta}{dr} \right|^2 + \frac{|\eta|^2}{r^2} \right) r\, dr \; . \tag{10.5.4}$$

The δW_v term is described by

$$\eta = \left(\frac{maB_\theta^a}{r} - krB_z^0 \right) \xi \; . \tag{10.5.5}$$

Here $B_\theta^a = B_\theta^0(a)$ and $\xi_a = \xi(a)$. Minimization of δW_{pot} and δW_v leads to Euler's equations

$$\frac{d}{dr}\left(f \frac{d\xi}{dr} \right) - g\xi = 0 \qquad (r \leq a) \tag{10.5.6}$$

and

$$\frac{d}{dr}\left(\frac{r}{m^2 + k^2 r^2} \frac{d\eta}{dr} \right) - \frac{\eta}{r} = 0 \qquad (r > a) \; . \tag{10.5.7}$$

In addition to meeting the zero condition for $r = 0$ and $r = b$, to solve these equations we must also satisfy the matching condition for $r = a$

$$\eta(a) = (mB_\theta^a - kaB_z^0)\xi_a \; . \tag{10.5.8}$$

The solution of Euler's equation for the vacuum region in the approximation $B_z^0 \gg B_\theta^0$ corresponding to $k^2 r^2 \ll m^2$ has the form

$$\eta = \frac{mB_\theta^a - kaB_z^0}{1 - (a/b)^{2m}} \left[\left(\frac{b}{r}\right)^m - \left(\frac{r}{b}\right)^m \right] \left(\frac{a}{b}\right)^m \xi_a \; . \tag{10.5.9}$$

Note that the approximation $B_z^0 \gg B_\theta^0$ corresponds to tokamak configurations where the resonant surface $q = m/n$ means $kr = mB_\theta^0/B_z^0 \ll m$. Substituting this value of η into δW_v, we obtain the potential energy in the vacuum region

$$\delta W_v = \frac{2\pi L}{\mu_0} \frac{(mB_\theta^a - krB_z^0)^2}{m} |\xi_a|^2 \lambda \; , \tag{10.5.10}$$

where

$$\lambda = \frac{1 + (a/b)^{2m}}{1 - (a/b)^{2m}} \; . \tag{10.5.11}$$

Thus, the total potential energy for plasmas with free boundaries is

$$\begin{aligned} \delta W \;\; = \;\; & \frac{2\pi L}{\mu_0}(B_\theta^a |\xi_a|)^2 \left[\left(1 - \frac{nq_a}{m}\right)^2 (1 + m\lambda) \right. \\ & \left. - 2\left(1 - \frac{nq_a}{m}\right) + \frac{\mu_0 \delta W_{pot}}{2\pi L (B_\theta^a)^2 |\xi_a|^2} \right] \end{aligned} \tag{10.5.12}$$

where $q_a = q(a)$. The instabilities correspond to $\delta W < 0$. Under the condition $k^2r^2/m^2 \ll 1$ only the second term of g in the potential energy $\delta W_{\rm pot}$ is important and

$$g \doteq (m^2 - 1)\frac{f}{r^2} + O(k^2r^2) . \qquad (10.5.13)$$

Then Euler's equation (10.5.6) for ξ is reduced to the simple form

$$\frac{d}{dr}\left[r^3\left(\frac{1}{q} - \frac{n}{m}\right)^2 \frac{d\xi}{dr}\right] - r\left(\frac{1}{q} - \frac{n}{m}\right)^2 (m^2 - 1)\xi = 0 . \qquad (10.5.14)$$

Also by using (10.5.6), $\delta W_{\rm pot}$ can be expressed as

$$\delta W_{\rm pot} = \frac{2\pi L}{\mu_0}\left(f\left|\xi\frac{d\xi}{dr}\right|\right)\Big|_{r=a} . \qquad (10.5.15)$$

Finally from (10.5.12,15), δW becomes

$$\begin{aligned} \delta W \ = \ & \frac{2\pi L}{\mu_0}(|\xi_a|B_\theta^a)^2\left[\left(1 - \frac{nq_a}{m}\right)^2(1 + m\lambda)\right. \\ & \left. -2\left(1 - \frac{nq_a}{m}\right) + (|\xi_a|^{-2}a)\left(1 - \frac{nq_a}{m}\right)^2\left(\left|\xi\frac{d\xi}{dr}\right|\right)\Big|_{r=a}\right] . \end{aligned} \qquad (10.5.16)$$

When the current density is constant or there is a uniform current profile, the safety factor q becomes constant in the plasma column. (Note that q must be constant when $B_\theta^0 \propto r$ for a uniform current profile.) The solution of (10.5.14) gives $r\xi'/\xi = m - 1$ and the stability criterion $\delta W > 0$ becomes

$$m\left(1 - \frac{nq_a}{m}\right)^2 - \left[1 - \left(\frac{a}{b}\right)^{2m}\right]\left(1 - \frac{nq_a}{m}\right) > 0 , \qquad (10.5.17)$$

or instability occurs for

$$m - 1 + (a/b)^{2m} < nq_a < m . \qquad (10.5.18)$$

For $m = 1/n = 1$ and $b \to \infty$ (without a conducting wall), the stability criterion becomes

$$q_a > 1 . \qquad (10.5.19)$$

This corresponds to *Kruskal-Shafranov limit* for the free-boundary kink mode. Thus in the case of uniform current profile the following conclusion can be drawn: the kink instability is possible only when magnetic surfaces resonant with the perturbation are located outside the highly conducting plasma. More generally, the stability criterion for $m = 1$ and $n = 1$ mode does not depend on the current density distribution, since the last term in (10.5.12) becomes small for $g \propto k^2r^2 \ll 1$.

10.6 MHD Waves

The MHD equations can also be used to study propagations such as the Alfven wave and the magnetosonic waves [10.2, 3]. Let us first study the normal modes of an infinite homogeneous plasma. Taking B in the z-direction the equilibrium state is specified by

$$\begin{aligned} &\boldsymbol{B}_0 = (0, 0, B_0) \\ &\varrho_0 = \text{const}\,, \quad P_0 = \text{const}\,, \quad B_0 = \text{const}\,. \end{aligned} \tag{10.6.1}$$

Since $\nabla P_0 = 0$, and $\nabla \times \boldsymbol{B}_0 = 0$, the linearized version of the equation of motion is given by [see (10.3.12, 13)]

$$\begin{aligned} -\sigma^2 \boldsymbol{\xi} &= \frac{\gamma P_0}{\varrho_0} \nabla\nabla \cdot \boldsymbol{\xi} + \frac{1}{\mu_0 \varrho_0} (\nabla \times \boldsymbol{B}_1) \times \boldsymbol{B}_0 \\ &= c_s^2 \nabla\nabla \cdot \boldsymbol{\xi} + v_A^2 \boldsymbol{b} \times [\nabla \times \nabla \times (\boldsymbol{b} \times \boldsymbol{\xi})]\,, \end{aligned} \tag{10.6.2}$$

where $c_s = \sqrt{\gamma P_0/\varrho_0}$, $v_A^2 = B_0^2/\mu_0\varrho_0$ and $\boldsymbol{b} = \boldsymbol{B}_0/B_0$. We use σ^2 instead of ω^2 to indicate that the eigenvalues we are looking for are real.

Since all equilibrium quantities are constant we may write $\boldsymbol{\xi}(\boldsymbol{r})$ as a Fourier integral (or a Fourier series if we consider a finite box) of plane wave solutions:

$$\boldsymbol{\xi}(\boldsymbol{r}) = (2\pi)^{-\frac{3}{2}} \int_{-\infty}^{\infty} \boldsymbol{\xi}(\boldsymbol{k}) \exp(\mathrm{i}\boldsymbol{k} \cdot \boldsymbol{r}) d^3\boldsymbol{k}\,. \tag{10.6.3}$$

We study the modes $\boldsymbol{\xi}(\boldsymbol{k}) \exp[\mathrm{i}(\boldsymbol{k} \cdot \boldsymbol{r} - \sigma t)]$ separately by making the substitution $\nabla \to \mathrm{i}\boldsymbol{k}$ in (10.6.2). This gives

$$\begin{aligned} -\sigma^2 \boldsymbol{\xi}(\boldsymbol{k}) &= -(v_A^2 + c_s^2)\boldsymbol{k}\boldsymbol{k} \cdot \boldsymbol{\xi}(\boldsymbol{k}) \\ &\quad - v_A^2 \boldsymbol{k} \cdot \boldsymbol{b}[\boldsymbol{k} \cdot \boldsymbol{b}\boldsymbol{\xi}(\boldsymbol{k}) - \boldsymbol{k}\boldsymbol{b} \cdot \boldsymbol{\xi}(\boldsymbol{k}) - \boldsymbol{b}\boldsymbol{k} \cdot \boldsymbol{\xi}(\boldsymbol{k})]\,, \end{aligned} \tag{10.6.4}$$

or broken down into x, y and z components:

$$\begin{pmatrix} -k_x^2(v_A^2 + c_s^2) - k_z^2 v_A^2 & -k_x k_y (v_A^2 + c_s^2) & -k_x k_z c_s^2 \\ -k_x k_y (v_A^2 + c_s^2) & -k_x^2(v_A^2 + c_s^2) - k_z^2 v_A^2 & -k_y k_z c_s^2 \\ -k_x k_z c_s^2 & -k_y k_z c_s^2 & -k_z^2 c_s^2 \end{pmatrix} \begin{pmatrix} \xi(\boldsymbol{k})_x \\ \xi(\boldsymbol{k})_y \\ \xi(\boldsymbol{k})_z \end{pmatrix}$$
$$= -\sigma^2 \begin{pmatrix} \xi(\boldsymbol{k})_x \\ \xi(\boldsymbol{k})_y \\ \xi(\boldsymbol{k})_z \end{pmatrix}\,. \tag{10.6.5}$$

Solutions are obtained by setting the determinant of the left hand side equal to zero. This gives:

$$(\sigma^2 - k_\parallel^2 v_A^2)[\sigma^4 - k^2(v_A^2 + c_s^2)\sigma^2 + k_\parallel^2 k^2 v_A^2 c_s^2] = 0\,, \tag{10.6.6}$$

where $k^2 = k_x^2 + k_y^2 + k_z^2$, $k_\parallel^2 = k_z^2$. Consequently, we obtain three solutions:

$$\begin{aligned} &\sigma^2 \equiv \sigma_A^2 = k_\parallel^2 v_A^2\,, \\ &\sigma^2 \equiv \sigma_{s,f}^2 = \frac{1}{2}k^2(v_A^2 + c_s^2)\left[1 \pm \sqrt{1 - \frac{4k_\parallel^2 v_A^2 c_s^2}{k^2(v_A^2 + c_s^2)^2}}\right]\,. \end{aligned} \tag{10.6.7}$$

The mode $\sigma^2 = \sigma_A^2$ is called the *shear Alfven wave* and the modes $\sigma^2 = \sigma_f^2$ and σ_s^2 the *fast* (+) and the *slow* (–) *magnetosonic waves*, respectively.

We also calculate the corresponding eigenvectors $\boldsymbol{\xi}$ and the associated magnetic field perturbations $\boldsymbol{B}_1$ and the pressure perturbation P_1 by substituting (10.6.7) into (10.6.5) and using the relation $\boldsymbol{B}_1 = \nabla \times (\boldsymbol{\xi} \times \boldsymbol{B}_0) = -\mathrm{i}B_0\boldsymbol{k} \times (\boldsymbol{b} \times \boldsymbol{\xi}(\boldsymbol{k}))$ and $P_1 = \gamma P_0 \nabla \cdot \boldsymbol{\xi} = -\mathrm{i}\varrho_0 c_s \boldsymbol{k} \cdot \boldsymbol{\xi}(\boldsymbol{k})$. Without loss of generality the $\boldsymbol{k}$ vector may be chosen to lie in the xz-plane, so that $k_y = 0$. We then obtain the following expressions for the Alfven eigenmodes:

$$
\begin{aligned}
&\xi(\boldsymbol{k})_x = \xi(\boldsymbol{k})_z = 0\,, \qquad \xi(\boldsymbol{k})_y \neq 0\,, \\
&B_{x1} = B_{z1} = 0\,, \qquad B_{y1} = -\mathrm{i}B_0 k_z \xi(\boldsymbol{k})_y\,, \\
&P_1 = 0
\end{aligned}
\qquad (10.6.8)
$$

Likewise we find for the slow and fast magnetosonic eigenmodes

$$
\begin{aligned}
&\xi(\boldsymbol{k})_y = 0, \quad \xi(\boldsymbol{k})_z = \alpha_{s,f}\frac{k_z}{k_x}\xi(\boldsymbol{k})_x\,, \\
&B_{y1} = 0,\ k_x B_{x1} + k_z B_{z1} = 0\,, \quad B_{z1} = -\mathrm{i}B_0 k_x \xi(\boldsymbol{k})_x\,, \\
&P_1 = -\mathrm{i}\varrho c_s k_x \left(1 + \alpha_{s,f}\frac{k_z^2}{k_x^2}\right)\xi(\boldsymbol{k})_x\,,
\end{aligned}
\qquad (10.6.9)
$$

where $\alpha_{s,f} = 1 - k^2 v_A^2/\sigma_{s,f}^2$. The relation between $\xi(\boldsymbol{k})_z$ and $\xi(\boldsymbol{k})_x$ in (10.6.9) is derived by first solving the top line of (10.6.5) for $\xi(\boldsymbol{k})_z$ and then replacing it by the expression $\xi(\boldsymbol{k})_z = [\sigma^2\xi(\boldsymbol{k})_z - k_x k_z c_s^2 \xi(\boldsymbol{k})_x]/k_z^2 c_s^2$ which is obtained from the bottom line of (10.6.5). Equation (10.6.7) gives $\alpha_s < 0$ and $\alpha_f > 0$, so that the spatial orientation of $\boldsymbol{\xi}(\boldsymbol{k})$ with respect to $\boldsymbol{B}_0$ is different for the fast and slow modes.

The Alfven waves are transverse waves with respect to both $\boldsymbol{\xi}(\boldsymbol{k})$ and $\boldsymbol{B}_1$, but they do not affect the pressure. The magnetosonic waves do affect the pressure and they have both transverse and longitudinal components. Together these three waves $\boldsymbol{\xi}_A$, $\boldsymbol{\xi}_s$, and $\boldsymbol{\xi}_f$ form an orthogonal triad in space. This indicates that any arbitrary displacement can be decomposed into the three different eigenmodes.

Ideal MHD waves are strongly anisotropic as is clear when one examines their phase velocity:

$$
\boldsymbol{v}_{ph} = \frac{\sigma \boldsymbol{k}}{k^2}. \qquad (10.6.10)
$$

Note that $v_{ph} = \sigma/k$ depends only on the angle θ between $\boldsymbol{k}$ and $\boldsymbol{B}$ and is independent of $\boldsymbol{k}$; $\sigma/k = f(\theta)$. Such waves are called non-dispersive and a plane wave packet constructed from them propagates without distortion. The group velocity of such a packet yields a flow of energy with the velocity:

$$
\boldsymbol{v}_g = \frac{\partial \sigma}{\partial \boldsymbol{k}}\,. \qquad (10.6.11)
$$

For the Alfven wave we find that the energy flow is always along the magnetic field

$$\boldsymbol{v}_{\mathrm{gA}} = v_{\mathrm{A}}\boldsymbol{b} \,. \tag{10.6.12}$$

We can now explain the physical mechanism of a shear Alfven wave. From the induction equation $\partial \boldsymbol{B}/\partial t = -\nabla \times \boldsymbol{E}$, the electric field $E_{1x} = (\sigma/k_z) \cdot B_{1y}$ appears in the presence of an oscillatory magnetic field $B_{1y} \cdot \exp[\mathrm{i}(k_z z - \sigma t)]$. In the ideal MHD model there is no electric field along the magnetic field line, that is, $E_{1z} = 0$ since $\boldsymbol{E} + \boldsymbol{v} \times \boldsymbol{B} = 0$. Electrons and ions drift in the y-direction with the same velocity of $\boldsymbol{E} \times \boldsymbol{B}/B^2 = -(E_{1x}/B_0)\hat{e}_y$. On the other hand, the phase velocity of the Alfven wave is σ/k_z and the oscillatory velocity of the magnetic field line in the y-direction is $-(B_{1z}/B_0)\sigma/k_z$. By using the relation between E_{1x} and B_{1y}, this velocity becomes equal to $-E_{1x}/B_0$. Therefore, the plasma oscillates as if it were frozen in the magnetic field line. For the concept of 'frozen-in' to be satisfied, the absence of an electric field along the magnetic field line is essential.

An analogy to the Alfven wave is a transverse vibration of a string. The plasma is frozen in the direction of the magnetic field line (the string) and there are B_0 field lines per unit area across it. Each field line can be regarded as having a mass density of ϱ_0/B_0 per unit length. On the other hand, the string also has a parallel magnetic stress B_0/μ_0. Therefore the transverse vibration of the field line, when regarded as a string of mass density ϱ_0/B_0 and tension B_0/μ_0, is characterized by the frequency $k\sqrt{(B_0/\mu_0)(\varrho_0/B_0)^{-1}} = k_{\|}v_{\mathrm{A}}$.

In the case of the low beta plasmas, the sound velocity is much smaller than the Alfven velocity, since $\beta = nT_{\mathrm{e}}/(B_0^2/2\mu_0) = 2c_{\mathrm{s}}^2/v_{\mathrm{A}}^2$. Under this approximation, $\xi(\boldsymbol{k})_z \doteq 0$ for the fast magnetosonic wave and $\xi(\boldsymbol{k})_z \doteq -(k^2 v_{\mathrm{A}}^2/k_x k_z c_{\mathrm{s}}^2)\xi(\boldsymbol{k})_x$ for the slow magnetosonic wave. This means that the dominant oscillation is in the x-direction in the fast mode and in the z-direction in the slow mode. This can be illustrated for the three MHD waves by plotting σ^2 as a function of $k_{\|}$ while keeping k_x fixed and vice versa for $v_{\mathrm{A}} > c_{\mathrm{s}}$ (Fig. 10.9). While the group

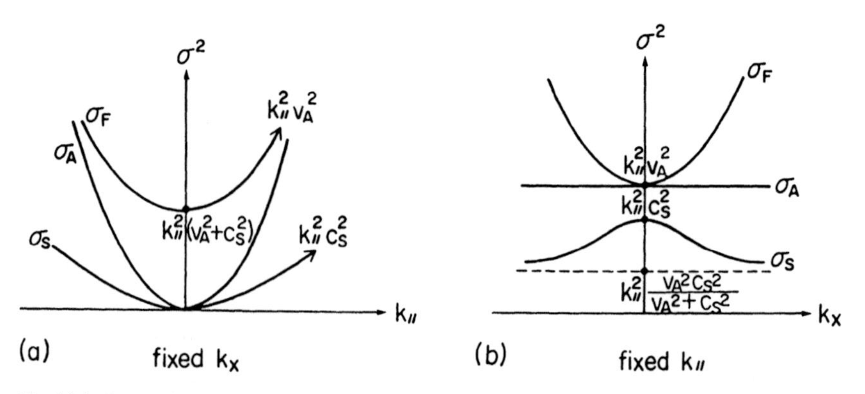

Fig. 10.9. Dispersion relations corresponding to MHD waves with $v_{\mathrm{A}} > c_{\mathrm{s}}$. (a) σ^2 as a function of $k_{\|}$ for a fixed k_x. (b) σ^2 as a function of k_x for a fixed $k_{\|}$

velocity in the parallel direction $\partial\sigma/\partial k_{||}$ is greater than zero for all three waves, the group velocities in the perpendicular direction $\partial\sigma/\partial k_\perp$ are quite different:

$$\begin{aligned} \frac{\partial\sigma}{\partial k_\perp} &> 0 \qquad \text{for fast waves} \\ \frac{\partial\sigma}{\partial k_\perp} &= 0 \qquad \text{for Alfven waves} \\ \frac{\partial\sigma}{\partial k_\perp} &< 0 \qquad \text{for slow waves}\ . \end{aligned} \tag{10.6.13}$$

To describe this behavior in inhomogeneous media we break $\boldsymbol{k}$ down into three components k_x, k_y, k_z, where k_z is in the direction of the magnetic field and k_x and k_y are in the perpendicular directions. The x-axis is chosen as the direction of the inhomogeneity. We now examine σ^2 as a function of k_x while keeping $k_{||}$ and k_y fixed.

In a slab of finite length in the x-direction with conducting plates at $x = \pm a$, the wavenumber k_x is quantized: $k_x = n\pi/2a$ and n is the number of nodes in the eigenfunction $\boldsymbol{\xi}(\boldsymbol{k})$ in the x-direction. Even though the medium is inhomogeneous, as long as the equilibrium quantities also vary in the x-direction, such a number to label the eigenvalues still makes physical sense. The essential features of the three discrete spectra of eigenvalue n are:

(1) The eigenvalue $\sigma_A^2 = k_z^2 v_A^2$ of the Alfven spectrum is infinitely degenerate;
(2) The slow wave eigenvalues have an accumulation point for $k_x \to \infty$ (or $n \to \infty$):

$$\sigma_s^2 \to \sigma_S^2 \equiv k_z^2 v_A^2 c_s^2/(v_A^2 + c_s^2) \tag{10.6.14}$$

which is obtained by noting that k_x^2 is included through k^2 in (10.6.7).
(3) For large wavenumbers $k_x \to \infty$ the fast wave eigenvalues behave as

$$\sigma_f^2 \doteq k_x^2(v_A^2 + c_s^2) \to \sigma_F^2 \equiv \infty\ , \tag{10.6.15}$$

so that the accumulation point of the fast wave spectrum is at infinity. We define two values of σ^2 denoted by σ_I^2 and σ_{II}^2 which correspond to the frequencies of the slow and the fast modes with $n = 0$:

$$\sigma_{I,II} = \frac{1}{2}k_0^2(v_A^2 + c_s^2)\left[1 \pm \sqrt{1 - 4k_z^2 v_A^2 c_s^2/k_0^2(v_A^2 + c_s^2)}\right]\ , \tag{10.6.16}$$

where $k_0^2 = k_y^2 + k_z^2$. The following sequence of inequalities is found:

$$0 \le \sigma_S^2 \le \sigma_s^2 \le \sigma_I^2 \le \sigma_A^2 \le \sigma_{II}^2 \le \sigma_f^2 \le \sigma_F^2\ . \tag{10.6.17}$$

Thus we now have obtained a clear separation of the three discrete subspectra.

This section is largely based on [10.18] which is by the author's mistake missing in the previous editions.

10.1. Derive (10.1.29).

10.2. Check (10.3.23) and derive (10.3.24).

10.3. Show a configuration satisfying the stability condition (10.3.27) everywhere.

10.4. Consider a cylindrical plasma with a vacuum region in $a < r < b$.

i) Show that the perturbed vacuum field energy can be written as

$$\begin{aligned}\frac{2\pi L}{\mu_0}\int_a^b |B_\perp|^2 r\,dr = \frac{2\pi L}{\mu_0}\int_a^b \Bigg[&\left(k^2+\frac{m^2}{r^2}\right)|\zeta|^2 \\ &+2\mathrm{Re}\left\{\left[-k\frac{d}{dr}(\xi B_\theta^0)+\frac{m^2}{r^2}\frac{d}{dr}(\xi r B_z^0)\right]\zeta^*\right\} \\ &+\left(\frac{m}{r}B_\theta^0+kB_z^0\right)^2|\xi|^2+\left|\frac{d}{dr}(\xi B_\theta^0)\right|^2 \\ &+\left|\frac{1}{r}\frac{d}{dr}(\xi r B_z^0)\right|^2\Bigg]\, r\,dr\,.\end{aligned}$$

ii)By minimizing with respect to ζ, derive (10.5.4).

10.5. Derive (10.5.15).

10.6. When the current density is homogeneous the solution of (10.5.14) shows that $r\xi'/\xi = m-1$. Check this statement.

10.7. Show that $\boldsymbol{\xi}_A, \boldsymbol{\xi}_s$, and $\boldsymbol{\xi}_f$ form an orthogonal triad in space.

10.8. Show that $\alpha_f > 0$ and $\alpha_s < 0$ where $\alpha_{s,f} = 1 - k^2 v_A^2/\sigma_{s,f}^2$.

10.9. Check the inequalities (10.6.17).

11. Resistive Magnetohydrodynamics

Although the plasma relevant to fusion research can normally be treated as being collisionless, the collisional resistivity, however small, often plays a crucial role in the macroscopic processes of MHD plasmas. Indeed, the resistivity causes a "reconnection" of different magnetic field lines, which never happens in an ideal MHD plasma because of the "frozen-in" condition of the fluid, and results in a topological change of the magnetic configuration accompanied by magnetic energy dissipation. Magnetic field line reconnection takes place in a very thin layer where the resistivity cannot be ignored because the parallel electric field locally vanishes, i.e., $\boldsymbol{E} \cdot \boldsymbol{B} = 0$. In this chapter, we shall be concerned mainly with this problem. First in Sect. 11.1, we derive the reduced MHD equations [11.1–5] which describe the plasma in the presence of a strong longitudinal magnetic field with shear. Then the *tearing mode instability* that occurs in a local resistive layer and the resulting magnetic island formation are described in Sect. 11.2. The next section is devoted to the resistive modification of the kink instability discussed in Sect. 10.5 [11.4]. Ballooning instability, which is a localized Rayleigh-Taylor instability [11.6], and resistive modification of the interchange instability are described in Sects. 11.4, 5 [11.1, 3.4]. Disruptive instabilities observed in tokamaks are briefly explained in Sect. 11.6. Section 11.7 deals with the nonlinear magnetic reconnection driven by an external plasma flow. The final section is devoted to the description of the plasma self-organization process due to the topological change via magnetic reconnection.

11.1 Reduced MHD Equations

11.1.1 Slab Geometry

We first consider the magnetohydrodynamics in a slab geometry. The magnetic field is assumed to be $\boldsymbol{B} = (B_x(x,y,t), B_y(x,y,t), B_z)$ in Cartesian coordinates. We further assume that $|B_z| = \text{const} \gg |B_x|$ and $|B_y|$ during the dynamic motion of the magnetofluid. This means that the dynamics takes place in the two-dimensional xy-space.

In the absence of perturbations the initial magnetic field is given by

$$B_x = 0\,, \qquad B_y = \bar{B}F(x)\,, \qquad B_z = B_{z0}\,, \tag{11.1.1}$$

where $F(x)$ is an odd function having the properties $F(x) \simeq x$ for $|x| \ll 1$ and

$F(x) \to 1$ for $|x| \gg 1$. The magnetic field B_y is produced by a sheet current in the z direction and the current is a function of x only.

The magnetic field given by (11.1.1) is called the sheared slab model, since $\boldsymbol{k} \cdot \boldsymbol{B}$ becomes zero at $x = 0$ for a two-dimensional perturbation with the wavenumber vector $\boldsymbol{k} = (k_x, k_y, 0)$. This model field is frequently used in stability analyses for drift waves as discussed in Sect. 5.5 and for resistive MHD instabilities shown in the following sections of this chapter.

A flux function $\Psi(x, y)$ is introduced by the relation

$$\boldsymbol{B} = \nabla\Psi \times \hat{\boldsymbol{z}} + B_{z0}\hat{\boldsymbol{z}} \tag{11.1.2}$$

which satisfies $\nabla \cdot \boldsymbol{B} = 0$ automatically. The function Ψ also satisfies $\boldsymbol{B} \cdot \nabla\Psi = 0$, which means that $\boldsymbol{B}$ is tangential to the surface defined by $\Psi(x, y) = \text{const}$. From $\nabla \times \boldsymbol{B} = \mu_0 \boldsymbol{J}$, we have

$$\nabla^2\Psi = -\mu_0 J_z \tag{11.1.3}$$

where J_z denotes the current in the z-direction. The simplified form of Ohm's law is

$$\boldsymbol{E} + \boldsymbol{u} \times \boldsymbol{B} = \eta \boldsymbol{J} \ , \tag{11.1.4}$$

where η is a resistivity (3.5.11). Here the fluid motion is assumed to be incompressible, $\nabla \cdot \boldsymbol{u} = 0$. This is justified when a strong magnetic field exists in one direction and the plasma pressure is low compared to the magnetic pressure, or $\beta \doteq 0$. By using $\partial \boldsymbol{B}/\partial t = -\nabla \times \boldsymbol{E}$ and noting that $\nabla\Psi \times \hat{\boldsymbol{z}} = \nabla \times (\Psi\hat{\boldsymbol{z}})$, we obtain

$$\frac{\partial\Psi}{\partial t} + (\boldsymbol{u} \cdot \nabla)\Psi = \frac{\eta}{\mu_0}\nabla^2\Psi + E \tag{11.1.5}$$

as the z-component of (11.1.4), where E is an arbitrary external electric field which satisfies $\nabla \times (E\hat{\boldsymbol{z}}) = 0$. The resistivity is assumed constant for simplicity.

For the velocity field, the equation of magnetohydrodynamic motion is used,

$$\varrho\frac{d\boldsymbol{u}}{dt} = -\nabla P + \boldsymbol{J} \times \boldsymbol{B} \ , \tag{11.1.6}$$

where d/dt is the convective derivative $d/dt = \partial/\partial t + \boldsymbol{u} \cdot \nabla$. The parallel component of (11.1.6) is $\varrho\, du_\parallel/dt = \nabla_\parallel P \doteq \partial P/\partial z = 0$, since P does not depend on z. This means that $u_\parallel \doteq 0$ if $u_\parallel|_{t=0} = 0$. We introduce the velocity stream function ϕ by the relation $\boldsymbol{u} = \nabla\phi \times \hat{\boldsymbol{z}}$, by noting that $\nabla \cdot \boldsymbol{u} = 0$ and $u_\parallel$ is small. By applying the operator $\hat{\boldsymbol{z}} \cdot \nabla\times$ to (11.1.6) (which means first operate with $\nabla\times$ and then separate out the z-component), we have

$$\varrho\frac{d}{dt}\nabla^2\phi = -\frac{1}{\mu_0}[\nabla\Psi \times \nabla(\nabla^2\Psi)] \cdot \hat{\boldsymbol{z}} \ . \tag{11.1.7}$$

Here the mass density ϱ is assumed to be constant, which is valid when the current

J_z is influential in the dynamics such as in current-driven MHD instabilities (kink or *tearing mode* which will be explained in Sect.11.2).

Thus two coupled second order partial differential equations are obtained for the two scalar functions ϕ and Ψ. The equilibrium state is described by $\Psi_0'(x) = -\bar{B}F(x)$ and $\phi_0 = 0$. The initial current density profile is given by $\mu_0 J_{z0} = -\Psi_0''$, where the primes denote derivatives with respect to x.

The equations (11.1.5, 7) are called the *reduced MHD equations*. These equations cannot describe all of the physics contained in the original MHD equations, but do describe the essential features in phenomena such as the linear stability of the tearing mode or the nonlinear evolution of unstable tearing modes corresponding to magnetic island formations (Sect. 11.2). Equations (11.1.5, 7) are also applicable to shear Alfvén waves in the slab geometry.

11.1.2 Toroidal Geometry

Similar reduced MHD equations in the three variables Ψ, ϕ, and P, where P is the scalar pressure, can be derived for toroidal plasmas for the specific case of a tokamak configuration. Usually the coordinates (r, θ, ζ) with the following metrics are used to describe the toroidal plasma:

$$(ds)^2 = (dr)^2 + (rd\theta)^2 + \left(1 + \frac{x}{R_0}\right)^2 (d\zeta)^2 , \qquad (11.1.8)$$

where $x = r\cos\theta$ and $\zeta = -R_0\varphi$. The poloidal angle is θ and the toroidal angle is φ.

A useful parameter to analyze the toroidal plasma is the inverse aspect ratio $\varepsilon = r/R_0$, where r is the minor radius and R_0 is the major radius of the torus. Table 11.1 shows that all of the plasma parameters can be expressed in powers of ε, where this "ordering" function is either $O(1)$, $O(\varepsilon)$, or $O(\varepsilon^2)$. The essential assumptions are that ε and β are small, i.e., $\varepsilon \ll 1$ and $\beta \sim O(\varepsilon)$, where $\beta = P/(B^2/2\mu_0)$ denotes the ratio of the plasma pressure to the magnetic pressure. These assumptions are reasonable in usual tokamaks.

The magnetic field is assumed to be of the form

$$\boldsymbol{B} = B_0\hat{\boldsymbol{z}} + \nabla\Psi \times \hat{\boldsymbol{z}} + B_1\hat{\boldsymbol{z}} , \qquad (11.1.9)$$

where the first and second terms are essentially the same as (11.1.2) but we now have $|\nabla\Psi \times \hat{\boldsymbol{z}}|/B_0 \sim O(\varepsilon)$. The third term represents the plasma effect on the magnetic field due to finiteness of the β value ($\beta \neq 0$); in other words, B_1 corresponds to the diamagnetic effect in the finite beta plasma of $\beta \sim O(\varepsilon)$. The second term corresponds to the poloidal magnetic field produced by the toroidal current. From the ordering function in Table 11.1 we see that $\varepsilon B_0/|\nabla\Psi \times \hat{\boldsymbol{z}}| \sim \varepsilon B_0/B_\theta = q(r) \sim 1$, where $q(r)$ is the safety factor in the cylindrical approximation.

The toroidal magnetic field is produced by the poloidal current I by $B_\varphi = I/R_0$. Two corrections to the zero-order constant toroidal field must be con-

Table 11.1. Plasma parameters in terms of an ordering function $O(\varepsilon^i)$, $i = 0, 1, 2$, used to derive the reduced MHD equations for a toroidal plasma

Ordering	O(1)	O(ϵ)	O(ϵ^2)
Beta value		β	
pressure		p	
poloidal flux		Ψ	
stream function		ϕ	
resistivity		η	
toroidal field	B_0	B_1	
poloidal field		$\nabla\Psi \times \hat{z}$	
perpendicular velocity		$\nabla\phi \times \hat{z}$	
parallel velocity			$u_{\parallel}$
time derivative		$\frac{\partial}{\partial t}$	
perpendicular derivative	$\nabla_{\perp}$		
parallel derivative		$\nabla_{\parallel}$	

sidered: one due to the toroidal curvature, $-B_0 x/R_0$, and the other due to the diamagnetic plasma current, I_1/R_0. By considering that the geometrical correction does not produce current, Ampere's law gives

$$\begin{aligned} \mu_0 \boldsymbol{J} &= \nabla \times (\nabla\Psi \times \hat{\boldsymbol{z}} + I_1/R_0 \hat{\boldsymbol{z}}) \\ &= -\nabla_{\perp}^2 \Psi \hat{\boldsymbol{z}} + \nabla I_1/R_0 \times \hat{\boldsymbol{z}} \,, \end{aligned} \tag{11.1.10}$$

where $\nabla_{\perp} = \nabla - \hat{\boldsymbol{z}}\, \partial/\partial z$ and I_1 represents the diamagnetic current effect. In deriving the second expression in (11.1.10), we used the fact that the derivative along the z-direction gives a quantity which is small (of order ε) compared to the derivative $\nabla_{\perp}$.

Another important ordering is $|\boldsymbol{u}_z|/|\boldsymbol{u}_{\perp}| \sim O(\varepsilon)$, where $\boldsymbol{u}_z$ and $\boldsymbol{u}_{\perp}$ are the velocity components parallel and perpendicular to the toroidal direction, respectively. The time evolution is related to the shear Alfvén waves and the characteristic time is $a/[B_\theta/(\mu_0 \varrho)^{1/2}]$, where B_θ is the poloidal magnetic field. This corresponds to $\partial/\partial t \sim O(\varepsilon)$, since the fastest time scale motion is suppressed by the incompressibility. When incompressibility is assumed, $\nabla \cdot \boldsymbol{u} = 0$, we find $dI_1/dt \sim O(\varepsilon^2)$ from $\partial B_z/\partial t = -B_z(\nabla \cdot \boldsymbol{u}) + (\boldsymbol{B} \cdot \nabla) u_z - (\boldsymbol{u} \cdot \nabla) B_z$.

By combining $|\boldsymbol{u}_z|/|\boldsymbol{u}_{\perp}| \sim O(\varepsilon)$ and $\nabla \cdot \boldsymbol{u} = 0$ we find that $\nabla_{\perp} \cdot \boldsymbol{u}_{\perp}$ is of order ε^3. A stream function ϕ can be introduced by the relation $\boldsymbol{u}_{\perp} = \nabla\phi \times \hat{\boldsymbol{z}}$. From Ohm's law and Faraday's induction law we get the equation

$$\frac{\partial \Psi}{\partial t} = \boldsymbol{B} \cdot \nabla\phi + \frac{\eta}{\mu_0} \nabla_{\perp}^2 \Psi + E \tag{11.1.11}$$

where again $\nabla \times (E\hat{\boldsymbol{z}}) = 0$ and $\boldsymbol{B}$ does not include $B_1 \hat{\boldsymbol{z}}$, since its contribution is only of order $O(\varepsilon^3)$ and hence is negligible. This is essentially the same as (11.1.5).

The same order analysis is applied to the equation of motion. Since $|\boldsymbol{u}| \sim O(\varepsilon)$ and $\partial/\partial t \sim O(\varepsilon)$, the inertial term becomes $O(\varepsilon^2)$. The terms of $O(\varepsilon)$ appear in the right hand side of the equation of motion perpendicular to the toroidal direction,

$$0 = -\nabla_\perp P + \boldsymbol{J} \times B_0 \hat{\boldsymbol{z}} \,. \tag{11.1.12}$$

In terms of I_1, the right hand side is written as

$$\nabla_\perp \left(P + \frac{I_1}{I_0} \frac{B_0^2}{\mu_0} \right) = 0 \,. \tag{11.1.13}$$

This equation means that the pressure balances with the diamagnetic variation of the toroidal magnetic field to confine plasmas of $\beta \sim O(\varepsilon)$. This situation is the same for both the tokamak and the theta pinch configuration. The parallel component of the equation of motion becomes $\boldsymbol{B} \cdot \nabla P = 0$ in the lowest order. Also $d(\boldsymbol{B} \cdot \nabla P)/dt = 0$ is consistent with incompressible magnetohydrodynamics (Problem 11.3). When $u_\parallel$ is zero and $\boldsymbol{B} \cdot \nabla P = 0$ in the initial MHD equilibrium, $u_\parallel$ remains small during the time evolution of the perturbation. If the perpendicular component of plasma current $\boldsymbol{J}_\perp$ found from the equation of motion is substituted into $\nabla \cdot \boldsymbol{J} = 0$, we get

$$\nabla \cdot \left[-(\varrho \frac{d\boldsymbol{u}}{dt} + \nabla P) \times \frac{\boldsymbol{B}}{B^2} + \sigma \boldsymbol{B} \right] = 0 \,. \tag{11.1.14}$$

The term $\sigma \boldsymbol{B}$ represents $\boldsymbol{J}_\parallel$, thus $\sigma = -\nabla_\perp^2 \Psi / \mu_0 B_0$. Equation (11.1.14) can be alternatively written as

$$\boldsymbol{B} \cdot \nabla \times \frac{\varrho}{B^2} \frac{d\boldsymbol{u}}{dt} = \boldsymbol{B} \cdot \nabla \sigma + \frac{(\nabla B^2 \times \nabla P) \cdot \boldsymbol{B}}{B^4} \,. \tag{11.1.15}$$

The left hand side can be rewritten as

$$\boldsymbol{B} \cdot \nabla \times \frac{\varrho}{B^2} \frac{d\boldsymbol{u}}{dt} \doteq \boldsymbol{B} \cdot \frac{\varrho}{B_0^2} \nabla \times \frac{d\boldsymbol{u}}{dt} = -\frac{\varrho}{B_0} \frac{d}{dt} \nabla_\perp^2 \phi \,, \tag{11.1.16}$$

where ϱ is assumed to be constant. The square of the magnetic field is

$$B^2 = B_0^2 \left[1 - 2 \left(\frac{\mu_0 P}{B_0^2} + \frac{x}{R_0} \right) \right] \,, \tag{11.1.17}$$

where we note that $B_1 = I_1/R_0 - B_0 x/R_0$ and $I_1/I_0 = -\mu_0 P/B_0^2$ from (11.1.13).

By substituting (11.1.17) into the right hand side of (11.1.15), we have

$$\varrho \frac{d}{dt} \nabla_\perp^2 \phi = \frac{1}{\mu_0} \boldsymbol{B} \cdot \nabla \nabla_\perp^2 \Psi + \left(\nabla \frac{2x}{R_0} \times \nabla P \right) \cdot \hat{\boldsymbol{z}} \,. \tag{11.1.18}$$

The pressure equation in the incompressible approximation becomes

$$\frac{\partial P}{\partial t} + (\nabla\phi \times \hat{z}) \cdot \nabla P = 0 , \qquad (11.1.19)$$

where we used the ordering $|u_\perp| \gg |u_z|$.

Usually the reduced MHD equation is normalized by using $t = at/v_{PA}$, $r = r/a, z = z/R_0 q(a)$, $\phi = \phi/av_{PA}$, $\Psi = \Psi/aB_\theta(a)$, $P = P/\{[B_\theta(a)]^2/2\mu_0\varepsilon\}$ and $\eta = \eta/av_{PA}\mu_0$, where $v_{PA} = (B_\theta^2/\mu_0\varrho)^{1/2}$. Here $q(a)$ is the safety factor at the plasma surface and $B_\theta(a)$ is the poloidal magnetic field at the plasma surface. Then the reduced MHD equation for ϕ, Ψ and P can be written

$$\frac{\partial}{\partial t}\nabla_\perp^2\phi = [\phi, \nabla_\perp^2\phi] + [\nabla_\perp^2\Psi, \Psi] + \frac{\partial}{\partial z}\nabla_\perp^2\Psi + [x, P] , \qquad (11.1.20)$$

$$\frac{\partial \Psi}{\partial t} = [\phi, \Psi] + \frac{\partial \phi}{\partial z} + \eta\nabla_\perp^2\Psi , \qquad (11.1.21)$$

$$\frac{\partial P}{\partial t} = [\phi, P] , \qquad (11.1.22)$$

where in (11.1.21) E [as in (11.1.5)] is neglected for simplicity. The Poisson bracket for two physical quantities $f(r, \theta, z)$ and $g(r, \theta, z)$ means $[f, g] = (\nabla f \times \nabla g) \cdot \hat{z}$. The Poisson bracket has the following properties:

$$[f, g] = -[g, f] \qquad (11.1.23)$$

$$[[f, g], h] + [[g, h], f] + [[h, f], g] = 0 \qquad (11.1.24)$$

$$[fg, h] = g[f, h] + f[g, h] \qquad (11.1.25)$$

$$[f(g), h] = \frac{df}{dg}[g, h] . \qquad (11.1.26)$$

Equation (11.1.24) is known as Jacobi's identity. When the relation $[f, g] = \nabla \cdot (f\nabla g \times \hat{z})$ is used, we have

$$\int [f, g]dV = 0 \qquad (11.1.27)$$

for the fixed boundary condition or $f|_{r=a} = g|_{r=a} = 0$. Equations (11.1.23, 25, 27) also give

$$\int f[g, h]dV = \int g[h, f]dV = \int h[f, g]dV . \qquad (11.1.28)$$

These relations are useful to derive the energy conservation relation for the reduced MHD equations. First, (11.1.20–22) are multiplied by $\phi, \nabla_\perp^2\Psi$ and x, respectively, and integrated in the plasma region. Then both sides of the three equations are summed up. Under the fixed boundary condition, we have

$$\frac{\partial}{\partial t}\int \left[\frac{1}{2}(\nabla_\perp\Psi)^2 + \frac{1}{2}(\nabla_\perp\phi)^2 - xP\right] dV = -\eta\int (\nabla_\perp^2\Psi)^2 dV . \qquad (11.1.29)$$

The quantity inside the square bracket corresponds to the energy density: the first term is the magnetic energy, the second term is the kinetic energy, and the third term is the internal energy. The right hand side shows Ohmic dissipation. When $\eta \to 0$, the total energy is conserved and the reduced MHD equations for an ideal MHD model are obtained.

11.2 Tearing Mode Instability and Magnetic Islands

Consider an initial field of the form given in (11.1.1). The plasma is initially stationary $\boldsymbol{u}_0 = 0$. The singular surface $\boldsymbol{k} \cdot \boldsymbol{B}_0 = 0$ exists at $x = 0$ for modes propagating in the y-direction. The zero order field is fixed in time and all perturbations are required to vanish far from the singular surface, since the free energy source for instability is localized in the neighborhood of the singular surface.

The reduced equations (11.1.5, 7) are now linearized and perturbations can be introduced in the form

$$\Psi(x,y) = \Psi_0(x) + \Psi_1(x)\cos ky \tag{11.2.1}$$

$$\phi(x,y) = \frac{\gamma}{k\bar{B}}\phi_1(x)\sin ky \tag{11.2.2}$$

where $\Psi_1(x)$ and $\phi_1(x)$ are assumed to vary in time as $e^{\gamma t}$. Here it is noted that the zero order state is completely described by $\Psi_0'(x) = -\bar{B}F(x)$ and $\phi_0 = 0$.

In Fig. 11.1 the resulting pattern of the magnetic surfaces Ψ = const given by (11.2.1) is shown. The region inside the separatrix is called the *magnetic island* [11.1–4]. The value of Ψ on the separatrix is found easily at the point $\Psi_s = \Psi_0(0) - \Psi_1(0)$. Setting $\Psi(x,y)$ equal to this value for $y = 0$ and expanding $\Psi_0(x)$ in a Taylor series around $x = 0$ we find the island width w as

$$w = 4\sqrt{-\Psi_1(0)/\Psi_0''}\ . \tag{11.2.3}$$

Now we can explain the linear stability of the tearing mode which produces the magnetic island in the vicinity of the singular surface.

Substituting (11.2.1, 2) into (11.1.5, 7) gives two coupled second order ordinary differential equations for $\Psi_1(x)$ and $\phi_1(x)$,

$$\Psi_1(x) - F(x)\phi_1(x) = \frac{1}{\gamma\tau_R}[\Psi_1''(x) - k^2\Psi_1(x)] \tag{11.2.4}$$

$$-\gamma^2\tau_A^2[\phi_1''(x) - k^2\phi_1(x)] = F(x)[\Psi_1''(x) - k^2\Psi_1(x)] - F''(x)\Psi_1(x)\ , \tag{11.2.5}$$

where the primes denote d/dx and $\tau_A = (\mu_0\varrho)^{1/2}/k\bar{B}$, $\tau_R = L^2/\eta$ are the characteristic Alfvén and resistive times. The layer width of the current carrying region around the singular surface is denoted by L which is also the unit length.

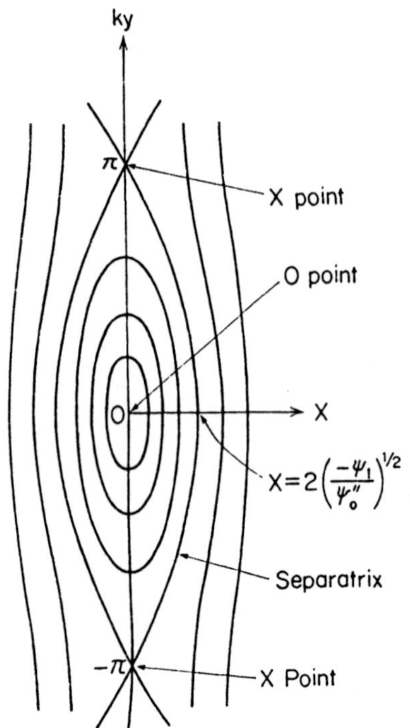

Fig. 11.1. Magnetic island configuration

At this point, a number of assumptions must be made which must be verified a posteriori. It is assumed that the growth rate is intermediate between the Alfvén time and the resistive time: $1/\tau_{\mathrm{R}} \ll \gamma \ll 1/\tau_{\mathrm{A}}$. Further, it is assumed that the resistivity is relevant only in a narrow layer $x_{\mathrm{T}} \ll 1$ where reconnection or tearing of magnetic field lines takes place.

Examine first the exterior region far from the singular surface $|x| > x_{\mathrm{T}}$ where the resistivity is assumed to be negligible. Assuming that the scale length for the solution are on the order of L and using $\gamma\tau_{\mathrm{A}} \ll 1$ we have

$$\Psi_1(x) = \phi_1(x)F \tag{11.2.6}$$

$$F(\Psi_1'' - k^2\Psi_1) = F''\Psi_1 \, . \tag{11.2.7}$$

Since it is not easy to find solutions for a general profile of $F(x)$, we choose a particular profile $F(x) = \tanh(x)$ such that $F(x) \doteq x$ for $|x| \ll 1$ and $F(x) \to 1$ for $|x| \gg 1$. For this profile (11.2.7) has the solution

$$\Psi_1(x) = \exp(\mp kx)\left[1 \pm \frac{1}{k}\tanh(x)\right] \, , \tag{11.2.8}$$

where the upper (lower) signs correspond to x positive (negative). The function $\Psi_1(x)$ has a discontinuous derivative at $x = 0$, and it is readily found that the discontinuity is given by

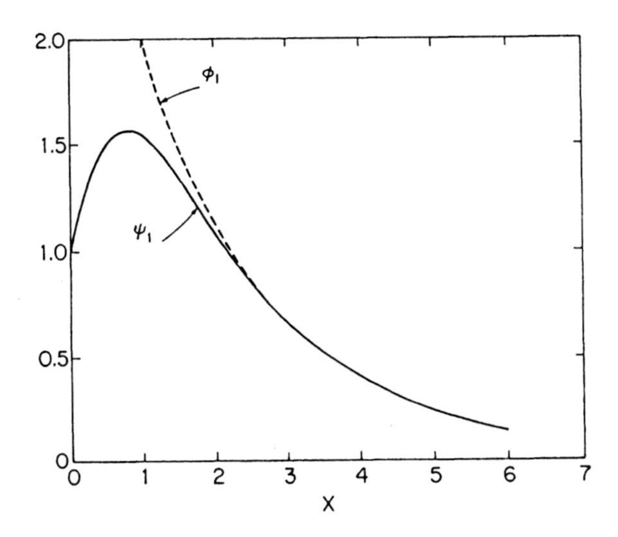

Fig. 11.2. Profiles of perturbations $\Psi_1(x)$ and $\phi_1(x)$ (*dotted line*)

$$\Delta' = \frac{\Psi_1'(0+) - \Psi_1'(0-)}{\Psi_1(0)} = 2\left(\frac{1}{k} - k\right) . \tag{11.2.9}$$

Note that Δ' is positive for $k < 1$, i.e., when the width of current carrying layer is short compared to the wavelength of the mode. The exterior solutions $\Psi_1(x)$ and $\phi_1(x)$ are shown in Fig. 11.2. The discontinuity of the derivative at $x = 0$ can be removed by considering the finite resistivity in the neighborhood of the singular surface or tearing layer. Now the resistive terms in (11.2.4, 5) must be retained to solve this region and the solutions must be matched at the boundary. We note that $x_T \ll 1$, so that within the tearing layer $F(x) \doteq x$. It is also assumed from the nature of the exterior solution that Ψ_1 is even, and that within the tearing layer $\Psi_1(x)$ is approximately constant (the "constant Ψ approximation") although $\Psi_1'(x)$ must be changing (Sect. 11.6). Since the higher derivatives with respect to x are large in this region, one can take $\Psi_1'' \gg k^2\Psi_1, \phi_1'' \gg k^2\phi_1$. Thus (11.2.4, 5) give

$$\Psi_1(0) - x\phi_1 = \Psi_1''/\gamma\tau_R \tag{11.2.10}$$

$$\gamma^2\tau_A^2\phi_1'' = -x\Psi_1'' . \tag{11.2.11}$$

Here (11.2.11) gives the condition for matching to the exterior solution:

$$-\frac{\gamma^2\tau_A^2}{\Psi_1(0)}\int_{-\infty}^{\infty}\frac{\phi_1''}{x}dx = \Delta' , \tag{11.2.12}$$

where in the extension of the integral to $\pm\infty$ an assumption has been made about the asymptotic behavior of ϕ_1. This assumption has been confirmed by numerical solution of (11.2.4, 5). The convergence of this integral guarantees that the interior solution Ψ_1 has constant slope asymptotically, and thus can be matched to the exterior solution.

Introducing the transformation

$$\chi(z) = -\left(\frac{\gamma\tau_A^2}{\tau_R}\right)^{1/4} \frac{\phi_1(x)}{\Psi_1(0)}, \qquad z = \left(\frac{\tau_R}{\gamma\tau_A^2}\right)^{1/4} x , \tag{11.2.13}$$

we have

$$\chi'' - z^2\chi = z . \tag{11.2.14}$$

The solution of this equation is odd and asymptotically equals $\chi \doteq -1/z$ for large $|z|$. The matching condition (11.2.12) becomes

$$\gamma^{5/4}\tau_R^{3/4}\tau_A^{1/2} \int_{-\infty}^{\infty} \frac{\chi''}{z} dz = \Delta' . \tag{11.2.15}$$

Apart from a constant the growth rate is now proportional to

$$\gamma \propto \tau_R^{-3/5}\tau_A^{-2/5}(\Delta')^{4/5} \tag{11.2.16}$$

when $\gamma\tau_R \propto S^{2/5} \gg 1$ and $\gamma\tau_A \propto S^{-3/5} \ll 1$ where S is the magnetic Reynolds number (10.1.3). We see that positive Δ' gives instability. Furthermore, it can be shown that the integration in (11.2.15) can be truncated at $z = 2$ with very little error, which was confirmed by a numerical calculation. Thus, the tearing layer width will be arbitrarily set as $x_T = 2(\gamma\tau_A^2/\tau_R)^{1/4}$ which is seen to scale as $S^{-2/5}$. In Fig. 11.3 the first-order fields and flow velocities estimated from Ψ_1 and ϕ_1 in the slab model are shown.

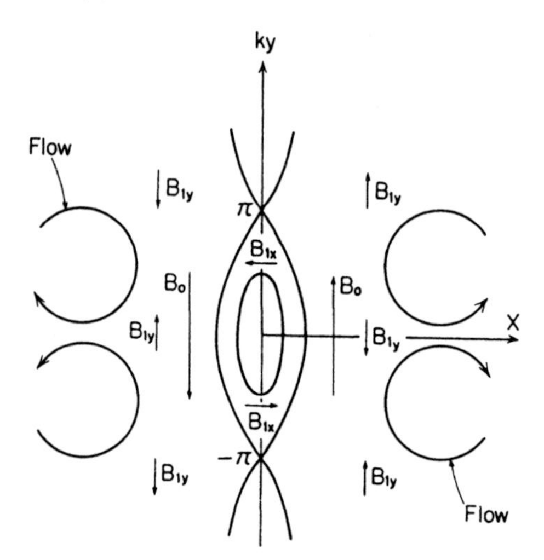

Fig. 11.3. Plasma flows in a magnetic island configuration in the slab model

Next we will examine the tearing mode in cylindrical geometry. We consider a fluid motion where all quantities are functions of τ, r, and t only, where $\tau = m\theta + kz$ and $k = -n/R_0$. This constraint corresponds to helical symmetry. Here m and n are the poloidal and toroidal mode numbers of the perturbation which has the form $f(r)\exp[i(m\theta + kz)]$. The helical symmetry allows the z coordinate to be eliminated through $\partial/\partial z = (k/m)\partial/\partial\theta$. With this symmetry in mind, we introduce a scalar function $\Psi^*(r,\tau) \equiv \Psi(r,\tau) - kr^2 B_{z0}/2m$. Considering a low β plasma and neglecting the toroidal correction term B_1 in (11.1.9), we can then write the magnetic field as

$$B = \nabla\Psi^* \times \hat{z} - \frac{kr}{m} B_{z0}\hat{\theta} + B_{z0}\hat{z} \ . \qquad (11.2.17)$$

It is easy to show $B \cdot \nabla\Psi^* = 0$, that is, Ψ^* is a flux function. We now use the reduced MHD equations, and assume low β plasmas or neglect the pressure term and the pressure equation. The equation for Ψ^* becomes

$$\frac{\partial\Psi^*}{\partial t} + \hat{z} \cdot \nabla_\perp\phi \times \nabla_\perp\Psi^* = -\eta J_z + E \ , \qquad (11.2.18)$$

by substitution of $\Psi = \Psi^* + kr^2 B_{z0}/2m$ into (11.1.11). The z-component of the current can be written by using Ψ^*,

$$\mu_0 J_z = -\nabla_\perp^2\Psi^* - \frac{2k}{m} B_{z0} \ . \qquad (11.2.19)$$

The equation of motion (11.1.18) gives

$$\varrho\frac{d}{dt}\nabla_\perp^2\phi = -\hat{z} \cdot \frac{1}{\mu_0}[\nabla_\perp\Psi^* \times \nabla_\perp(\nabla_\perp^2\Psi^*)] \ , \qquad (11.2.20)$$

by neglecting the pressure term. These three equations obtained assuming single helicity are useful for investigating the nonlinear development of the unstable tearing mode especially for $(m,n) = (1,1)$ and $(m,n) = (2,1)$ in the cylindrical geometry. We note that the form of (11.2.20) is similar to (11.1.7) in the slab geometry.

In order to determine the stability criterion for the tearing mode Δ' and the radial dependence of perturbations, we linearize (11.2.18, 20) by writing

$$\begin{aligned} \Psi^* &= \Psi_0^*(r) + \Psi_1(r)\cos\tau \\ \phi &= \phi_1(r)\sin\tau \\ J_z &= J_0(r) + J_1(r)\cos\tau \ . \end{aligned} \qquad (11.2.21)$$

The perturbations are assumed to vary as $e^{\gamma t}$. By eliminating all but Ψ_1 and ϕ_1, we obtain

$$\gamma\varrho\nabla_\perp^2\phi_1 = \frac{m}{\mu_0 r}\left(\frac{d\Psi_0^*}{dr}\nabla_\perp^2\Psi_1 - \Psi_1\frac{d^3\Psi_0^*}{dr^3}\right) , \qquad (11.2.22)$$

$$\gamma\Psi_1 + \frac{m}{r}\frac{d\Psi_0^*}{dr}\phi_1 = \frac{\eta}{\mu_0}\nabla_\perp^2\Psi_1 \qquad (11.2.23)$$

with the perpendicular Laplacian

$$\nabla_\perp^2 \equiv \frac{1}{r}\frac{\partial}{\partial r}r\frac{\partial}{\partial r} - \frac{m^2}{r^2} \ . \qquad (11.2.24)$$

By using the assumption of $\gamma\tau_A \ll 1$, we find

$$\nabla_\perp^2 \Psi_1 + \frac{\mu_0 \, dJ_0/dr}{d\Psi_0^*/dr} \Psi_1 = 0 \,, \tag{11.2.25}$$

away from the singular surface $d\Psi_0^*(r_s)/dr = 0$.

By numerically solving (11.2.25) for a given current profile, Δ' can be calculated. In the cylindrical tokamak model, usually there is at most one singular surface. Two numerical solutions Ψ_{1i} and Ψ_{1e} are obtained, one in the region from $r = 0$ to r_s and the other in the region from the plasma boundary to r_s, respectively. The stability criterion is given by calculating

$$\Delta'(\varepsilon) = \left. \frac{d\Psi_{1e}}{dr} \frac{1}{\Psi_{1e}} \right|_{r=r_s+\varepsilon} - \left. \frac{d\Psi_{1i}}{dr} \frac{1}{\Psi_{1i}} \right|_{r=r_s-\varepsilon} \,, \tag{11.2.26}$$

where $\Delta'(\varepsilon)$ becomes constant asymptotically for small ε. The stability is determined by $\lim_{\varepsilon \to 0} \Delta'(\varepsilon) < 0$. The stability behavior estimated by (11.2.25, 26) is consistent with tokamak experiments and it predicts that the $(m, n) = (2, 1)$ tearing mode is the most unstable.

11.3 Current Driven Instability (II)

Consider a dense plasma at equilibrium having a constant current density $J_0 \hat{z}$ extending from $r = 0$ to $r = r_1$ and a very low density, resistive plasma with zero current for $r_1 < r < a$. With B_θ at $r = r_1$, from (11.2.19, 24) and continuity of the flux function, we have

$$\begin{aligned}
\Psi_0^* &= -\left(\frac{\mu_0}{4} J_0 + \frac{k}{2m} B_0 \right) r^2 && \text{for } r < r_1 \\
&= -\frac{k}{2m} B_0 r^2 - \frac{\mu_0}{4} J_0 r_1^2 [1 + 2 \ln (r/r_1)] && \text{for } r > r_1
\end{aligned} \tag{11.3.1}$$

which gives

$$\begin{aligned}
B_\theta &= \frac{\mu_0}{2} J_0 r, && q = B_0 r / B_\theta R = \text{const} && \text{for } r < r_1 \\
&= \frac{\mu_0}{2} J_0 \frac{r_1^2}{r}, && q \propto r^2 && \text{for } r > r_1
\end{aligned} \tag{11.3.2}$$

and singular surface $r = r_s$ is assumed to be in the low density resistive plasma $(r_1 < r_s < a)$.

We begin with (11.2.25, 26). For a constant current profile, $d^3\Psi_0^*/dr^3 = 0$. Then for $\gamma \tau_A \ll 1$, (11.2.25) gives

$$\left(\frac{1}{r} \frac{d}{dr} r \frac{d}{dr} - \frac{m^2}{r^2} \right) \Psi_1 = 0 \,. \tag{11.3.3}$$

The problem is then to solve the Laplace equation in the three regions $0 < r < r_1, r_1 < r < r_s$ and $r_s < r < a$, and match the solutions such that Ψ_1 is continuous at $r = r_1$ and $r = r_s$ and vanishes at $r = a$. Such a solution can be written in the form

$$\Psi_1 = \begin{cases} (r/r_1)^m & \text{for} \quad r < r_1 \\ C_1(r/r_1)^m + C_2(r/r_1)^{-m} & \text{for} \quad r_1 < r < r_s \\ C_3[(r/a)^m - (r/a)^{-m}] & \text{for} \quad r_s < r < a \end{cases} \tag{11.3.4}$$

with

$$C_1 = 1 - C_2 \tag{11.3.5}$$

$$C_3[(r_s/a)^m - (r_s/a)^{-m}] = C_1(r_s/r_1)^m + C_2(r_s/r_1)^{-m} \,. \tag{11.3.6}$$

The coefficient C_2 can be determined from the discontinuity of Ψ_1' at $r = r_1$ where the equilibrium quantities J_z or $\Psi_0^{*''}$ and ϱ_0 are discontinuous,

$$\begin{aligned} \delta\Psi_0^{*''} &= \Psi_0^*(r_1+0) - \Psi_0^*(r_1-0) = \mu_0 J_0 = \frac{2B_0}{Rq} \\ \delta\varrho_0 &= \varrho_0^+ - \varrho_0^- \,, \end{aligned} \tag{11.3.7}$$

where $\varrho_0^{\pm} \equiv \varrho_0(r_1 \pm 0)$ and we substitute in (11.3.1, 2).

The discontinuity in Ψ_1' and ϕ_1' at $r = r_1$, denoted by $\delta\Psi_1'$ and $\delta\phi_1'$, can be determined from (11.2.22, 23). First, by integrating (11.2.22) across the surface $r = r_1$, we have the relation

$$\begin{aligned} \delta[\varrho_0\phi_1'] &= \varrho_0^+\delta\phi_1' + \delta\varrho_0\phi_1'(r_1 - 0) \\ &= \frac{m}{\mu_0\gamma r_1}(\Psi_0^{*'}\delta\Psi_1' - \Psi_1\delta\Psi_0^{*''}) \,. \end{aligned} \tag{11.3.8}$$

Next from (11.2.23), we have

$$\gamma\Psi_1 - \frac{mB_0}{R}Q\phi_1 = 0 \tag{11.3.9}$$

where the resistivity is neglected, which is justified away from the singular surface, and

$$Q = q^{-1} - \frac{n}{m} \tag{11.3.10}$$

which can be obtained from the relation $d\Psi_0^*/dr = -B_\theta + (kr/m)B_0$ [see (11.2.17)]. Differentiating (11.3.9) and noting that q =const for $r < r_1$, we get

$$\gamma\Psi_1' - \frac{mB_0}{R}Q\phi_1' = 0 \tag{11.3.11}$$

and for $r > r_1$;

$$\gamma\Psi_1' - \frac{mB_0}{R}Q\phi_1' - \frac{mB_0}{R}Q'\phi_1 = 0 \,. \tag{11.3.12}$$

We calculate $\delta[\varrho_0\phi_1']$ from the above two equations and set the result equal to (11.3.8) to obtain

$$\left(\frac{m}{\mu_0\gamma r_1}\Psi_0^{*\prime} - \frac{R\gamma}{mB_0Q}\varrho_0^{+}\right)\delta\Psi_1'$$

$$= \frac{m}{\mu_0\gamma r_1}\Psi_1\delta\Psi_0^{*\prime\prime} + \frac{R\gamma}{mB_0Q}\delta\varrho_0\Psi_1'(r_1-0) + \frac{2R\gamma}{mB_0Q^2qr_1}\varrho_0^{+}\Psi_1\ , \qquad (11.3.13)$$

where use was made of the relations (11.3.9) and $Q' = -2/rq\,(r > r_1)$. We can calculate $\Psi_0^{*\prime}$ from (11.3.1, 2) to get $\Psi_0^{*\prime} = -B_0r_1Q/R$ and $\delta\Psi_0^{*\prime\prime}$ as shown in (11.3.7). Substituting these results into (11.3.13) and using the solution (11.3.4) with (11.3.5), which gives $\delta\Psi_1' = -2mC_2/r_1$, we obtain

$$C_2 = (m-nq)^{-1}\left[1+\frac{n^2\gamma^2\tau_+^2}{m^2Q^2}\right]^{-1}\left[1+\frac{n^2q\gamma^2(\tau_+^2-\tau_-^2)}{2mQ}+\frac{n^2\gamma^2\tau_+^2}{m^2Q^2}\right] \qquad (11.3.14)$$

where $\tau_{\pm} = (\mu_0\varrho_0^{\pm})^{1/2}R/nB_0$ are the characteristic Alfven times of the inner (−) and outer (+) regions.

As in the tearing mode theory, Ψ_1' has a discontinuity at the singular surface, $\Delta' = [\Psi_1'(r_s+0) - \Psi_1'(r_s-0)]/\Psi_1(r_s)$, which is related to the growth rate. Using (11.3.4, 6) we can calculate Δ'

$$\Delta' = \frac{m}{r_s}\left\{\frac{1+(r_s/a)^{2m}}{1-(r_s/a)^{2m}} - \frac{1-C_2[1+(r_1/r_s)^{2m}]}{1-C_2[1-(r_1/r_s)^{2m}]}\right\}\ . \qquad (11.3.15)$$

Combining (11.3.14, 15), we can determine γ in terms of Δ'.

Two limits are readily obtained. First, we consider the limit $\tau_+ \to 0$, that is the low density limit in the outer region. We can infer from (11.3.15) that for fixed γ, Δ' is equal to 0, which upon combining (11.3.14) with $\tau_+ = 0$ and (11.3.15) with $\Delta' = 0$ gives

$$\gamma^2\tau_-^2 = \left(\frac{m}{nq}-1\right)^2\frac{2}{(r_1/a)^{2m}-1} + \frac{2}{nq}\left(\frac{m}{nq}-1\right)\ . \qquad (11.3.16)$$

This gives the growth rate of the free boundary kink mode discussed in Sect. 10.5. In this limit the island growth is impeded by the inertia of the central core region against the resistive tearing. The other limit corresponds to the tearing mode limit which is found by setting the density discontinuity $\delta\varrho_0$ equal to zero in (11.3.14). In this case, $C_2 = (m-nq)^{-1}$ independent of γ gives the tearing mode stability criterion in a step-like current profile.

The free boundary kink mode and the tearing mode can be viewed as two different limits of the same instability. In the case of the ideal kink the resonant surface must be located outside the plasma in the surrounding vacuum region, and is perturbed by a distortion of the plasma-vacuum boundary. If a resistive plasma is introduced in the vacuum region, magnetic reconnection occurs and the growth rate of the mode is modified accordingly by the resistivity.

A limiting result is obtained from (11.3.16) for $a \to \infty$:

$$\gamma^2\tau_-^2 = -2\left(\frac{m}{nq}-1\right)\left(\frac{m-1}{nq}-1\right)\ . \qquad (11.3.17)$$

This result is the same as (10.5.18) except for the conducting wall position and the instability occurs for

$$m - 1 < nq < m \, . \tag{11.3.18}$$

Especially for the $(m, n) = (1, 1)$ mode, for the plasma to be stable $q > 1$ is required, which is the Kruskal-Shafranov limit.

11.4 Ballooning Instability

The linearized reduced MHD equations are written as

$$\varrho_0 \frac{\partial}{\partial t} \nabla_\perp^2 \phi_1 = \frac{1}{\mu_0} \boldsymbol{B}_0 \cdot \nabla \nabla_\perp^2 \Psi_1 + \frac{1}{\mu_0} \boldsymbol{B}_1 \cdot \nabla \nabla_\perp^2 \Psi_0 + \nabla \frac{2x}{R_0} \times \nabla P_1 \cdot \hat{\boldsymbol{z}} \, , \tag{11.4.1}$$

$$\frac{\partial P_1}{\partial t} + (\nabla \phi_1 \times \hat{\boldsymbol{z}}) \cdot \nabla P_0 = 0 \tag{11.4.2}$$

$$\frac{\partial \Psi_1}{\partial t} = (\boldsymbol{B}_0 \cdot \nabla) \phi_1 + \frac{\eta}{\mu_0} \nabla_\perp^2 \Psi_1 \, , \tag{11.4.3}$$

where the quantities with subscript 1 are perturbations. Now we consider the small region outside of the toroidal plasma. The local coordinates of this region are (x, y, z), where x is the major radius direction, y is the direction parallel to the axis of the torus and z is the toroidal direction. Here it is noted that (11.4.1–3) are derived for tokamak type devices where the poloidal field is smaller than the toroidal field. Therefore, the magnetic field is assumed to be along the z-direction in the local coordinates. Then, the perturbations are expanded as $\exp(\mathrm{i}k_x x + \mathrm{i}k_y y + \mathrm{i}k_z z - \mathrm{i}\omega t)$. There is a pressure gradient in the x-direction $\nabla P_0 = \hat{\boldsymbol{x}} \partial P_0 / \partial x$. However, here we consider the case that the wavelength of the perturbation along the x-direction is much smaller than the scale length of the pressure gradient, and expand the perturbation as $\exp(\mathrm{i}k_x x)$ in the x-direction.

By applying the above Fourier expansion to (11.4.1–3) and neglecting the resistivity and the ∇J_0 term , that is, the second term on the right hand side of (11.4.1) for simplicity, we find

$$\varrho_0 \omega k_\perp^2 \tilde{\phi} = -\frac{B_0}{\mu_0} k_z k_\perp^2 \tilde{\Psi} + \frac{2}{R_0} k_y \tilde{P} \, , \tag{11.4.4}$$

$$-\omega \tilde{P} + k_y \tilde{\phi} \frac{dP_0}{dx} = 0 \, , \tag{11.4.5}$$

$$-\omega \tilde{\Psi} = B_0 k_z \tilde{\phi} \, , \tag{11.4.6}$$

where $k_\perp^2 = k_x^2 + k_y^2$. From (11.4.5, 6), $\tilde{P}$ and $\tilde{\Psi}$ are expressed in terms of $\tilde{\phi}$ and then substituted into (11.4.4) to obtain

$$\omega^2 = \frac{k_z^2 B_0^2}{\mu_0 \varrho_0} + \frac{2}{\varrho_0 R_0} \frac{k_y^2}{k_\perp^2} \frac{dP_0}{dx} . \tag{11.4.7}$$

When $k_x \doteq k_y$ and $k_z = 0$, ω^2 becomes

$$\omega^2 \doteq \frac{1}{\varrho_0 R_0} \frac{dP_0}{dx} . \tag{11.4.8}$$

In the outer region of the toroidal plasma, where $dP_0/dx < 0, \omega$ becomes imaginary and gives an unstable solution. The major radius of the torus R_0 in (11.4.8) corresponds to R_c defined in (3.3.7) and is positive. The dispersion relation (11.4.8) is the same as that for Rayleigh-Taylor instability discussed in Sect. 10.4. The instability is localized to the region of the outside of the toroidal plasma and is called the *ballooning mode*. The outside region has a positive curvature in the magnetic field line concave to the plasma with positive $\boldsymbol{g}$ (Sect. 10.4). This region is called the bad curvature region due to the instability with respect to ballooning. The dispersion relation (11.4.7) predicts that for finite k_z or $k_z \neq 0$ and $k_x \doteq k_y$ modes, the plasma is stable when the condition

$$k_z^2 > \frac{\beta}{2R_0 L_P} \tag{11.4.9}$$

is satisfied, where L_P is the scale length of the pressure gradient. The inner region of a toroidal plasma has a magnetic field line curved convex to the plasma and gives negative $\boldsymbol{g}$. In other words, the Rayleigh-Taylor instability localized to the inner region never occurs; this region is called the good curvature region. In a toroidal plasma, however, the magnetic field line connects the positive $\boldsymbol{g}$ region and the negative $\boldsymbol{g}$ region since a poloidal component is required to maintain MHD equilibrium. The length between the bad curvature region (positive $\boldsymbol{g}$) and the good curvature region (negative $\boldsymbol{g}$) is called the *connection length* and is estimated as $L_c \sim qR_0$, where q is the safety factor which is related to the ratio between the toroidal magnetic field component and the poloidal magnetic field component.

Thus ballooning mode localized to the bad curvature region has $k_z \neq 0$ and is more unstable than the mode with constant amplitude along the magnetic field line at $k_z \simeq 0$. Typical wavelengths of ballooning modes are estimated at $k_z \sim 1/L_c \sim 1/qR_0$. If this is substituted into (11.4.9), then

$$\beta < \frac{2L_P}{R_0 q^2} \tag{11.4.10}$$

is required for stability against the ballooning mode. The expression in (11.4.10) gives an upper limit, called the β *stability-limit*, to the stably confined plasma pressure determined by MHD stability. This is one of the most important parameters used to evaluate the quality of the magnetic confinement system.

In order to obtain a more precise β-limit it is necessary to perform a more elaborate analysis. The most difficult problem is to define the mode structure in a toroidal system with a magnetic shear. The mode structure in a tokamak with circular flux surfaces is described by

$$\phi_1(\Delta r,\theta,\zeta) = \sum_{l=-\infty}^{\infty} \hat{\phi}(\theta+2\pi l)\exp\left[inq'\Delta r(\theta+2\pi l)+in(q_0\theta-\zeta)\right], \tag{11.4.11}$$

where $\Delta r = r - r_{mn}, r_{mn}$ is the value of r at the rational surface $q(r_{mn}) = q_0 = m/n, q' = dq/dr$, the coordinates r, θ and ζ are used as defined in (11.1.8). Here the radial coordinate is denoted by $nq(r)$ instead of r, and q is expanded in the neighborhood of r_{mn}. We note that a large θ region corresponds to a small Δr region in (11.4.11). The expression (11.4.11) is called the *ballooning representation* [11.6]. It is normally used to study the ballooning mode in a sheared magnetic field ($dq/dr \neq 0$). By substituting the ballooning representation to (11.4.1–3) under the assumption of marginal stability, $\gamma = 0$, we have

$$\frac{1}{\mu_0}(\boldsymbol{B}_0\cdot\nabla)\nabla_\perp^2(\boldsymbol{B}_0\cdot\nabla)\phi_1 - \nabla\frac{2x}{R_0}\times\nabla\left[(\nabla\phi_1\times\hat{\boldsymbol{z}})\cdot\nabla P_0\right]\cdot\hat{\boldsymbol{z}} = 0. \tag{11.4.12}$$

Using the approximations $\boldsymbol{B}_0\cdot\nabla \doteq (B_\theta/r)\partial/\partial\theta$, $\nabla_\perp^2 \doteq -(n^2q_0^2/r^2)[1+(rq_0'\theta/q_0)^2]$, $\partial/\partial\theta \doteq inq_0$, $\partial/\partial r \doteq inq_0'\theta$ and $\nabla P_0 \doteq -(P_0/L_P)\hat{\boldsymbol{r}}$, we obtain

$$\frac{d}{d\theta}\left(1+\frac{q_0'^2r^2}{q_0^2}\theta^2\right)\frac{d\hat{\phi}}{d\theta} + \hat{\beta}\left(\cos\theta+\frac{rq_0'}{q_0}\theta\sin\theta\right)\hat{\phi} = 0, \tag{11.4.13}$$

where $\hat{\beta} = (2\mu_0P_0/B_\theta^2)(r^2/R_0L_P) = \beta q_0^2R_0/L_P$. If we introduce the relation $s = rq_0'/q_0$, then (11.4.13) becomes

$$\frac{d}{d\theta}(1+s^2\theta^2)\frac{d\hat{\phi}}{d\theta} + \hat{\beta}(\cos\theta+s\theta\sin\theta)\hat{\phi} = 0. \tag{11.4.14}$$

This equation is the ballooning mode equation for $n\to\infty$ and $\Delta r\to 0$. The numerical solution of this equation with the boundary conditions $\hat{\phi}(\pm\pi) = 0$ for $l = 0$ gives the eigenvalue $\hat{\beta}$ as a function of s, which is plotted in Fig. 11.4. For

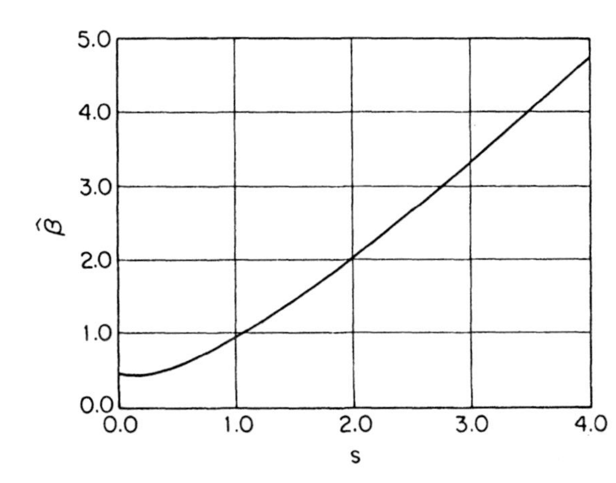

Fig. 11.4. Marginal beta values as a function of shear parameter

$s > 1.0, \hat{\beta} \doteq 2s$. The upper region of the curve is unstable with respect to the ballooning mode. Recently by more elaborate analyses it has been found that the ballooning mode becomes stable again in the region of higher $\hat{\beta}$; this is called the *second stability region* [11.7].

When the resistivity term in (11.4.3) is retained we have *resistive ballooning mode* [11.8] which becomes unstable for the beta below the curve in Fig. 11.4. Stability analysis of this mode is more complicated than for the ideal ballooning mode, but is similar to the resistive interchange mode discussed in Sect. 11.5.

A recent numerical analysis which takes into account both the free boundary kink mode and the ballooning mode shows that the beta stability limit in a tokamak is given by

$$\beta_{\text{lim}} = C\frac{\mu_0 I}{aB_{\text{T}}} \tag{11.4.15}$$

where I is the total current and C is a constant. This limit is called the *Troyon limit* [11.9] and has been experimentally confirmed. Physically, it can be understood as being the ratio of the plasma radius a to the connection length L_{c}, since $\mu_0 I/aB_{\text{T}} \sim B_{\text{p}}/B_{\text{T}} = a/R_0 q = a/L_{\text{c}}$. It implies a balance of the plasma pressure that tends to deform the field line with the magnetic pressure that tends to resist the deformation. Another explanation is the balance of the average centrifugal force density, $n\langle mv^2\rangle/R_{\text{c}} \sim nT/L_{\text{c}}$ and the magnetic restoring force density, $(B_0^2/2\mu_0 L_{\text{c}})(a/L_{\text{c}})$, against a perpendicular field line displacement.

11.5 Resistive Interchange and Rippling Modes

First we consider the full resistive MHD model discussed in Sect. 4.5. From Maxwell's equations $\partial \boldsymbol{B}/\partial t = -\nabla \times \boldsymbol{E}$, $\nabla \times \boldsymbol{B} = \mu_0 \boldsymbol{J}$ and Ohm's law we have

$$\frac{\partial \boldsymbol{B}}{\partial t} = (\boldsymbol{B}\cdot\nabla)\boldsymbol{u} - (\boldsymbol{u}\cdot\nabla)\boldsymbol{B} - \boldsymbol{B}(\nabla\cdot\boldsymbol{u}) - \nabla\times\frac{\eta}{\mu_0}(\nabla\times\boldsymbol{B}) \tag{11.5.1}$$

which for $\eta = 0$ describes the convection (the first and the second term on the right hand side) and compression (the third term) of the magnetic field with the fluid.

To illustrate the basic consequence of (11.5.1) in magnetohydrodynamic stability problems, we examine the interchange mode which takes place in the presence of a mass density gradient and an effective gravitational force across the magnetic field (Sect. 10.4). Perturbations of the form $f_1 = f_1(x)\exp[\gamma t + \mathrm{i}(k_y y + k_z z)]$ are considered in the slab equilibrium with $\boldsymbol{u}_0 = 0$ and

$$\boldsymbol{B}_0 = B_{y0}(x)\hat{\boldsymbol{y}} + B_{z0}\hat{\boldsymbol{z}}\ , \tag{11.5.2}$$

which is the same as (11.1.1). In addition we have $d\varrho_0/dx$ and $\boldsymbol{g} = g\hat{\boldsymbol{x}}$.

The linearized equation of (11.5.1) with $\eta = 0$ becomes

$$\frac{\partial \boldsymbol{B}_1}{\partial t} = (\boldsymbol{B}_0\cdot\nabla)\boldsymbol{u}_1 - (\boldsymbol{u}_1\cdot\nabla)\boldsymbol{B}_0 - \boldsymbol{B}_0(\nabla\cdot\boldsymbol{u}_1)\ . \tag{11.5.3}$$

The x-component of (11.5.3) is then written as

$$\gamma B_{x1} = \mathrm{i}(\boldsymbol{k} \cdot \boldsymbol{B}_0) u_{x1} \ . \tag{11.5.4}$$

If $\boldsymbol{B}_0$ is unidirectional over all space, i.e., B_{y0} = const, then $\boldsymbol{k} = (0, k_y, k_z)$ can be oriented so that $\boldsymbol{k} \cdot \boldsymbol{B}_0 = 0$, and the perturbation does not then distort magnetic field lines since $B_{x1} = 0$. A displacement without distortion of the magnetic field lines does not change the energy. This means that a plasma in a unidirectional magnetic field is unaffected by the interchange mode.

Now consider the more general magnetic field, where the direction of $\boldsymbol{B}_0$ is x-dependent. For any choice of $\boldsymbol{k}$, one can divide $\boldsymbol{B}_0$ into a component $\boldsymbol{B}_{0\perp}$ orthogonal to $\boldsymbol{k}$ and a component $\boldsymbol{B}_{0\parallel}$ parallel to $\boldsymbol{k}$. The behavior of the component $\boldsymbol{B}_{0\perp}$ during an interchange perturbation is identical with the behavior of the unidirectional $\boldsymbol{B}_0$. Therefore, this component has no influence on stability. The component $\boldsymbol{B}_{0\parallel}$ appears in (11.5.4) and its existence implies that any interchange perturbation increases the magnetic energy of the system via B_{x1} generation. In this way finite magnetic shear produced by the slab equilibrium (11.5.2) tends to stabilize the system against interchange modes. This is due to the fact that the magnetic field is frozen in the plasma fluid in the absence of dissipation. The situation can change if a finite resistivity exists.

Let us now examine to what extent resistivity can affect decoupling of the magnetic field from the fluid. Since $\gamma \boldsymbol{B}_1 = -\boldsymbol{k} \times \boldsymbol{E}_1$, we assume $\boldsymbol{E}_{\perp 1} = 0$ to suppress $\boldsymbol{B}_{\perp 1}$. In this case, from Ohm's law a perpendicular current

$$\boldsymbol{J}_{\perp 1} = (\boldsymbol{u}_1 \times \boldsymbol{B}_{0\parallel})/\eta_0 \tag{11.5.5}$$

is produced associated with the fluid motion $\boldsymbol{u}_1$. This current yields a restoring force $\boldsymbol{F}_1 = \boldsymbol{J}_{\perp 1} \times \boldsymbol{B}_{0\parallel}$ with the x-component

$$F_{x1} = u_{x1} B_{0\parallel}^2 / \eta_0 \ . \tag{11.5.6}$$

For $\eta_0 \to 0, -F_{x1}/u_{x1}$ becomes infinite everywhere except at the nulls of $B_{0\parallel}$ and the fluid is forced to follow the magnetic field by an infinite restoring force; this fact is consistent with the fact that for $\eta_0 = 0, B_{x1} \equiv 0$ is not possible. For any finite η_0, however, there is a finite region δL about each null of $B_{0\parallel}$ where the restoring force is arbitrarily weak. In this region the magnetic field is decoupled from the fluid and slips through it, thereby inducing an instability. The width of this reconnection region must be calculated for each mode, and is related to the growth rate. The driving power that balances the work done by the restoring force is

$$P = -\boldsymbol{u}_{x1} \cdot (\boldsymbol{J}_1 \times \boldsymbol{B}_0) = u_{x1}^2 (B_{0\parallel}')^2 (\delta L)^2 / \eta_0 \ , \tag{11.5.7}$$

where $B_{0\parallel} \sim B_{0\parallel}' \delta L$. In general, the instability wavelength is much larger than δL and thus, for an incompressible fluid the kinetic energy is largest in the $\boldsymbol{k}$-direction. Equating the rate of change of this energy to the driving power gives

$$\gamma \varrho_0 u_\parallel^2 = \gamma \varrho_0 \frac{u_{x1}^2}{k^2 (\delta L)^2} = \frac{u_{x1}^2 (B_{0\parallel}')^2 (\delta L)^2}{\eta_0} \tag{11.5.8}$$

where the incompressibility condition $\nabla \cdot \boldsymbol{u} = 0$ was used. We then get

$$\delta L = (\gamma \varrho_0 \eta_0 / k^2 (B'_{0\|})^2)^{1/4} . \tag{11.5.9}$$

There are two time scales associated with resistive reconnection: the ideal magnetohydrodynamic time τ_A proportional to $\varrho_0^{1/2}$ and the resistive time τ_R proportional to $1/\eta_0$. The ratio $S = \tau_R/\tau_A \propto 1/\eta_0 \varrho_0^{1/2}$ is the magnetic Reynolds number and is very large in most plasmas of interest.

The resistive interchange mode occurs in the presence of a mass density gradient and $\boldsymbol{F} = \varrho \boldsymbol{g} = \varrho g\ \hat{\boldsymbol{x}}$. The linearized driving force is of the form

$$\boldsymbol{F}_1 = \varrho_1 g \hat{\boldsymbol{x}} = -u_{x1} \frac{d\varrho_0}{dx} \frac{\boldsymbol{g}}{\gamma} \tag{11.5.10}$$

which is destabilizing if g points toward decreasing density (Sect. 10.4). Here ϱ_1 is calculated by using the continuity equation and the incompressibility condition. Equating $u_{x1} F_1$ to (11.5.7) gives

$$\delta L = \frac{\sqrt{-\eta_0 g (d\varrho_0/dx)}}{B'_{0\|} \gamma^{1/2}} . \tag{11.5.11}$$

From (11.5.9, 11)

$$\gamma = \left[-\frac{kg}{B'_{0\|}} \frac{d\varrho_0}{dx} \left(\frac{\eta_0}{\varrho_0} \right)^{1/2} \right]^{2/3} \propto S^{2/3} \eta_0 . \tag{11.5.12}$$

This resistive interchange mode becomes more unstable as the wavelength gets smaller. When the growth rate (11.5.12) is substituted into (11.5.9), it is found that the reconnection region is very narrow since $\delta L \propto S^{-1/3} \ll 1$.

As an example, let us consider the interchange mode in a stellarator field. To this end we modify the reduced MHD equation (11.1.17) to include the stellarator field [11.10, 11]. A simplified model for a stellarator field is the magnetic field $\boldsymbol{B} = -\nabla \varphi$ with the potential φ given by

$$\varphi = 2\varphi_l I_l(hr) \sin(l\theta + hz) . \tag{11.5.13}$$

Here I_l is the modified Bessel function and $h^{-1} = 2\pi R_0/N$; N corresponds to the pitch number and l is the pole number of the stellarator field. By taking the average

$$\bar{f}(r,\theta) = \frac{N}{2\pi R_0} \int_0^{h^{-1}} f(r,\theta,z) dz \tag{11.5.14}$$

to calculate the curvature term, we can show that the right hand side of the second term of (11.1.18) changes from $2x/R_0$ to

$$\frac{2x}{R_0} + \frac{\overline{(\nabla\varphi)^2}}{B_0^2} \tag{11.5.15}$$

and the second term of (11.5.15) becomes

$$\overline{(\nabla\varphi)^2} = 2\varphi_l^2 h^2 \left\{ [I_l'(hr)]^2 + \left(1 + \frac{l^2}{h^2 r^2}\right) I_l^2(hr) \right\} \tag{11.5.16}$$

where the prime denotes the derivative with respect to the argument. The average curvature due to the stellarator field is nonzero even for the straight cylindrical approximation $R_0 \to \infty$ which should be substituted into g of (11.5.12) to study the resistive interchange mode in the stellarator. Usually (11.5.15) gives a bad curvature, making g positive which corresponds to instability. However, this instability might be suppressed by adjusting a vertical field to change bad curvature into good curvature.

The rippling mode is also categorized as a resistive mode. This is caused by the local channeling of the current through ripples in the resistivity. Ohm's law is modified by the resistivity change to

$$\eta_0 \boldsymbol{J}_1 = -\eta_1 \boldsymbol{J}_0 + \boldsymbol{u}_1 \times \boldsymbol{B}_0 + \boldsymbol{E}_1 \,, \tag{11.5.17}$$

where η_1 is given by the convective law

$$\frac{d\eta_1}{dt} = \frac{\partial \eta_1}{\partial t} + (\boldsymbol{u}_1 \cdot \nabla)\eta_0 = 0 \,, \tag{11.5.18}$$

and $\eta_1 = -\boldsymbol{u}_1 \gamma^{-1} \cdot \nabla \eta_0$. Within the reconnection layer where we assume $\boldsymbol{E}_{1\perp} = 0$, the $\boldsymbol{J}_1$ term in (11.5.17) gives rise to a driving force

$$\boldsymbol{F}_1 = \boldsymbol{J}_1 \times \boldsymbol{B}_0 = \frac{\boldsymbol{u}_1 \cdot \nabla \eta_0}{\gamma \eta_0} (\boldsymbol{J}_0 \times \boldsymbol{B}_0) \,. \tag{11.5.19}$$

In the slab model, F_{x1} changes sign at the singular surfaces since u_{x1} is directed toward the singular surface. Hence, this $\boldsymbol{F}_1$ is stabilizing on the side of higher resistivity and destabilizing on the side of lower resistivity. An unstable mode results if the region of decoupled flow lies on the lower resistivity side and has a width δL such that the driving power

$$u_{x1} F_{x1} \doteq \frac{u_{x1}^2 \eta_0' (B_{y0}')^2 \delta L}{\gamma \eta_0 \mu_0} \tag{11.5.20}$$

dominates the work done by the restoring force on the fluid. Here $B_{y0}(x) \doteq B_{y0}'|_{x=0} \delta L$ and $\mu_0 J_{z0} = dB_{y0}/dx \doteq B_{y0}'|_{x=0}$. Comparing (11.5.7) and (11.5.20) gives

$$\delta L \sim \eta_0' / \gamma \mu_0 \tag{11.5.21}$$

when $k_z = 0$ or $B_{y0}' = B_{\|0}'$. From (11.5.9, 20), and assuming $\eta_0' = \eta_0 / L_\eta$, where L_η is the scale length of the resistivity gradient,

$$\gamma = \left[\frac{(\eta_0')^2 k_y B_{y0}'}{(\eta_0 \varrho_0 \mu_0^4)^{1/2}} \right]^{2/5} \propto S^{2/5} \eta_0 \,. \tag{11.5.22}$$

This mode thus grows more rapidly at decreasing wavelengths.

The linear stability of the rippling mode and its nonlinear behavior can be studied more rigorously by the following reduced equations

$$\frac{\partial \Psi}{\partial t} = \boldsymbol{B} \cdot \nabla \Phi + \frac{\eta}{\mu_0} \nabla^2 \Psi + E \,, \tag{11.5.23}$$

$$\varrho \frac{d}{dt} \nabla_\perp^2 \Phi = \frac{1}{\mu_0} \boldsymbol{B} \cdot \nabla \nabla_\perp^2 \Psi \,, \tag{11.5.24}$$

$$\frac{d\eta}{dt} = \nabla_\parallel \cdot (\chi_\parallel \nabla_\parallel \eta) \,. \tag{11.5.25}$$

Here the pressure term of (11.1.18) is neglected and in (11.5.18) a parallel heat transport term is added to the right hand side. It is known that a parallel heat conductivity $\chi_\parallel$ has a stabilizing tendency on the rippling mode.

11.6 Disruptive Instability

In tokamaks there exists a deteriorative instability known as the *disruptive instability* which is a rapidly developing process involving a redistribution of current over the column cross section with a decrease in poloidal field energy and an expulsion of part of the poloidal flux beyond the boundary of the plasma column [11.12–18]. This process is interpreted as a sequence of reconnections of magnetic field lines inside the plasma column.

The case under consideration involves reconnection of the magnetic field component transverse to a helical perturbation of the form $f(r, \theta + kz/m)$ in the cylindrical coordinates (r, θ, z). This component has the form $\boldsymbol{B}_* = \boldsymbol{B}_\perp + (krB_z/m)\hat{\boldsymbol{\theta}}$, where $\boldsymbol{B}_\perp$ is the transverse component of the magnetic field, B_z is the longitudinal component. At a point where $B_\theta = -krB_z/m$, i.e., where the field line pitch is the same as the pitch of the helical instability, $\boldsymbol{B}_* \cdot \hat{\theta}$ vanishes (see Fig. 11.6a and set $m = 1$). We can write the expression for $\boldsymbol{B}_*$ in terms of the safety factor q when we note that $-k/m = 1/R_0 q_s$, where q_s is the value of q at the point where the pitch of the perturbation coincides with that of the field line.

Here we consider the case $q_s = 1$ where the $(m, n) = (1, 1)$ mode has a singular surface. The nonlinear behavior of the $(m, n) = (1, 1)$ tearing mode induces the internal disruptive instability or *internal disruption*. Before discussing the nonlinear behavior, we demonstrate the linear stability of the $(m, n) = (1, 1)$ tearing mode.

First the reduced MHD equations for a low β tokamak plasma are linearized as in (11.2.22, 23). Then we introduce the variables $\gamma \xi_r = -(m/r)\phi_1$ and $\tilde{\Psi} = (m/r)\Psi_1$ and the equations in the neighborhood of the resonant surface $r = r_s$ become

$$\eta \tilde{\Psi}'' - \gamma(\tilde{\Psi} - k_\parallel B_0 \xi_r) = 0 \tag{11.6.1}$$

$$\gamma^2 \varrho_0 \xi_r'' + \frac{1}{\mu_0} k_\parallel B_0 \tilde{\Psi}'' = 0 , \tag{11.6.2}$$

where only the highest derivative with respect to r is retained in $\nabla_\perp^2$ and the scale length of the inhomogeneous current density is assumed to be larger than that of the mode localization.

To estimate the growth rate γ and the tearing layer width λ one assumes that both terms in (11.6.1, 2) are comparable and expand $k_\parallel \simeq k'(r - r_s) \sim k'\lambda$. One further estimates that $\xi_r'' \sim \xi_r/\lambda^2$ and $\tilde{\Psi}'' \sim \tilde{\Psi}/\lambda^2$. Then (11.6.1) yields $\gamma \sim \eta/\lambda^2$ at $r \doteq r_s$ corresponding to localized diffusion, while (11.6.2) gives $\gamma \sim k_\parallel v_A \sim k'\lambda v_A$ by using $\tilde{\Psi} \sim k_\parallel B_0 \xi_r$. After elimination of λ with $\tau_A = a/v_A$ and $\tau_s = a^2/\eta$,

$$\gamma^3 \sim \frac{(k_\parallel' a^2)^2}{(\tau_s \tau_A^2)} \tag{11.6.3}$$

is obtained. This instability having $\gamma \propto \eta^{1/3}$ is different from the usual tearing mode discussed in Sect. 11.2, which shows $\gamma \propto \eta^{3/5}$. The point is that the assumption $\tilde{\Psi}'' \sim \tilde{\Psi}/\lambda^2$ corresponds to a discontinuity in $\tilde{\Psi}$ in the radial field perturbation for $\lambda \ll a$, while in the usual tearing mode theory $\tilde{\Psi}$ itself is continuous and only its gradient $\tilde{\Psi}'$ changes appreciably across the tearing layer λ. The apparent discontinuity in slope is measured by using (11.2.26) where $\lambda \ll \varepsilon \ll a$ and the second derivative of $\tilde{\Psi}$ becomes $\tilde{\Psi}'' \sim \Delta' \tilde{\Psi}(r_s)/\lambda$. [Note that $\tilde{\Psi}(r_s) = m/r_s \Psi_{1i}(r_s) = m/r_s \Psi_{1e}(r_s)$.] Without making any assumption regarding the magnitude of Δ', one may repeat the manipulations leading to (11.6.3). Then (11.6.1) implies $\gamma \sim \eta \Delta'/\lambda$, while (11.6.2) yields $k_\parallel B_0 \Delta' \tilde{\Psi}/\lambda \sim \varrho_0 \gamma^2 \xi_r'' \sim (B_0/v_A)^2 \gamma^2 \xi_r/\lambda^2$. Assuming $\tilde{\Psi} \sim k_\parallel B_0 \xi_r$, one obtains $\gamma \sim k_\parallel v_A (\Delta' \lambda)^{1/2}$. After eliminating λ, we find

$$\gamma^5 \sim (k_\parallel' a^2)^2 (\Delta' a)^4 \tau_s^{-3} \tau_A^{-2} , \tag{11.6.4}$$

where the tearing instability occurs only if $\Delta' > 0$ [11.19]. Usually for tearing modes $\Delta' a \sim 1$ is valid which implies that $\tilde{\Psi}(r) \sim \tilde{\Psi}(r_s)$ through the tearing layer width λ. This is commonly referred to as the "constant Ψ approximation" [11.1]. If $\Delta' \lambda \sim 1$, then the relation $\gamma \propto \eta^{1/3}$ is recovered. An example of this type of tearing instability is the $(m, n) = (1, 1)$ mode in tokamaks for $q(0) < 1$ [11.20].

The reconnection process in an internal disruption is pictured in Fig. 11.5. Figure 11.5a shows the pattern of field lines for the auxiliary field $\boldsymbol{B}_*$ in the initial state of equilibrium. We see that $\boldsymbol{B}_*$ has opposite signs on the two sides of the $q = 1$ line. The flux function Ψ^* corresponding to the field $\boldsymbol{B}_*$ is defined by rewriting (11.2.17) as

$$\boldsymbol{B}_* = \nabla \Psi^* \times \hat{\boldsymbol{z}} . \tag{11.6.5}$$

When the internal region shifts, the oppositely directed fields become tangent to one another and begin to reconnect (Fig. 11.5b, c). A new plasma column formed

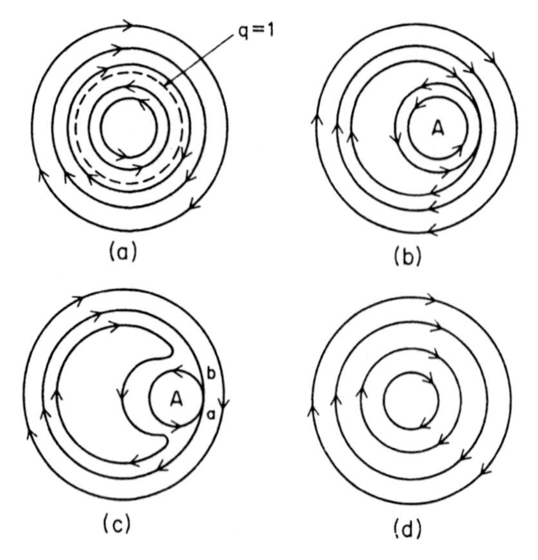

Fig. 11.5. Changes of flux surface topologies during internal disruption. (**a**) initial state of equilibrium; (**b**, **c**) shift of inner region and reconnection of field lines; (**d**) completed reconnection, new equilibrium state

in region A pushes the inner column along the line ab towards the oppositely directed field until complete reconnection occurs (Fig. 11.5d). After reconnection, which covers the region from 0 to r_0, where $\Psi^*(r_0) = \Psi^*(0)$ (Fig. 11.6b), the energy of the magnetic field B_* decreases substantially, so that this process is energetically favorable. It is easy to see from Fig. 11.6 that the lines of force with the same values of Ψ^* are involved in the reconnection process, so that the total helical flux between these two lines is zero. (It should be noted that the sign of B_* changes at $r = r_s$.) The two regions, inner and outer, with the same $d\Psi^*$ after the reconnection merge into a single annular region so that after the

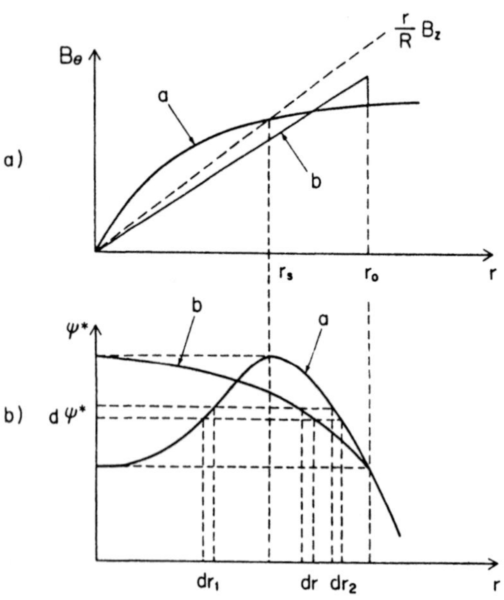

Fig. 11.6. (**a**) Profile of poloidal magnetic field before, a, and after, b, the internal disruption. For reference, the position of the singular surface rB_z/R is shown. (**b**) Profile of helical flux function before, a, and after, b, the internal disruption

reconnection Ψ^* has the form shown by curve b in Fig. 11.6b. The fact that B_θ is proportional to r after the reconnection (curve b in Fig. 11.6a) means that the current profile is uniform.

In incompressible flow, two small rings with areas $2\pi r_1 dr_1, 2\pi r_2 dr_2$ merge into a single ring with an area $2\pi r dr = 2\pi r_1 dr_1 + 2\pi r_2 dr_2$, which immediately makes it possible to find $\Psi_1^*(r)$ after the reconnection. It also follows that the value of $\int_s \Psi^* dS$, where S is the cross section area with radius r_0, is conserved.

For the simple special case of a parabolic current density $J = J_0(1 - r^2/a^2)$, the function $\iota = 1/q = RB_0/rB_z$ also turns out to be parabolically dependent on r, and all the calculations are simplified. Namely, before the reconnection, $1/q = (1/q_0)(1 - r^2/2a^2)$, so that $r_s^2 = (1 - q_0)2a^2$ for $m = 1$ and $n = 1$, where q_0 is the value of q at the center of the column. The reconnection region is given by $r = r_0$, with $r_0^2 = 2r_s^2$ (Problem 11.11), so that at the boundary of this region

$$q = q_{r0} = \frac{q_0}{2q_0 - 1} \; . \tag{11.6.6}$$

The internal disruption explained here was first observed in the central region of tokamaks as sawtooth oscillations of soft X-ray emission. It is interesting that this kind of internal disruption was also found in current carrying stellarator/heliotrons and reversed field pinches.

The internal disruption model shown by Figs. 11.5, 6 was introduced by Kadomtsev and explained observations in many small and medium size tokamaks. But the experimental results on the Joint European Torus JET (which is the largest tokamak in the world) deviated from the Kadomtsev model [11.14] in that the reconnection process depends on the nonlinear evolution of the $(m, n) = (1, 1)$ tearing mode and the collapse time at the internal disruption becomes longer in the larger tokamak. The collapse time is $\sim 100\,\mu$s in JET [11.21] which is comparable to that in much smaller tokamaks.

A recently developed theoretical model reveals that the deviation from the Kadomtsev model comes about from the safety factor profile. The introduction of the concept of a flattened q-profile leads to a completely different picture of the behavior of the plasma during the internal disruption. Consider the idealized case where $q = 1$ everywhere inside the central region. The absence of shear makes an interchange of flux tubes induce no bending of the magnetic field line (Sect. 10.3). In reality there will be weak shear in the central region and we expect a *quasi-interchange mode*. To see this we consider the potential energy given by (10.3.28),

$$\delta W_{mk} = \frac{4\pi^2}{\mu_0}\frac{(B_z^0)^2}{R}\int_0^{2\pi R}\left(1 - \frac{1}{q}\right)\left(\frac{d\xi}{dr}\right)^2 r^3 dr + \mathrm{O}\left(\left(\frac{r}{R}\right)^4\right) , \tag{11.6.7}$$

where we assumed $|kr| = r/R \ll 1$ and $L = 2\pi R$. For the usual q-profile it is necessary that the large contribution of the first term in (11.6.7), which is positive definite, be reduced to zero. This is achieved by taking the leading order displacement within the $q = 1$ surface as $\xi =$ const and decreasing it to zero at the narrow region in the neighborhood of the $q = 1$ surface. In other words

$d\xi/dr = 0$ where $q = 1$. However, if $(1 - q)$ is sufficiently small inside the $q = 1$ surface, the separation of the potential energy into two terms as in (11.6.7) fails and ξ is no longer required to be constant inside the $q = 1$ surface.

On the time scale of the initial phase of the internal disruption the plasma behaves as a perfect conductor and the magnetic topology is unchanged. The fast temperature collapse at the magnetic axis is simply due to the rapid sideway motion of the plasma core. Any reconnection takes place in a later phase.

In tokamaks, another disruption which sometimes terminates plasma current thereby abruptly losing MHD equilibrium is called a *major disruption*. This phenomenon can be explained by a mechanism based on multi-helical tearing modes. In tokamaks usually $q(0) < 1$ and $q(a) > 3$ and q increases monotonically towards the edge. There are resonant surfaces at $q = 2$ and $q = 3/2$ inside the plasma column. In some cases tearing modes with $(m, n) = (2, 1)$ and $(m, n) = (3, 2)$ become unstable simultaneously. When the magnetic perturbation of an (m, n) tearing mode is expressed through $\Psi_{mn}(r)\cos(m\theta - nz/R)$, the magnetic field line equations in the cylindrical geometry become

$$\frac{dr}{dz} = \sum_{m,n} \frac{m\Psi_{mn}}{rB_0} \sin\left(m\theta - \frac{n}{R}z\right)$$

$$\frac{d\theta}{dz} = \frac{B_\theta^0(r)}{rB_0} + \sum_{m,n} \frac{1}{B_0} \frac{d\Psi_{mn}}{dr} \cos\left(m\theta - \frac{n}{R}z\right) . \tag{11.6.8}$$

We note that without the perturbation the magnetic field line stays on the $r =$ const surface.

The tearing mode produces a magnetic island in the neighborhood of the resonant surface, which can be confirmed by following the magnetic field lines by (11.6.8) for a single Ψ_{mn}. The width of the island is denoted by w_{mn} and the resonant surface is r_{mn}. The parameter

$$K \equiv \frac{w_{mn} + w_{m'n'}}{|r_{mn} - r_{m'n'}|} \tag{11.6.9}$$

is defined for two magnetic islands due to the tearing mode (m, n) and (m', n'). For $K \ll 1$, magnetic islands with small width are clearly seen around the resonant surface; however, when $K \gg 1$ we have magnetic island overlapping, and the magnetic field lines followed by (11.6.8) including Ψ_{mn} and $\Psi_{m'n'}$ show ergodic behavior in the region between the two resonant surfaces r_{mn} and $r_{m'n'}$. In other words, the magnetic surface is destroyed. In the multi-helical tearing mode model the magnetic surfaces of the whole confinement region are destroyed by many tearing modes with different (m, n) at different resonant surfaces. When this occurs, high temperature plasma confinement is lost through parallel heat transport since particles following magnetic field lines can reach the edge region by parallel motion alone (Sect. 13.6). This scenario of a major disruption has been confirmed by solving the reduced MHD equations for ϕ, Ψ numerically in cylindrical and toroidal geometries.

11.7 Driven Magnetic Reconnection

Let us consider under what condition reconnection takes place. At first, a tokamak equilibrium is established. As discussed in Sect. 11.6, by subtracting the helical field in resonance with an (m, n) mode (m and n are the poloidal and toroidal mode numbers, respectively), the tokamak field reduces to $\boldsymbol{B}_*$ which changes its sign across the resonance (rational) surface $q(r_s) = m/n$, where q is the safety factor. In a resistive plasma it is known that the tearing mode instability occurs at the resonance surface and the anti-parallel fields get reconnected spontaneously.

Spontaneous reconnection is commonly observed in magnetically confined plasmas. In natural plasmas, however, it is often observed that an anti-parallel field configuration is locally formed as a result of a configurational change or a collision of two plasmas with different origins. The solar wind-magnetosphere interaction [11.22] is a typical example. In order to distinguish it from the spontaneous process, we call a reconnection that occurs as a result of a collision of two plasmas a *driven reconnection*. Since spontaneous reconnection can occur with or without an external disturbance, it can alternatively be called a *linear reconnection*. In contrast, a driven reconnection requires an external force that drives a tearing free system to a system with a locally anti-parallel field configuration. In this sense a driven reconnection is essentially 'nonlinear'.

Let us describe the difference between spontaneous and driven reconnections. For simplicity, we shall consider the resistive tearing instability a spontaneous reconnection. Since the instability occurs independently of the external condition, the reconnection rate must be a function of internal variables. The electric field induced in the magnetically neutral sheet, which is a measure of the reconnection rate, is given by $E = \eta J$ where η is the resistivity and J is the neutral sheet current. Since the neutral sheet current exists initially, the linear reconnection rate is given by ηJ_0 where J_0 is the initial sheet current. Normalizing it by $v_\mathrm{A} B_0$, we obtain

$$\varepsilon_L = \frac{\eta J_0}{v_\mathrm{A} B_0} = \frac{\eta}{\mu_0 L v_\mathrm{A}} = S^{-1} \qquad (11.7.1)$$

where B_0 is the characteristic magnetic field, L is the characteristic length perpendicular to the magnetic field, v_A is the characteristic Alfvén velocity and S is the magnetic Reynolds number. This relation indicates that the reconnection rate of the tearing instability is roughly on the order of S^{-1}.

On the other hand, a driven reconnection is rather independent of the initial configuration, since the anti-parallel configuration is achieved as a result of nonlinear deformation due to a driving force. Thus, the neutral sheet current is not a causal variable for reconnection but rather a resulting variable, in clear contrast with the spontaneous case. The induced neutral sheet current is given by

$$J = \frac{E}{\eta} = \frac{u_\mathrm{d} B_0}{\eta} \qquad (11.7.2)$$

where u_d is the driving plasma flow. Normalizing ηJ by $v_\mathrm{A} B_0$, one obtains

$$\varepsilon_{NL} = \frac{E}{v_A B_0} = \frac{u_d}{v_A} . \tag{11.7.3}$$

Comparing (11.7.1) and (11.7.3), it can be seen that depending on the strength of the driving flow the reconnection rate of driven reconnection can take an arbitrary value between 0 and 1, while that of the spontaneous process is in general very small since $S \gg 1$. Thus it can be said that the driven reconnection process is usually a much more active process than the spontaneous process. It should also be noted that the spontaneous reconnection process is highly dependent on the resistivity, while the driven process is rather independent of it. This can be explained as follows. As far as the driving flow persists, the neutral sheet current increases until the induced electric field matches the driving field. In other words, when the resistivity is small, the induced current becomes large, whereas when the resistivity is large, the current does not have to increase very much. Thus, in the analysis of driven reconnection one does not have to be so careful about the effect of resistivity, but in the analysis of resistivity-triggered reconnection, the presence of the tearing mode induces a very sensitive dependence on the resistivity, since the growth rate depends on the resistivity as in (11.6.3) or (11.6.4).

Driven reconnection is expected to exhibit a variety of activities in a magnetized plasma depending on the magnetic field configuration and the driving flow condition. One of the attractive features of magnetic reconnection is its effectiveness in converting magnetic to kinetic energy. For example, one sets up an initial condition having an anti-parallel magnetic field configuration as shown in Fig. 11.7. From the left and right boundaries, plasmas intrude into the region shown in the figure, so that the neutral sheet is compressed and the neutral sheet current is enhanced. The other boundaries are assumed to be free boundaries. The resistivity is assumed to be

$$\eta = \begin{cases} \alpha(J - J_c)^2 & \text{for} \quad J > J_c \\ 0 & \end{cases} \tag{11.7.4}$$

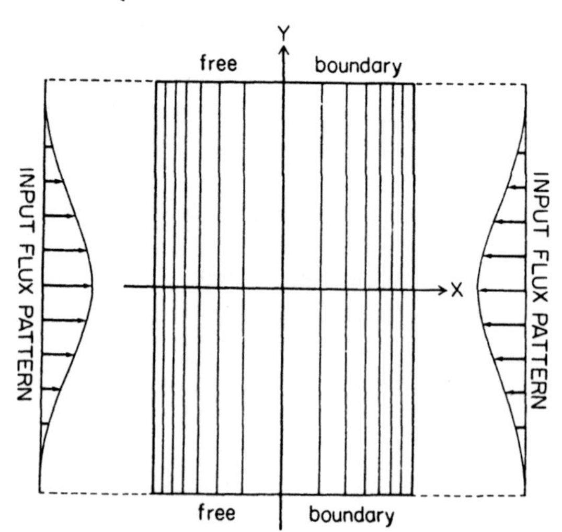

Fig. 11.7. A configuration for driven reconnection

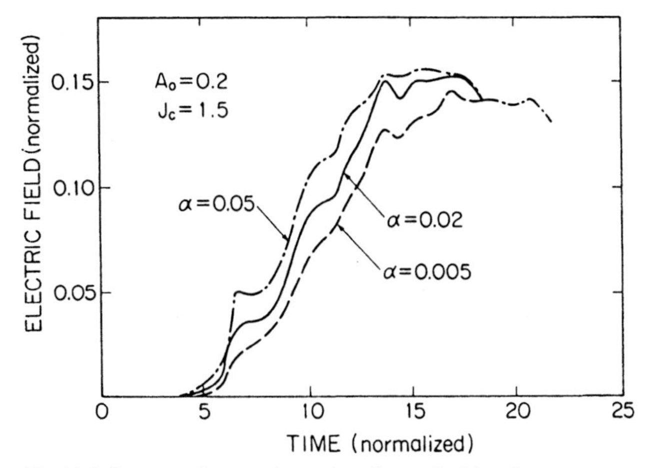

Fig. 11.8. Reconnection rate for various flow velocities A_0

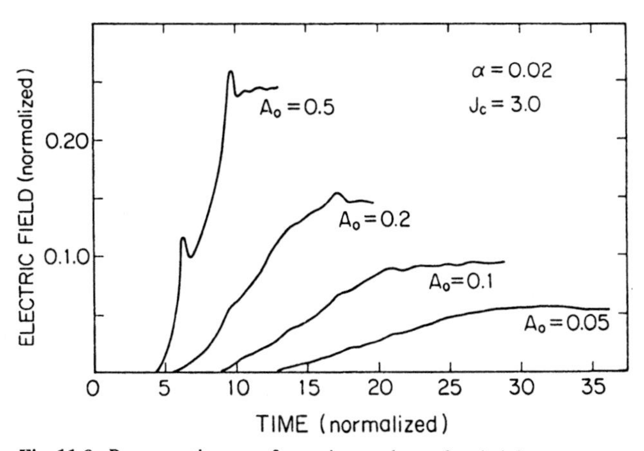

Fig. 11.9. Reconnection rate for various values of resistivity α

which simulates an anomalous resistivity driven by a current driven instability. It is noted that J_c is chosen to be larger than the initial peak current J_0. This model was discussed by *Sato* and *Hayashi* [11.23, 24] and Fig. 11.8, 9 were obtained by them.

Figure 11.8 shows the time evolution of the reconnection rate (the electric field measured at the reconnection point in Fig. 11.7) for different resistivities (different α) with a fixed driving flow speed (denoted by $A_0 \propto u_d$). Figure 11.9 shows the time evolution of the reconnection rate for different driving flows with a fixed resistivity. From these results one can definitely conclude that the reconnection rate is strongly dependent on the driving flow, but only very weakly on the resistivity. It is certain that there is a reconnection process which is fundamentally different from the tearing mode type reconnection.

In astronomical media, in general the energy input plays an essential role in driving reconnection. Once reconnection is driven, then the magnetic energy is converted into kinetic energy, whereby the internal energy is carried away from the reconnection region [11.22]. The macroscopic energy thus carried away could in turn cause configurational changes in the neighboring region to drive another reconnection. Similar phenomena may be observed in laboratory plasmas. Suppose that an equilibrium exists in a current carrying plasma but the state is not stable magnetohydrodynamically. Then part of the magnetic energy is converted to kinetic energy by a current driven kink instability and consequently, the magnetic configuration is changed. Suppose further that a lower energy equilibrium state exists. Then reconnection is inevitably driven as a result of nonlinear deformation due to the kink instability, even if the initial equilibrium is stable with respect to the tearing mode. In the driven reconnection process, current peaking occurs locally in the region where instability driven flows converge, and ohmic dissipation is enhanced there. Due to this ohmic dissipation, the system can promptly relax towards the lower energy state. The self-reversal process in the reversed field pinch is a typical example of the relaxation process due to magnetic reconnection, specifically, driven magnetic reconnection [11.24].

11.8 Self-Organization

Here we regard a plasma as a conducting fluid having small resistivity and small viscosity. Even in this simple model interaction of the plasma with magnetic fields leads to extremely complex behavior, especially when turbulence occurs. It is remarkable, therefore, that one can make quantitative predictions about the plasma configuration resulting from such turbulence. This is possible because the turbulence, which has a small resistivity, allows the plasma rapid access (in a time short compared with the usual resistive diffusion time) to a particular minimum energy state. This process, known as plasma relaxation, involves the reconnection of magnetic field lines and is a remarkable example of *self-organization* of a plasma [11.25].

A plasma resembles an infinite number of interlinked flexible conductors and the problem is to identify the appropriate constraints. If there were no constraints the state of minimum energy would be a vacuum field with no plasma current. This is indeed the eventual state of an isolated resistive plasma but is clearly not what we are concerned with here. At the other extreme, if the plasma is perfectly conducting then there are an infinite number of constraints. These arise because the fluid moves precisely with the magnetic field, each field line maintains its identity and the flux through any closed curve moving with the fluid is constant.

To express these constraints mathematically we introduce the vector potential, $\boldsymbol{B} = \nabla \times \boldsymbol{A}$. For every infinitesimal flux tube surrounding a closed line of force the quantity

$$K = \int \boldsymbol{A} \cdot \boldsymbol{B} \, d\tau \tag{11.8.1}$$

is an invariant. This is confirmed by using the relation (10.1.23)

$$\frac{dK}{dt} = \int \left(\frac{\partial \boldsymbol{A}}{\partial t} \cdot \boldsymbol{B} + \boldsymbol{A} \cdot \frac{\partial \boldsymbol{B}}{\partial t} \right) d\tau + \int_s (\boldsymbol{A} \cdot \boldsymbol{B})(\boldsymbol{n} \cdot \boldsymbol{v}) dS \,. \tag{11.8.2}$$

By noting $\partial \boldsymbol{B}/\partial t = \nabla \times (\boldsymbol{v} \times \boldsymbol{B})$,

$$\int \boldsymbol{A} \cdot \frac{\partial \boldsymbol{B}}{\partial t} d\tau = -\int \nabla \cdot [\boldsymbol{A} \times (\boldsymbol{v} \times \boldsymbol{B})] d\tau + \int (\boldsymbol{v} \times \boldsymbol{B}) \cdot (\nabla \times \boldsymbol{A}) d\tau \tag{11.8.3}$$

is obtained. Here the second term of the right hand side vanishes identically. By using the Gauss theorem,

$$\int \nabla \cdot [\boldsymbol{A} \times (\boldsymbol{v} \times \boldsymbol{B})] d\tau = \int_s [(\boldsymbol{A} \cdot \boldsymbol{B})(\boldsymbol{v} \cdot \boldsymbol{n}) - (\boldsymbol{A} \cdot \boldsymbol{v})(\boldsymbol{B} \cdot \boldsymbol{n})] dS \tag{11.8.4}$$

and from the definition of the infinitesimal flux tube $\boldsymbol{B} \cdot \boldsymbol{n} = 0$. Then, (11.8.2–4) give

$$\frac{dK}{dt} = \int \left(\frac{\partial \boldsymbol{A}}{\partial t} \cdot \boldsymbol{B} \right) d\tau \,. \tag{11.8.5}$$

Here $E = -\nabla \phi - \partial \boldsymbol{A}/\partial t$ is substituted into (11.8.5) to yield

$$-\int (\boldsymbol{E} \cdot \boldsymbol{B} + \nabla \phi \cdot \boldsymbol{B}) d\tau = -\int \nabla \cdot (\phi \boldsymbol{B}) d\tau = -\int_s \phi (\boldsymbol{B} \cdot \boldsymbol{n}) dS \,, \tag{11.8.6}$$

where $\boldsymbol{E} \cdot \boldsymbol{B} = 0$ is valid for ideal MHD. By noting $\boldsymbol{B} \cdot \boldsymbol{n} = 0$ for the flux tube surrounding a closed line of force, we confirm

$$\frac{dK}{dt} = 0 \,. \tag{11.8.7}$$

The quantity K is called the *magnetic helicity*.

Note that this invariant is essentially topological – it involves the identification of lines of force and represents the linkage of lines of force with one another. If one closed field line initially links another n-times then in a perfectly conducting plasma, the two loops must remain linked n-times during any plasma motion.

If we minimize the magnetic energy,

$$W = \frac{1}{2\mu_0} \int B^2 d\tau \tag{11.8.8}$$

subject to the constraint of magnetic helicity conservation described above, then for a plasma confined by a perfectly conducting toroidal shell the equilibrium satisfies

$$\nabla \times \boldsymbol{B} = \lambda \boldsymbol{B} \,, \tag{11.8.9}$$

as will be shown below.

From the first variation of $W - \lambda K/2$,

$$\delta\left(W - \lambda\frac{K}{2}\right) = \int \frac{\boldsymbol{B}\cdot\delta\boldsymbol{B}}{\mu_0}d\tau - \frac{\lambda}{2}\int(\delta\boldsymbol{A}\cdot\boldsymbol{B} + \boldsymbol{A}\cdot\delta\boldsymbol{B})d\tau\,, \tag{11.8.10}$$

where λ is a Lagrange multiplier. The first term is written as

$$\begin{aligned}\frac{1}{\mu_0}\int \boldsymbol{B}\cdot\delta\boldsymbol{B}d\tau &= \frac{1}{\mu_0}\int \boldsymbol{B}\cdot\nabla\times\delta\boldsymbol{A}\,d\tau \\ &= -\frac{1}{\mu_0}\left(\int \boldsymbol{B}\times\delta\boldsymbol{A}\cdot\boldsymbol{n}\,dS - \int \delta\boldsymbol{A}\cdot\nabla\times\boldsymbol{B}\,d\tau\right),\end{aligned} \tag{11.8.11}$$

and the second term is

$$\begin{aligned}\int(\delta\boldsymbol{A}\cdot\boldsymbol{B} + \boldsymbol{A}\cdot\delta\boldsymbol{B})d\tau &= \int(\delta\boldsymbol{A}\cdot\boldsymbol{B} + \boldsymbol{A}\cdot\nabla\times\delta\boldsymbol{A})d\tau \\ &= 2\int(\delta\boldsymbol{A}\cdot\boldsymbol{B})d\tau - \int(\boldsymbol{A}\times\delta\boldsymbol{A})\cdot\boldsymbol{n}\,dS\,.\end{aligned} \tag{11.8.12}$$

Here we assume that the wall is perfectly conducting. Then $\boldsymbol{n}\times\delta\boldsymbol{A} = 0$ is imposed on the wall, which corresponds to $\boldsymbol{n}\times\boldsymbol{E} = 0$, and the surface integrals of (11.8.11, 12) vanish. Therefore, (11.8.10) becomes

$$\delta\left(W - \lambda\frac{K}{2}\right) = \int \delta\boldsymbol{A}\cdot[(\nabla\times\boldsymbol{B}) - \lambda\boldsymbol{B}]d\tau\,. \tag{11.8.13}$$

For an arbitrary choice of $\delta\boldsymbol{A}$, (11.8.13) becomes zero when (11.8.9) is satisfied. When (11.8.13) is zero, the magnetic energy is minimized with the helicity being held constant. Thus the state of minimum magnetic energy is a force free equilibrium.

However, the above discussion cannot be the appropriate description of the quiescent state. In order to determine λ, one would have to calculate the invariant K for each closed field line and relate it to its initial value. Hence, far from being universal and independent of initial conditions, the state defined by (11.8.9) depends on every detail of the initial state.

To resolve this difficulty we must recognize that real plasmas, especially turbulent ones, are never perfectly conducting in the sense discussed above. In the presence of resistivity, however small, the topological properties of lines of force are no longer preserved. Lines of force may break and reconnect even though the resistive diffusion time may be very long and there is insignificant flux dissipation. Mathematically the situation is one of non-uniform convergence; when $\eta = 0$ the equations do not permit changes in the topology of field lines, whereas such changes may occur when $\eta \neq 0$, even in the limit of small η. Physically, as $\eta \to 0$ the regions over which resistivity acts get smaller but the field gradients get larger and the rate of reconnection does not diminish as fast as η or may not diminish at all. This situation is similar to driven reconnection discussed in Sect. 11.7.

One concludes that in a turbulent resistive plasma, flux tubes have no continuous independent existence. Consequently all the topological invariants K cease to be relevant, not because the magnetic flux changes significantly but because it is no longer possible to identify the field by which the flux changes. However, the sum of all the invariants, that is, the integral of $\boldsymbol{A} \cdot \boldsymbol{B}$ over the total plasma volume V_0, is independent of any topological considerations and of the need to identify particular field lines. Therefore, it remains a good invariant so long as the resistivity is small.

To obtain the relaxed state of a slightly resistive plasma, therefore, we must minimize the energy subject to the single constraint that the total magnetic helicity

$$K_0 = \int_{V_0} \boldsymbol{A} \cdot \boldsymbol{B}\, d\tau \tag{11.8.14}$$

be invariant. When $\eta \neq 0$ in the plasma and $\boldsymbol{B} \cdot \boldsymbol{n} = 0$ at the conducting boundary,

$$\frac{dK_0}{dt} = -\int \eta \boldsymbol{J} \cdot \boldsymbol{B}\, d\tau \tag{11.8.15}$$

and

$$\frac{dW}{dt} = -\int \eta J^2 d\tau \, . \tag{11.8.16}$$

By noting $\mu_0 \boldsymbol{J} = \nabla \times \boldsymbol{B}$,

$$\frac{|dK_0/dt|}{|dW/dt|} \doteq \frac{|\eta J B|}{|\eta J^2|} = \frac{B}{J} = \mu_0 L \, . \tag{11.8.17}$$

When L is small compared to the system size, K_0 changes more slowly than W.

From the first variation $\delta(W - \lambda\, K_0/2) = 0$,

$$\nabla \times \boldsymbol{B} = \lambda \boldsymbol{B} \tag{11.8.18}$$

is again obtained for equilibrium of a plasma enclosed by a perfectly conducting toroidal shell where λ is a constant. This relaxed state depends only on a single parameter λ.

For a circular cross section torus of large aspect ratio one may take the cylindrical limit in which the solution of (11.8.18) is

$$B_r = 0 \, , \quad B_\theta = B_0 J_1(\lambda r) \, , \quad B_z = B_0 J_0(\lambda r) \, . \tag{11.8.19}$$

This equilibrium is known as the *Bessel function solution.* The field profiles given by (11.8.19) agree well with those observed in the quiescent phase of many toroidal pinch discharges. The onset of the spontaneous reversed toroidal field at the wall can also now be determined to occur when $\lambda a > 2.4$. This result is also in good agreement with many observations in reversed field pinches (Sect. 9.2). It should be noted that K_0 is gauge invariant. Under a gauge transformation $\boldsymbol{A} \rightarrow \boldsymbol{A} + \nabla \chi$, the change in the helicity K_0 is

$$\int \boldsymbol{B} \cdot \nabla \chi\, d\tau = \int_s \chi \boldsymbol{B} \cdot \boldsymbol{n}\, dS \, . \tag{11.8.20}$$

With the boundary condition $\boldsymbol{B} \cdot \boldsymbol{n} = 0$ the surface integral vanishes and K_0 is gauge invariant.

A comment is also necessary on the role of plasma pressure. Relaxation proceeds by reconnection of lines of force. During this reconnection, the plasma pressure can equalize itself so that the fully relaxed state is also a state of uniform pressure. Hence the inclusion of plasma pressure does not change our conclusion about the relaxed or the self-organized state. Of course, one may argue that pressure relaxation might be slower than field relaxation so that upon field equilibration some pressure gradients might remain.

The energy relaxation process due to driven reconnection explained in Sect. 11.7 can also be regarded as a self-organization process. It was shown by a numerical simulation done by *Sato* and *Horiuchi* that the magnitude of the local helicity $|\boldsymbol{A} \cdot \boldsymbol{B}|$ takes on a minimum value, or $\boldsymbol{A} \cdot \boldsymbol{B} \doteq 0$, at the reconnection point where the magnetic energy is dissipated [11.26].

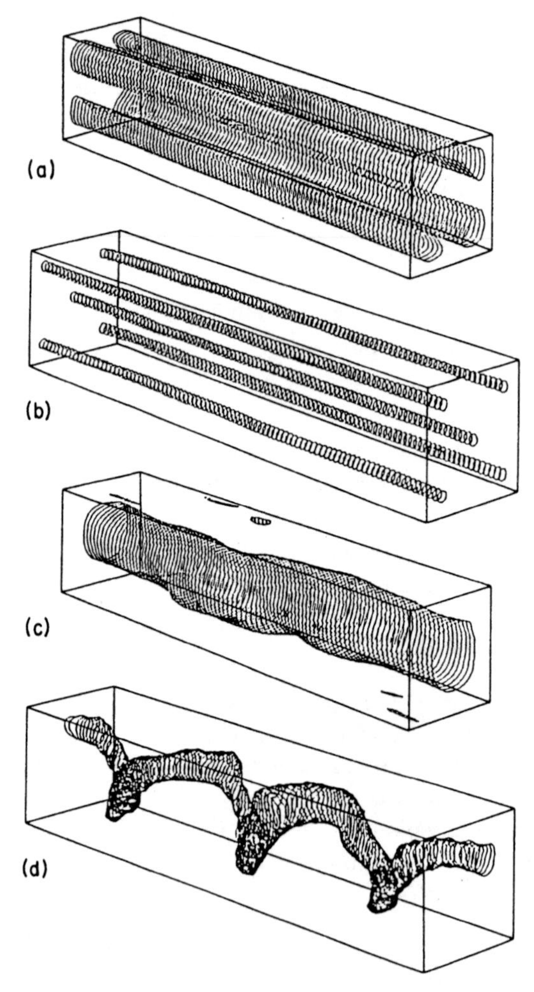

Fig. 11.10. Three dimensional display of the contours of the toroidal magnetic flux, (a) and (b) are initial states, (c) and (d) are final quasi-stationary states corresponding to (a) and (b), respectively

Thus the configurational change associated with the magnetic energy relaxation due to driven reconnection conserves the magnetic helicity. As an example, consider the initial force free equilibria consisting of parallel current channels as shown in Fig. 11.10a,b. These equilibria are not in minimum energy state and are kink unstable. The plasma motion associated with the kink instabilities drives a series of magnetic reconnections resulting in the formation of final equilibrium states which consist of a single current channel as shown in Fig. 11.10c,d. Both final states are minimum energy states, but are clearly different from each other. For the case of Fig. 11.10c, the current channel has an axisymmetric cylindrical structure, while for the case of Fig. 11.10d it has a helically twisted structure. The difference arises from the different initial values of the helicity which are preserved throughout the reconnection or energy relaxation processes. It was shown by *Reiman* [11.27] and by *Horiuchi* and *Sato* [11.26] that the minimum energy state becomes either an axisymmetric or a helically twisted state depending on the helicity value. The results of the full MHD numerical simulation shown in Fig. 11.10 are consistent with this theory [11.26].

PROBLEMS

11.1. Derive (11.1.5, 7).
11.2. Show that $dI_1/dt \sim O(\varepsilon^2)$ when $\nabla \cdot \boldsymbol{u} = 0$.
11.3. Confirm $d(\boldsymbol{B} \cdot \nabla P)/dt = 0$ in incompressible magnetohydrodynamics.
11.4. Derive (11.1.13) from (11.1.12).
11.5. Derive (11.1.15) from (11.1.14).
11.6. Confirm the properties of Poisson bracket (11.1.23–26).
11.7. Confirm that $d\Psi^*/dr = 0$ gives the singular surface.
11.8. Show (11.5.9).
11.9. Derive (11.6.1, 2).
11.10. Derive $\gamma \sim \eta^{1/3}$ from (11.6.4) when $\Delta'\lambda \sim 1$.
11.11. Confirm that the reconnection radius is given by $r_0^2 = 2r_s^2$.

12. Wave-Plasma Interactions

When an electromagnetic wave is incident on an inhomogeneous plasma from the vacuum, a collective oscillation, or wave, is excited inside the plasma. The excited wave then propagates inside the non-uniform plasma and dissipates due to wave-particle interactions and collisions. This type of problem is generally called the wave-plasma interaction and has a variety of applications to fusion research such as plasma heating, stability control, and current-drive. This chapter deals with such problems. Because of their complex nature, we shall restrict ourselves to only giving an outline of the problems. First, in Sect. 12.1 we describe the basic notion of wave propagation in non-uniform plasma. Then in Sect. 12.2 we consider wave propagation in a magnetically confined plasma and describe how the externally excited wave can reach the point where the wave can deposit energy and/or momentum via wave-particle resonances. The next section discusses the possibility of current control via radio-frequency waves. Finally, in Sect. 12.4, we briefly describe the wave-plasma interaction when intense laser light is used as an energy driver for inertial confinement fusion.

12.1 Waves in Non-uniform Plasma

Real plasmas are spatially non-uniform. For the moment, we consider the situation in which a wave is generated at a steady local source and propagates inside a weakly non-uniform plasma at constant frequency ω.

12.1.1 The Eikonal Approximation

As the wave propagates in a non-uniform plasma, the amplitude and the propagation direction change slowly, although the phase itself changes rapidly. In such a case, the eikonal representation is useful which for one-dimensional propagation is

$$\phi'(x,t) = \exp\left[\mathrm{i}\int^{x} dx' k(x') - \mathrm{i}\omega t\right] \tag{12.1.1}$$

and in three dimensions, by taking into account of the change of the polarization vector it is

$$\boldsymbol{E}'(\boldsymbol{r},t) = \boldsymbol{\varepsilon}(\boldsymbol{r}) \exp\left[\mathrm{i}\int^{\boldsymbol{r}} \boldsymbol{k}(\boldsymbol{r}') \cdot d\boldsymbol{r}' - \mathrm{i}\omega t\right] , \tag{12.1.2}$$

where $\boldsymbol{\varepsilon}(\boldsymbol{r})$ is the unit vector in the direction of the wave electric field [12.1].

12.1.2 The Local Dispersion Relation

If the spatial non-uniformity is sufficiently weak that it can be ignored in the scale length on the order of the wavelength, we can use the local dispersion relation given by

$$\overleftrightarrow{D}(\boldsymbol{k},\omega,\boldsymbol{r})\cdot\boldsymbol{\varepsilon}(\boldsymbol{r})=0 \tag{12.1.3}$$

or

$$\det\overleftrightarrow{D}(\boldsymbol{k},\omega,\boldsymbol{r})=0\ . \tag{12.1.4}$$

We denote the solution of (12.1.4) by

$$\boldsymbol{k}=\boldsymbol{k}(\boldsymbol{r}) \tag{12.1.5}$$

for a given ω, which describes how the wavevector changes as the wave propagates in a spatially non-uniform medium.

12.1.3 Geometrical Optics

The three-dimensional wave propagation characteristics in the local approximation are described by the geometrical optics equations

$$\frac{d\boldsymbol{r}}{dt}=\left(\frac{\partial\omega}{\partial\boldsymbol{k}}\right)_{\boldsymbol{r}}\ , \tag{12.1.6}$$

$$\frac{d\boldsymbol{k}}{dt}=-\left(\frac{\partial\omega}{\partial\boldsymbol{r}}\right)_{\boldsymbol{k}}\ , \tag{12.1.7}$$

where ω is the formal solution of (12.1.4)

$$\omega=\omega(\boldsymbol{k},\boldsymbol{r})\ . \tag{12.1.8}$$

Equation (12.1.6) simply describes the fact that the wave packet propagates with the group velocity of the wave. Equation (12.1.7) can be derived by using (12.1.6):

$$0=\frac{\partial\omega}{\partial\boldsymbol{k}}\cdot\frac{d\boldsymbol{k}}{dt}+\frac{\partial\omega}{\partial\boldsymbol{r}}\cdot\frac{d\boldsymbol{r}}{dt}=\frac{d\boldsymbol{r}}{dt}\cdot\left[\frac{d\boldsymbol{k}}{dt}+\frac{\partial\omega}{\partial\boldsymbol{r}}\right]\ .$$

Usually (12.1.6, 7) are used in the ray tracing calculation (Sect. 12.2).

12.1.4 The Eigenvalue Problem

The local dispersion relation describes the refraction and reflection of the wave, but it does not give the eigenfrequency ω. To determine the eigenfrequency, we have to solve an eigenvalue problem.

Here we give simple examples for determining the eigenfrequency of an electrostatic wave propagating in one dimension. The basic equation is the Poisson equation which reads

$$\varepsilon_0 \frac{\partial^2 \phi(x)}{\partial x^2} = -\int dx' Q(x,x',\omega)\phi(x') , \tag{12.1.9}$$

where we denoted the potential as $\phi(x)\mathrm{e}^{-\mathrm{i}\omega t}$ and included only the induced charge and neglected the source charge. By the transformation

$$x - x' = \Delta x , \qquad \frac{x+x'}{2} = x - \frac{\Delta x}{2} = X , \tag{12.1.10}$$

we write

$$Q(x,x',\omega) = Q(\Delta x, X, \omega) , \tag{12.1.11}$$

where the X-dependence is due to the non-uniformity of the plasma. We use the eikonal representation (12.1.1) and assume

$$\left| k^{-2}\frac{dk}{dx} \right| , \qquad \left| \frac{1}{Qk}\,\frac{dQ}{dx} \right| \sim \lambda \ll 1 . \tag{12.1.12}$$

We expand in powers of Δx as

$$\int^{x'} k(x'')dx'' = \int^{x} k(x'')dx'' - \Delta x\, k(x) + \frac{(\Delta x)^2}{2}\,\frac{dk(x)}{dx} + \cdots$$

$$Q(\Delta x, X, \omega) = Q(\Delta x, x, \omega) - \frac{\Delta x}{2}\,\frac{\partial Q(\Delta x, x, \omega)}{\partial x} + \cdots . \tag{12.1.13}$$

Then to first order in λ, the right hand side of (12.1.9) becomes

$$-\int d\Delta x \left(1 - \frac{\Delta x}{2}\,\frac{\partial}{\partial x}\right) Q(\Delta x, x, \omega)\mathrm{e}^{-\mathrm{i}k(x)\Delta x}\left[1 + \mathrm{i}\frac{(\Delta x)^2}{2}\,\frac{dk(x)}{dx}\right]\phi(x) . \tag{12.1.14}$$

We define the local coefficient $\tilde{Q}(k,x,\omega)$ as

$$\tilde{Q}(k,x,\omega) = \int d\Delta x \mathrm{e}^{-\mathrm{i}k\Delta x} Q(\Delta x, x, \omega) . \tag{12.1.15}$$

Using the relations

$$\int d\Delta x \mathrm{e}^{-\mathrm{i}k\Delta x} Q(\Delta x, x, \omega)(\Delta x)^n = \left(\mathrm{i}\frac{\partial}{\partial k}\right)^n \tilde{Q}(k,x,\omega) \tag{12.1.16}$$

$$\frac{d}{dx} = \frac{\partial}{\partial x} + \frac{dk(x)}{dx}\,\frac{\partial}{\partial k} , \tag{12.1.17}$$

we can reduce (12.1.9) to the form

$$\varepsilon_0 \frac{d^2\phi(x)}{dx^2} = -\left\{ \tilde{Q}(k,x,\omega) - \frac{\mathrm{i}}{2}\left[\frac{d}{dx}\,\frac{\partial}{\partial k}\tilde{Q}(k,x,\omega)\right]\right\}\phi(x) . \tag{12.1.18}$$

This equation has the form of the Schrödinger equation in quantum mechanics with the expansion parameter λ playing the role of the Planck constant. We can therefore use the quasi-classical approximation (WKB approximation) to solve the equation. Namely, we expand $k(x)$ in powers of λ as

$$k(x) = k_0(x) + k_1(x) + \cdots, (k_0 = O(1), k_1 = O(\lambda), \cdots) . \tag{12.1.19}$$

To zero order in λ we obtain the local dispersion relation

$$\varepsilon_0 - k_0^{-2}\tilde{Q}[k_0(x), x, \omega] = 0 \tag{12.1.20}$$

which determines $k_0(x)$. To first order (12.1.18) becomes

$$-2\varepsilon_0 \left[k_1(x) - \frac{\mathrm{i}}{2}\frac{d}{dx}\right] k_0(x) = -\left[k_1(x) - \frac{\mathrm{i}}{2}\frac{d}{dx}\right] \frac{\partial}{\partial k_0}\tilde{Q}(k_0, x, \omega) ,$$

which can be solved for $k_1(x)$ as

$$k_1(x) = \frac{\mathrm{i}}{2}\frac{d}{dx}\log\left[\varepsilon_0 k_0(x) - \frac{\mathrm{i}}{2}\frac{\partial}{\partial k_0}\tilde{Q}(k_0, x, \omega)\right] . \tag{12.1.21}$$

Substituting (12.1.19) into (12.1.1), we obtain

$$\phi(x) = \left[\varepsilon_0 k_0(x) - \frac{1}{2}\frac{\partial}{\partial k_0}\tilde{Q}(k_0, x, \omega)\right]^{-1/2} \exp\left[\mathrm{i}\int^x k_0(x')dx'\right] . \tag{12.1.22}$$

The WKB approximation breaks down at the reflection point $k_0 = 0$. In this case, we introduce analytical continuation to the complex k-plane and smoothly connect the solutions on both sides of the reflection point. When the eigenmode is localized between two reflection points at $x = a$ and $x = b$ the Bohr-Sommerfeld quantization condition determines the eigenfrequency [12.2]

$$\int_a^b k_0(x')dx' = (n + 1/2)\pi, (n = 0, 1, 2, \cdots) . \tag{12.1.23}$$

For a long wavelength low frequency mode, the local charge neutrality condition is satisfied and in the left hand side of the Poisson equation (12.1.9) $\partial^2\phi/\partial x^2$ is negligible. In this case, an eigenvalue equation can be obtained by expanding $\phi(x')$ as

$$\phi(x') = \phi(x) + (x' - x)\phi'(x) + \frac{(x' - x)^2}{2}\phi''(x) + \cdots . \tag{12.1.24}$$

Substituting (12.1.24) into (12.1.9) and neglecting the left hand side, we get

$$\frac{d^2\phi}{dx^2} + P_1(x)\frac{d\phi(x)}{dx} + P_2(x)\phi(x) = 0 , \tag{12.1.25}$$

where

$$P_1(x) = 2\,\frac{\int dx' Q(x, x', \omega)(x' - x)}{\int dx' Q(x, x', \omega)(x' - x)^2} , \tag{12.1.26}$$

$$P_2(x) = 2\,\frac{\int dx' Q(x,x',\omega)}{\int dx' Q(x,x',\omega)(x'-x)^2}\,. \tag{12.1.27}$$

In many cases, $P_1(x)$ vanishes. Then (12.1.25) is reduced to eigenvalue equation similar to (12.1.18). For a more general case, we have an integral equation in wavenumber space which can be solved by an eikonal approximation in wavenumber space. [12.3, 4]

12.1.5 Absolute vs Convective Instabilities

When the dispersion relation has a solution whose imaginary part is positive ($-\gamma > 0$) in (7.2.25), the wave grows exponentially. This is an instability. There are two types of instabilities, one is the *convective instability* in which the wave amplitude grows as seen from the frame moving with the group velocity of the wave as the wave propagates. The other is the *absolute instability* in which the wave amplitude grows in the rest frame. Analysis shows that an absolute instability results when there is an unstable solution ($-\gamma > 0$) of the dispersion relation with zero group velocity [12.5, 6].

The difference between the convective instability and the absolute instability becomes important in an inhomogeneous plasma. In a non-uniform plasma, the instability occurs in a spatially localized region. If the instability is *convective*, the wave energy excited at the local region passes out of that region with an effective damping rate of v_g/L, where v_g is the group velocity and L is the width of the unstable region. If the growth rate is less than v_g/L the wave amplitude does not grow in that local region. In other words, the convective instability can be suppressed by a spatial non-uniformity of scale length $L \lesssim v_g/(-\gamma)$. If the instability is *absolute*, the wave energy never convects out of the unstable region, so that the wave amplitude grows locally and the instability cannot be suppressed by convective damping.

A temporally growing mode in a spatially localized region can occur even with convective instability, either when the convecting wave energy is reflected back to the unstable region by some effect, or when the convective damping is small compared to the growth rate $-\gamma$. In the latter case, we can find an unstable localized eigenmode under the *outgoing wave boundary condition* which requires that outside the growing region the solution describes outgoing waves [12.7].

12.2 Accessibility of Waves in Magnetized Plasmas

In most laboratory plasmas for wave heating experiments, the antenna system of the radio-frequency (RF) generator is located in a region of zero or low plasma density. The plasma is inhomogeneous, and a resonant absorption of wave energy by the plasma takes place in a region of relatively high density which is deep inside the plasma. One may then ask whether the radio-frequency wave launched from the antenna system will in fact reach the high density region, or whether the wave will be reflected at some region of intermediate density. If no such intermediate reflection occurs, we shall say that the resonance is accessible.

Only ray trajectories are involved in the question of accessibility. In general, the group velocity in the direction of the density gradient becomes slower and slower, and the direction of energy flow turns increasingly sideways. In the immediate vicinity of resonance, however, this question must be analyzed by wave equations, and decisive roles are played by finite temperature and dissipative effects.

We shall concern ourselves now only with the question of accessibility up to the region where the WKB approximation or the ray tracing based on a cold loss-free plasma model starts to be inappropriate. We shall assume that the variation of plasma parameters is sufficiently slow within a wavelength so that at each point the homogeneous plasma dispersion relation (using the values of plasma parameters at that point i.e., the local dispersion relation) gives an accurate description of the propagation by ray trajectories as discussed in Sect. 12.1.

In general, the solution of the ray trajectory problem requires the prior solution of the phase trajectory problem, or more accurately, the determination of the surfaces of constant phase. If we assume that the components of electric field vary as $Ae^{i\phi}$, in which the amplitudes A are slowly varying functions of position, we may develop sets of equations which are ordered according to the magnitude of terms like $|\nabla \ln A|/|\nabla\phi|$. The lowest-order set of equations turns out to be

$$\left(\frac{\partial\phi_0}{\partial x}\right)^2 + \left(\frac{\partial\phi_0}{\partial y}\right)^2 + \left(\frac{\partial\phi_0}{\partial z}\right)^2 = |\boldsymbol{k}_0|^2 \tag{12.2.1}$$

and $\boldsymbol{k}_0(x,y,z)$ is the wavenumber vector found by the local dispersion relation. The solution given in (12.2.1) determines, in this lowest order approximation, the surfaces of constant phase in the medium. In an isotropic nondispersive medium, the ray trajectories are orthogonal to these constant phase surfaces.

The most tractable case is the propagations of a plane wave in a plane-stratified medium. Let $\boldsymbol{k}_0$ be a function of x alone (not necessarily perpendicular to the direction of the static magnetic field). Then a solution of (12.2.1) is

$$\begin{aligned}
\frac{\partial\phi_0}{\partial x} &= \hat{\boldsymbol{x}}\cdot\boldsymbol{k}_0 && \text{(depends on } x\text{)}\\
\frac{\partial\phi_0}{\partial y} &= \hat{\boldsymbol{y}}\cdot\boldsymbol{k}_0 && (=\text{const})\\
\frac{\partial\phi_0}{\partial z} &= \hat{\boldsymbol{z}}\cdot\boldsymbol{k}_0 && (=\text{const})
\end{aligned} \tag{12.2.2}$$

The last two equations are the consequence of the plane symmetry and the periodicity of the plane wave. The first equation may be integrated as

$$\phi_0 = \int^x \hat{\boldsymbol{x}}\cdot\boldsymbol{k}_0(x')dx' \,. \tag{12.2.3}$$

Reflection occurs when the square of the propagation vector in the direction of the stratification k_x^2 passes through zero. The other components k_y and k_z are constant in the above example. Accessibility is then attained if k_x^2 (computed from the local dispersion relation) is positive for all values of density less than the resonant density, which is defined as the value of plasma density at which resonance occurs with the given parameters ω, k_y, k_z, and B_0.

Under the condition that the plasma is fully ionized and quasistatic, the density gradient will necessarily be perpendicular to the magnetic field. We shall therefore adopt the notation that the $\hat{x}$ is the direction of increasing density, and that $B_0 = B_0\hat{z}$, as before. Moreover, for simplicity, we shall only carry through the calculations for the case $k_y = 0$. Analysis of the hybrid resonances requires the full cold plasma dispersion relation which we write

$$aN_x^4 - bN_x^2 + c = 0 \tag{12.2.4}$$

where

$$\begin{aligned}
a &= \chi_1 \,, \\
b &= RL + \chi_1\chi_3 - \chi_3 N_z^2 - \chi_1 N_z^2 \,, \\
c &= \chi_3(RL - 2\chi_1 N_z^2 + N_z^4) \,, \\
R &= 1 - \sum_s \omega_{ps}^2/[\omega(\omega + \Omega_s)] \,, \\
L &= 1 - \sum_s \omega_{ps}^2/[\omega(\omega - \Omega_s)] \,, \\
\chi_1 &= (R+L)/2 \,, \qquad \chi_3 = 1 - \sum_s \omega_{ps}^2/\omega^2 \\
N_x &= k_x c/\omega, \quad N_z = k_z c/\omega \,.
\end{aligned}$$

Equation (12.2.4) is the same as (7.3.10) with changed notation.

The hybrid resonance occurs when the coefficient of N_x^4 goes to zero, allowing N_x^2 to become infinite. Therefore, the relevant solution of (12.2.4) can be written as

$$N_x^2 = \frac{b + \sqrt{b^2 - 4ac}}{2a} \,, \tag{12.2.5}$$

since we now know that we must select the root which approaches $N_x^2 = b/a$ as $a \to 0$. We see that $\chi_1 = 1$ at zero density and also that it is linear in the density. Since χ_1 is zero at resonance, we can conclude that χ_1 drops linearly from 1 to 0 as the density increases to the resonant density.

For sufficiently low densities, we may therefore approximate $\chi_1 \simeq 1, RL \simeq 1$. In this case, which applies to a thin region of low density at the edge of the plasma, (12.2.4) can be factored as

$$[N_x^2 - \chi_3(1 - N_z^2)][N_x^2 - (1 - N_z^2)] = 0 \,. \tag{12.2.6}$$

It can be shown that the accessibility to the hybrid resonances described by (7.3.29, 30) requires $N_z^2 > 2$ (12.2.9); therefore the vanishing of the second factor of (12.2.6) corresponds to an evanescent wave, and in fact it gives that root of (12.2.4) which is conjugate to (12.2.5). The vanishing of the first factor, which corresponds to (12.2.5), starts as an evanescent wave at zero density ($\chi_3 = 1$) which then becomes a propagating wave after χ_3 becomes negative. For the lower hybrid wave of (7.3.30), the $\chi_3 = 0$ layer occurs at an extremely low value of electron density. The evanescent region in the presence of plasma is therefore very thin, and the wave attenuation will be very slight.

We shall now consider the region of plasma away from the edge where we are allowed to use the approximation $|\chi_3| \gg 1$ and hence $|\chi_3| \gg |\chi_1|$. From (12.2.5), we can see that accessibility will be attained (N_x^2 will be positive) in the case that $b, b^2 - 4ac$, and a are all positive. We first notice that $a = \chi_1$ is positive. With our assumption $|\chi_3| \gg |\chi_1|$, b will be positive if

$$N_z^2 > \left|\frac{RL}{\chi_3}\right| + \chi_1 \tag{12.2.7}$$

This condition is satisfied automatically if $b^2 - 4ac > 0$: with $|\chi_3| \gg |\chi_1|$, this inequality may be written

$$(RL - \chi_1\chi_3)^2 + \chi_3 N_z^2(\chi_3 N_z^2 - 2RL - 2\chi_1\chi_3 + 8\chi_1^2 - 2\chi_1 N_z^2) > 0\,. \tag{12.2.8}$$

If we use the relations $RL = \chi_1^2 - \chi_2^2, |\chi_3| \gg |\chi_1|$ and $|\chi_1| < 1$, a sufficient condition for accessibility is given by

$$N_z^2 > \left|\frac{2\chi_2^2}{\chi_3}\right| + 2\,, \tag{12.2.9}$$

where

$$\chi_2 = \frac{1}{2}(R - L) = \sum_s \frac{\omega_{ps}^2}{\omega^2}\,\frac{\omega\Omega_s}{\omega^2 - \Omega_s^2}\,.$$

Here we note that $\chi_3 < 0$.

Next the accessibility of electron cyclotron range frequency (ECRF) waves is shown. To understand the properties of wave propagation, use is made of a *Clemmow-Mullaly-Allis* (CMA) *diagram* (Fig. 12.1) [12.8]. The vertical axis of the diagram shows the change in the magnetic field normalized to ω_{ce}/ω and the horizontal axis shows the change in density normalized to ω_{pe}^2/ω^2. The conditions $R = 0$ and $L = 0$ give the so-called right hand and left hand cut-offs,

$$\omega = \tfrac{1}{2}\left[\pm\omega_{ce} + \sqrt{\omega_{ce}^2 + 4\omega_{pe}^2}\right]\,. \tag{12.2.10}$$

Here the plus and the minus signs correspond to right and left cut-offs, respectively. The marks R, L, O, and X in the diagram denote right hand, left hand, ordinary and extraordinary waves (7.3.25,26), respectively. Waves can be classified both in terms of their polarization at $\theta = 0$ and electric field orientation at $\theta = \pi/2$, i.e., RX-waves, LO-waves, etc., where θ is the angle between the wave propagation and the magnetic field. The closed lines are the wave normal surfaces, oriented with respect to the magnetic field that is assumed to be pointed

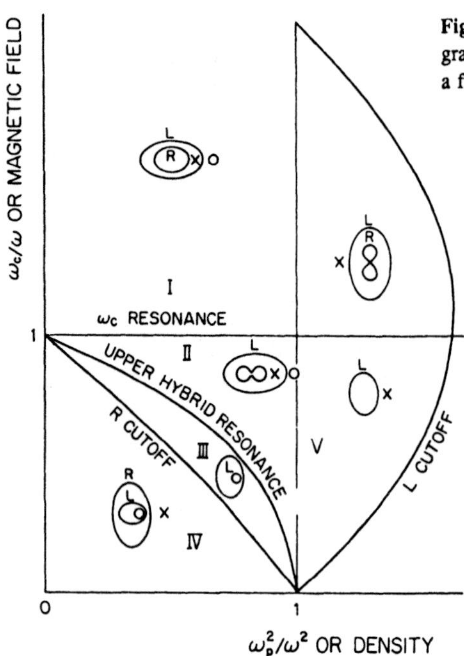

Fig. 12.1. Clemmow-Mulloly-Allis (CMA) diagram showing the change in magnetic field as a function of plasma density for ECRF waves

in the positive vertical direction. A wave normal surface is a surface that is traced out by the tip of the phase velocity vector. The relative sizes of the wave normal surfaces distinguish fast and slow waves. Resonances and cut-offs are indicated by captioned curves.

In the ECRF there are five principal regions, labeled in Fig. 12.1 by Roman numerals. Region I is the high magnetic field region, thus termed because $\omega < \omega_{ce}$. Here both RX and LO waves propagate. Region II is the region between the electron cyclotron resonance and the upper hybrid resonance. Again both RX and LO waves exist, but the RX wave does not propagate at $\theta = 0$. Region III is the evanescent region for the RX wave. Region IV is the low magnetic field edge region where both RX and LO waves exist. Region V is beyond $\chi_3 = 0$ or the plasma cut-off, and the RO wave does not propagate in this region. In summary, the X wave sees the upper hybrid resonance, the right and left hand cut-offs, while the O wave sees only the plasma cut-off.

The CMA diagram can be used to schematicaly illustrate accessibility conditions. As the wave propagates from the edge into higher density regions, the point on the diagram moves from somewhere on the vertical axis to the right. An increase in the magnetic field corresponds to an upward movement, and a decrease to a movement downward. As an example, consider the accessibility to the upper hybrid resonance of an extraordinary wave. Path 1 in Fig. 12.2 shows a wave that is able to access the resonance. This wave starts from a high field region and propagates to a lower field region. Path 2 in the same figure indicates a wave that is unable to access the resonance. Here the wave encounters a right hand cut-off as it propagates from a low density region to a higher density

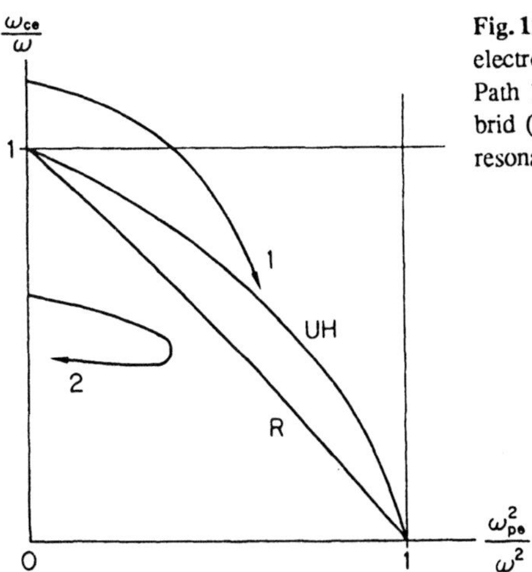

Fig. 12.2. CMA diagram showing accessibility of an electromagnetic wave to electron cyclotron resonance. Path 1: an extraordinary wave accesses the upper hybrid (UH) resonance. Path 2: wave cannot access the resonance due to right hand cut-off (R)

region. These results are strictly for a cold loss-free plasma. For a finite temperature plasma, the wave energy is absorbed by Landau and cyclotron damping, and electron cyclotron resonances at the fundamental and the second harmonic become the dominant resonances for both the ordinary and extraordinary waves.

Let us consider an electromagnetic wave propagating in the y-direction perpendicular to the magnetic field which is directed along the z-axis. We shall restrict ourselves to the O mode so that the wave electric field is linearly polarized in the direction of the magnetic field. The wave magnetic field will then be in the x-direction. The linear dispersion relation for the O mode is

$$\frac{k^2c^2}{\omega^2} = \varepsilon_{zz} , \qquad (12.2.11)$$

where ε_{zz} is the zz-component of the dielectric tensor ($\varepsilon_{zz} = \varepsilon_0\chi_3$ for the cold plasma model) and we have assumed the wave variables vary as $\exp[\mathrm{i}(ky - \omega t)]$. For a hot plasma in a uniform magnetic field the dispersion relation takes the form

$$\frac{k^2c^2}{\omega^2} = 1 - \frac{\omega_{pe}^2}{\omega}\exp(-b_e)\sum_{n=-\infty}^{\infty}\frac{I_n(b_e)}{(\omega - n\Omega_e)} , \qquad (12.2.12)$$

where $b_e = k^2v_{Te}^2/\Omega_e^2$, and I_n is a modified Bessel function. Notice that in the cold plasma approximation $b_e = 0$, there is no resonance in the dispersion relation. However, it is clear that in the hot plasma dispersion relation there are resonances at multiples of the cyclotron frequency. One interpretation of these resonances is given in terms of a *mode conversion* mechanism. For this purpose, it is sufficient to consider the resonance at the fundamental.

For frequencies in the vicinity of Ω_e and for long wavelength such that $b_e \ll 1$, the dispersion relation can be approximated by including only the $n = 0$ and $n = -1$ terms giving

$$\frac{k^2c^2}{\omega^2} = 1 - \frac{\omega_{pe}^2}{\omega^2} - \omega_{pe}^2 \frac{\exp(-b_e)I_1(b_e)}{\omega(\omega - |\Omega_e|)} \tag{12.2.13}$$

Equation (12.2.13) describes two distinct modes of the plasma. First, there is the usual cold plasma branch given by $\omega^2 = \omega_{pe}^2 + c^2k^2$ and second, a hot plasma mode given by $\omega \simeq |\Omega_e|$. This mode is called the cyclotron wave. It has the same polarization as the cold plasma O mode and only propagates for a narrow range of frequency around $\omega = |\Omega_e|$. The dispersion relation is illustrated in Fig. 12.3. The two modes are physically distinct provided their frequencies are well separated. However, when their frequencies and wavenumbers are very close, the two modes couple and lose their identity.

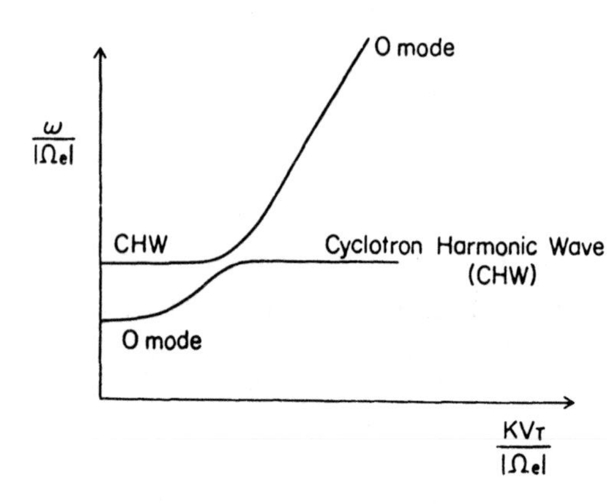

Fig. 12.3. Dispersion relation for the O mode and the cyclotron harmonic wave in the vicinity of $\omega = |\Omega_e|$

Now consider the case of an inhomogeneous plasma where both the density and magnetic field are functions of position. The electromagnetic wave propagates into the plasma from the outside so that we can take B_0 and n_0 to be functions of y. We now write (12.2.13) in the form

$$[\omega - \omega_1(y,k)][\omega - \omega_2(y,k)] = \eta(y,k)$$

or

$$\left[\omega - \sqrt{\omega_{pe}^2 + k^2c^2}\right](\omega - |\Omega_e|) = \frac{b_e}{4}\,\omega_{pe}^2\,, \tag{12.2.14}$$

where the right hand side is the coupling term. Imagine a fixed frequency O mode propagating from the low magnetic field side of the plasma into higher field regions. At some point $y = y_c$ the frequencies of the O mode and the cyclotron harmonic wave will be in resonance and some of the incident wave energy will be mode converted to the cyclotron harmonic wave. The fraction of the energy incident in the first mode that is converted to the second mode is related to $\eta(y_c, k)$ and $\partial\omega_1/\partial k, \partial\omega_1/\partial y, \partial\omega_2/\partial k, \partial\omega_2/\partial y$ evaluated at $y = y_c$. Mode conversion occurs frequently in the problem of wave propagation in inhomogeneous plasmas [12.9].

12.3 RF Control of Magnetized Plasmas

One important aspect of plasma-RF interaction is that RF power may be used for controlling the plasma properties. Some of these roles are discussed here.

The plasma-RF interaction is essentially based on the wave-particle resonance. Resonant transfer of momentum and energy from the wave to the particles opens the possibility of controlling the plasma properties. When the resonant phase velocity is much larger than the thermal velocity, particles in the tail of the Maxwell distribution function are accelerated. These particles form a plateau in the distribution function by quasilinear effects and as a result DC current can be driven. Such a current drive by an electromagnetic wave is possible via a lower hybrid wave. In tokamaks, RF current drive has been produced by launching the lower hybrid wave with adjusted frequency and wave number to couple with supra-thermal electrons.

Another important characteristic of the plasma-RF interaction is that the resonance region, e.g., satisfying $\omega = \omega_{ce}, \omega = \omega_{ci}$ or $\omega = \omega_{LH}$ (ω_{LH} is the lower hybrid frequency), is localized spatially in an inhomogeneous medium. In a magnetically confined system, the magnetic field, the density profile, and the temperature profile are intrinsically inhomogeneous. For an electron cyclotron range frequency (ECRF) or an ion cyclotron range frequency (ICRF) electromagnetic wave the resonant position can be chosen arbitrarily by adjusting the magnetic field and frequency ω of the external electromagnetic power source to satisfy $\omega = \omega_{ce}$ or $\omega = \omega_{ci}$.

When wave energy is absorbed by the plasma only at the resonant position, the power deposition to the plasma is localized there. The temperature profile is related to both the power deposition profile, or the source term of the heat conduction equation, and the heat transport coefficient. The temperature profile is also related to stability of the magnetically confined plasma through the pressure gradient, e.g., stability against the interchange mode (Sect. 10.3) or the ballooning mode (Sect. 11.4). Therefore, by using the property that the power deposition profile changes according to the resonant position, the temperature profile may be controlled such that the stability criterion is satisfied.

In most tokamaks the current profile is directly related to the electron temperature profile through the Spitzer resistivity. No anomalous resistivity has so far been observed in high density plasmas. The plasma current profile is the most important parameter in controlling the plasma against current driven instability (Sect. 10.5 and Sect. 11.3). Since local electron heating is possible by ECRF wave heating, the current profile is also controllable by adjusting the resonant position. It is also possible to control the current profile directly by using the RF current drive technique, when the region of the current drive can be determined externally.

One interesting application of the local heating is to measure the heat conductivity. We first modulate the local temperature by a rapid local heating. By measuring the propagation of the heat pulse, heat conductivity can be estimated by comparing the pulse propagation to the solution of the heat diffusion equation.

We know that the ponderomotive force (Sects. 3.4, 5.6 and 8.5) is produced in the plasma by a large amplitude RF wave. In a mirror system, the ponderomotive force is used for end plugging or reducing particle loss flux through the loss cone in the velocity space. The ponderomotive force is also useful to suppress flute instability in an axisymmetric mirror, which is intrinsically unstable to the flute mode as shown in Sect. 10.3. The applications of RF wave-plasma interactions discussed above are quite new and have recently been intensively investigated [12.10].

Here we discuss the mechanism of the current drive by the RF wave. The interest in driving currents arises from the possibility of operating tokamak reactors at steady state by replacing the inherently pulsed ohmic transformer current. The crucial quantity by which the practicability of a reactor incorporating any of the current drive schemes may be assessed is J/P_d, the amount of current generated per power dissipated. This quantity, which we will determine presently, should be maximized for useful application of the current drive to tokamaks.

Consider the displacement of a small number of electrons δf in velocity space from coordinates subscripted 1 to subscripted 2. The energy expended to produce this displacement is given by

$$\Delta E = (E_2 - E_1)\delta f \ , \tag{12.3.1}$$

where E_1 is the kinetic energy associated with velocity space location 1. Electrons at different coordinates will scatter at different rates. Suppose the displaced electrons would have lost their momenta parallel to the magnetic field, which is in the z-direction, at a rate ν_1. At the new location 2 in the velocity space they now lose it at a rate ν_2. The z-directed current density is then given by

$$\delta J(t) = -e(v_{z1}e^{-\nu_1 t} - v_{z2}e^{-\nu_2 t})\delta f \ , \tag{12.3.2}$$

where v_z is the velocity parallel to the magnetic field and $-e$ is the electron charge.

Consider the quantity J, the time smoothed current over an interval Δt which is large compared to both the $1/\nu_1$ and $1/\nu_2$ so that

$$J \equiv \frac{1}{\Delta t}\int_0^{\Delta t} \delta J(t)dt \simeq -\frac{e\delta f}{\Delta t}\left(\frac{v_{z1}}{\nu_1} - \frac{v_{z2}}{\nu_2}\right) . \tag{12.3.3}$$

The term in the integral may be interpreted as the time integrated current attributable to an energy input ΔE. Substituting now for δf from (12.3.1) and identifying $\Delta E/\Delta t$ as the dissipated power P_d, we find the crucial parameter

$$\frac{J}{P_d} = e\left(\frac{v_{z1}/\nu_1 - v_{z2}/\nu_2}{E_1 - E_2}\right) . \tag{12.3.4}$$

By considering the limit $\boldsymbol{v}_2 \to \boldsymbol{v}_1$,

$$\frac{J}{P_d} \to \frac{e\hat{\boldsymbol{s}} \cdot \nabla_v(v_z/\nu)}{\hat{\boldsymbol{s}} \cdot \nabla_v E} \ , \tag{12.3.5}$$

where we have also assumed that location 1 and 2 are separated infinitesimally (which allows the dropping of the subscript) in the direction of the velocity displacement vector $\hat{s}$.

Suppose that $\nu \propto v^{-3}$ as in the Coulomb collision frequency for $v > v_T$ (Sect. 3.5), where v is the speed of the resonant electrons. It follows that for lower hybrid waves where $\hat{s} /\!/ \hat{z}$,

$$\frac{J}{P_\mathrm{d}} \propto [v_z^{-1}(v_z^2 + v_\perp^2)^{3/2} + 3v_z(v_z^2 + v_\perp^2)^{1/2}] , \tag{12.3.6}$$

where $v_\perp$ is the perpendicular velocity of the resonant electrons and is far less than v_z. Then $J/P_\mathrm{d} \propto 4v_z^2$. We may consider the case in which energy is given only to the resonant electrons by choosing $\hat{s}$ parallel to $\hat{v}_\perp$. This may be accomplished, for example, by heating in the perpendicular direction with a wave that resonates with the selected electrons. An example of such a wave is the electron cyclotron wave (Sect. 12.2). The associated J/P_d is

$$\frac{J}{P_\mathrm{d}} \propto 3v_z\sqrt{v_z^2 + v_\perp^2} . \tag{12.3.7}$$

By comparing (12.3.7) with (12.3.6), the J/P_d for the electron cyclotron wave current drive will be about 3/4 of that for the lower hybrid current drive.

12.4 Laser-Plasma Interaction

In inertial confinement fusion, an intense laser light shines on a solid target to produce an expanding plasma (12.11, 12). The interaction of the laser light with the plasma is quite a complicated physical problem. The most widely used laser at present is the neodymium glass laser whose frequency ω_0 is about $1.8 \times 10^{15}\,\mathrm{s}^{-1}$ at a wavelength of 1.06 μm, while the plasma frequency at solid density is typically one order of magnitude higher than ω_0. Therefore the incident laser light cannot propagate inside the solid target, but first ionizes the target surfaces. For a power density of $10^{15}\,\mathrm{W/cm^2}$, the associated electric field is about 15 V/Å. The ionized gas or plasma expands outward with approximately the ion sound speed which is about 10^6 m/s. During the $\sim$ 1 ns laser pulse the plasma expands about 1 mm. This is much longer than the laser wavelength, so the subsequent laser light pulses interact with the expanding low density plasma of density gradient length L much longer than the wavelength. In the classical theory, laser light absorption takes place by a process which is the inverse of bremsstrahlung, that is, absorption of a photon accompanied by an electron-ion collision. For $\omega_0 \gg \omega_\mathrm{pe} \gg \bar{\nu}_\mathrm{ei}$ the absorption length is given by

$$l_{ab} = \left(\frac{\omega_0}{\omega_\mathrm{pe}}\right)^2 \sqrt{1 - \frac{\omega_\mathrm{pe}^2}{\omega_0^2}}\,\frac{c}{\bar{\nu}_\mathrm{ei}} , \tag{12.4.1}$$

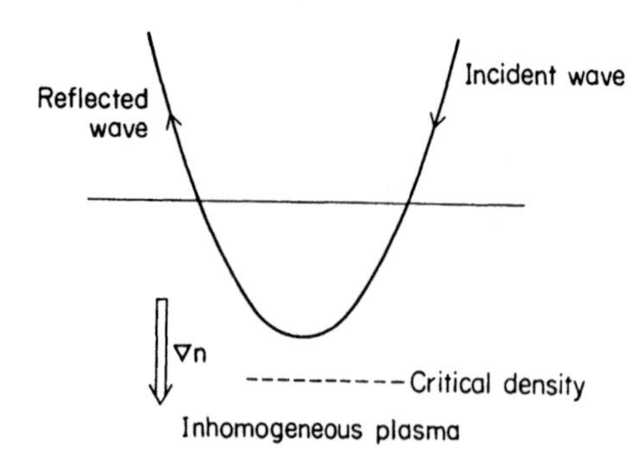

Fig. 12.4. Path of laser light incident on an inhomogeneous plasma at an angle to the density gradient

where ω_{pe} is the electron plasma frequency, $\bar{\nu}_{ei}$ is the electron-ion collision frequency (3.5.8). Since $\omega_{pe}^2\bar{\nu}_{ei}$ is proportional to $n^2T_e^{3/2}$, l_{ab} depends sensitively on the plasma density and temperature. We choose n to be one half the critical density, i.e., $(\omega_0/\omega_{pe})^2 = 2$. The temperature depends on the incident laser power density. At high power density T_e becomes a few keV. Then l_{ab} becomes several centimeters which is much larger than the plasma depth L. The classical inverse bremsstrahlung is then ineffective and most of the laser energy arrives at the critical density. There the above formula is no longer valid. When the laser light is incident at finite angle to the direction of the density gradient, the laser light is reflected as shown in Fig. 12.4. The oscillating electromagnetic field tunnels into the region of the critical density where it excites an electrostatic electron plasma wave (linear mode conversion) which in turn is absorbed by the Landau damping. This is called the resonance absorption of the electromagnetic wave and is very effective near the critical density. The laser energy is then absorbed by the electrons in a thin layer at the critical density producing energetic electrons which expand outward. This expanding hot electron flow produces two important effects. One is a current which produces a magnetic field. The strength of the magnetic field becomes on the order of megagauss. The other is to produce fast ions by the space charge field. If the laser energy is carried by a small fraction of ions, then the resultant ablation pressure is reduced. The hot electrons are brought back to the target by the space charge field and penetrate into the high density region. Since the mean free path of the hot electrons is long, they penetrate into the central core region and deposit their energy there, causing preheating of the central core plasma. Experiments show that the heat conduction due to the hot electrons in the region of over critical density is substantially reduced below the classical value. Various mechanisms have been proposed to explain this reduction of the hot electron heat conductivity, such as a self-generated magnetic field, two-stream instability due to the cold electron return current, nonlocal heat transport due to the steep temperature gradient, etc. In any case, reduction of the heat conductivity further enhances the hot electron energy at the critical region.

In addition to these effects, various parametric instabilities (Chap. 8) can take place when the laser power density is enhanced. Parametric instabilities occurring near the critical density usually cause anomalous absorption of laser energy but

mainly by hot electrons, while parametric instabilities at lower density, such as the stimulated Brillouin scattering, enhance the reflection and hence reduce the absorption efficiency.

PROBLEMS

12.1. Show that the wave intensity $|\phi(x)|^2$ obtained by the WKB approximation (12.1.22) is proportional to the inverse of the group velocity of the wave.

12.2. Consider an electromagnetic wave incident on a magnetic field free plasma which is weakly non-uniform in the x-direction, from the low density side, at an angle θ with the x-axis. Find the plasma density at the reflection point of this electromagnetic wave.

12.3. Derive (12.2.4) from (7.3.10).

12.4. Derive the inequality (12.2.9).

12.5. Confirm that (12.2.7) is satisfied when (12.2.9) is true.

12.6. Consider a cylindrical tokamak. Calculate the safety factor profile $q(r) = RB_\theta/rB_z$ for the case when the electron temperature profile is Gaussian $T_e(r) = T_e(0)\exp[-r^2/a^2]$.

13. Transport Processes

The first and most fundamental problem in thermonuclear fusion research is to heat a plasma to the ignition temperature and to confine such a plasma such that the product of the plasma density n and the confinement time τ satisfies the Lawson criterion (9.1.12). For magnetic confinement, the confinement time τ is determined by the transport process. It is a simple problem if the magnetic field is straight and the particles can be transported across the magnetic field only by collisions. However, if one wants to confine a plasma in a reasonably sized vessel such that the MHD equilibrium and stability are guaranteed, one inevitably has to use a curved magnetic field geometry, as explained in Chap. 9, so that the transport processes have to be treated in a complex configuration such as in tokamaks. The problem then becomes extremely complicated, partly due to the geometrical effects on the particle orbits and more seriously due to electromagnetic fluctuations that result from microscopic instabilities. In this chapter, we first describe in Sects. 13.1–3 the geometrical effects, considering only the collisional displacement of the particle orbits across magnetic surfaces. We then briefly describe the so-called anomalous transport due to electromagnetic fluctuations in Sect. 13.4–6. At present, the physical mechanism of the anomalous transport as observed in tokamaks and other advanced devices is not well understood, but remarkable progress has been seen in recent years. These developments are beyond the scope of the present book. Several recent seminal papers are cited, however [13.1–5].

13.1 General Properties of Collisional Transport

Consider a plasma immersed in a uniform magnetic field directed along the z-axis, $B_0\hat{e}_z$. In the absence of collisions, a charged particle executes a gyrating motion with its guiding center along a magnetic line of force. Collisions can displace the guiding center across the field line. We assume that collisions are sufficiently rare and the sequence of the displacements of the guiding center in the xy-plane (the plane perpendicular to the magnetic field) can be treated as independent events (Markoff random process) and introduce the probability $g(\Delta\boldsymbol{r},\tau)$ for the particle to be displaced by $\Delta\boldsymbol{r}$ in time τ. Then the particle density $n(\boldsymbol{r},t)$ at position $\boldsymbol{r}$ and time t can be written as

$$n(\boldsymbol{r},t) = \int d(\Delta\boldsymbol{r})n(\boldsymbol{r}-\Delta\boldsymbol{r},t-\tau)g(\Delta\boldsymbol{r},\tau)\,. \tag{13.1.1}$$

Here τ is the mean time between collisions and $\boldsymbol{r}$ is the position vector in the xy-plane, $\boldsymbol{r} = (x, y) = (r\cos\theta, r\sin\theta)$. Due to axisymmetry, we can assume that $g(\Delta\boldsymbol{r}, \tau)$ is independent of the angular variable θ, i.e., $g = g(\Delta r, \tau)$. Then we have

$$\int d(\Delta\boldsymbol{r}) g(\Delta r, \tau)(\Delta\boldsymbol{r}) = 0 .$$

We further assume that the dispersion of the radial displacement is proportional to τ:

$$\int d(\Delta\boldsymbol{r}) g(\Delta r, \tau)(\Delta r)^2 = 2D\tau , \tag{13.1.2}$$

where D is a constant. We assume that the density variation is sufficiently slow in space and time and approximate the right hand side of (13.1.1) by

$$\int d(\Delta\boldsymbol{r}) \left[n(\boldsymbol{r}, t) - \tau \frac{\partial}{\partial t} n(\boldsymbol{r}, t) - \Delta\boldsymbol{r} \cdot \nabla n(\boldsymbol{r}, t) + \frac{1}{2}(\Delta\boldsymbol{r} \cdot \nabla)^2 n(\boldsymbol{r}, t) \right] g(\Delta r, \tau) . \tag{13.1.3}$$

Using this approximation we get from (13.1.1, 2) the diffusion equation

$$\frac{\partial n}{\partial t} = D \frac{1}{r} \frac{\partial}{\partial r} r \frac{\partial n}{\partial r} . \tag{13.1.4}$$

If we denote the left hand side of (13.1.2) by

$$\int d(\Delta\boldsymbol{r}) g(\Delta r, \tau)(\Delta r)^2 = \left\langle (\Delta r)^2 \right\rangle ,$$

we get

$$D = \left\langle (\Delta r)^2 \right\rangle / 2\tau . \tag{13.1.5}$$

If the relation $\Delta r = \Delta v \tau$ holds, D can be alternatively written as

$$D = \frac{\tau}{2} \left\langle (\Delta v)^2 \right\rangle . \tag{13.1.6}$$

The expressions (13.1.5, 6) are useful for a wide class of problems to estimate a transport coefficient in the context of random walk. In the example of cross-field diffusion in a uniform magnetic field,

$$\frac{\langle (\Delta r)^2 \rangle}{2} \doteq \varrho_{\mathrm{L}}^2 , \qquad \tau \doteq 1/\nu , \tag{13.1.7}$$

from which the classical diffusion coefficient $D_{\mathrm{c}} = \nu \varrho_{\mathrm{L}}^2$ is obtained. Here ϱ_{L} is the Larmor radius and ν is the electron-ion collision frequency.

We will explain the classical diffusion by using the two-fluid equations of Sect. 4.4 [13.1]. When the inertia terms are neglected, the equations of motion for the electron and ion fluids become

$$-en(\boldsymbol{E} + \boldsymbol{u}_{\mathrm{e}} \times \boldsymbol{B}) - \nabla P_{\mathrm{e}} + n m_{\mathrm{e}} \nu_{\mathrm{ei}} (\boldsymbol{u}_{\mathrm{i}} - \boldsymbol{u}_{\mathrm{e}}) = 0 \tag{13.1.8}$$

$$en(\boldsymbol{E} + \boldsymbol{u}_\mathrm{i} \times \boldsymbol{B}) - \nabla P_\mathrm{i} - nm_\mathrm{i}\nu_\mathrm{ie}(\boldsymbol{u}_\mathrm{i} - \boldsymbol{u}_\mathrm{e}) = 0 \; , \tag{13.1.9}$$

under the quasi-neutrality condition $n = n_\mathrm{e} \doteq n_\mathrm{i}$. In the cylindrically symmetric plasma column with uniform magnetic field $B_0\hat{\boldsymbol{z}}$, only a radial electric field is produced, since density and pressure profiles depend only on r. The θ-components of (13.1.8, 9) are written as

$$enu_{\mathrm{e}r}B_0 + nm_\mathrm{e}\nu_\mathrm{ei}(u_{\mathrm{i}\theta} - u_{\mathrm{e}\theta}) = 0 \tag{13.1.10}$$

$$-enu_{\mathrm{i}r}B_0 - nm_\mathrm{i}\nu_\mathrm{ie}(u_{\mathrm{i}\theta} - u_{\mathrm{e}\theta}) = 0 \; . \tag{13.1.11}$$

Summing up (13.1.10, 11) and noting the relation $m_\mathrm{e}\nu_\mathrm{ei} = m_\mathrm{i}\nu_\mathrm{ie}$, we have $u_{\mathrm{i}r} = u_{\mathrm{e}r}$. This means that the electron and ion radial particle fluxes across the magnetic field $\Gamma_{\mathrm{e}r} = nu_{\mathrm{e}r}$ and $\Gamma_{\mathrm{i}r} = nu_{\mathrm{i}r}$ are equal. The diffusion satisfying $\boldsymbol{\Gamma}_\mathrm{e} = \boldsymbol{\Gamma}_\mathrm{i}$ is called the *ambipolar diffusion*. In fully ionized plasmas in cylindrical geometry, the ambipolar diffusion is automatically established at steady state due to momentum conservation in electron-ion collisions. From the momentum conservation, like-particle collisions such as ion-ion collisions or electron-electron collisions do not contribute to the particle flux. The radial components of (13.1.8, 9) are written as

$$-en(E_r + u_{\mathrm{e}\theta}B_0) - \frac{dP_\mathrm{e}}{dr} = 0 \tag{13.1.12}$$

$$en(E_r + u_{\mathrm{i}\theta}B_0) - \frac{dP_\mathrm{i}}{dr} = 0 \; . \tag{13.1.13}$$

The azimuthal velocity is given by

$$u_{\mathrm{e}\theta} = -\frac{E_r}{B_0} - \frac{1}{enB_0}\frac{dP_\mathrm{e}}{dr} \; , \tag{13.1.14}$$

$$u_{\mathrm{i}\theta} = -\frac{E_r}{B_0} + \frac{1}{enB_0}\frac{dP_\mathrm{i}}{dr} \; . \tag{13.1.15}$$

Combining (13.1.14, 15) with (13.1.10) or (13.1.11), we have

$$\Gamma_{\mathrm{e}r} = -\frac{m_\mathrm{e}\nu_\mathrm{ei}}{e^2B_0^2}\frac{d}{dr}(P_\mathrm{e} + P_\mathrm{i}) \; . \tag{13.1.16}$$

The radial electric field causes a drift motion in the θ-direction.

In axisymmetric toroidal plasmas the *ambipolar condition* is also satisfied automatically. However, in nonaxisymmetric systems, particle diffusion is intrinsically non-ambipolar (Sect. 13.3) and the condition $\boldsymbol{\Gamma}_\mathrm{e} = \boldsymbol{\Gamma}_\mathrm{i}$ determines the local radial electric field. Note that if $\boldsymbol{\Gamma}_\mathrm{e} \neq \boldsymbol{\Gamma}_\mathrm{i}$, charge separation occurs which produces a local electric field which imposes $\boldsymbol{\Gamma}_\mathrm{e} = \boldsymbol{\Gamma}_\mathrm{i}$.

The deviation of the particle orbit from the flux surfaces is determined by the drift velocity which is a superposition of gradient B and curvature drifts,

$$\boldsymbol{v}_\mathrm{d} = \boldsymbol{v}_\mathrm{G} + \boldsymbol{v}_\mathrm{c} = \frac{W_\perp(\boldsymbol{B} \times \nabla B)}{eB^3} + \frac{2W_\parallel(\boldsymbol{R}_\mathrm{c} \times \boldsymbol{B})}{eB^2R_\mathrm{c}^2} \; , \tag{13.1.17}$$

where $W_\perp$ and $W_\parallel$ are perpendicular and parallel energy of the particle, respectively.

The predominant magnetic field variation in an axisymmetric toroidal system results from $B \doteq B_t = B_0(1 - r/R\cos\theta)$, where R is the major radius and θ is the poloidal angle. The parallel velocity can be expressed as

$$v_\parallel = \pm\sqrt{\frac{2}{m}(W - W_\perp)} = \pm\sqrt{\frac{2}{m}(W - \mu B)}\,, \tag{13.1.18}$$

where μ denotes the magnetic moment and it is conserved as an adiabatic invariant (Sect. 3.3).

Equation (13.1.18) shows that when $\mu B = W$ is satisfied or when $\mu B_0/W > 1 - \varepsilon$, the parallel velocity vanishes and the particle will be trapped in the local mirror field where $\varepsilon = r/R$. This implies that the drift in the guiding center motion due to magnetic field inhomogeneity is approximately vertical: positively (negatively) charged particles drift downward (upward). These facts determine the character of guiding center (g.c.) orbits in a torus. Note in particular that when projected onto the ϕ = const plane (or poloidal plane), they appear to be closed as shown in Fig. 13.1. The net radial drift after one bounce for *trapped particles*, or one poloidal circuit around the magnetic axis for *untrapped particles*, is zero because in the presence of a poloidal magnetic field, the lowest order motion of the g.c. causes it to spend equal times above the magnetic axis (where an electron, for example, drifts radially outward) and below the magnetic axis (where an electron drifts inward). The guiding center orbit of trapped particles on the poloidal plane as shown in Fig. 13.1a is called the *banana orbit*.

Next, we estimate the thickness d of the banana orbit. For gradient B drift, the drift speed is on the order of $mv_\perp^2/eB_0R$, while the bounce time is $\tau_B \sim qR/v_\parallel$, where qR corresponds to the connection length (Sect. 13.2). It is clear from (13.1.18) that, for trapped particles with $\mu B_0/W \doteq 1 - r/R$,

$$v_\parallel \doteq \sqrt{r/R}\, v_T \tag{13.1.19}$$

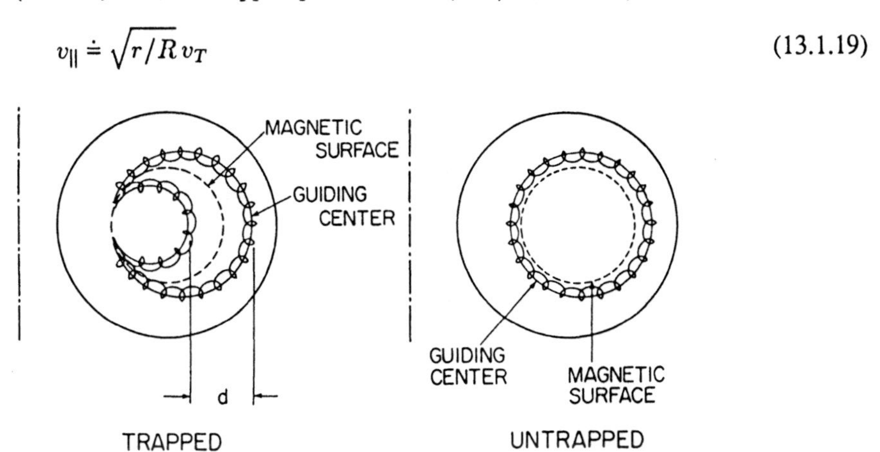

Fig. 13.1. Particle orbits in a torus projected onto a ϕ = const plane for a trapped particle (a) and an untrapped particle (b). The width of the "banana orbit" is denoted by d

and the banana width becomes

$$d \doteq v_d \tau_B \doteq \varrho_L \frac{B_0}{B_\theta}\sqrt{\frac{r}{R}} . \tag{13.1.20}$$

Since in tokamaks $B_\theta/B_0 \ll (r/R)^{1/2} \ll 1$, where B_θ is the poloidal magnetic field, d becomes larger than the Larmor radius ϱ_L.

From (13.1.18), a particle is trapped if at $\theta = 0$ its pitch angle α satisfies

$$\sin\alpha = \sqrt{\frac{W_\perp(\theta=0)}{W}} \geq \sqrt{\frac{B_{min}}{B_{max}}} = \sqrt{\frac{1-\varepsilon}{1+\varepsilon}} . \tag{13.1.21}$$

The *fraction of trapped particles* can be estimated from

$$f_{trap} = \left(1 - \int_0^{\alpha_0} \sin\alpha \, d\alpha\right) = \sqrt{1 - \frac{1-\varepsilon}{1+\varepsilon}} \doteq \sqrt{2\varepsilon} \tag{13.1.22}$$

where α_0 is given by $\alpha_0 = \arcsin\left[\sqrt{B_{min}/B_{max}}\right]$, and we have assumed an isotropic particle distribution.

One can also calculate an *effective collision frequency* ν_{eff} for scattering a trapped particle out of the trapped region:

$$\nu_{eff} \doteq \frac{R}{r}\nu_c , \tag{13.1.23}$$

where ν_c is the collision frequency for 90 degree scattering (Sect. 3.5). For banana diffusion, we have shown that the significant scattering angle is $(r/R)^{1/2}$ times smaller than the usual $\frac{\pi}{2}$.

We may now estimate the diffusion coefficient in the presence of trapped particles. In this case the characteristic length is d given by (13.1.20) and the characteristic time is $1/\nu_{eff}$ of (13.1.23). We also consider the diffusion as resulting from only a fraction $\sqrt{r/R}$ of the total number of particles (13.1.22). Therefore,

$$D \doteq \nu_{eff} d^2 \sqrt{\frac{r}{R}} = \left(\frac{B_0}{B_\theta}\right)^2 \sqrt{\frac{r}{R}} D_c , \tag{13.1.24}$$

is obtained. Expression (13.1.24) is called the *neoclassical diffusion coefficient in the banana regime.* Details are discussed in the next section [13.2].

13.2 Neoclassical Transport in Axisymmetric Toroidal Configurations

The most widely used axisymmetric confinement system is a tokamak. The principle of tokamak confinement is explained in Chap. 9. In the tokamak, the angle between the toroidal direction and the direction of magnetic field line on the circular magnetic surface of radius r is given by

$$\delta = \arctan\left(\frac{B_\theta}{B_0}\right) = \arctan\left(\frac{r}{R_0 q}\right) . \tag{13.2.1}$$

As before, B_θ is the poloidal magnetic field, B_0 is the toroidal magnetic field and q is the safety factor. If we define the *connection length* L_c by the length of the magnetic field line starting from the outermost region and ending at the innermost region of the cross section, then δ is also equal to $\arcsin(\pi r/L_c)$. Since $L_c \gg r$ and $R_0 \gg r$, L_c is given by $L_c \doteq \pi q R_0$.

Let us first consider the extent of the increase of particle diffusion for a tokamak plasma in the case of relatively short mean free paths, that is, short as compared with the connection length, so that the influence of trapped particles in local magnetic mirrors has been neglected. The mean free path is given by the thermal velocity multiplied by the collision time, $l_{mfp} = v_T \tau$.

To obtain the diffusion coefficient D we have to determine the characteristic displacement of particles in the toroidal magnetic field. We know that the particle drifts across the magnetic surface with a velocity on the order of $v_\perp = v_T^2/\omega_c R_0$, where v_T is the thermal velocity of the particle, ω_c is the cyclotron frequency, and R_0 is the radius of the torus. The maximum displacement from the magnetic surface is obtained when we multiply $v_\perp$ by the time of flight along a section of a line of force in which the curvature does not change its sign. This time is on the order of $t = (qR_0)^2/D_\parallel$, where $D_\parallel$ is the coefficient of longitudinal diffusion given by $D_\parallel = v_T^2/\nu$ and characteristic length is estimated as on the order of the connection length $L_c \sim qR_0$. Thus, $\Delta r \sim v_\perp t \sim v_\perp (qR_0)^2/D_\parallel$. Substituting this expression into the estimate $D \doteq (\Delta r)^2/t = (\Delta r)v_\perp$ (13.1.5) we see that the toroidal contribution to the diffusion coefficient is on the order of $\nu\varrho^2 q^2$ and the total diffusion coefficient including the classical diffusion, is equal to

$$D_\perp = D_c(1 + \alpha_n q^2) , \tag{13.2.2}$$

where α_n is a constant near unity and $D_c = \nu\varrho^2$ is the classical diffusion coefficient in the uniform magnetic field.

Expression (13.2.2) has been obtained with the help of a simple estimate using the analogy of a random walk of a particle in the diffusion process. However, in a collisional plasma, the toroidal contribution to $D_\perp$ is due to a regular motion of particles which, at the outer contour of the torus, is directed outward from the magnetic surfaces while at the inner contour it is directed toward the center of the plasma column. Therefore, we are actually concerned with a regular laminar convection caused by the toroidal curvature. To obtain a more complete expression for $D_\perp$, we need to consider this convection. To describe the convection it is sufficient to use the equilibrium equation

$$\nabla P = \boldsymbol{J} \times \boldsymbol{B} , \tag{13.2.3}$$

Ohm's law,

$$\boldsymbol{E} + \boldsymbol{u} \times \boldsymbol{B} = \frac{1}{\sigma_\perp}\boldsymbol{J}_\perp + \frac{1}{\sigma_\parallel}\boldsymbol{J}_\parallel \tag{13.2.4}$$

and the condition of quasi-neutrality

$$\nabla \cdot \boldsymbol{J} = 0 . \tag{13.2.5}$$

In Ohm's law we took into account the difference between the longitudinal and transverse conductivity.

We shall consider the following axisymmetric torus with a magnetic field of the form $B = (0, B_\theta(r), B_0(1+\varepsilon\cos\theta)^{-1})$. This model is applicable to a tokamak with a small inverse aspect ratio $\varepsilon = a/R \ll 1$. The toroidal coordinates used here are similar to the cylindrical system and the length element is described by $ds^2 = dr^2 + r^2 d\theta^2 + (1+\varepsilon\cos\theta)^2 d\zeta^2$, where r is the radius of the magnetic surface under consideration, θ is the azimuthal angle, and ζ is the coordinate along the torus. Letting $\varepsilon = r/R_0 \ll 1$ we can solve (13.2.3–5) by the method of successive approximations.

In the zeroth approximation with respect to ε, we have a straight plasma cylinder, where the parameters are functions of r only. We note that the zeroth order electric field is vanishing under the MHD equilibrium (13.2.3). In the first approximation the quantities become functions of the azimuthal angle θ as well. We are interested in the particle flux across the magnetic surface. From (13.2.3, 4) we obtain

$$u_r = \frac{E_\theta^{(1)}}{B_0} - \frac{1}{\sigma_\perp B_0^2}\frac{dP}{dr} . \tag{13.2.6}$$

Here, the first term represents the drift of plasma under the influence of the electric field, which is a first order quantity, and the second term describes the classical diffusion in a straight cylinder. The electric field $E_\theta^{(1)}$ can be obtained from the longitudinal component of Ohm's law. Noting $E_\| = \boldsymbol{E}\cdot\boldsymbol{B}/B \doteq E_\theta^{(1)} B_\theta/B_0$, we have

$$E_\theta^{(1)} = \frac{B_0}{B_\theta}\frac{J_\|}{\sigma_\|} . \tag{13.2.7}$$

The component $J_\|$ can be expressed in terms of ∇P by using the equation of quasi-neutrality (13.2.5) and $B_\theta \ll B_z$,

$$\begin{aligned} \nabla_\| \cdot \boldsymbol{J}_\| &= \frac{B_\theta}{B_0}\frac{1}{r}\frac{\partial J_\|}{\partial\theta} = -\nabla_\perp \cdot \frac{\boldsymbol{B}\times\nabla P}{B^2} \\ &= -\varepsilon\frac{2}{B_0}\frac{dP}{dr}\frac{1}{r}\frac{\partial}{\partial\theta}\cos\theta . \end{aligned} \tag{13.2.8}$$

Hence it follows that $J_\| = -\varepsilon(2/B_\theta)(dP/dr)\cos\theta$, which is called the *Pfirsch-Schlüter current*, and the poloidal electric field is

$$E_\theta^{(1)} = -2\varepsilon\frac{B_0}{\sigma_\| B_\theta^2}\frac{dP}{dr}\cos\theta . \tag{13.2.9}$$

The mean flux through the magnetic surface is obtained when $n_0 u_r$ is multiplied by $(1+\varepsilon\cos\theta)$ to take into account the variation of the surface element in the system and then averaged over θ:

$$\Gamma = \frac{1}{2\pi}\int_0^{2\pi} n_0 u_r(1+\varepsilon\cos\theta)d\theta = -\frac{n_0}{\sigma_\perp B_0^2}\frac{dP}{dr}\left(1+\frac{\sigma_\perp}{\sigma_\parallel}q^2\right) . \qquad (13.2.10)$$

This expression corresponds exactly to the diffusion coefficient obtained by Pfirsch and Schlüter. The coefficient $\sigma_\perp/\sigma_\parallel$ is approximately 1 or $\alpha_n \doteq 1$ in (13.2.2).

From the derivation of (13.2.2) we see that it applies only if the mean free path l_{mfp} is smaller than qR_0 or when the time t the particle spends on the outer contour ($\cos\theta > 0$) is determined by the longitudinal diffusion.

In a rarefied plasma with $l_{\text{mfp}} > qR_0$, the time t will simply correspond to the free traversal time $t \sim qR_0/v_\parallel$. As the collision frequency ν decreases such that $\nu < \nu_2 = v_T/qR_0$, particles trapped in the local mirror field appear. At some $\nu < \nu_1$, even the trapped particles will no longer undergo collisions during their time of flight between the turning points. It is obvious that only for $\nu < \nu_1$ does the effect of trapped particles become realistic. Let us first consider the frequency interval $\nu_1 < \nu < \nu_2$ (we shall specify ν_1 later) where no particles are trapped and the plasma consists only of transit particles. We use the approximation where the Fokker-Planck collision term is simply replaced by a term of the form $-\nu_{\text{eff}} f$, with an effective collision frequency ν_{eff}.

The following kinetic equation can be given for the distribution function $f(v_\parallel)$ averaged over the transverse velocities:

$$\frac{v_\parallel}{qR_0}\frac{\partial f_1}{\partial\theta} + v_{\text{d}}\sin\theta\frac{\partial f_0}{\partial r} = -\nu_{\text{eff}} f_1 , \qquad (13.2.11)$$

where $v_{\text{d}} = T/m\Omega R_0$ is the mean drift velocity. For simplicity, we have neglected the curvature drift term in v_{d}. The first term on the left hand side of (13.2.11) corresponds to the rotation of particles in the θ-direction through their motion along the helical magnetic field lines of force in tokamaks. In the second term we kept only the θ-dependent part arising from the radial derivative of the zeroth distribution function f_0. The first order distribution function f_1 denoting a small θ-dependent correction is easily determined from (13.2.11) as

$$f_1 = -\frac{\nu_{\text{eff}}\sin\theta - (v_\parallel/qR_0)\cos\theta}{\nu_{\text{eff}}^2 + (v_\parallel/qR_0)^2} v_{\text{d}}\frac{\partial f_0}{\partial r} \qquad (13.2.12)$$

and the radial flux averaged over θ is given by

$$\begin{aligned} \langle n_0 v_r\rangle &= \frac{1}{2\pi}\iint v_{\text{d}}\sin\theta f_1 d\theta\, dv_\parallel \\ &= -\frac{1}{2}\int \frac{\nu_{\text{eff}}}{\nu_{\text{eff}}^2 + (v_\parallel/qR_0)^2} v_{\text{d}}^2\frac{\partial f_0}{\partial r} dv_\parallel , \end{aligned} \qquad (13.2.13)$$

where f_0 is given by $[n_0/(\sqrt{\pi}v_T)]\exp(-mv_\parallel^2/2T)$. In the case of small ν_{eff} the first factor in the integrand can be replaced by $\pi\delta(v_\parallel/qR_0)$, where $\delta(x)$ denotes a delta function. Then we obtain

$$\langle n_0 v_r \rangle \doteq -\frac{\sqrt{\pi}}{2}\frac{v_T}{qR_0}\varrho^2 q^2 \frac{dn_0}{dr}, \tag{13.2.14}$$

which differs from the more exact expression only by a numerical factor of order one. From (13.2.14) we see that in the collision frequency range $\nu_1 < \nu < \nu_2$, the diffusion coefficient is independent of the collision frequency and has the form of a Pfirsch-Schlüter coefficient with the frequency $\nu = \nu_2$.

When the collision frequency decreases further, a situation will be reached in which the condition of applicability of the kinetic equation (13.2.11) is no longer satisfied. In a collisionless plasma, the trapped particles dominate the plasma characteristics. By virtue of the differential character of the Coulomb collision terms, the effective collision frequency is $\nu_{\text{eff}} = \nu v_T^2/(\Delta v)^2$, where Δv is the velocity interval in which the distribution function is substantially changed. For the trapped particles, $\Delta v \sim v_T\sqrt{\varepsilon}$ and the effective collision frequency becomes $\nu_{\text{eff}} = \nu/\varepsilon$. Thus, the value of ν beyond which particle trapping must be taken into account is equal to $\nu_1 = \varepsilon^{3/2} v_T/qR_0$ which is determined by the condition that $\nu_{\text{eff}} = v_T\sqrt{\varepsilon}/qR_0$, where the quantity $qR_0/v_T\sqrt{\varepsilon}$ is the trapped particle excursion time over the connection length qR_0.

For $\nu < \nu_1$ the diffusion is determined by the trapped particles whose number is $n_{\text{trap}} \sim n_0\sqrt{\varepsilon}$ (13.1.22). Since the mean displacement, or the width of the banana, is on the order of $\Delta x \sim q\varrho/\sqrt{\varepsilon}$ and the effective collision frequency $\nu_{\text{eff}} = \nu/\varepsilon$, the corresponding diffusion coefficient by (13.1.5) will be on the order of

$$D_\perp \doteq \sqrt{\varepsilon}(\Delta x)^2 \nu_{\text{eff}} \doteq \varepsilon^{-3/2} q^2 \varrho^2 \nu \,. \tag{13.2.15}$$

This expression exceeds that given by Pfirsch-Schlüter diffusion coefficient by a factor of $\varepsilon^{-3/2}$.

The complete dependence on the collision frequency of the diffusion coefficient in a toroidal configuration was first obtained by *Galeev* and *Sagdeev* [13.3] and is shown in Fig. 13.2 [13.3]. If $\nu > \nu_2$, the Pfirsch-Schlüter coefficient holds, while in the range $\nu_1 < \nu < \nu_2$ we have a 'plateau' and the diffusion in this regime is called the *plateau diffusion*, and finally with $\nu < \nu_1$ we are in the banana regime and the dependence on ν is again linear with nonzero slope.

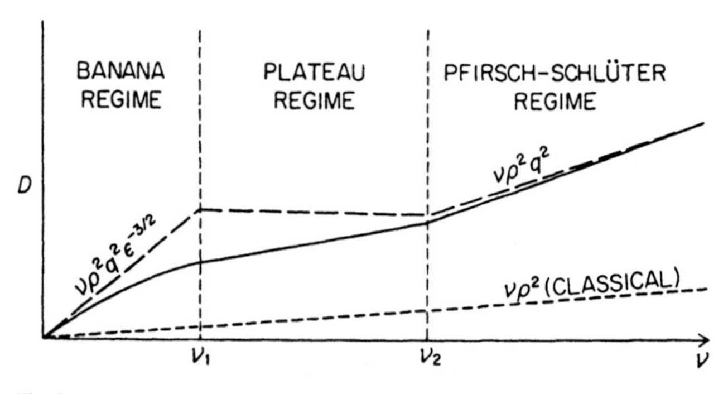

Fig. 13.2. Neoclasssical diffusion coefficient for a tokamak as a function of collision frequency

Similar behavior is observed for the ion thermal transport coefficient χ_i. An approximate value is given by (13.2.14, 15) if ϱ is understood to be the Larmor radius of the ions and ν and v_T are the ion collision frequency and ion thermal velocity, respectively. Similarly, for the electron thermal transport coefficient χ_e, ϱ is electron Larmor radius, ν electron-electron collision frequency and v_T electron thermal velocity. Since the ambipolar condition is not required for thermal transport, $\chi_e \sim \sqrt{m_e/m_i}\chi_i$ for $T_e \sim T_i$. Usually the ion thermal transport is larger.

It should be noted that in (13.2.15) the quantity ν is a sum of the electron-ion collision frequency ν_{ei} and the electron-electron collision frequency ν_{ee}. This is understandable since in collisions between trapped and untrapped electrons the sum of the displacements of the two colliding particles are different from zero, resulting in a net drift of the center of mass. Thus, on the average, there is net diffusion caused by collisions between like-particles of different drift motions.

A characteristic feature of diffusion in axisymmetric systems is its strong dependence on the magnetic field: for a given q, it decreases in proportion to B^{-2}. This is connected with the fact that an increase of the field reduces the Larmor radius and the width of the bananas.

In tokamaks, usually the plasma current is driven by an axisymmetric magnetically induced electric field

$$E_\varphi = -\frac{\partial A_\varphi}{\partial t}, \tag{13.2.16}$$

where A_φ is the φ-component of the magnetic vector potential. Since tokamaks are axisymmetric, the canonical angular momentum of a charged particle will be conserved. Hence, averaging over a cyclotron period we have

$$\frac{d}{dt}\left[R\left(\frac{mv_\parallel B_\varphi}{B} + eA_\varphi\right)\right] = 0\,. \tag{13.2.17}$$

Now we consider a trapped particle motion with two successive mirror reflection points at the minor radii r and $r + \Delta r$ which occur at times t and $t + \Delta t$. Since $v_\parallel$ is zero at these points (or v_θ is zero in the orbit projected onto the φ = const plane), (13.2.17) gives

$$\Delta r \frac{\partial}{\partial r}(RA_\varphi) + \Delta t R \frac{\partial}{\partial t} A_\varphi = 0\,. \tag{13.2.18}$$

The radial variation of RA_φ is directly related to B_θ:

$$\frac{1}{R}\frac{\partial}{\partial r}(RA_\varphi) = -B_\theta\,. \tag{13.2.19}$$

Thus (13.2.16, 18, 19) yield

$$\frac{\Delta r}{\Delta t} = -\frac{E_\varphi}{B_\theta}\,. \tag{13.2.20}$$

This relation suggests that banana orbits of all trapped particles drift towards the magnetic axis with a velocity of E_φ/B_θ. This inward drift motion of trapped particles induces a particle flux $\Gamma \sim \sqrt{\varepsilon} n E_\varphi / B_\theta$ and it is called the *Ware pinch.* It is expected that the Ware pinch is a factor in the density increase after gas puffing.

Another interesting feature of the neoclassical transport theory is the *bootstrap current.* It depends on the presence of trapped particles but it is not itself carried by the trapped particles. In tokamaks the trapped electrons carry a diamagnetic current in the toroidal direction associated with their banana orbits:

$$J_z \sim -\sqrt{\varepsilon} \frac{\varepsilon}{B_\theta} \frac{dP}{dr} , \tag{13.2.21}$$

Here collisions between trapped and passing particles transfer momentum to the untrapped electrons at a rate of $(\nu_{ee}/\varepsilon) J_z$ at an effective collision frequency ν_{ee}/ε. This must be balanced by the loss of momentum due to collisions between electrons and ions, so that

$$(\nu_{ee}/\varepsilon) J_z \sim \nu_{ei} J_\varphi^{\mathrm{BS}} \tag{13.2.22}$$

leading to a toroidal current

$$J_\varphi^{\mathrm{BS}} \sim \varepsilon^{1/2} \frac{1}{B_\theta} \frac{dP}{dr} , \tag{13.2.23}$$

with $\nu_{ee} \sim \nu_{ei}$. This is the neoclassical bootstrap current J_φ^{BS}. It has been remarked that the Ware pinch and the bootstrap current are complementary effects connected by the Onsager relations for off-diagonal transport coefficients [13.2].

13.3 Neoclassical Transport in Nonaxisymmetric Magnetic Confinement

Another toroidal magnetic confinement system, the stellarator/heliotron, is discussed in Sect. 9.2. Usually it has helical symmetry in the straight approximation, i.e., without toroidal curvature. In the toroidal system, unlike tokamaks a stellarator/heliotron can confine plasmas without toroidal plasma current since the helical magnetic field required for toroidal MHD equilibrium is produced by external helical coils or helical windings. Though they have a great advantage for steady state operation of magnetic fusion reactors, the geometrical simplicity of axisymmetry is lost. In this section we consider how the loss of axisymmetry affects the confinement properties.

First consider the axisymmetric torus again in the small inverse aspect ratio limit ($\varepsilon = r/R \ll 1$). The field lines satisfy

$$\frac{dr}{ds} = \frac{B_r(r,\theta)}{B(r,\theta)} , \tag{13.3.1}$$

$$r\,\frac{d\theta}{ds}=\frac{B_\theta(r,\theta)}{B(r,\theta)}\,, \tag{13.3.2}$$

where s is the length along the magnetic field line. These equations can be combined to obtain

$$\frac{dr}{d\theta}=r\,\frac{B_r(r,\theta)}{B_\theta(r,\theta)}\,. \tag{13.3.3}$$

When we make the approximation of cylindrical geometry or the limit $\varepsilon = r/R \to 0, B_\theta$ depends only on r and B_r vanishes identically so that the magnetic field line moves on a constant r surface $r = r_0 =$ const. The situation changes in a nonaxisymmetric torus. The nonaxisymmetric perturbation can be expressed by

$$\frac{dr}{d\theta}=g_1(r,\theta,\phi)\,. \tag{13.3.4}$$

We will represent g_1 by a Fourier expansion

$$g_1(r,\theta,\phi)=\sum_{m,n} g_{1mn}\,(r)\mathrm{e}^{\mathrm{i}(m\theta+n\phi)}\,. \tag{13.3.5}$$

Equation (13.3.4) can be used to examine the radial displacement of the field lines as they spiral about the cylindrical plasma. Recall that the cylinder has a length $2\pi R$ and the angle ϕ is given by z/R, where z is the axial coordinate. For simplicity, consider a single Fourier component g_{1mn}. The displacement in r, when the poloidal angle advances through 2π, is

$$\delta r_1 = g_{1mn}(r_0)\int_{\theta_0}^{\theta_0+2\pi}\mathrm{e}^{\mathrm{i}(m\theta+n\phi)}d\theta\,. \tag{13.3.6}$$

By using the definition of the rotational transform $\iota, \phi = \theta/\iota$. When this is substituted into (13.3.6) we get

$$\delta r_1 = g_{1mn}(r_0)\frac{\mathrm{e}^{\mathrm{i}(m\theta_0+n\phi_0)}(\mathrm{e}^{2\pi\mathrm{i}n/\iota}-1)}{\mathrm{i}(m+n/\iota)}\,, \tag{13.3.7}$$

where the radial variation of ι is assumed to be negligibly small. Similarly, when θ advances another $2\pi, \delta r_2 = \delta r_1 \exp(2\pi\mathrm{i}n/\iota)$. In general, $\delta r_j = \delta r_1 \exp[2\pi\mathrm{i}(j-1)n/\iota]$. The cumulative radial displacement after an infinite number of poloidal revolutions can be determined from

$$\begin{aligned}\Delta r &= \lim_{n\to\infty}\sum_{j=1}^{n}\delta r_j=\delta r_1\sum_{j=1}^{\infty}\exp[2\pi\mathrm{i}(j-1)n/\iota]\\ &= \mathrm{i}\frac{g_{1mn}(r_0)}{m+n/\iota}\exp[\mathrm{i}(m\theta_0+n\phi_0)]\end{aligned} \tag{13.3.8}$$

Thus the cumulative displacement is comparable to the magnitude of the non-axisymmetry g_{1mn} unless $n/m = -\iota$, in which case the cumulative displacement grows without bound. If there are field lines with a rational rotational transform and the nonaxisymmetry has a corresponding Fourier component, then the magnetic field line excursion from the $r = r_0$ surface becomes large. This discussion is related to the non-existence of MHD equilibrium in a three dimensional geometry. The details of this problem are beyond the scope of this book. To simplify the situation, we assume that $\iota \doteq$ const and does not satisfy the resonance condition $m/n = -\iota$ except for very large m and n. Then the existence of flux surfaces is true asymptotically. Hereafter our concern is limited to such systems.

We can distinguish between two types of inhomogeneities of the magnetic field: one owing to the helical winding and another due to the curvature that arises when the system is bent to a torus. The magnetic field along the line of force will vary as shown in Fig. 13.3. Here the deep and more frequent oscillations of the field correspond to the helical windings and the slow modulation corresponds to the toroidal curvature. The inhomogeneity caused by the helical windings can be characterized by a modulation depth of the longitudinal field ε_h given by (13.3.10). The inhomogeneity caused by the toroidal effect is characterized by the parameter ε_t. Usually $\varepsilon_t < \varepsilon_h < 1$ is valid in a stellarator/heliotron.

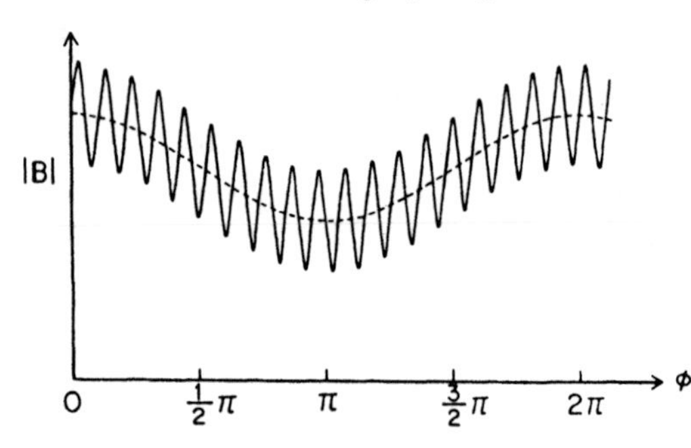

Fig. 13.3. Magnitude of a magnetic field along the toroidal direction in a nonaxisymmetric system. Slow oscillation due to toroidal curvature; fast oscillation due to helical windings

To analyze the particle motion in such systems, especially the motion of trapped particles, it is convenient to make use of the longitudinal adiabatic invariant J_2 (Sect. 3.2). In strong magnetic fields where the width of the bananas (or projected orbits of the trapped particles onto the poloidal plane) is considerably smaller than the characteristic dimensions of the magnetic field inhomogeneity, the longitudinal adiabatic invariant is conserved. This means that the trapped particles (or more precisely, the centers of the bananas) move along the surfaces $J_2 =$ const.

Here we consider the case where a straight helically symmetric field is bent to a torus. If the perturbation destroying the helical symmetry is small, the longitudinal invariant can be written as $J_2 = J_2^0(r) + \delta J(r, \theta)$ (Problem 13.3), where J_2^0 is the longitudinal invariant of the helically symmetrical system which, with given energy E and magnetic moment μ, depends only on the magnetic surface. The perturbation due to the bending of the straight system to a torus δJ depends not only on the coordinate r corresponding to the magnetic surface, but also on the azimuthal angle θ. We expand J_2^0 with respect to the small deviation $r - r_0$ of the bananas from a given surface r_0. Keeping terms up to second order in the deviation, we can write the condition $J_2 = \text{const}$ in the following form:

$$(r - r_0)\frac{\partial J_2^0(r_0)}{\partial r} + \frac{1}{2}(r - r_0)^2 \frac{\partial^2 J_2^0(r_0)}{\partial r^2} + \delta J(r_0, \theta) - \delta J(r_0, \theta_0) = 0 \;. \quad (13.3.9)$$

Here the constant equal to J_2 has been chosen such that the banana passes through the point (r_0, θ_0). The first two terms of (13.3.9) are determined by the inhomogeneity of the magnetic field owing to the helical windings and are on the order of ε_h, while the last two terms are on the order of ε_t. If $\partial J_2^0/\partial r \neq 0$ (and is not small), the second term in (13.3.9) can be neglected. In this case $\delta r = r - r_0$ is on the order of $\delta r \sim \delta J/(\partial J_2^0/\partial r) \sim r\varepsilon_t/\varepsilon_h$. This indicates that deviations of the corresponding banana motions from $r = r_0$ are proportional to ε_t. They circumvent the torus with respect to the azimuthal angle θ. The drift velocity of these transit bananas is on the order of $\varepsilon_h T/eBr$ or gradient B drift due to the helical ripple.

If $\partial J_2^0/\partial r$ tends to zero, the deviation δr increases and near the point at which $\partial J_2^0/\partial r = 0$ we must take into account the second derivative term in (13.3.9). Near the point $\partial J_2^0/\partial r = 0$, the bananas can be subdivided into transit and trapped bananas. The first group can be compared with the transit particles as they freely circumvent the torus with respect to θ. The trapped bananas can move within a limited interval of θ-values as shown in Fig. 13.4. We note that the centers of the bananas form a closed trajectory, since the toroidal effect has up-down symmetry. The orbit of the trapped banana is also called a *superbanana*. The displacement of the trapped bananas which is estimated from (13.3.9) at $\partial J_2^0/\partial r = 0$ is on the order of $\delta r \doteq \sqrt{\delta J/\partial^2 J_2^0/\partial r^2} \sim r\sqrt{\varepsilon_t/\varepsilon_h}$, i.e., it is much larger than that of the transit bananas. Unlike the displacement of the guiding centers, the displacement of the bananas is independent of the magnetic field strength. Only the magnetic drift velocity drops as B is increased, but the displacement remains unchanged.

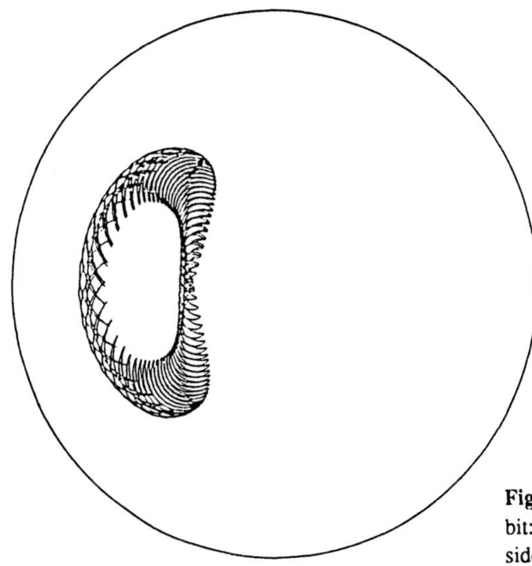

Fig. 13.4. Superbanana (trapped banana) orbit: The axis of the torus is in the left hand side

Superbananas exist only in the presence of drifting bananas with $\partial J_2^0/\partial r = 0$. If a radial electric field E_r exists in the plasma, $\boldsymbol{E} \times \boldsymbol{B}$ drift must be added to the magnetic drift. With a sufficiently strong radial electric field, the former may exceed the latter. In this case, $\partial J_2^0/\partial r$ does not vanish and the superbananas will disappear. For this case the electric field satisfies $E_r > (T_e/ea)\delta B/B \sim \varepsilon_h T/ea$, where T_e is the electron temperature, δB is the modulation of the magnetic field and a is an averaged minor radius of the nonaxisymmetric plasma.

Now we will discuss the diffusion coefficient more quantitatively [13.3, 4]. In a nonaxisymmetric system, the model magnetic field varies as

$$B = B_0 \left[1 - \varepsilon_h \cos\left(l\theta - \frac{N}{R}z\right) - \varepsilon_t \cos\theta\right] , \tag{13.3.10}$$

where θ is the azimuthal angle, z is the coordinate along the torus, l is the pole number of the helical windings and N is the pitch number of the helical windings. For example, for a corrugated torus $l = 0$, for a stellarator with three helical windings $l = 3$. The heliotron has $l = 2$. In (13.3.10), $\varepsilon_t < \varepsilon_h < 1$ is valid for a stellarator/heliotron.

The main role in the transport is played by the localized particles trapped by a field inhomogeneity caused by the helical windings. Their fraction is $\sqrt{\varepsilon_h}$. These particles, which oscillate between the mirrors of the helical winding, drift with respect to θ by a certain angular velocity $\omega(\alpha)$ which is proportional to the particle energy and depends on the pitch angle of the velocity space denoted by α. When ω is averaged with respect to the energy, for a given α, we obtain the approximation

$$\omega(\alpha) \doteq \frac{T}{eBr^2}\varepsilon_h f(\alpha) = \omega_h f(\alpha) , \tag{13.3.11}$$

where T is the temperature, r is the minor radius, $f(\alpha)$ is a function of α of order unity and $\omega_h = \varepsilon_h T/eBr^2$. Recall that a variation of $f(\alpha)$ in the trapped particle range $0 < \alpha < \sqrt{\varepsilon_h}$ entails a change of sign of the function f such that it vanishes at a certain α value ($\alpha = \alpha_0 < \sqrt{\varepsilon_h}$) (Problem 13.4).

The presence of a toroidal curvature ε_t causes a drift with respect to r so that similarly to (13.2.11), we obtain a drift kinetic equation

$$\omega(\alpha)\frac{\partial f_1}{\partial\theta} + \langle v_\perp\rangle \sin\theta\frac{\partial f_0}{\partial r} = -\nu_{\text{eff}} f_1 , \tag{13.3.12}$$

where $\langle v_\perp\rangle = T/eBR_0$, $f_0 = f_0(r)\exp(-E/2T)$ and $f_1 = f_1(r, \theta, \alpha)$. Here E is the particle energy. The drift kinetic equation (13.3.12) is only applicable in the case of not very small collision frequencies. Then all bananas may be considered to be of transit type, that is, they circumvent the stellarator/heliotron with respect to θ.

From (13.3.12) we obtain f_1 by the same calculation as in Sect. 13.2 and then the radial flux averaged over θ:

$$\Gamma_\perp = -\frac{1}{2}\int \frac{\nu_{\text{eff}}\langle v_\perp\rangle^2}{\nu_{\text{eff}}^2 + [\omega(\alpha)]^2}\frac{\partial f_0}{\partial r}d\alpha . \tag{13.3.13}$$

If $\nu_{\text{eff}} \doteq \nu/\varepsilon_{\text{h}} > \omega_{\text{h}} = (T/eBr^2)\varepsilon_{\text{h}}$, it is possible to neglect ω in (13.3.13) so that we obtain the diffusion coefficient in front of $\partial f_0/\partial r$ as

$$D_{\perp} \doteq \frac{1}{2}\frac{\varepsilon_{\text{h}}\omega_{\text{h}}}{\nu}\frac{\varepsilon_{\text{t}}^2}{\varepsilon_{\text{h}}^{1/2}}\frac{T}{eB}\,, \tag{13.3.14}$$

noting that the fraction of trapped particles is $\sqrt{\varepsilon_{\text{h}}}$. If $\nu_{\text{eff}} < \omega_{\text{h}}$, the integrand in (13.3.13) can be replaced by $\pi\delta(\omega)$ and we obtain the approximate expression

$$D_{\perp} \doteq \frac{\varepsilon_{\text{t}}^2}{\varepsilon_{\text{h}}^{1/2}}\frac{T}{eB}\,. \tag{13.3.15}$$

If ν decreases, the main contribution in (13.3.13) will be made by bananas with α close to α_0 and $\omega = 0$. Bananas with sufficiently small $\delta\alpha = \alpha - \alpha_0$ are no longer transit bananas, they become superbananas. This takes place at $\delta\alpha \sim \sqrt{\varepsilon_{\text{t}}}$ since in the case of a superbanana a banana becomes trapped in an inhomogeneity of ε_{t}. Thus the frequency of drift of the bananas to form a superbanana is on the order of $\omega_{\text{t}} \sim \sqrt{\varepsilon_{\text{t}}/\varepsilon_{\text{h}}}\,\omega_{\text{h}}$. Accordingly, the transition to superbanana diffusion will take place at $\nu_{\text{eff}} \doteq \nu/\varepsilon_{\text{t}} \sim \omega_{\text{t}}$. If $\nu < \nu_{\text{s}} = (\varepsilon_{\text{t}}/\varepsilon_{\text{h}})^{3/2}\varepsilon_{\text{h}}\omega_{\text{h}}$, the corresponding diffusion coefficient is of the order

$$D_{\perp} \doteq \sqrt{\varepsilon_{\text{t}}}\,\nu_{\text{eff}}(\delta r)^2 \doteq \frac{\nu}{\nu_{\text{s}}}\frac{\varepsilon_{\text{t}}^2}{\varepsilon_{\text{h}}^{1/2}}\frac{T}{eB}\,, \tag{13.3.16}$$

for $\nu_{\text{eff}} = \nu/\varepsilon_{\text{t}}$ and $\delta r \doteq T/eBR_0\omega_{\text{t}}$. This diffusion decreases with ν.

It is an interesting peculiarity of diffusion in nonaxisymmetric toroidal systems that each component (the electron and the ion components) diffuses independently of the others. The formulas given above apply to both electrons and ions provided that ν is properly defined as the corresponding collision frequency. The connection between the electrons and ions exists via the electric field, that is, the field E_r established in the system must balance the particle flux of electrons and ions. Consequently, in that range of plasma parameters where the ion diffusion coefficient D_{i} is larger than the electron diffusion coefficient D_{e}, the plasma will be charged negatively and when $D_{\text{e}} > D_{\text{i}}$ it will have a net positive charge.

13.4 Convective Cell and Associated Transport

The term 'convective cell' has several meanings. Our concern here is with two connotations: a specific low frequency eigenmode of the two-dimensional fluid equations; any low frequency, large scale cross-field motion of magnetized plasma induced by instabilities such as MHD modes or drift waves [13.6, 7, 8].

The equation for the standard convective cell mode with $k_z = 0$ in a homogeneous magnetized plasma where the magnetic field is in the z-direction is

$$\frac{\partial}{\partial t}\nabla_\perp^2\phi + \frac{\hat{z}\times\nabla\phi}{B}\cdot\nabla\nabla_\perp^2\phi = \mu\nabla_\perp^4\phi\,, \tag{13.4.1}$$

which can be derived from the two-dimensional reduced MHD equations (11.1.7) by neglecting the magnetic fluctuation Ψ and adding the viscosity term in the right hand side. Here ϕ is the electrostatic potential corresponding to $B_0\phi$ in Sect. 11.1, the second term represents the $\boldsymbol{E}\times\boldsymbol{B}$ drift motion, and the viscosity term is added to the right hand side of (13.4.1). Thus a convective cell is the plasma analogue of a vortex in fluid mechanics where ϕ corresponds to the stream function. By linearizing (13.4.1), the dispersion relation

$$\omega = -\mathrm{i}\mu k^2 \tag{13.4.2}$$

is obtained. In turbulent plasmas, it should be pointed out that the viscosity μ is increased above the value obtained when only Coulomb collisions are assumed to occur and that it is related to enhanced diffusion due to the turbulence. Usually, the collisional viscosity is small in high temperature plasmas and the damping rate given by (13.4.2) is negligible. Since the convective cells have a quasistatic vortex nature, they may lead to rapid particle diffusion. The diffusion coefficient can be estimated from the test particle diffusion coefficient in the homogeneous plasma explained in Sect. 13.1,

$$D \doteq \frac{(\Delta x)^2}{\Delta t}\,, \tag{13.4.3}$$

if we can identify the correlation length Δx and the correlation time Δt. Suppose that there arises some density fluctuation δn, uniform along the field, with a characteristic perpendicular space scale Δx. There are two kinds of perpendicular velocities associated with δn, a possible translational motion of the fluctuation as a whole or a flow across the magnetic field (not of interest presently), and an internal velocity of Δu which deforms the fluctuation. A characteristic time of the deformation is estimated as

$$\Delta t \sim \frac{\Delta x}{\Delta u}\,. \tag{13.4.4}$$

In the model where particles move with the fluid, the particle diffusion due to the fluctuations of scale Δx, from (13.4.3, 4) is of order

$$D \sim \Delta x \Delta u\,. \tag{13.4.5}$$

Since Δu is the velocity fluctuation across Δx, it can be determined from the electric field gradient over the convective cell:

$$\Delta u \sim \frac{1}{B}\left(\frac{\partial \bar{E}}{\partial x}\Delta x\right) \sim \frac{1}{B}\frac{\Delta\bar{\phi}}{\Delta x}\,, \tag{13.4.6}$$

where $\Delta\bar{\phi}$ is the root mean square potential fluctuation across the cell and $\bar{E} = \Delta\bar{\phi}/\Delta x$. The unknown scale Δx cancels out in (13.4.5), leaving

$$D \sim \frac{\Delta\bar{\phi}}{B} = \frac{T}{eB}\frac{e\Delta\bar{\phi}}{T} . \tag{13.4.7}$$

The result obeys the *Bohm diffusion* scaling which is proportional to the temperature and inversely proportional to the magnetic field, provided that $e\Delta\bar{\phi}/T$ is independent of T and B. Here it is important to stress that we assumed the fluctuation to be two-dimensional. If the fluctuation has structure along the field, as would be the case if instabilities with finite k_z were excited, then parallel streaming motion would give an additional rapid decorrelation mechanism and the estimate of the correlation time (13.4.4) is likely to be different.

Low frequency drift waves can act as convective cells and cause rapid particle transport if they are excited to a large amplitude. Then a particle can be convected from one side of the cell to another within a wave period as shown in Fig. 13.5.

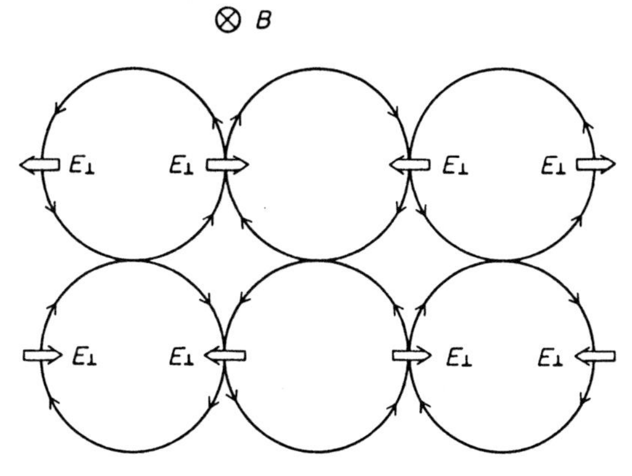

Fig. 13.5. Convective flows excited by $\boldsymbol{E} \times \boldsymbol{B}$ motion in the plane perpendicular to the magnetic field

For a mode with $k_x \sim k_y$ (the xy-plane is perpendicular to the magnetic field direction), this condition becomes $\omega/k_\perp < \tilde{u}_\perp$, where $k_\perp$ is the perpendicular wavenumber and $\tilde{u}_\perp$ is the perturbed $\boldsymbol{E} \times \boldsymbol{B}$ drift velocity. In this case the drift wave has a large amplitude potential $\delta\phi$ such that $\tilde{u}_\perp$ is greater than the diamagnetic velocity

$$\frac{k_\perp \delta\phi}{B} > \frac{\omega_*}{k_\perp} = v_* = \frac{T}{eBL_n} , \tag{13.4.8}$$

(5.5.14) or equivalently,

$$\frac{e\delta\phi}{T} > \frac{1}{k_\perp L_n} , \tag{13.4.9}$$

where $L_n^{-1} = |dn_0/dx|/n_0$ when the equilibrium density has an inhomogeneity in the x-direction.

Nonlinear excitation of convective cells due to drift instabilities can be expected since the dispersion relation of drift waves has a real frequency

$$\omega_r = \omega_* \frac{e^{-b_i} I_0(b_i)}{2 - e^{-b_i} I_0(b_i)} \tag{13.4.10}$$

with $b_i = k_\perp^2 \varrho_{Li}^2$ and $k_\parallel \ll k_\perp$, where $k_\parallel$ and $k_\perp$ are the parallel and perpendicular components of the wavenumber with respect to the magnetic field direction. The frequency given in (13.4.10) shows that the drift wave is dispersive for $k_\perp \varrho_{Li} \geq 1$. Parametric mode coupling takes place between two drift waves with the same frequency but different $k_\perp$, so that $\omega_1 - \omega_2 \doteq 0$ and $k_{\parallel 1} - k_{\parallel 2} \doteq 0$. Here the subscripts 1 and 2 denote two different waves. Thus, a convective cell is produced. This process is important because the drift instabilities have a maximum growth rate at $k_\perp \varrho_{Li} \doteq 1$.

13.5 Drift Wave Turbulence and Associated Transport

Small fluctuations in the electric and magnetic fields lead to small fluctuations in the particle's velocity and radial positions. This can lead to transport of both particles and energy across the confining magnetic field [13.5, 11]. For low frequency fluctuations ($\omega \ll \omega_{ci}$) in magnetically confined plasma such as a tokamak plasma, a particle's radial velocity fluctuation can be written as

$$\tilde{v}_r \doteq \frac{\tilde{E}_p}{B} + \frac{v_\parallel \tilde{B}_r}{B} , \tag{13.5.1}$$

where the subscripts p and r denote the poloidal and radial component respectively, and $v_\parallel$ is the particle velocity along the unperturbed magnetic field B_0. (Here the coordinate system is that of a tokamak with circular cross section flux surfaces. In a tokamak, the poloidal field is much smaller than the toroidal field $B_p \ll B_t \doteq B_0$.) The fluctuations in the particle's radial motion can lead to both anomalous particle transport and associated thermal transport. However, the net transport depends not just on the level of the fluctuations, but also on the correlation between various fluctuating quantities. Thus, to determine the fluctuation-induced transport, correlations of fluctuations must be examined.

First, consider the radial particle flux $\Gamma = nu_r$ where u_r denotes the fluid velocity. In a turbulent plasma, each quantity can be written as a time averaged part plus a fluctuating part, e.g., $n = \langle n \rangle + \tilde{n}$, where $\langle \ \rangle$ denotes an average over a time scale which is long compared to the frequency of the fluctuations, so that $\langle \tilde{n} \rangle = 0$ and $\langle n \rangle$ is the average, or macroscopic density. The net particle flux can be written as

$$\Gamma = \langle (\langle n \rangle + \tilde{n}) \, (\tilde{u}_r + \langle u_r \rangle) \rangle = \langle n \rangle \langle u_r \rangle + \langle \tilde{n} \tilde{u}_r \rangle \, . \tag{13.5.2}$$

The contribution from the turbulence thus depends on the correlation between the density and radial velocity fluctuations,

$$\Gamma = \langle n u_r \rangle = \int_{-\infty}^{\infty} n^*(\omega) u_r(\omega) d\omega \, , \tag{13.5.3}$$

where $\tilde{n}(\omega)$ and $\tilde{u}_r(\omega)$ are Fourier components with respect to the angular frequency ω.

On the assumption that the flux is ambipolar, or that the electron and ion fluxes are equal (Sect. 13.1), and using (13.5.1), the particle flux resulting from the turbulence can be written as

$$\Gamma = \frac{\langle \tilde{n}\tilde{E}_{\rm p}\rangle}{B} - \frac{\langle \tilde{J}_{\|}\tilde{B}_r\rangle}{eB} , \tag{13.5.4}$$

where

$$\tilde{J}_{\|} = -e\int_{-\infty}^{\infty} v_{\|}\tilde{f}_{\rm e} dv_{\|} \tag{13.5.5}$$

is the fluctuation in the parallel electron current and $\tilde{f}_{\rm e}$ is the fluctuating part of the electron distribution function. The particle diffusion coefficient is then defined by $D = -\Gamma/(d < n > /dr)$. When this significantly exceeds the classical value $D = \nu\varrho_{\rm L}^2$, it is called the *anomalous diffusion*. It should be noted that this definition is idealized, since usually the turbulent flux Γ can depend on many different quantities, such as $dT_{\rm e}/dr$.

For low frequency, electrostatic ($\tilde{B} = 0$) microturbulence such as drift wave turbulence, the heat flux depends on the correlation between the fluctuating radial velocity $\tilde{u}_r = \tilde{E}_{\rm p}/B$ and the pressure fluctuations. The heat flux can be written as

$$Q_r = \frac{5}{2B}\langle \tilde{E}_{\rm p}\tilde{P}\rangle , \tag{13.5.6}$$

where Q_r is the total heat flux (convective plus conductive), or $Q_r = 5/2T\Gamma_r + q_r$. Here q_r is the conductive heat flux. In addition to (13.5.6), the radial magnetic field fluctuations in finite β plasmas generate another heat flux which results from that part of the large classical parallel conductivity that is now directed radially. The classical conductive heat flux is

$$\boldsymbol{q} = -K_{\|}\nabla_{\|}T - K_{\perp}\nabla_{\perp}T , \tag{13.5.7}$$

where $K_{\|}$ and $K_{\perp}$ are the classical conductivity parallel and perpendicular to the magnetic field, respectively. The contribution to the radial heat flux from the magnetic fluctuations is obtained from (13.5.7) when the unit vector along the magnetic field $\boldsymbol{b}$, as well as the temperature, are written as an average plus a fluctuating part, i.e.,

$$\boldsymbol{b} = \langle \boldsymbol{b}\rangle + \frac{\tilde{B}_r}{B}\hat{\boldsymbol{r}} , \tag{13.5.8}$$

where $\langle \boldsymbol{b}\rangle$ is in the direction of the unperturbed magnetic field. The radial component of the heat flux is obtained by multiplying $\tilde{B}_r\hat{\boldsymbol{r}}/B$ by (13.5.7)

$$q_r = -K_{\|}\left\langle \left|\frac{\tilde{B}_r}{B}\right|^2\right\rangle \frac{d\langle T\rangle}{dr} - K_{\perp}\left\langle \left|\frac{\tilde{B}_r}{B}\right| \frac{d\tilde{T}}{dr}\right\rangle \tag{13.5.9}$$

where we set $\langle \nabla_{\|}\rangle\langle T\rangle = 0$ and third order correlations which can arise from the fluctuation of $K_\parallel$ have been neglected. From (13.5.9), it is estimated that very low levels of fluctuation $|\tilde{B}_r/B| \sim 10^{-4}$ can lead to anomalous heat conductivity comparable to that observed in tokamak experiments because the classical parallel conductivity $K_\parallel$ is so large at typical plasma parameters.

Now we explain quasilinear calculations of transport due to electrostatic drift waves. One underlying assumption of quasilinear theory is that there exists a spectrum of linearly unstable normal modes whose amplitude is small enough so that the interaction of the modes with each other can be neglected. Quasilinear theory also assumes that the particles respond linearly to the wave.

Assuming quasi-neutrality and that the radial velocity fluctuation is due to the fluctuating $\boldsymbol{E} \times \boldsymbol{B}$ drift, the fluctuation-induced particle flux can be written from the first term of (13.5.4) as

$$\Gamma = -D\frac{d\langle n\rangle}{dr} = -\frac{1}{B}\langle \tilde{n}\tilde{E}_{\rm p}\rangle \ . \tag{13.5.10}$$

In low frequency drift waves, the electrons reach a nearly Boltzmann distribution, and the density perturbation with the wavenumber $\boldsymbol{k}$ can often be approximated as

$$\tilde{n}_{\boldsymbol{k}} \doteq \langle n\rangle \frac{e\tilde{\phi}_{\boldsymbol{k}}(1 - {\rm i}\delta_{\boldsymbol{k}})}{T_{\rm e}} \ , \tag{13.5.11}$$

where $\tilde{\phi}_{\boldsymbol{k}}$ is the fluctuation in the potential and $\delta_{\boldsymbol{k}}$ is a real small quantity which stands for the phase difference between $\tilde{n}_{\boldsymbol{k}}$ and $\tilde{\phi}_{\boldsymbol{k}}$. Then using (13.5.10) the particle flux becomes

$$\Gamma = \langle n\rangle \frac{T_{\rm e}}{eB} \sum_{\boldsymbol{k}} k_{\rm p}\delta_{\boldsymbol{k}} \left|\frac{\tilde{n}_{\boldsymbol{k}}}{\langle n\rangle}\right|^2 \ , \tag{13.5.12}$$

where $k_{\rm p}$ is the poloidal wavenumber.
The growth rate of the drift wave is determined by $\delta_{\boldsymbol{k}}$. Usually $\delta_{\boldsymbol{k}} \sim \gamma(\boldsymbol{k})/\omega_{\rm r}(\boldsymbol{k}) \sim \gamma(\boldsymbol{k})/\omega_{*{\rm e}}$ as will be explained later. Here $\omega_{*{\rm e}} = k_{\rm p}T_{\rm e}/eBL_{\rm n}$ (5.5.14) is the drift frequency. Then the diffusion coefficient can be approximated as

$$D \sim L_{\rm n}^2 \sum_{\boldsymbol{k}} \gamma(\boldsymbol{k}) \left|\frac{\tilde{n}_{\boldsymbol{k}}}{\langle n\rangle}\right|^2 \ , \tag{13.5.13}$$

where $L_n = \langle n\rangle/(d\langle n\rangle/dr)$ is the scale length of the density gradient. This expression is often used as a basis for estimating the amount of transport to be expected from observed or estimated levels of density fluctuations. In making such estimates, usually it is assumed that $\gamma \sim 0.1\omega_{*{\rm e}}$ or $\gamma \sim \Delta\omega$, where $\Delta\omega$ is an average width of the wavenumber spectrum of the fluctuation.

From (13.5.12), it can be seen that the flux is directly proportional to $\delta_{\boldsymbol{k}}$. Thus this quasilinear calculation of the fluxes is very sensitive to the assumption of linear electron response. Moreover, the fluxes are also directly proportional to the fluctuation level $|\tilde{n}/\langle n\rangle|$, which is not determined in the quasilinear theory.

One frequently used estimate of the fluctuation level is the 'mixing length' level

$$\left|\frac{\tilde{n}}{\langle n\rangle}\right| \doteq \frac{1}{k_\perp L_n} \,. \tag{13.5.14}$$

At this level, the perturbed density gradient $k_\perp \tilde{n}$ is comparable to the mean density gradient, which is the free energy source of the drift wave instability where $k_\perp$ is the perpendicular wavenumber. Using this level, (13.5.13) gives

$$D \doteq \frac{\gamma(k)}{k_\perp^2} \,, \tag{13.5.15}$$

which is a commonly occurring estimate of the diffusion coefficient. By comparing this with the random walk argument in Sect. 13.1, we can interpret the result such that the correlation length is $1/k_\perp$ and the correlation time is $1/\gamma$.

As explained in Sect. 5.5, the drift wave is a quasi-electrostatic wave which propagates perpendicularly to both the magnetic field and the density gradient. Electrons move along the field line to compensate for the charge separation by the ion $\boldsymbol{E} \times \boldsymbol{B}$ drift along the density gradient, i.e., to keep the charge neutrality via Debye shielding. As far as the Debye shielding is secured, the Boltzmann relation $e\tilde{\phi}_{\boldsymbol{k}}/T_\mathrm{e} \sim \tilde{n}_{\boldsymbol{k}}/\langle n\rangle$ is maintained. However, if the Debye shielding is prevented by some means, the potential propagation is delayed behind the propagation of the density perturbation. Then the resulting $\boldsymbol{E} \times \boldsymbol{B}$ drift tends to enhance the density perturbation, since the drift from the high (or low) density side takes place at the crest (or trough) of the perturbation. The growth rate is naturally expected to be proportional to the phase delay $\gamma(\boldsymbol{k}) \propto \delta_{\boldsymbol{k}}$. We have seen a mechanism for this delay concerning the finite ion Larmor radius for a collisionless drift wave in Sect. 6.5. As explained above, inhibition of the Debye screening can also cause a delay. For instance, electron-ion collision can reduce the Debye shielding efficiency. The instability caused by such a collision is called the *resistive drift wave instability*.

The reduced MHD equations (11.1.18) can be written as

$$\frac{d}{dt}\frac{\nabla_\perp^2 \tilde{\phi}}{B_0 \Omega_\mathrm{i}} = \frac{1}{e n_0}\frac{\partial}{\partial z} J_z \tag{13.5.16}$$

by noting $\mu_0 J_z = -\nabla_\perp^2 \Psi$ and neglecting the pressure term. Here J_z is the perturbed current density in the magnetic field direction $\boldsymbol{B} = B_0 \hat{\boldsymbol{z}}$ and Ω_i is the ion cyclotron frequency. The convective derivative is written as

$$\frac{d}{dt} = \frac{\partial}{\partial t} + \frac{\nabla \tilde{\phi} \times \hat{\boldsymbol{z}}}{B_0} \cdot \nabla \,. \tag{13.5.17}$$

We note that the electrostatic potential $\tilde{\phi}$ in (13.5.16, 17) corresponds to $B_0 \phi$ in Sect. 11.1. The continuity equation relates the number density to the electron current density

$$\frac{dn}{dt} = \frac{1}{e}\frac{\partial J_z}{\partial z} \,. \tag{13.5.18}$$

Here we assumed that ion motion is two-dimensional and that the ion contribution to the parallel current is negligible. We also assume that the electrons are isothermal. Then the electron equation of motion (4.5.1) in the z-direction gives

$$J_z = \frac{T_e}{e\eta}\frac{\partial}{\partial_z}\left(\frac{n_1}{n_0} - \frac{e\tilde{\phi}}{T_e}\right), \tag{13.5.19}$$

where $n = n_0 + n_1$ and n_0 is the equilibrium density which does not depend on the z-coordinate, η is the resistivity which is proportional to the electron collision frequency. By eliminating J_z from (13.5.16, 19) we can construct coupled nonlinear equations for $\tilde{\phi}$ and n_1. If we use the normalization $e\tilde{\phi}/T_e \equiv \phi, n_1/n_0 \equiv n$, $\Omega_i t \equiv t$ and $x/\varrho_s \equiv x$ (or $y/\varrho_s \equiv y$) where $\varrho_s = \sqrt{T_e/m_i}/\Omega_i$, the coupled equations become

$$\left(\frac{\partial}{\partial t} - \nabla\phi \times \hat{z}\cdot\nabla\right)\nabla_\perp^2\phi = -\frac{T_e}{e^2 n_0 \eta \Omega_i}\frac{\partial^2}{\partial z^2}(\phi - n) \tag{13.5.20}$$

and

$$\left(\frac{\partial}{\partial t} - \nabla\phi \times \hat{z}\cdot\nabla\right)(n + \ln n_0) = -\frac{T_e}{e^2 n_0 \eta \Omega_i}\frac{\partial^2}{\partial z^2}(\phi - n)\,. \tag{13.5.21}$$

These equations are useful to study resistive drift wave turbulence and the related anomalous particle transport. We note further that after (13.5.21) is subtracted from (13.5.20) and the Boltzmann relation $n = \phi$ is substituted in,

$$\frac{\partial}{\partial t}(\nabla_\perp^2\phi - \phi) - (\nabla\phi \times \hat{z})\cdot\nabla(\nabla_\perp^2\phi - \phi) + \nabla \ln n_0 \times \nabla\phi\cdot\hat{z} = 0 \tag{13.5.22}$$

or

$$\frac{d}{dt}(\nabla_\perp^2\phi - \phi) + \nabla \ln n_0 \times \nabla\phi\cdot\hat{z} = 0 \tag{13.5.23}$$

is obtained. This equation is called the *Hasegawa-Mima equation* for nonlinear drift waves or drift wave turbulence. An inverse cascade process in two-dimensional turbulence in which wave energy is carried from short wavelengths to long wavelenghts resulting in a formation of convective cell was observed after numerical solution of (13.5.20, 21) or (13.5.23) [13.7, 12].

13.6 MHD Turbulence and Associated Transport

The nonlinear incompressible MHD equations for a plasma with a uniform density are frequently used to discuss MHD turbulence:

$$\frac{\partial \boldsymbol{B}}{\partial t} = \nabla \times (\boldsymbol{u} \times \boldsymbol{B}) + \eta\nabla^2\boldsymbol{B} \tag{13.6.1}$$

$$\left(\frac{\partial}{\partial t} + \boldsymbol{u}\cdot\nabla\right)\boldsymbol{u} = -\frac{\nabla P}{\varrho} + \frac{1}{\varrho}(\nabla \times \boldsymbol{B}) \times \boldsymbol{B}\,, \tag{13.6.2}$$

where ϱ is the mass density.

In the presence of an infinitely strong externally applied magnetic field $\boldsymbol{B}_0$ without net current, the three-dimensional equations (13.6.1, 2) reduce exactly to the two-dimensional version of the same equations, where now fluctuations in $\boldsymbol{B}$ and $\boldsymbol{u}$ have components and variations only in the plane perpendicular to $\boldsymbol{B}_0 = B_0\hat{\boldsymbol{z}}$ [13.5, 6, 7]. the significance of the reduction of the three-dimensional system to the two-dimensional isotropic system in the perpendicular plane is that much is known about two-dimensional fluid turbulence. In particular, a 'dual cascade' has been observed in MHD turbulence. The nonlinearities lead to an *inverse cascade* of the vector potential $\boldsymbol{A} = A_z\hat{\boldsymbol{z}}$ from large k to small k where $\boldsymbol{B} = \nabla \times \boldsymbol{A}$. Note that the magnetic helicity is given by $\int A_z B_0 dV$ (Sect. 11.7). In other words, the helicity cascades inversely and a simultaneous cascade of spectral energy (kinetic plus magnetic) takes place from small k to large k. Generalizing to the case of finite external field B_0, the spectrum splits into two components: a dominant part which is nearly two-dimensional isotropic MHD turbulence and a spectrum of shear Alfven waves $\omega = k_z v_A$ (Sect. 10.6). The correlation length along $\boldsymbol{B}_0$ is much longer than those transverse to $\boldsymbol{B}_0$. This is consistent with experimental observations of magnetic fluctuations in tokamaks, namely $\tilde{B}_\perp \gg \tilde{B}_\parallel$, and the correlation length is much greater in the parallel direction.

In the absence of radial magnetic fluctuations, a magnetic field line in a tokamak will remain in its equilibrium flux surface as it winds around the torus (Sect. 13.2). When radial magnetic fluctuations B_r are present, the field line trajectory will deviate from the original flux surface [13.9, 10]. As a result of such radial magnetic fluctuations or 'magnetic flutter', each electron now has a radial velocity perturbation given by $\tilde{v}_r = v_\parallel \tilde{B}_r/B$, and thus part of the rapid electron motion along the field lines is converted into radial motion. This mechanism of anomalous thermal transport is given in Sect. 13.5. Heat transport associated with an internal disruption explained in Sect. 11.6 is a typical example. When magnetic islands due to magnetic fluctuations centered at different radii or rational surfaces begin to overlap, the magnetic field lines begin to make a braid. If the overlap is large enough, the magnetic flux surfaces are essentially destroyed and each field line trajectory wanders stochastically away from its original radial location as it winds around the torus. The electrons follow these wandering field lines and thus the radial transport can be enhanced as in the case of the major disruption described in Sect. 11.6. The average square of the radial displacement of the magnetic field line Δr moving a distance s along the field line can be written as

$$\langle(\Delta r)^2\rangle = 2sD_\mathrm{m} , \tag{13.6.3}$$

with

$$D_\mathrm{m} = \int_0^\infty dz\langle b_r(0)b_r(z)\rangle = L_0 b_0^2 , \tag{13.6.4}$$

and $b_r = \tilde{B}_r/B$ and $b_0^2 = \langle|b_r(0)|^2\rangle$. Here D_m is identified as the magnetic field line diffusion coefficient, L_0 is the parallel auto-correlation length of the radial magnetic fluctuations and the unperturbed field has been taken to be in the z-

direction. When the auto-correlation function is calculated by integrating along the unperturbed magnetic field line trajectory, this is referred to as the quasilinear limit of D_m [13.10].

Using the fact that electron orbits in a tokamak follow the field lines, we can calculate the electron test particle diffusion coefficient which results from the diffusion of the stochastic magnetic field lines. Particle transport, however, cannot take place at this rate, since the ion test particle diffusion is much slower than the electron diffusion. If the electrons attempt to diffuse faster than the ions, an ambipolar electric field arises which reduces the electron diffusion. However, thermal electron transport is not inhibited. Thus, in the anomalous thermal transport due to the stochastic magnetic field fluctuations the electron test particle diffusion coefficient may be equated to the electron thermal diffusivity. In these test particle diffusion calculations, the magnetic fluctuation spectrum is assumed to be specified and it is difficult to calculate it self-consistently with the electron motion.

The physical picture underlying these test particle diffusion calculations is the following: an electron follows a wandering field line for a distance s until, by one of a number of physical processes, it moves to another field line uncorrelated with the first. In the distance s along the field line the electron has moved radially a distance Δr given by (13.6.3). The movement of the electron between uncorrelated field lines is necessary to make sure that each radial step is independent of the previous one.

The collisionless regime is defined by $l_{mfp} \gg L_k$, where $l_{mfp} = v_{Te}/\nu_{ei}$ is the electron mean free path and L_k is the characteristic length over which two neighboring field line trajectories diverge exponentially

$$|d\boldsymbol{r}(s)| = |d\boldsymbol{r}(0)| \exp(s/L_k) \,. \tag{13.6.5}$$

Here $|d\boldsymbol{r}|$ is the separation of two field lines and s is the distance along the field line. For this case, in a given collision time an electron streams along the field line a distance l_{mfp}. At this distance, neighboring trajectories have diverged significantly and the collision causes the electron to move to an uncorrelated field line. Then by using (13.6.3) the radial step in the collision time is $\langle(\Delta r)^2\rangle^{1/2} = (2 l_{mfp} D_m)^{1/2}$ and the test particle diffusion coefficient becomes

$$D = \frac{\nu_{ei}}{2}\langle(\Delta r)^2\rangle = v_{Te} D_m \,. \tag{13.6.6}$$

Under the assumption of a collisionless plasma this test particle diffusion coefficient is equal to the thermal diffusivity in the stochastic magnetic field. Expression (13.6.6) is often used to estimate the anomalous transport in tokamaks based on the observed magnetic fluctuations or MHD turbulence.

PROBLEMS

13.1. In the model magnetic field for tokamaks $B = (0, B_\theta(r), B_0)$ in the toroidal coordinates (r, θ, φ), (i) show the guiding center equation of motion; (ii) show that the guiding center orbit in the poloidal plane is closed in the shape of a banana for trapped particles.

13.2. Derive the linearized drift kinetic equation (13.2.11) from (4.3.21) by adding a collision term to the right hand side.

13.3. By using the model magnetic field

$$B = B_0(r,\theta) - B_1(r,\theta)\cos(l\theta - N\phi)\ , \tag{13.P.1}$$

the longitudinal adiabatic invariant J_2 is written as

$$J_2(k) = \begin{cases} (2B_1/B_0)k^{1/2}E(1/k) & \text{for } k > 1 \\ (2B_1/B_0)[E(k) + (k-1)K(k)] & \text{for } k < 1\ , \end{cases} \tag{13.P.2}$$

where K and E are complete elliptic integrals of the first and second kind, respectively, and $B_1 \ll B_0$ is assumed. Here

$$k = (E - \mu B_0 + \mu B_1)\,/\,(2\mu B_1)\ . \tag{13.P.3}$$

Confirm the relation (13.P.2).

13.4. The orbit period of the banana in the model magnetic field (13.P.1) is given by

$$T(k) = \begin{cases} 4(\mu B_1/m)^{1/2}K(1/k)/k & \text{for } k > 1 \\ 4(\mu B_1/m)^{1/2}K(k) & \text{for } k < 1\ . \end{cases} \tag{13.P.4}$$

The banana moves along the surface $J_2(r,\theta) = \text{const}$, which gives $\Delta\theta = -\partial J_2/\partial r$ and $\Delta r = \partial J_2/\partial\theta$. The angular velocity of the banana in the poloidal direction is given by

$$\frac{d\theta}{dt} = \frac{\Delta\theta}{T} = -\frac{\partial J_2}{\partial r}\frac{1}{T(k)}\ . \tag{13.P.5}$$

Show that the angular velocity can be zero.

13.5. Confirm the diffusion coefficient (13.3.14, 15).

13.6. Confirm the expression of the diffusion coefficient (13.3.16).

13.7. Derive (13.5.9) from (13.5.7, 8).

14. Progress in Fusion Research

We have described the basic theory of plasma physics relevant to thermonuclear fusion research. In this last chapter, we briefly outline the recent progress in this field. First, in Sect. 14.1, we describe progress in tokamak research which has been the mainstream of the thermonuclear fusion in the past quarter century. Then in Sect.14.2, we outline recent progress in other magnetic confinement systems. Section 14.3 is devoted to the present status of inertial confinement fusion research. Finally, in Sect. 14.4 we give a brief presentation of plasma physics with nuclear reaction effects taken into account.

14.1 A Brief History of Tokamak Research

As shown in Fig. 14.1, in a tokamak the axisymmetric plasma is confined in a toroidal vacuum chamber by a strong magnetic field (toroidal magnetic field) running along the long circumference of the torus together with a current flowing through the ring of plasma itself (which is required for MHD equilibrium). In order to obtain stability, i.e., $q > 1$, where q is the safety factor (Sect. 10.4) the magnitude of the toroidal magnetic field must be larger than the poloidal magnetic field produced by the plasma current. A small vertical magnetic field serves as a $\boldsymbol{J} \times \boldsymbol{B}$ force to maintain the position of the plasma ring at a suitable place inside the chamber. The plasma current is induced inductively since the plasma acts as a single turn secondary winding of a transformer. It also heats the plasma resistively. The ohmic heating is the simplest and most efficient method of creating a relatively hot ($T_e \sim 1\,\mathrm{keV}$) and dense plasma ($n_e \gtrsim 10^{19}\,\mathrm{m}^{-3}$), but

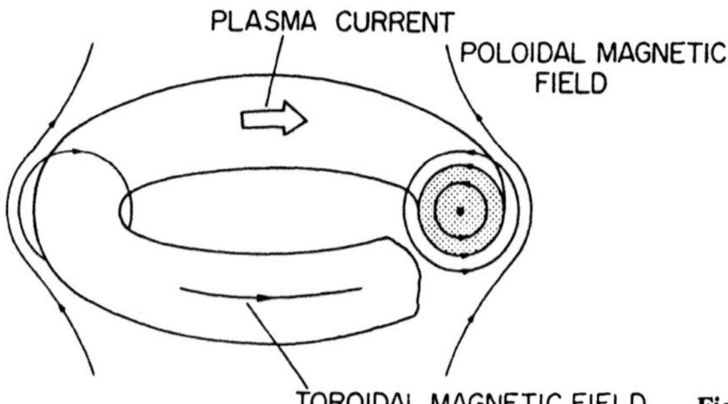

Fig. 14.1. Tokamak configuration

it is generally thought to be necessary to provide supplementary heating to bring the plasma to thermonuclear temperatures.

Initial tokamak research was started in the USSR before 1960. At the 1968 IAEA Conference on Plasma Physics and Controlled Fusion Research in Novosibirsk, a group from the Kurchatov Institute in Moscow led by L.A. Artimovich presented T-3 tokamak results showing promising data, i.e., $T_e \gtrsim 600\,\mathrm{eV}$ and $n_e \sim 2 \times 10^{19}\,\mathrm{m}^{-3}$. The diagnostic team of the Culham Laboratory (England) confirmed this data by the ruby laser Thomson scattering measurement. The name for their early machine, the tokamak, has now become the generic name for all such devices. After the conference, the C-Stellarator at the Princeton Plasma Physics Laboratory (USA) was modified to the ST-tokamak. The Russian success led to a rapid expansion of tokamak research and many tokamaks were constructed in the US, Europe, and Japan. The total number of tokamaks now exceeds one hundred. Russian tokamak research has been continued but has gradually lost its leading position.

The success of the T-3 tokamak may be attributed to the development of high vacuum chambers and the improvement of equilibrium control using vertical magnetic fields whose theoretical analyses were also pursued during the 1960s. In the 1970s progress was made in macroscopic stability, transport and scaling law, additional heating, and impurity control. These items are explained briefly.

There are three major forms of macroscopic disturbances; disruptive instability, Mirnov oscillations and internal disruption (Sect. 11.6). The disruptive instability is an abrupt and generally unpredictable expansion of the plasma column accompanied by a large negative spike of the loop voltage around the plasma column, which is the voltage driving the plasma current. Minor disruptions may occur repeatedly in a given discharge, whereas a major disruption generally terminates the current. The disruptive instability limits both the plasma current and the attainable density for given discharge conditions.

Under normal tokamak operating conditions without disruptive instabilities, small helical perturbations in the poloidal magnetic field are usually observed at the plasma edge. These are the Mirnov oscillations whose frequencies are in the range of 10–30 kH$_z$. As the Mirnov oscillations become larger, the energy confinement time drops indicating that the helical perturbations increase the rate at which energy is transported from the central region to the edge of the plasma column.

The internal disruptions were first observed in measurements of soft X-ray emission in the ST-tokamak and later in other tokamaks. They appear as a relaxation or as sawtooth-type oscillations in the plasma temperature. The effects of the internal disruptions are concentrated near the center of the discharge, and these effects on the overall confinement are usually modest in tokamaks using ohmic heating. However, sometimes large scale sawtooth oscillations are observed at high power levels in tokamaks with additional heating. According to MHD theory, the plasma can be unstable to helical perturbations or kink and tearing modes when their pitch angle is the same as that of the magnetic field line or when a resonant surface exists. This can occur when the safety factor q is equal to m/n. If the safety factor drops below unity in the central region, the plasma becomes unstable to the $m = 1/n = 1$ mode. This is related to the internal disruption (Sect. 11.6).

Even if $q > 1$ everywhere, instabilities of higher mode numbers can set in, provided that the resonant surface lies in a region of the plasma with finite conductivity. In this case the tearing modes grow, and the magnetic field lines break and reconnect to form magnetic islands around the resonant surface (Sect. 11.2). Because of rapid transport along the field lines (Sect. 13.6), the main effect of the magnetic islands is to short circuit the transport across the islands, resulting in the deterioration of plasma confinement.

In the neoclassical theory discussed in Sect.13.2, collisional transport in a uniform magnetic field is modified by particle orbits specific to the non-uniform tokamak field geometry. A key question is how close experimentally observed values of ion and electron energy transport are to the value predicted by the neoclassical theory. Many tokamak experiments have studied ion energy loss and their results generally lie within a factor of 2 to 5 of the neoclassical theory. Although these results are roughly consistent with the neoclassical theory this does not mean the electron transport is also neoclassical. The neoclassical theory predicts that the energy loss via electrons is smaller by about the square root of the mass ratio $\sqrt{m_e/m_i}$ than the thermal conduction due to ions (Sect. 13.2). In practice, however, the opposite has been seen in most tokamaks. Experimentally observed electron thermal transport is larger than the neoclassical predictions by a factor ranging from 10 to 500.

In view of the theoretical uncertainties in making reliable predictions of energy loss, tokamak physicists have turned to empirical scaling laws to extrapolate from the established experimental data to yet unexplored regimes. Because the ranges of parameter variation available on individual tokamaks are limited, derivation of scaling laws involves comparison of the data from various tokamaks. Such scaling laws are particularly important with regard to electron energy transport, where the underlying physical mechanism of the anomalous transport has not yet been completely identified (Sects. 13.5–7). The important scaling laws are those that relate electron energy confinement time to parameters such as the density, the temperature, or the size of the devices. The gross energy confinement time for ions plus electrons τ_E increases linearly with density by the neo-Alcator scaling law

$$\tau_E \propto naR^2 , \qquad (14.1.1)$$

which was established in ohmic heated plasmas with medium density. The previously accepted scaling law was $\tau_E \propto na^2$, called the Alcator scaling. These scaling laws indicate that better confinement should occur with higher density; however, due to the disruptive instability or increase of radiation loss the magnitude of the density is limited. The empirical density limit is called the *Murakami limit*.

Controlling impurities in plasmas has been a significant technical concern in recent tokamaks. Low power discharge cleaning has been employed to condition the chamber wall surfaces and to pump away low-Z impurities. The low power of the discharge is intended to dissociate hydrogen while minimizing dissociation of water vapor and hydrocarbons. A thin film of titanium is deposited to bury impurities adsorbed on the chamber wall and to enhance the adsorption of hydrogen or deuterium emerging from the boundary of the plasma. This procedure is called the titanium gettering. The high-Z impurities have been reduced from the discharge volume by selecting low-Z limiter materials such as carbon or stainless steel. As another technique for controlling particle supply, "gas puffing" is being used in many tokamaks. This involves injecting additional gas into the vacuum chamber to fuel the plasma after a discharge has been started in a low density gas. The edge cooling that accompanies gas puffing appears to be effective in reducing sputtering and plasma-limiter interaction. A recently developed method calls for the injection of a small solid pellet composed of hydrogen or deuterium at velocities of 500 to 1500 m/s to increase the density. This pellet injection may be useful for profile control of density and temperatures. Some tokamaks have been equipped with more active devices for controlling impurities called divertors.

At the present time, the most successful, best understood, and most widely used auxiliary heating method is neutral beam injection (NBI) of atomic hydrogen isotopes. These are produced by the acceleration of ions into a charge exchange region where they are neutralized by electron attachment from unaccelerated atoms. The technology of producing intense neutral beams is evolving rapidly. Systems are now available that produce megawatts of neutral power at particle energies of 50 to 100 keV.

Several successful auxiliary heating studies have been carried out using radio-frequency microwaves in the ion cyclotron frequency range (tens of MH_z) and lower hybrid frequencies (hundreds of MH_z). Recent development of high power gyrotrons (about 200 kW) operating at 30 to 75 GH_z also makes electron cyclotron heating a possible auxiliary heating method. The production and transport of multimegawatts of RF power in the ion cyclotron and lower hybrid frequency ranges make these very promising.

In 1978, the PLT tokamak confined a plasma with $T_i \approx 7$ keV by neutral beam heating. Subsequently, the D-III tokamak reached an average β of 4.5% and the PBX tokamak improved this to $\bar{\beta} \approx 5.3\%$, where $\bar{\beta}$ is the β value (10.2.12) averaged over the plasma volume. This value is consistent with the Troyon limit (11.4.15) with the coefficient $C\mu_0 = 2.5$ in the units of I [MA], a [m] and B_T [T], as shown in Fig. 14.2. The PDX tokamak equipped with the divertor configuration was built at the Princeton Plasma Physics Laboratory as well as the PLT and PBX tokamaks [14.1, 2]. Recently, the D-III-D tokamak attained $\bar{\beta} \approx 11\%$ [14.3, 4]. These results confirm the advantage of noncircular tokamaks to improve the $\bar{\beta}$ limit. The Alcator C tokamak reached $n\tau_E \approx 3 \times 10^{19}$ m^{-3}s for high-density plasmas [14.5]. These values are almost consistent with the Lawson criterion when they are obtained simultaneously.

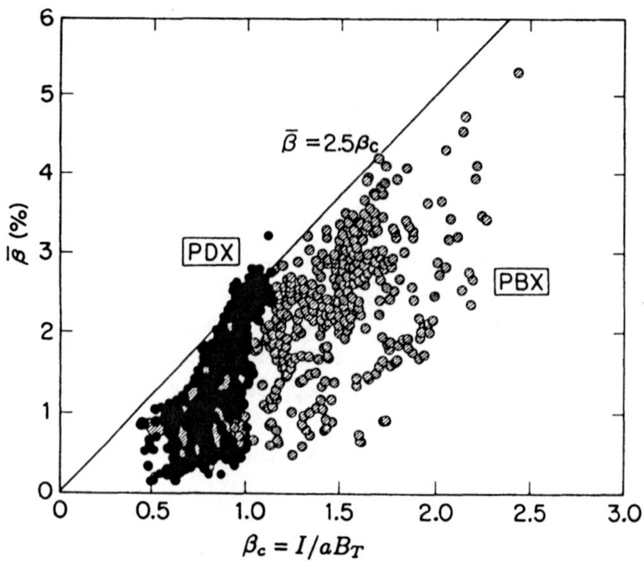

Fig. 14.2. Experimental β versus $\beta_c = I\,[\mathrm{MA}]/a\,[\mathrm{m}]\;B_T\,[\mathrm{T}]$ for PDX and PBX tokamaks. The line corresponding to $C\mu_0 = 2.5$ shows the Toroyon limit

When a high-power Neutral-Beam Injection (NBI) is used in tokamaks without divertors or with limiters for defining the plasma boundary, τ_E usually degrades through a phenomena called the *L mode*. In contrast, in tokamaks with divertors it has been found that the plasma confinement time does not decrease significantly with high-power neutral-beam heating, although the confinement-time scaling $\tau_E \propto I$ is different from the neo-Alcator scaling. This operating mode is called the *H mode* by the ASDEX tokamak group [14.6, 7]. In spite of the accumulation of experimental data, the origin of the anomalous transport in tokamaks is not completely understood. Experimental data analysis showed that the electron temperature profile in the *L mode* is resistant to changes in the power deposition profiles and the plasma density. This phenomena is called the *'profile consistency'*.

In order to overcome the pessimistic confinement behavior of the *L mode*, many tokamak experiments were concentrated on *H mode* discharges. These gave a confinement improvement with an H factor about 2 – 2.5, where the H factor is the ratio of τ_E between the L and *H modes*.

Recently, the energy confinement time of the *L mode* was evaluated with a regression analysis as the empirical scaling, called the *L mode* ITER-89P [14.8]:

$$\tau_E^{\text{ITER-89P}}[\mathrm{s}] = 0.048\; a^{0.8}\; R^{1.2}\; B^{0.2}\; n^{0.1}\; P_h^{-0.5}\; I^{0.85}\; \kappa^{0.5}\; A^{0.5}\;,$$

where $a\,[\mathrm{m}]$ is the minor radius in the horizontal direction, $R\,[\mathrm{m}]$ is the major radius, $B\,[\mathrm{T}]$ is the magnitude of the magnetic field, $n\;[10^{20}\,\mathrm{m}^{-3}]$ is the average density, $P_h\,[\mathrm{MW}]$ is the heating power, $I\,[\mathrm{MA}]$ is the total plasma current, κ is the ellipticity of magnetic surface near the separatrix, and A is the ion atomic number ($A = 1$ for a proton and $A = 2$ for a deuteron). In many tokamaks *H mode* plasmas were obtained; however, it is difficult to maintain the *H mode* with a *H* factor comparable to 2, since the plasma density increases continuously due to the improved particle confinement behind the *L-H transition*. One characteristic behavior of the *H mode* is the formation of sharp density and temperature gradients at the edge plasma region, i.e. the *edge transport barrier* associated with a sheared poloidal flow due to the $\boldsymbol{E} \times \boldsymbol{B}$ drift. Increase of the pressure gradient above a threshold value degrades the plasma stability. When the particle and energy losses across the separatrix to the divertor are enhanced by the pressure-driven *edge localized modes* (ELMs) intermittently, the *H mode* can be sustained. This type of confinement mode is called the ELMy *H mode*. It is shown that the ELMy *H mode* plasmas are governed by the *gyro-Bohm* transport given by χ_e^{GB}(electron thermal diffusivity) $= (\rho_s/a)\chi_e^{\mathrm{B}}$, where $\chi_e^{\mathrm{B}} \propto T_e/eB$ is the *Bohm* diffusion coefficient (Sect. 13.4), and ρ_s is the ion Larmor radius calculated with the electron temperature T_e [14.9]. It is suggested that the drift-wave turbulence in a sheared magnetic field is responsible for the *gyro-Bohm* transport.

The *H mode* is obtained, when P_h exceeds a threshold value given by the empirical scaling [14.10]

$$P_{\mathrm{th}}^{\mathrm{H}}\,[\mathrm{MW}] = 0.45\, n^{0.75}\, B\, R^2 \;,$$

where n [$10^{20}\,\mathrm{m}^{-3}$], B [T] and R [m]. For establishing a scaling of $P_{\mathrm{th}}^{\mathrm{H}}$, it is necessary to study a physical model for the *L-H transition* [14.11, 12]. It is generally easier to realize the *H mode* in deuterium plasmas than hydrogen plasmas [14.13]. It has recently been shown that the *H mode* in JET can be obtained with a lower threshold power for the D-T plasmas than the deuterium plasmas [14.14].

The current drive using electromagnetic waves has been studied intensively (Sect.12.3 and Appendix A.15) in the pursuit of a steady-state tokamak. In particular, lower-hybrid-current drives are useful in the low-density plasma regime. In the TRIAM-1M tokamak with super-conducting toroidal coils [14.15], the longest discharge of more than 1 hour was realized. It was also shown that the momentum input due to NBI makes it possible to drive a plasma current. An interesting idea is the use of a bootstrap current or diffusion-driven current, proportional to the plasma pressure P. If the bootstrap current is produced in reactor-relevant plasmas, steady-state operation may become easier, and the requirement for a RF-current drive may be reduced. From this point of view high-β_p discharges that exhibit a substantial confinement improvement are attractive, where $\beta_p = \int P dV/(B_p^2/2\mu_0)$ and B_p is the poloidal magnetic field at $r = a$.

Another interesting discharge revealing a confinement improvement appears in the *negative* (or *enhanced reverse*) *shear* configuration [14.16, 17]. The q profile has a minimum at the outer half-radius region, q being the safety factor. The region between $q(0)$ and $q_{\min}$ has a negative shear, since $q(0) > q_{\min}$. Here, the *internal transport barrier* (ITB) is formed near the $q_{\min}$ surface. The sheared plasma flow is also essential for obtaining ITB. A confinement improvement with an H factor comparable to 3 is possible. It is expected that the negative shear configuration is sustained continuously by both the bootstrap current and an additional current generated by the current drive. However, empirical scaling of the energy-confinement time has not yet been clarified; it is left as a near-term subject.

Recent experiments on large tokamaks TFTR (USA), JET (EC) and JT-60 (Japan) have yielded data which exceed parameters expected by the designers. D-T plasma experiments were already started at JET and TFTR. A-high-fusion-power plasma at JET recorded $T_i(0) \approx 23\,\mathrm{keV}$, $T_e(0) \approx 14\,\mathrm{keV}$ and $n_e(0) \approx 3.8 \times 10^{19}\,\mathrm{m}^{-3}$, which gave a fusion power of 13 MW ($> 10\,\mathrm{MW}$ for 0.5 s). The ratio of the fusion power to the heating is $Q \simeq 0.6$, and the energy confinement time reached to $\tau_E \simeq 0.9\,\mathrm{s}$.

It should be pointed out that α-driven MHD instabilities may take place in D-T tokamak reactors when the pressure of α particles becomes substantial. The dangerous instability is the Toroidicity-induced Alfvén Eigenmode (TAE) (Appendix A.17). Recently, in NBI- or ICRF-heated deuterium plasmas in large tokamaks, the existence of TAEs was shown, and the TAE-induced loss of high-energy ions was observed [14.18, 19]. For D-T plasmas in TFTR and JET, attempts were made to observe α-driven TAEs [14.20]; however, weak TAEs were detected only for a transient phase with slowing-down α's after the external heating power was shut down.

A new direction of tokamak research is the *Low-Aspect-Ratio Tokamak* (LART) with $R/a < 2$ [14.21]. Theoretically, the β limit set by the ballooning instability becomes higher than 10% for the LART (Sect. 11.4). The expression for the Troyon limit (11.4.15) also suggests a higher β for a lower R/a, since $\mu_0 I/aB_T \approx (a/R)(1/q)$. Recently, the small spherical tokamak START attained a high-β plasma with $\bar{\beta} \gtrsim 30\%$ for an neutral-beam-heated plasma. Two medium-size low-aspect-ratio tokamaks, MAST (UK) and NSTX (USA), are under construction.

14.2 Progress in Other Magnetic Confinement Research

In the 40-year history of magnetic confinement research, many types of magnetic configurations were proposed and many small devices were constructed to prove their principal ideas. From the initial stages, there were two different approaches to confining high temperature plasmas. One is the toroidal plasma confinement system and the other is the linear or mirror-based system. The former can be divided into axisymmetric and nonaxisymmetric systems and the essential difference comes from the requirement of a toroidal plasma current to maintain MHD equilibrium. In addition to tokamaks, reversed field pinch (RFP) and spheromaks belong to the axisymmetric systems (Chap. 9). Spheromaks can confine almost spherical plasmas or toroidal plasmas of extremely small inverse aspect ratio $R/a \sim 1$. From the topological point of view spheromaks are essentially the same as a linear system, since there is no conductor at the center of the toroidal plasma ring. This gives an advantage to spheromaks from the point of view of reactor design.

The stellarator/heliotron/torsatron devices are representatives of nonaxisymmetric systems. The stellarator-type device was the first one proposed, and experimental research was started in the mid-1950s. Since the theoretical studies of equilibrium, stability and transport for nonaxisymmetric systems were so difficult at the time, and because the production of a toroidal plasma without a net toroidal-plasma current requires an efficient wave-heating method, stellarator research lagged behind work on tokamaks. Around 1982, high-density toroidal plasmas without a net toroidal current were produced successfully in the W VII-A stellarator and the Heliotron E [14.22].

The term "stellarator" is employed to describe toroidal-confinement devices that produce closed-flux surfaces entirely by means of external magnetic fields. The classical stellarator has ℓ pairs of helical conductors with alternative antiparallel current flow that produces a poloidal field in the plasma. The net toroidal field in the plasma from these helical conductors is nullified, so that a set of toroidal field coils is also required.

The heliotron/torsatron configurations have ℓ helical conductors with a parallel current flow. These helical windings also provide the toroidal field. In general, a separate set of coils is required to provide a vertical field to form closed magnetic surfaces. Stellarators have a promising, but limited experimental basis in comparison to tokamaks. A recent topic is that of the *Large Helical Device* (LHD) which started its operation in 1998 at National Institute for Fusion Science (Toki, Japan). LHD has the device parameters: R (major radius) = 3.9 m, $\bar{a}$ (average minor radius) = 0.55 m, and B (at the center of plasma column) = 3 T produced by $\ell = 2$ helical coils with a pitch number of 10. The main object of LHD is to study confinement properties of currentless toroidal plasmas in the temperature range of (5–10) keV. If the energy-confinement time follows the LHD scaling derived for the design of LHD [14.23, 24], $\tau_E \approx 0.1$ s will be expected, where

$$\tau_E^{\mathrm{LHD}}[\mathrm{s}] = 0.17\, R^{0.75}\, a^2\, n^{0.69}\, B^{0.84}\, P_h^{-0.58} \left(\frac{A}{1.5}\right)^{0.5} .$$

Here, the units of R, a, n, B and P_h are the same as in the ITER-89P scaling. If this scaling is considered as $\tau_E \propto n^{0.6} B^{0.8} P_h^{-0.6}$, it is consistent with the *gyro-Bohm* transport discussed in the case of the ELMy *H mode*. However, for designing a compact fusion reactor based on the heliotron configuration, an improved confinement is required with about a factor of two compared to the LHD scaling.

The concept of an advanced stellarator has been developed in 1980s. The W7-AS stellarator was designed to reduce the Pfirsch-Schlüter current (Sect. 13.2), which corresponds to reducing the neoclassical transport in a stellarator (Sect. 13.3) [14.25]. This was successfully demonstrated with the reduction of the Shafranov shift of the magnetic axis (Sect. 10.2). Recently, W7-AS experiments also showed the *L-H transition* and a *H mode* similar to those in tokamaks, although the *H* factor has only been 1.2–1.3 [14.26]. The confinement improvement with a H factor of about two was obtained for the hot-ion mode.

A highly optimized stellarator configuration using modular coils was developed at the Max-Planck Institute for Plasma Physics (Germany), which is called helias [14.27]. As an experimental device based on the concept of the helias, the W7-X stellarator was designed and the construction was started in 1997 [14.28]. W7-X has the device parameters: $R = 5.5\,\mathrm{m}$, $\bar{a} = 0.5\,\mathrm{m}$, and $B = 3\,\mathrm{T}$ with superconducting modular coils. The target of the plasma parameters are similar to that of LHD. It is expected that a high-β plasma with $\bar{\beta} \simeq 5\%$ can be confined with good high-energy-particle confinement and reduced neoclassical ripple transport (Sect. 13.3). This configuration was found by selecting the magnetic spectrum in the specific magnetic coordinates called Boozer coordinates. The main components are $\ell = 1$ helical field, $\ell = 0$ bumpy field, and the toroidal correction component. In the W7-X or helias configuration, the toroidal effect is reduced in the Boozer coordinates compared to the geometrical aspect ratio, which contributes to the improvement of the high-energy-ion confinement and a reduction of the neoclassical ripple transport. For stable confinement of $\bar{\beta} \simeq 5\%$ plasmas, a sufficient magnetic well is generated, although the Shafranov shift is smaller than that in W7-AS. An interesting property is that the high-energy-particle confinement is improved with an increase of the plasma pressure or $\bar{\beta}$.

After the freedom to select the magnetic spectrum in the Boozer coordinates had widely been recognized, several interesting stellarator configurations were proposed. One is the quasi-helically symmetric stellarator with the dominant $\ell = 1$ helical field. The other is the quasi-axisymmetric stellarator with substantially smaller helical components compared to the toroidal correction component [14.29]. This is considered as a weakly non-axisymmetric tokamak without net toroidal plasma current. Experimental results for these configurations, in addition to the W7-X, will be available at the beginning of next century.

The RFP is an axisymmetric, toroidal confinement concept that is like the tokamak in many respects, but different in several essential ways. The plasma is confined in both systems by a combination of poloidal B_θ and toroidal B_ϕ magnetic fields. In both configurations the poloidal field is created by a toroidal plasma current I_ϕ which has been induced by transformer action, and the toroidal field is provided primarily by external coils. Both designs require an external vertical field to achieve force balance equilibrium.

The RFP differs from the tokamak primarily in that strongly sheared magnetic fields in the outer plasma region are used to suppress local MHD instabilities (Sect. 9.2). The highly sheared fields are created by reversal of the toroidal field in the outer plasma region. The shear stabilization of the RFP can be understood most readily by using the Suydam criterion for the necessary condition for local stability of a linear cylindrical pinch. In an RFP the magnetic field shear vanishes near the center and (10.3.36) must be satisfied by an off-axis peaking of the pressure profile leading to $dP/dr > 0$. Near the outer edge of the discharge $dP/dr < 0$ and a large value of shear or $|dq/dr|q^{-1}$ is required for stabilization of local modes.

For stabilization of globally unstable MHD modes with wavelength longer than the chamber radius, a conducting shell or a set of closely fitting external conductors are required in the RFP. This is a disadvantage for the RFP as a reactor concept relative to the tokamak.

The differing stability requirements in an RFP and a tokamak lead to relaxed engineering requirements which make the RFP overall more attractive. The stability constraint of $q > 1$ everywhere and $q(a) > 2 \sim 3$ in a tokamak is replaced by $dq/dr < 0$ everywhere in an RFP. The RFP can operate with $q < 1$ and this increases the achievable value of B_θ or I_ϕ for a fixed B_ϕ by a factor of about 3 over that in a tokamak. This situation implies that the toroidal field strength requirements for an RFP could be more modest than for a tokamak and that it may be possible to ohmically heat an RFP plasma to ignition, which is not the case for a tokamak.

The greatest theoretical uncertainty in RFP design had been the toroidal field reversal itself. That self-reversal can occur was not in question; the reversed field pinch state has been observed in many experiments. However, relaxation mechanisms and associated energy loss for setting up RFP are not well understood. But recently, with the aid of computer simulation, the relaxation mechanism has been disclosed (Sect. 11.8).

In the *simple mirror* configuration, conservation of kinetic energy requires

$$\tfrac{1}{2}m\left[v_{\|}^2(l) + v_{\perp}^2(l)\right] = E\ , \tag{14.2.1}$$

where l is the direction along the magnetic field line, since the stationary magnetic field does not do work on a charged particle. Conservation of the magnetic moment (3.2.13) yields

$$\tfrac{1}{2}mv_{\|}^2(l) = E - \mu B(l)\ . \tag{14.2.2}$$

This indicates that particles for which $E/\mu = B(l)$ for $B_{\text{min}} < B(l) < B_{\text{max}}$ will be trapped. Evaluating the constant E and μ from the relationship at $l = l_0$ corresponding to the position of B_{min} yields

$$\frac{1}{2}mv_{\|}^2(l) = \left[\frac{1}{2}mv_{\|}^2(l_0) + \frac{1}{2}mv_{\perp}^2(l_0)\right] - \frac{1}{2}mv_{\perp}^2(l_0)\frac{B(l)}{B_{\text{min}}}\ . \tag{14.2.3}$$

Thus, the condition for a particle to be trapped depends upon $v_{\perp}(l_0)/v_{\|}(l_0)$ and $B(l)/B_{\text{min}}$. The boundary in the velocity space $(v_{\|}, v_{\perp})$ between trapped and untrapped particles can be determined by evaluating the above equation for $v_{\|}(l_{\text{max}}) = 0$,

$$v_{\perp}(l_0) = \pm v_{\|}(l_0)\left(\frac{B_{\text{max}}}{B_{\text{min}}} - 1\right)^{-1/2}\ . \tag{14.2.4}$$

Noting that $v_{\perp}$ is the two-dimensional velocity component in the plane perpendicular to the magnetic field, (14.2.4) defines a cone, as shown in Fig. 9.6.

Particles with $v_{\perp}(l_0)/v_{\|}(l_0)$ which fall within the loss cone are lost immediately. Other particles can be scattered into the loss cone and are then lost. Thus, the confinement time is proportional to the time required for a particle to scatter into the loss cone. The time is related to the 90° deflection time τ:

$$\tau_P = \tau \log_{10}\left(\frac{B_{\mathrm{max}}}{B_{\mathrm{min}}}\right) . \tag{14.2.5}$$

Since $\tau_{\mathrm{ii}} \doteq \sqrt{m_{\mathrm{i}}/m_{\mathrm{e}}}\ \tau_{\mathrm{ee}}$ where τ_{ii} denotes the ion-ion collision time and τ_{ee} denotes the electron-electron collsion time, the electrons scatter into the loss cone and escape much faster than the ions. This creates a positive net charge in the plasma which results in a positive electrostatic potential that acts to confine electrons in the loss cone until ions can escape. Thus, the particle confinement time for ions and electrons in a simple mirror is proportional to the ion-ion 90° deflection time

$$\tau_P \doteq \tau_{\mathrm{ii}} \log_{10}\left(\frac{B_{\mathrm{max}}}{B_{\mathrm{min}}}\right) . \tag{14.2.6}$$

This confinement time is not enough for fusion to be able to occur in the reactor.

The simple mirror is unstable against flute instabilities (Sect. 9.2). They can be suppressed if the field increases, rather than decreases away from the plasma, that is, if the plasma is confined in a three-dimensional magnetic well. This is called the minimum-B stabilization of mirror plasma (Sect. 9.2). The principles of confinement in a *minimum-B mirror* are essentially the same as those described above for the simple mirror, namely, confinement is governed by ion-ion scattering into the loss cone. The estimate of the confinement time by (14.2.6) is still valid. Even the most favorable performance for a minimum-B mirror leads to a plasma power amplification factor $Q \doteq 1 \sim 2$ (Sect. 9.1). Because Q >10–15 is required for positive power balance in a reactor, the prospects of a minimum-B mirror configuration for electricity production are poor.

The most promising use of minimum B mirrors, however, is as end plugs to confine ions electrostatically in a central solenoidal cell. The basic idea of the *tandem mirror*, illustrated in Fig. 9.7, is to create a potential difference ϕ_{i} between the end plug and the central cell by creating a density difference. According to the Boltzmann distribution

$$n_{\mathrm{e}}(l) = n_0 \exp\left(\frac{e\phi}{T_{\mathrm{e}}}\right) . \tag{14.2.7}$$

Using the subscripts p and c to refer to the end plug and the central cell, respectively, this relation can be used to obtain

$$e\phi_{\mathrm{i}} = e(\phi_{\mathrm{p}} - \phi_{\mathrm{c}}) = T_{\mathrm{e}}\ln\left(\frac{n_{\mathrm{p}}}{n_{\mathrm{c}}}\right) = T_{\mathrm{e}}\ln\left(\frac{\beta_{\mathrm{p}}B_{\mathrm{p}}^2}{\beta_{\mathrm{c}}B_{\mathrm{c}}^2}\right) , \tag{14.2.8}$$

where β is defined as in (10.2.12) and the temperature is assumed to be constant. The basic tandem mirror concept has already been confirmed experimentally.

Since most of the plasma volume in the tandem mirror is in the central cell, the power density is proportional to $n^2\langle\sigma v\rangle$ which in turn is proportional to $(\beta_{\mathrm{c}}B_{\mathrm{c}}^2)^2$. Thus in order to achieve a large confining potential and an adequate

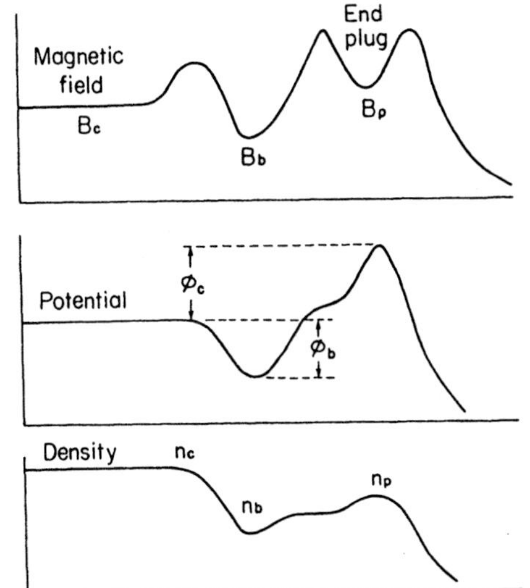

Fig. 14.3. Magnetic field, potential, and density in the thermal barrier of a tandem mirror: B_c-central cell magnetic field; B_b-barrier magnetic field; B_p-plug magnetic field; ϕ_c-central cell potential; ϕ_b-barrier potential; n_c-central cell density; n_b-barrier density; n_p-plug density

power density, it is necessary to have a large $\beta_p B_p^2$. Calculations indicate that magnetic fields in excess of 20 T may be required in the end plugs. Penetration of the neutral beam to heat the dense end plug plasmas would require neutral beam energies on the order of 1 MeV. These place severe demands on the technology.

A reduction of technology requirements can be achieved, in principle, by the addition of so-called *thermal barriers* shown in Fig. 14.3. It is produced by adding a single mirror at each end of the solenoid, partially isolating the end plugs from the solenoid. The intention is to thermally insulate hotter electrons in the end plug from the relatively colder electrons in the central cell. If a higher electron temperature can be sustained in the plug by intense electron heating (e.g., with microwaves at the electron cyclotron frequency as discussed in Sect. 12.2), then the potential barriers necessary to confine the ions escaping from the solenoid can be generated with a much lower density n_p in the end plugs. A large reduction in the plug density reduces the power consumed in the plugs and opens up options for reasonable magnets and neutral beam technology.

Another problem in the tandem mirror is the increased particle loss due to the nonaxisymmetric magnetic field, which is similar to the transport in nonaxisymmetric toroidal plasma explained in Sect. 13.3. The Gamma-10 device was designed to improve the particle loss by reducing the asymmetry in the magnetic field [14.30].

14.3 Progress in Inertial-Confinement Research

The basic concept of the Inertial Confinement Fusion (ICF) was described in Sect. 9.3. It is based on the compression of the fuel to a very high density, namely 10^3–10^4 times the solid density, such that a positive energy output can be gained by the fusion reaction within a short time less than the time needed for the fuel particles fly apart. Thus, ICF is based on a concept essentially different from the magnetic confinement fusion. As mentioned in Sect. 9.3, there are two fundamental problems which must be overcome in order for ICF to succeed: one is to develop a high power-density energy driver with a relatively high gain efficiency and reliability, and the other is to achieve a compression of the fuel to the required density levels. These two problems can, to a certain extent, be studied independently of each other, apart from a minor difference in the compression methods which depend on the choice of the driver. The most widely used high power energy driver at present is the neodymium-glass laser. Although its efficiency is considered to be too low ($< 1\%$) to be useful as a driver for the actual ICF reactor, it is capable of producing high-power densities in a small target (pellet) region. These lasers have been used for studying the physics associated with the high compression of the pellet material with great success. We therefore start with a description of the recent progress in the research of the fuel compression and ignition.

Compression

In 1988, the GEKKO XII laser system at Osaka University (Japan) has achieved a compression of the fuel target up to 600 times the solid density [14.31], by reducing the non-uniformity of the laser illumination and by employing a highly spherical pellet as the target [14.32]. In the density measurement of the compressed core, novel diagnostic techniques applying nuclear reaction products were employed [14.33]. The results were confirmed by US ICF visiting research group in 1990 and produced a great impact on the following ICF research activities.

According to the 1-D code simulation [14.34], the central part of the D-T fuel, when compressed up to such a high density, must be heated to a high temperature (a few keV) and a substantial amount of neutron yield as fusion products should be obtained. However, the observed neutron yield is orders of magnitude less than that expected from the 1-D simulation. This implies that the core compression would not be spherically symmetric. One can immediately infer a Rayleigh-Taylor instability at the acceleration and deceleration (stagnation) phase of the implosion, which causes a mixing of the compressed core with the cold fuel outside the compressed core. In order to investigate this possibility, an X-ray imaging diagnostic technique of high spatial ($\approx 10\,\mu$m) and temporal (≈ 10 ps) resolution was developed, and the result clearly showed an irregular compression of the core, in qualitative agreement with the prediction of 2-D simulation [14.35].

The Osaka group has made substantial efforts to improve the implosion symmetry, such as the use of amplified spontaneous emission, partially coherent light and an improved power balance of twelve laser beams etc., and some improvement was achieved for the neutron yield which is still far below the values expected from the onset of ignition [14.36].

It is mainly due to the illumination asymmetry caused by a restricted number of laser beams (twelve for the case of GEKKO XII). An analysis indicated the necessity of more than 100 beams to realize the high spherical symmetry sufficient for ignition.

NIF

In the US, a new project called *National Ignition Facility* (NIF) was proposed [14.37]. The construction started in 1997 and its full operation is expected to begin in 2003. The NIF-laser system consists of 192 high-power laser beams with 1.8 MJ of beam energy in the third harmonics of the neodymium glass laser, and both direct and indirect drive targets will be used to demonstrate the high compression and ignition with more than 20 MJ fusion energy output. A similar project was approved in France.

Fast Ignition Method

As an alternative approach to ignition, a new concept of fast ignition of a compressed core using an ultrashort high-intensity laser was proposed [14.38]. Such a method has become realistic since the discovery of the *Chirped-Pulse Amplification* (CPA) of laser light [14.39], which made it possible to compress the laser light to an ultrashort (≈ 1 ps) pulse of very high intensity (> 100 TW). The idea is to first compress the central core to a density sufficient for trapping of α-particles by the existing technique and then heat the core by an additional application of an ultrashort laser pulse of high intensity. In order for this method to work successfully, the picosecond high-intensity laser pulse has to penetrate deeply into the high-density region, say by laser drilling of the high-density plasma and/or by relativistic effects, and produce high-energy electron beams to deposit energy on the compressed core to heat it up to the ignition temperature. Development of an ultrashort high-intensity laser pulse is now in progress in the US, Japan and EU, and some preliminary experiments are going on for laser drilling and laser-plasma interaction in over-critical density plasma. Various types of parametric instabilities are expected to take place by such a high-intensity laser in an intricate manner. For instance, the relativistic electron motion induced by the high electromagnetic field causes an electron mass oscillation and the resulting parametric instability of the electron plasma wave [14.40]. Stimulated Raman scattering, which can occur only in the low-density region of a density less than one quarter of the critical density in the case of relatively weak electromagnetic field, takes place at the density well above the critical density due to the relativistic electron motion, and is mixed with the modulational instability [14.41]. Harmonic generation of laser light due to the coupling with the electron plasma waves causes a generation of energetic electrons [14.42]. It is expected that all these effects may take place simultaneously and/or cooperatively. At even higher intensity of laser light, electron-positron pair creation can take place, which further complicates the phenomena. At the same time, laser-plasma interaction studies of such a high-intensity laser with ultrahigh-density plasmas will open a new frontier of scientific research in connection with astrophysical plasmas.

ICF Drivers

As for energy drivers that have the potential of becoming a candidate for the reactor use, the Nd:YAG solid-state laser pumped by laser diodes is receiving a strong interest and is under development to demonstrate the potential of high-intensity (4 MJ in 20 ns), high-efficiency (more than 10%) and high-repetition-rate (more than 10 pulses per second) lasers of extremely compact size. The solid-state laser, when developed for the reactor use, will be applicable to a direct drive compression because of its good focusing on a small target, but independently of the ICF use it will be developed for use in industry applications as well, and this will lead to a considerable cost reduction in the future. On the other hand, for the indirect drive compression, heavy-ion beams can be employed as the energy driver because of their high efficiency, repetition rate and reliability and are being pursued as high-efficiency and large-energy ($\approx$ 1 MJ) drivers.

14.4 Plasma Physics Toward Fusion Reactors

Many of the macroscopic properties of a plasma are governed by the power balance [14.43, 44]. The power balance is affected by a variety of phenomena such as transport, heating instabilities, and so on. The relative importance of these various phenomena varies from location to location within the plasma and from phase to phase over the plasma burn cycle.

The plasma burn cycle is usually initiated by filling the plasma chamber with a cool gas, with the fuel in molecular or atomic form. The plasma is formed by ionizing this filling gas with a voltage pulse or by heating.

Once the breakdown phase is completed and a fully ionized plasma is established, the atomic processes become relatively less important except in the boundary region, and bulk transport effects dominate the plasma behavior. If we treat the boundary region separately, we can write a set of particle and power balance equations for the plasma core that couple to the boundary region through continuity conditions of particle and heat fluxes at the interface between the core and boundary regions.

A set of global balance equations for averaged n and T are given by

$$\frac{dn_{\mathrm{i}}}{dt} = S_{\mathrm{i}} - \frac{1}{2} n_{\mathrm{i}}^2 \langle \sigma v \rangle - \frac{n_{\mathrm{i}}}{\tau_{\mathrm{i}}} , \tag{14.4.1}$$

$$\frac{d\eta_\alpha}{dt} = S_\alpha + \frac{1}{4} n_{\mathrm{i}}^2 \langle \sigma v \rangle - \frac{n_\alpha}{\tau_\alpha} , \tag{14.4.2}$$

$$\frac{dn_Z}{dt} = S_Z - \frac{n_Z}{\tau_Z} , \tag{14.4.3}$$

for fuel ions n_{i}, alpha particles n_α, and impurity ions n_Z with charge Z, respectively, and

$$\frac{3}{2} \frac{d}{dt} (n_{\mathrm{e}} T_{\mathrm{e}}) = P_{\mathrm{Ohm}} + P_{\mathrm{A}}^{\mathrm{e}} + \frac{1}{4} n_{\mathrm{i}}^2 \langle \sigma v \rangle U_{\alpha \mathrm{e}} - Q_{\mathrm{ie}} - P_{\mathrm{rad}} - \frac{3}{2} \frac{n_{\mathrm{e}} T_{\mathrm{e}}}{\tau_{\mathrm{E}}^{\mathrm{e}}} , \tag{14.4.4}$$

$$\frac{3}{2} \frac{d}{dt} (n_{\mathrm{i}} T_{\mathrm{i}}) = P_{\mathrm{A}}^{\mathrm{i}} + Q_{\mathrm{ie}} + \frac{1}{4} n_{\mathrm{i}}^2 \langle \sigma v \rangle U_{\alpha i} - \frac{3}{2} \frac{n_{\mathrm{i}} T_{\mathrm{i}}}{\tau_{\mathrm{E}}^{\mathrm{i}}} , \tag{14.4.5}$$

for the electron and ion temperatures T_{e} and T_{i} and other notations will be explained below. The electron density is given by $n_{\mathrm{e}} = n_{\mathrm{i}} + 2n_\alpha + Zn_Z$. In the fusion reaction term in (14.4.2) we assume $n_{\mathrm{D}} = n_{\mathrm{T}} = n_{\mathrm{i}}/2$, where $\langle \sigma v \rangle$ is the fusion reaction rate. The quantities in (14.4.1–5) are obtained by appropriate spatial averages. These equations are applicable to all types of magnetically confined plasmas. The specifics of the confinement configuration enter through the specification of the particle confinement times $(\tau_{\mathrm{i}}, \tau_\alpha, \tau_Z)$ and the energy confinement times $(\tau_{\mathrm{E}}^{\mathrm{e}}, \tau_{\mathrm{E}}^{\mathrm{i}})$.

The particle sources $(S_{\mathrm{i}}, S_\alpha$, and $S_Z)$ depend on the particle and heat fluxes leaving the core and on the specifics of the boundary region, plasma chamber wall, and any impurity control mechanism. Any external fuel source such as pellet or neutral beam injection into the core also contributes to S_{i}. A simple order-of-magnitude estimate of these sources can be made. When there is no impurity removal mechanism and the boundary region is sufficiently tenuous that charge exchange is unimportant, then

$$S_{\mathrm{i}} = R_{\mathrm{i}} \frac{n_{\mathrm{i}}}{\tau_{\mathrm{i}}} , \tag{14.4.6}$$

$$S_\alpha = R_\alpha \frac{n_\alpha}{\tau_\alpha} , \tag{14.4.7}$$

and

$$S_Z = (R_Z + Y_Z)\,\frac{n_Z}{\tau_Z} + Y_i\,\frac{n_i}{\tau_i} + Y_\alpha\,\frac{n_\alpha}{\tau_\alpha}\,, \tag{14.4.8}$$

where R and Y are the reflection coefficient and sputtering yield, respectively.

Equations (14.4.1–8) are useful for examining the dynamic power balance during a tokamak burn cycle. During the early startup phase, ohmic heating of the electrons P_{Ohm} is the dominant heating source, but this heating mechanism saturates when the electron temperature reaches about 1 keV. During ohmic heating the ions are heated only by the collisional energy transfer from electrons Q_{ie} except in a very low density regime. At this stage, about several tens of MW of additional heating power to electrons $P_{\mathrm{A}}^{\mathrm{e}}$ and to ions $P_{\mathrm{A}}^{\mathrm{i}}$ are applied to the plasma to increase the temperature into the thermonuclear regime where alpha heating becomes dominant. Once a D-T plasma, or any part of it, is heated to several keV by external means, the fusion reaction becomes a significant self-heating mechanism. The fusion rate is $n_{\mathrm{D}}\, n_{\mathrm{T}}\langle\sigma v\rangle$. Each fusion event produces a 14.1 MeV neutron which immediately escapes from the plasma and a 3.52 MeV alpha particle which transfers its energy to the ions $U_{\alpha\mathrm{i}}$ and electrons $U_{\alpha\mathrm{e}}$ via Coulomb collisions.

The analysis of the alpha particle heating process is similar to that for neutral beam heating. The birth distribution of alpha particles is just the fusion distribution. The virgin alpha particles are isotropically distributed in velocity space. The fast alpha particle orbits can be calculated by using the guiding center equations (Chap. 3). Those orbits which leave the plasma confinement region can be considered lost. The fraction of lost alpha particles depends on the proximity of the point at which the alpha particle is born to the plasma surface and on the magnitude of the excursion Δr of the alpha particle from the flux surface on which it is born. In a tokamak, Δr is related to the Larmor radius measured with the poloidal magnetic field $\Delta r \propto 1/B_\theta(r)$, and $B_\theta(r)$ is related to the toroidal plasma current by Ampere's law (Sect. 13.1). Thus the loss fraction at a given radial location decreases as the total plasma current increases or as the plasma current density distribution becomes more centrally peaked. Since the loss of alpha particles implies the loss of the corresponding fraction of the alpha heating, one criterion for a self-sustaining tokamak fusion plasma is that it maintains enough plasma current to confine most of the alpha particles. When the current profile is roughly parabolic in a tokamak with an aspect ratio of about three, approximately I_{p} = 5 MA is sufficient to confine more than 90% of the alpha particles.

Alpha particles may be lost from the plasma before giving up their energy by other mechanisms. The discrete nature of the toroidal field coils in tokamaks introduces nonaxisymmetric toroidal field ripples, which lead to particle trapping and other particle orbit modifications that can enhance the loss of alpha particles (Sect. 13.3).

It is possible that the alpha particles may drive instabilities in the plasma that has an effect on enhancement of the alpha particle loss during the slowing down process, since the distribution function deviates from the Maxwellian distribution.

During a plasma burn, the number of fuel ions is depleted by the fusion reactions, which corresponds to the second term in (14.4.1). These fusion losses are generally anticipated to be only a few percent of the transport losses. The fuel ions that are transported out of the plasma may be reflected from the chamber wall and recycled into the center of the plasma. In this case, a small number (sufficient to offset the ions lost in fusion) of fuel ions will be supplied at the edge of the plasma; in this manner constant plasma fuel ion density could be maintained. The plasma density could also be built up by adding more fuel atoms than are necessary to replace the fuel ions lost in fusion. These phenomena of recycling and density build-up by gas injection have been experimentally observed.

Injection of pellets of a few millimeters in diameter or clusters of smaller pellets of fuel atoms is a means of fueling the center of the plasma directly. As the pellet moves into the plasma its surface ablates. The injection velocity and size of the pellet can be determined so that most of the pellet is ablated in the center of the plasma. For pellets on the order of a few millimeters, velocities on the order of 5 to 10 $\times 10^3$ m/s are required to penetrate the fusion plasmas.

The presence of a relatively small concentration of impurity ions can lead to large radiative power losses, P_{rad} in (14.4.4), that are sufficient to either prevent a plasma from being heated to thermonuclear temperatures or to maintain such temperatures without substantial supplemental heating. There are methods for studying impurity control in the plasma boundary region. Impurities created by deterioration of the chamber walls can be reduced by a dense neutral or partially ionized gas blanket in the boundary region which insulates the plasma from the wall by reducing the electron and ion temperatures. The energy of charge exchange neutrals that strike the wall also decreases below the sputtering threshold. It is possible to create magnetic field configurations in which all the field lines inside a given separatrix surface remain within the confinement volume and all, or some, of the field lines outside of the separatrix leave the confinement volume. Ions diffusing out to the plasma or returning from the wall are swept along the diverted field lines out of the plasma chamber into a divertor chamber. Impurity ions from the wall are likewise swept out of the plasma chamber before they can diffuse across the boundary region and enter the plasma. Heat flow along the diverted field lines is quite large compared to cross-field heat flow, so that the diverted field lines act to cool the plasma in the boundary region as well as to remove ions.

We now examine some of the problems that will be encountered in maintaining a steady state power balance in a thermonuclear plasma and consider some control mechanisms that can be used to achieve this objective. The problems fall into two categories, composition changes and dynamical instabilities. The desired overall power balance is given by

$$\frac{dW}{dt} = (P_\alpha + P_A) - (P_{rad} + P_{tr}) \,, \tag{14.4.9}$$

where $W = 3nkT$ is the plasma internal energy, the first two terms are the alpha P_α and supplemental power sources P_A and the last two terms are the radiation P_{rad} and transport power losses P_{tr}. A power balance is achieved when $dW/dt = 0$.

Assume that a power balance has been achieved at the beginning of a burn cycle, and consider the effects of the eroding wall particles and fusion alpha impurities upon the power balance. The most obvious effect is the increase in P_{rad}, which is usually the principal effect. If the maximum value of the average beta $\bar{\beta}$ is limited by MHD stability requirements, then the accumulation of impurity ions must be compensated by a reduction in the fuel ion concentration, so that $\bar{\beta} \propto (n_i + n_\alpha + n_Z + n_e)T$ remains constant. This reduces the alpha heating $P_\alpha \propto n_i^2 \langle\sigma v\rangle_f U_\alpha$.

Again, assume that a power balance has been achieved, and now consider the effect of a small temperature perturbation. A linear expansion of (14.4.9) about the power balance condition yields

$$\frac{dW}{dt} = \left[\left(\frac{\partial P_\alpha}{\partial T}\right)_T \Delta T + \left(\frac{\partial P_A}{\partial T}\right)_T \Delta T\right] - \left[\left(\frac{\partial P_{rad}}{\partial T}\right)_T \Delta T + \left(\frac{\partial P_{tr}}{\partial T}\right)_T \Delta T\right] . \tag{14.4.10}$$

If the right hand side of (14.4.10) is positive then the response to the perturbation is to increase the temperature perturbation. Here $(\;)_T$ means a quantity at a temperature T. Thus the power balance is unstable if

$$\left[\left(\frac{\partial P_\alpha}{\partial T}\right)_T + \left(\frac{\partial P_A}{\partial T}\right)_T\right] - \left[\left(\frac{\partial P_{rad}}{\partial T}\right)_T + \left(\frac{\partial P_{tr}}{\partial T}\right)_T\right] > 0 \,. \tag{14.4.11}$$

Because the fusion cross section increases sharply with temperature (up to about $T = 80\,\mathrm{keV}$ for D-T fusion), the alpha heating term is destabilizing. The supplemental power term may be zero for neutral beam injection since P_A depends very weakly on T, and may depend on the absorption mechanism for wave heating. The radiative loss term is generally positive and stabilizing for low-Z impurities which are fully stripped. However, it can be negative and destabilizing for high-Z impurities which are partially ionized. The transport loss term can be stabilizing or destabilizing depending on whether τ_E varies inversely or directly (e.g., according to the neoclassical theory with $\tau_E \propto T^{1/2}$ as shown in Sect. 13.2) with temperature.

When power balance is maintained by supplemental heating, the destabilizing P_α term in (14.4.11) is smaller relative to the other terms than when there is no supplemental heating. Thus a power balance that is maintained with supplemental heating is more stable than an ignition power balance ($P_\alpha = P_{rad} + P_{tr}$).

A. Appendix

A.I Problems to Part I Basic Theory

A.1. Ponderomotive Force in a Magnetic Field

Consider a particle of mass m and charge q in the presence of a uniform static magnetic field $\boldsymbol{B}_0$ and an oscillating electromagnetic field whose electric field $\boldsymbol{E}_1$ is given by $\boldsymbol{E}_1 = \boldsymbol{E}_1(\boldsymbol{r})\cos(\omega t)$ and magnetic field by $\boldsymbol{B}_1 = \boldsymbol{B}_1(\boldsymbol{r})\sin(\omega t)$ with $\boldsymbol{B}_1(\boldsymbol{r}) = -\omega\nabla \times \boldsymbol{E}_1(\boldsymbol{r})$. We assume that the spatial dependence of $\boldsymbol{E}_1(\boldsymbol{r})$ is sufficiently slow that its scale length is much larger than the excursion length $|q\boldsymbol{E}_1/(m\omega^2)|$.

(i) Show that if the spatial variation of $\boldsymbol{E}_1(\boldsymbol{r})$ is neglected the solution of the equation of motion for the part oscillating at frequency ω can be written as (Sect. 3.4.2)

$$\boldsymbol{v}_1(t) = \left[\frac{\Omega^2}{(\Omega^2-\omega^2)}\right]\left[\frac{\boldsymbol{E}_1 \times \boldsymbol{B}_0}{B_0^2}\right] + \left(\frac{q}{m}\right)\frac{\{\boldsymbol{B}_0 \times [(d\boldsymbol{E}_1/dt) \times \boldsymbol{B}_0]/B_0^2\}}{(\Omega^2-\omega^2)} - \frac{\boldsymbol{b}[\boldsymbol{b}\cdot d\boldsymbol{E}_1/dt]}{\omega^2} \tag{A.1}$$

$$\boldsymbol{r}_1(t) = \left[\frac{\Omega^2}{(\Omega^2-\omega^2)}\right]\{-[d\boldsymbol{E}_1/dt] \times \boldsymbol{B}_0/(\omega^2 B_0^2)\} + \left(\frac{q}{m}\right)\frac{\{\boldsymbol{B}_0 \times [(\boldsymbol{E}_1 \times \boldsymbol{B}_0]/B_0^2\}}{(\Omega^2-\omega^2)} - \frac{\boldsymbol{b}[\boldsymbol{b}\cdot \boldsymbol{E}_1]}{\omega^2}\ . \tag{A.2}$$

(ii) Approximating $\boldsymbol{E}_1(\boldsymbol{r}+\boldsymbol{r}_1)$ by $\boldsymbol{E}_1(\boldsymbol{r})+\boldsymbol{r}_1\cdot\nabla\boldsymbol{E}_1(\boldsymbol{r})$, show that the second/order average force (ponderomotive force)

$$\boldsymbol{F}_2 = q\overline{\boldsymbol{r}_1\cdot\nabla\boldsymbol{E}_1} + q\overline{\boldsymbol{v}_1 \times \boldsymbol{B}_1}$$

where the bar denotes the time average over the oscillating period, can be written as

$$\boldsymbol{F}_2 = q\nabla\overline{(\boldsymbol{r}_1\cdot\boldsymbol{E}_1)}/2\ . \tag{A.3}$$

Use the relation $\overline{(dA/dt)B} = \overline{-A(dB/dt)}$.

(iii) Substituting (A.1, 2) into (A.3), show that the ponderomotive force in the presence of a magnetic field changes its direction depending on the frequency of the oscillating field. Also discuss the effect of the cyclotron resonance. These properties can be used for the rf stabilization of the flute mode.

A.2. Weibel Instability

Consider a cold, uniform electron plasma imbedded in a uniform neutralizing positive charge. We assume that static magnetic field is absent. The electrons are streaming in the y-direction with the constant speed U_0.

(i) Consider a linear electromagnetic wave propagating in the x-direction with electric field in the y-direction and magnetic field in the z-direction. Show that this wave is accompanied by a density perturbation due to a velocity perturbation in the x-direction.

(ii) Show that the above electromagnetic wave satisfies the following linear dispersion relation:

$$k^2c^2\left[1+(\omega_p/\omega)^2(U_0/c)^2\right]=\omega^2-\omega_p^2\,. \tag{A.4}$$

(iii) Show that this dispersion relation has a growing (unstable) solution. Calculate the growth rate and find the maximum growth condition.

(iv) Discuss the physical mechanism of the instability.

A.3. Envelope Soliton and Hole

Consider the nonlinear Schrödinger equation (5.6.27) for one-dimensional propagation of a strongly dispersive wave. For simplicity, we assume that the coefficients p and q are real and constant.

(i) Galilean invariance
Show that if $u_0(x,t)$ is a solution of (5.6.27), then

$$u(x,t)\equiv u_0(x-x_0-Vt,t)\exp\{\mathrm{i}Vx/(2p)-\mathrm{i}[V^2t/(4p)+\theta]\} \tag{A.5}$$

is also a solution of (5.6.27), where x_0, V and θ are arbitrary constants.

(ii) Sagdeev potential
Consider a stationary solution, $u(x,t)=w(x)\exp(-\mathrm{i}\nu t)$, with ν being real and constant. Setting

$$w(x)\equiv R\exp(\mathrm{i}\phi)\,,\quad \boldsymbol{R}\equiv\{X=R\cos\psi\,,\ Y=R\sin\psi\}$$

with R and ψ being real, show that the equation for $\boldsymbol{R}(x)$ can be reduced to a two-dimensional equation of motion,

$$\partial^2\boldsymbol{R}/\partial^2x=-\partial U(R)/\partial\boldsymbol{R}\,, \tag{A.6}$$

in the presence of a cylindrically symmetric Sagdeev potential $U(R)$ given by

$$U(R) = (\nu R^2 + qR^4/2)/(2p) + \text{const.} \,. \tag{A.7}$$

Draw $U(R)$ for four cases of (a) $p\nu > 0$, $pq > 0$, (b) $p\nu > 0$, $pq < 0$, (c) $p\nu < 0$, $pq > 0$, and (d) $p\nu < 0$, $pq < 0$.

(iii) Finite-amplitude plane wave
Show that (A.6) has two constants of motion:

$$M = R^2 d\psi/dx\,, \quad \text{and} \quad E = (1/2)(dR/dx)^2 + U(R) + M^2/(2R^2)\,.$$

Find the dispersion relation for the finite-amplitude plane-wave solution at the minimum and the maximum of the effective one-dimensional potential, $U(R) + M^2/(2R^2)$, and discuss the frequency shift due to finite amplitude.

(iv) Envelope soliton
Show that for the case $p\nu < 0$ and $pq > 0$, (A.6) has a soliton-like solution with fixed phase ($M = 0$) being of the form

$$R(x) = R_0 \operatorname{sech}(\alpha x) \quad (R_0,\ \alpha = \text{constant})\,.$$

Determine R_0 and α.

(v) Envelope hole
Show that for the case $p\nu > 0$ and $pq < 0$, (A.6) has a solitary hole solution of the form

$$R(x) = R_M[1 - \alpha \operatorname{sech}^2(\beta x)]\,,$$

where R_M is the value of R that gives a maximum of the effective potential $U(R) + M^2/(2R^2)$.
Determine α and β. Discuss the special case of $M = 0$ where a phase jump takes place at $x = 0$.

A.4. Coupled Nonlinear Electron-plasma and Ion-acoustic Waves

Consider an electron-plasma wave in the presence of a density perturbation δn which propagates with the constant speed V close to the ion acoustic speed c_s. The basic equation for the electron-plasma wave is (5.6.17). Choosing ω_{Pe}^{-1}, $\lambda_D = v_{Te}/\omega_{Pe}$ and n_0 as units of the time, length and density, this equation can be reduced to the form

$$i\partial u/\partial t + (3/2)\partial^2 u/\partial x^2 - (1/2)\delta n_e u = 0\,. \tag{A.8}$$

(i) For low-frequency density perturbation δn_e, we can use (5.6.22), but since the density perturbation moves with the finite speed V, we can no longer neglect the ion inertia. Replacing $\partial/\partial t$ and $\partial/\partial x$ in the ion equation by $-V\partial/\partial\xi$ and $\partial/\partial\xi$, respectively, where $\xi = x - Vt$, and using a linear approximation for ions, and assuming charge neutrality $\delta n_e = \delta n_i$, show that the normalized electron-density perturbation can be written as $\delta n_e = \varepsilon|u|^2(V^2/\varepsilon^2 - 1)^{-1}$, where ε^2 is the electron-to-ion mass ratio.

(ii) Show that if V is very close to the ion acoustic speed, i.e. $V \doteq \varepsilon$, the linear approximation and local charge neutrality breaks down. Using the calculation similar to that in Sect. 5.6.1, derive the following equation for $\delta n_e \doteq \delta n_i \equiv v(\xi)$:

$$\partial^2 v/\partial\xi^2 - 2\lambda v + v^2 + |u|^2 = W = \text{const.}\,, \tag{A.9}$$

where λ is the excess Mach number, $\lambda = (V - \varepsilon)/\varepsilon$.

(iii) Show that if we set $u(x,t) = R(\xi)\exp(-\mathrm{i}\nu t + \mathrm{i}\psi)$ with R real and ν constant, (A.8) has the two integrals

$$R^2 d\psi/d\xi \equiv M = \text{constant}\,,$$
$$(dR/d\xi)^2 + (2/3)(\nu - v/2)R^2 + M^2/R^2 \equiv E = \text{constant}\,.$$

(iv) Show that for $M = 0$, a stationary solution of the coupled system of equations (A.8, 9) can be written in the form

$$v(\xi) = v_0 + A cn^2(\alpha\xi; k)\,, \quad u(\xi) = B cn(\alpha\xi; k) sn(\alpha\xi; k)$$

where A, B, v_0, α are constants, and $cn(x; k)$, $sn(x; k)$ are Jacobi's elliptic functions of modulus k. Derive the following relations:

$$A = -18k^2\alpha^2\,, \quad B = (432)^{1/2}k^2\alpha^2\,, \quad \alpha^2 = (2/3)(5k^2 - 4)^{-1}[(v_0/2) - \nu]\,,$$
$$k^2 - 1 = [(v_0 - 2\lambda)v_0 - W]/(2A\alpha^2)\,, \quad -\lambda + v_0 = 2\alpha^2(1 + 4k^2)\,.$$

(v) Show that for $k^2 = 1$ and $v_0 = 0$, the above solution can be reduced to a solitary-wave solution being of the form

$$v(\xi) = C\mathrm{sech}^2(\mu\xi)$$
$$u(\xi) = D\mathrm{sech}(\mu\xi)\tanh(\mu\xi)\,.$$

Determine C, D, and μ, which are all constant.

A.5. Lotka-Volterra Model for Relaxation Oscillation

Consider a coupling of two wave modes whose actions are given by N_1 and N_2. We assume that mode 1 is linearly unstable with the linear growth rate γ, while mode 2 is linearly damped with the linear damping rate Γ, and that they obey the following system of nonlinear mode-coupling equations:

$$\partial N_1/\partial t = 2\gamma N_1 - \alpha N_1 N_2\,,$$
$$\partial N_2/\partial t = -2\Gamma N_2 + \beta N_1 N_2\,, \tag{A.10}$$

where γ, Γ, α, β are all constants.

(i) Find the stationary solutions of (A.10) and discuss the linear stability of the stationary solutions.

(ii) Prove the following conservation relation:

$$I \equiv \beta N_1 + \alpha N_2 - 2\Gamma \log N_1 - 2\gamma \log N_2 = \text{constant} .$$

(iii) By a numerical analysis, show that the solutions of (A.10) with an initial condition that $N_1(t = 0) \neq 0$ and $N_2(t = 0) = 0$ displays an indefinite relaxation oscillation.

(iv) Show that if we add a small constant source term S on the right-hand side of the first equation of (A.10) the above relaxation oscillation damps and the solution approaches to a stationary solution.

A.6. Initial Memory of the Vlasov Equation

Consider the linearized Vlasov equation for the Fourier component $F_k(v,t)$ in one-dimensional plasma without magnetic field, see (6.6.12).

(i) Find the solution under the initial condition $F_k(v, t = 0) = g_k(v)$ in the form of (6.8.9). This solution shows that the memory of the initial perturbation $g_k(v)$ is kept forever.

(ii) Show that this initial memory dies out with time for the potential $\phi_k(t)$ which is determined by the Poisson equation (6.2.1), provided that the initial distribution $g_k(v)$ is a smooth function of v.

(iii) Show that if the initial distribution $g_k(v)$ has a form,

$$g_k(v) = A \exp(ikv\tau)/(v - \zeta) \tag{A.11}$$

with $A = \text{const.}$, $\tau > 0$, and $\text{Im}\{k\zeta\} > 0$, the potential $\phi_k(t)$ grows with the time in the region $t < \tau$ in contradiction with the Landau damping. This result demonstrates that the Landau damping is a result of a statistical average over various initial conditions and is not valid for *any* initial conditions.

(iv) Show that for $t > \tau$ $\phi_k(t)$ damps by the Landau damping. Discuss why $\phi_k(t)$ grows until $t = \tau$ by considering the oscillation of the distribution function due to the velocity-dependent phase factor.

A.7. Plasma-wave Echo

Consider a one-dimensional plasma and apply two successive pulse fields of the wave numbers $\pm k$ and $\pm k'$ ($k' > k$) at time $t = 0$ and $t'(> 0)$, respectively. Then at time $t = t'' \equiv k't'/k''$ ($k'' \equiv k' - k$), a third wave of $\pm k''$ appears spontaneously. This is called the *plasma-wave echo*. It is a result of the combined effect of the initial memory kept in the distribution function $F_{\pm k}$ and the reversal of the sign of the phase of the distribution function $F_{\pm k''}$ due to the application of the second pulse field at $t = t'$.

(i) Assuming that there is no perturbation at $t < 0$ and writing the first pulse field as

$$E_k(x,t) = [V_k \exp(ikx) + V_{-k} \exp(-ikx)]\delta(t) ,$$

calculate the distribution function $F_{\pm k}(v, t = +0)$ by solving (6.6.12).

(ii) Calculate the Fourier component $E(\pm k, \omega)$ of the induced electric field in the linear approximation.

(iii) Derive the expression for $F_{\pm k}(v, t = t')$ in terms of $F_{\pm k}(v, t = +0)$ and $E(k, \omega)$.

(iv) Assuming that t' is sufficiently large compared to the Landau damping time of the wave excited by the first pulse field and that the electric field of wave number $\pm k'$ at time $t = t'$ can be written as

$$E_{k'}(x,t) = [V_{k'} \exp(ik'x) + V_{-k'} \exp(-ik'x)]\delta(t - t') ,$$

show that the distribution function $F_{\pm k''}(v, t = t' + 0)$ produced by the second-order coupling of the perturbations at the wave numbers $\mp k$ and $\pm k'$ is given by

$$F_{\pm k''}(v, t = t' + 0)$$
$$= (e/m)^2 V_{\pm k'} \cdot \partial/\partial v[\exp(\pm ikvt')E(\mp k, \omega = \mp kv)(d/dv)F_0(v)] ,$$

where we neglected the contribution of the initial perturbation $F_{\pm k}(v, t = +0)$ to $F_{\pm k}(v, t = t')$.

(v) Show that the above distribution function, as an initial distribution for the perturbation at the wave number $\pm k''$ has a form of (A.11) with $\tau = t'' - t'$. This yields a plasma-wave echo.

(vi) Calculate the line shape of the above plasma-wave echo for the case $k' = 2k$.

A.8. H-theorem for the Quasilinear Equations

(i) Show that the quasilinear equation (6.6.14) satisfies the following H-theorem:

$$(d/dt) \int dv F_0(v,t) \log[F_0(v,t)] \leqq 0 ,$$

and that the equality in the above equation holds only when $\partial F_0(v,t)/\partial v = 0$ (i.e., a plateau in the velocity space) in the region $D(v) \neq 0$.

(ii) Show that the three-dimensional version of the quasilinear equations in the absence of the magnetic field can be written in the form

$$\partial F_0(\boldsymbol{v},t)/\partial t = \int d^3\boldsymbol{k}/(2\pi)^3 \boldsymbol{k} \cdot \partial/\partial \boldsymbol{v} \alpha(\boldsymbol{k},\boldsymbol{v}) W_k(t) \boldsymbol{k} \cdot \partial F_0(\boldsymbol{v},t)/\partial \boldsymbol{v} ,$$

$$\partial W_k(t)/\partial t = \int d^3\boldsymbol{v} m\omega_k \alpha(\boldsymbol{k},\boldsymbol{v}) W_k(t) \boldsymbol{k} \cdot \partial F_0(\boldsymbol{v},t) \partial \boldsymbol{v} , \qquad \text{(A.12)}$$

where $W_k(t)$ and ω_k are the energy and frequency of the wave of wave number $\boldsymbol{k}$, and $\alpha(\boldsymbol{k},\boldsymbol{v})$ is given by

$$\alpha(\boldsymbol{k}, \boldsymbol{v}) = (q^2/mk^2)|\boldsymbol{E}_{\boldsymbol{k}}(t)|^2 \pi\delta(\omega_{\boldsymbol{k}} - \boldsymbol{k}\cdot\boldsymbol{v}) ,$$

where, for simplicity, only one component of the plasma particles is considered. The first equation of (A.12) is the quasilinear diffusion equation in the velocity space, and the second equation describes the Landau damping of the wave. Verify the H-theorem for the three-dimensional case.

(iii) Show that if the spontaneous emission of the wave is taken into account, the term $W_{\boldsymbol{k}}(t)\boldsymbol{k}\cdot\partial F_0(\boldsymbol{v},t)/\partial\boldsymbol{v}$ on the right-hand sides of (A.12) is to be replaced by the expression

$$W_{\boldsymbol{k}}(t)\boldsymbol{k}\cdot\partial F_0(\boldsymbol{v},t)/\partial\boldsymbol{v} + m\omega_{\boldsymbol{k}}F_0(\boldsymbol{v},t) .$$

In the first equation of (A.12), the additional term, $m\omega_{\boldsymbol{k}}F_0(\boldsymbol{v},t)$, comes from the plasma-wave contribution to $J_1(\boldsymbol{v},t)$ which arises from the pole at $\boldsymbol{k}\cdot\boldsymbol{v} \doteq \omega_{\boldsymbol{k}}$ in the inverse of the dielectric function in (6.8.19).

(iv) Show that for the above generalized quasilinear set of equations, the entropy defined by the relation

$$S(t) \equiv \int d^3\boldsymbol{v}\, F_0(\boldsymbol{v},t)\log[F_0(\boldsymbol{v},t)] + \int d^3\boldsymbol{k}/(2\pi)^3 \log[W_{\boldsymbol{k}}(t)]$$

satisfies the following H-theorem

$$dS(t)/dt \geqq 0 ,$$

in which the equality holds only when

$$\boldsymbol{k}\cdot\partial\log[F_0(\boldsymbol{v},t)]/\partial\boldsymbol{v} = -m\omega_{\boldsymbol{k}}/W_{\boldsymbol{k}}(t) .$$

Verify that a Maxwellian distribution of temperature T for $F_0(\boldsymbol{v},t)$ with the thermal level of the wave energy, $W_{\boldsymbol{k}} = T$, satisfies the above condition for $dS/dt = 0$.

A.9. High-frequency Conductivity

Consider a situation in which a dipole oscillating electric field (8.3.5) is applied to a collisionless plasma in the absence of the magnetic field.

(i) Show that if we denote the linear-response electron-current density as $\boldsymbol{J}\exp(-\mathrm{i}\omega_0 t) + \mathrm{c.c.}$, we have the relation

$$\boldsymbol{J} = \boldsymbol{J}_0 + \boldsymbol{J}' ,$$

with $\boldsymbol{J}_0 = -[e^2/(\mathrm{i}m\omega_0)]n_{e0}\boldsymbol{E}_0$, and $\boldsymbol{J}' = -[e^2/(\mathrm{i}m\omega_0)]\langle n_e'\boldsymbol{E}'\rangle$, where n_e' and $\boldsymbol{E}'$ are the fluctuating parts of the electron density and the electric field, respectively, and the angled bracket denotes the spatial average. Note that the ion contribution to the current density is ignored because of the large ion mass.

(ii) Assuming that the fluctuations are electrostatic, show that $\boldsymbol{J}'$ can be written as

$$\boldsymbol{J}' = \left[Ze^3/(m\varepsilon_0\omega_0)\right] \sum_{\boldsymbol{k}} (\boldsymbol{k}/k^2) \int d\omega/(2\pi^2) \langle n_e'(\boldsymbol{k},\omega+\omega_0) n_i'(-\boldsymbol{k},-\omega)\rangle , \tag{A.13}$$

where Z is the valency of the ion. This relation reveals that electron-density fluctuations alone have no contribution to the current density $\boldsymbol{J}$, ion-density fluctuations being necessary as well as electron-density fluctuation. Discuss the physical reason for this.

(iii) In the presence of a dipole oscillating field, all particles of the same species oscillate in phase with the same amplitude. For electrons, this oscillation is represented by $\boldsymbol{r}_0(t) \equiv [e\boldsymbol{E}_0/(m\omega_0^2)] \exp(-\mathrm{i}\omega_0 t) + \mathrm{c.c.}$. Then by the transformation to the oscillating frame, $\boldsymbol{r} - \boldsymbol{r}_0(t)$, we can eliminate the external field from the basic equations for the electrons. Denoting this transformation by an operator Δ, i.e., $\Delta A(\boldsymbol{r},t) \equiv A(\boldsymbol{r}-\boldsymbol{r}_0(t),t)$, and using (6.4.11), show the following relations for the Fourier transform in the rest frame and the oscillating frame

$$(\Delta A)(\boldsymbol{k},\omega) = \int d\omega' \Delta(\boldsymbol{k},\omega') A(\boldsymbol{k},\omega-\omega') = \sum_{n=-\infty}^{\infty} \mathrm{i}^n \mathrm{J}_n(\varrho) A(\boldsymbol{k},\omega+n\omega_0)$$

$$A(\boldsymbol{k},\omega) = \sum_{n=-\infty}^{\infty} \mathrm{i}^n \mathrm{J}_n(-\varrho)(\Delta A)(\boldsymbol{k},\omega+n\omega_0)$$

where $\varrho = 2\boldsymbol{k}\cdot\boldsymbol{E}_0/(m\omega_0^2)$, which is the ratio of the excursion length of the electron due to the applied oscillating field to the wavelength, and $\mathrm{J}_n(\varrho)$ is the Bessel function.

(iv) In the electron oscillating frame, the small-amplitude electron fluctuations can be treated by the linear theory in the absence of the applied field. For ions, we use the rest frame neglecting the oscillation due to the applied field. Derive the following relations

$$(\Delta n_e')(\boldsymbol{k},\omega) = [\varepsilon_0\chi_e(\boldsymbol{k},\omega)/\varepsilon_e(\boldsymbol{k},\omega)] Z(\Delta n_i')(\boldsymbol{k},\omega) ,$$
$$Z n_i'(\boldsymbol{k},\omega) = [\varepsilon_0\chi_i(\boldsymbol{k},\omega)/\varepsilon_i(\boldsymbol{k},\omega)] n_e'(\boldsymbol{k},\omega)$$

where $\chi_s(\boldsymbol{k},\omega)$ is the electric susceptibility (6.2.12) of the s-th species of particle and $\varepsilon_s(\boldsymbol{k},\omega) \equiv \varepsilon_0[1+\chi_s(\boldsymbol{k},\omega)]$.

(v) Using the linear approximation with respect to ϱ and the relations derived in (iii) and (iv), derive the following relation

$$n_e'(\boldsymbol{k},\omega+\omega_0) = (\mathrm{i}Z\varrho/2)[\varepsilon_0/\varepsilon_e(\boldsymbol{k},\omega) - \varepsilon_0/\varepsilon_e(\boldsymbol{k},\omega+\omega_0)] n_i'(\boldsymbol{k},\omega) .$$

(vi) Substituting the above expression into (A.13) and neglecting ω in the electron dielectric function ε_e, as ω is a frequency of the ion response, derive the following formula

$$J' = \left[iZ^2e^4/\left(m^2\omega_0^3\right)\right]\sum_k \left(\boldsymbol{kk}\cdot\boldsymbol{E}_0/k^2\right)[1/\varepsilon_e(\boldsymbol{k},0) - 1/\varepsilon_e(\boldsymbol{k},\omega_0)]S_i(k)\,, \tag{A.14}$$

where $S_i(\boldsymbol{k})$ is the ion form factor in the absence of the applied field

$$S_i(\boldsymbol{k}) \equiv \int d\omega/(2\pi)^2\langle|n_i'(\boldsymbol{k},\omega)|^2\rangle\,.$$

The Formula (A.14) is called the *Dawson-Oberman formula* for the high-frequency conductivity of plasma. Discuss the effect of enhanced ion-density fluctuations on the absorption constant which is the imaginary part of the conductivity, $\boldsymbol{J}/\boldsymbol{E}_0$.

A.10. Exact Kinetic Theory for Electrostatic Waves in the Presence of a Dipole Field

As mentioned in the proceeding problem, in the presence of an externally applied dipole field given by (8.3.5), all particles of the same species oscillate in phase with the same amplitude.

(i) Show that in the presence of a uniform static magnetic field $\boldsymbol{B}_0$ in the z-direction, the oscillating motion of the particle of mass m and charge q is given by

$$\boldsymbol{r}_0(t) = \{x_0\cos(\omega_0 t), y_0\sin(\omega_0 t), z_0\cos(\omega_0 t)\}$$

with

$$\begin{aligned} x_0 &= -(2q/m)E_{0\perp}/\left(\omega_0^2 - \Omega^2\right)\,, \\ y_0 &= (2q/m)E_{0\perp}\Omega/\left[\omega_0\left(\omega_0^2 - \Omega^2\right)\right]\,, \\ z_0 &= -\left[2q/\left(m\omega_0^2\right)\right]E_{0\parallel} \end{aligned}$$

where $\boldsymbol{E}_0$ is assumed to be in the x, z-plane, and $E_{0\perp}$ and $E_{0\parallel}$ are the electric field components perpendicular (x-direction) and parallel (z-direction) to the magnetic field B_0, respectively.

(ii) Representing the transformation to the oscillating frame, $\boldsymbol{r}\to\boldsymbol{r}-\boldsymbol{r}_0(t)$, by an operator Δ, as in the preceding problem, derive the following relation for the Fourier transform in the oscillating frame and the rest frame

$$(\Delta A)(\boldsymbol{k},\omega) = \sum_{n-\infty}^{\infty}\exp(in\theta)J_n(a)A(\boldsymbol{k},\omega+n\omega_0)\,,$$

$$(\Delta^{-1}A)(\boldsymbol{k},\omega) = \sum_{n-\infty}^{\infty}\exp(-in\theta)J_n(a)A(\boldsymbol{k},\omega-n\omega_0)\,,$$

where

$$a^2 \equiv (k_x x_0 + k_z z_0)^2 + (k_y y_0)^2 , \quad \theta \equiv \tan^{-1}[(k_x x_0 + k_z z_0)/(k_y y_0)] .$$

(iii) In the linear approximation with respect to the fluctuations which are assumed to be electrostatic, i.e., $\boldsymbol{E}(\boldsymbol{k},\omega) = -\mathrm{i}\boldsymbol{k}\phi(\boldsymbol{k},\omega)$, the relation corresponding to (6.2.11) holds in the oscillating frame. Using (6.2.6) we have

$$-e(\Delta_e n_e)(\boldsymbol{k},\omega) = -e\chi_e(\boldsymbol{k},\omega)[Z(\Delta_e n_i)(\boldsymbol{k},\omega) - (\Delta_e n_e)(\boldsymbol{k},\omega)] ,$$
$$Ze(\Delta_i n_i)(\boldsymbol{k},\omega) = -e\chi_i(\boldsymbol{k},\omega)[Z(\Delta_i n_i)(\boldsymbol{k},\omega) - (\Delta_i n_e)(\boldsymbol{k},\omega)]$$

where $\chi_s(\boldsymbol{k},\omega)$ is the electric susceptibility (6.4.19) of the s-th species of particle in the presence of a magnetic field, and Δ_s is the transformation operator to the oscillating frame of the s-th species of particle. Using these relations, derive the following relation

$$\left\{1 + \chi_i(\boldsymbol{k},\omega)\Delta_i\Delta_e^{-1}[1+\chi_e]^{-1}\Delta_e\Delta_i^{-1}\right\}(\Delta_i n_i)(\boldsymbol{k},\omega) = 0 . \qquad \text{(A.15)}$$

(iv) Using the Bessel function relation

$$\sum_{n=-\infty}^{\infty} \exp(in\theta)\mathrm{J}_{n+p}(a)\mathrm{J}_n(b) = \exp(ip\psi)\mathrm{J}_p(\lambda) ,$$

where $\lambda = (a^2+b^2-2ab\cos\theta)^{1/2}$, and $\psi = \tan^{-1}[b\sin\theta/(a-b\cos\theta)]$, derive the following relation

$$\Delta_i\Delta_e^{-1}[1+\chi_e]^{-1}\Delta_e\Delta_i^{-1}(\Delta_i n_i)(\boldsymbol{k},\omega)$$
$$= \sum_p\sum_q \exp[\mathrm{i}q(\pi - \phi + \theta_e)]\mathrm{J}_p(\mu)J_{p-q}(\mu)[1+\chi_e(\boldsymbol{k},\omega+p\omega_0)]^{-1}$$
$$\times (\Delta_i n_i)(\boldsymbol{k},\omega + q\omega_0) , \qquad \text{(A.16)}$$

where

$$\mu = \left[a_e^2 + a_i^2 - 2a_e a_i\cos(\theta_i - \theta_e)\right]^{1/2} ,$$
$$\phi = \tan^{-1}\{a_i\sin(\theta_i - \theta_e)/[a_e - a_i\cos(\theta_i - \theta_e)]\} .$$

Substitution of (A.16) into (A.15) yields the set of mode-coupling equations driven by an external dipole field which are valid to all orders in E_0.

A.11. Dispersion Relation for Parametric Instability due to a Weak Dipole Pump Field

In the weak pump case, i.e. $\mu^2 \ll 1$, we have $J_0(\mu) \doteq 1, J_{\pm 1}(\mu) \doteq \pm\mu/2$ and $J_{\pm n}(\mu) \doteq 0$ ($|n| > 1$).

(i) Derive a coupled system of equations for $(\Delta_i n_i)(\boldsymbol{k},\omega)$ and $(\Delta_i n_i)(\boldsymbol{k},\omega \pm \omega_0)$.
(ii) In the low-frequency and long-wavelength case, i.e. $|\omega| \ll \omega_{pi}, \omega_{ci}$, and $k^2\lambda_D^2 \ll 1$, $\varepsilon_e(\boldsymbol{k},\omega) \doteq \varepsilon_0\chi_e(\boldsymbol{k},0) = \varepsilon_0(k^2\lambda_D^2)^{-1} \gg \varepsilon_0$, so that we can ignore $\varepsilon_e(\boldsymbol{k},\omega)^{-1}$ as compared with $\varepsilon_e(\boldsymbol{k},\omega\pm\omega_0)^{-1}$, provided that ω_0 is in

the electron resonance frequency range, i.e. $|\varepsilon_e(\boldsymbol{k},\omega\pm\omega_0)| \ll \varepsilon_0(k^2\lambda_D^2)^{-1}$. Derive the following dispersion relation for parametric instabilities in this case

$$1+\frac{\mu^2\varepsilon_0^2\chi_i(\boldsymbol{k},\omega)}{4k^2\lambda_D^2\varepsilon(\boldsymbol{k},\omega)}\left[\frac{1}{\varepsilon(\boldsymbol{k},\omega+\omega_0)}+\frac{1}{\varepsilon(\boldsymbol{k},\omega-\omega_0)}\right]=0\,, \qquad \text{(A.17)}$$

where $\varepsilon(\boldsymbol{k},\omega)=\varepsilon_0[\chi_e(\boldsymbol{k},\omega)+\chi_i(\boldsymbol{k},\omega])$, the dielectric function of the electron-ion system.

(iii) In the absence of the static magnetic field, i.e. $\Omega=0$, $y_0=0$ and $\theta=\pi/2$. Neglecting the ion response near the pump frequency, i.e. $\varepsilon(\boldsymbol{k},\omega\pm\omega_0)\doteq\varepsilon_e(\boldsymbol{k},\omega\pm\omega_0)$, derive the flowing dispersion relation independent of the value of $k^2\lambda_D^2$:

$$1+\frac{\varrho^2\varepsilon_0\chi_i(\boldsymbol{k},\omega)\varepsilon_e(\boldsymbol{k},\omega)}{4\varepsilon(\boldsymbol{k},\omega)}\left[\frac{1}{\varepsilon_e(\boldsymbol{k},\omega+\omega_0)}+\frac{1}{\varepsilon_e(\boldsymbol{k},\omega-\omega_0)}\right]=0 \qquad \text{(A.18)}$$

where $\varrho=2e\boldsymbol{k}\cdot\boldsymbol{E}_0/(m\omega_0^2)$.

(iv) Using (A.18), verify (8.3.23).

(v) Using (A.17), discuss the parametric excitation of a couple of upper hybrid mode and a lower hybrid mode by an electromagnetic pump.

A.12. Kinetic Theory for Electric Susceptibility in Weakly Inhomogeneous Plasma

Consider a plasma in a uniform static magnetic field B_0 in the z-direction. The plasma is assumed to have a density gradient in the x-direction with the scale length much longer than the wavelength.

(i) Show that in this case the unperturbed distribution function F_0 is a function of $v_z, v_\perp$ and $X\equiv x-v_y/\Omega$.

(ii) Show that (6.4.6) is to be replaced by

$$\boldsymbol{k}\cdot\partial/\partial\boldsymbol{v}'=k_z\partial/\partial v_z+(\boldsymbol{k}_\perp\cdot\boldsymbol{v}_\perp'/v_\perp)\partial/\partial v_\perp-(k_y/\Omega)\partial/\partial X\,.$$

(iii) Carrying out a calculation similar to that in Sect. 6.4, show that the only effect of the density gradient is to replace $k_z\partial/\partial v_z$ in (6.4.17) and (6.4.19) by $k_z\partial/\partial v_z-(k_y/\Omega)\partial/\partial X$.

(iv) Assuming a Maxwellian distribution,

$$F_0(\boldsymbol{v},X)=n_0(X)(2\pi v_T^2)^{-3/2}\exp\left[-v^2/\left(2v_T^2\right)\right]\,,$$

and denoting $-n_0^{-1}\partial n_0/\partial X\equiv\kappa$, derive the following expression for the electric susceptibility

$$\chi(\boldsymbol{k},\omega)=\frac{1}{k^2\lambda_D^2}\left[1+\frac{\omega-\omega^*}{\sqrt{2}k_zv_T}\sum_{n=-\infty}^{\infty}Z(\zeta_n)\Lambda_n(b)\right]$$

where $\omega^* = \kappa k_y v_T^2/\Omega$, the drift frequency, $Z(\zeta)$ is the plasma-dispersion function defined by (6.3.19) with ζ_n being given by

$$\zeta_n = (\omega + n\Omega)/(\sqrt{2}k_z v_T)$$

and $\Lambda_n(b) \equiv \exp(-b)I_n(b)$ (Problem 6.5).

(v) Consider an electron-ion system and assuming $k_z v_{Ti} \ll |\omega| \doteq \omega^* \ll \omega_{ci}$, $k_z v_{Te}$ derive the dispersion relation for the collisionless drift wave. Use the following expansion formulas for $Z(\zeta)$:

$$Z(\zeta) = \mathrm{i}\pi^{1/2}\exp(-\zeta^2) - 2\zeta(1 - 2\zeta^2/3 + \ldots) \quad (|\zeta| \ll 1)\,,$$
$$Z(\zeta) = \mathrm{i}\pi^{1/2}\exp(-\zeta^2) - \zeta^{-1}(1 + 1/(2\zeta^2) + \ldots) \quad (|\zeta| \gg 1)\,.$$

A.2 Problems to Part II Applications to Fusion Plasmas

A.13. Bernstein Waves

Consider a spatially uniform plasma in a static uniform magnetic field $\boldsymbol{B}_0$. There exist electrostatic waves propagating perpendicular to $\boldsymbol{B}_0$ called *Bernstein waves.*

(i) Assuming a Maxwellian distribution of the form (3.5.22) for both the electron and the ion, and applying the calculation similar to Problem 6.5 to (6.4.19), derive the following dispersion relation for the Bernstein wave

$$1 + \sum_s \left(k^2\lambda_{Ds}^2\right)^{-1} \sum_{n=0}^{\infty} \frac{2I_n(b_s)\exp(-b_s)n^2\Omega_s^2}{(\omega^2 - n^2\Omega_s^2)} = 0\,,$$

where the notation is the same as in Problem 6.5 with suffix s denoting the particle species. Clearly, the Bernstein wave consists of many modes whose frequencies are separated by integer multiple of cyclotron frequency.

(ii) The Bernstein wave is not subject to Landau or cyclotron damping. Discuss the physical reason for this.

(iii) At high frequencies, i.e. $\omega^2 > \Omega_e^2$, the ion contribution can be neglected, provided that there exist a small but finite collisional effect to smooth out the ion cyclotron resonance. The resulting waves are called the electron Bernstein waves. Study the dispersion characteristics of the electron Bernstein waves in the frequency range $\omega_{ce} < \omega < 2\omega_{ce}$.

(iv) Discuss the relation of the electron Bernstein waves to the upper hybrid mode in the low temperature limit.

(v) At low frequencies where $\omega \sim \omega_{ci} \ll \omega_{ce}$ and at long wavelength $k\varrho_e \ll 1$, show that the electron susceptibility can be approximated as $\chi_e \doteq \omega_{pe}^2/\omega_{ce}^2$. The resulting waves are called the *ion Bernstein waves*.

(vi) Discuss the dispersion relation, i.e. the ω versus $k\varrho_i$ curves, for the ion Bernstein waves.

A.14. Magnetosonic Waves in Tokamaks

We model the tokamak plasma by a cylinder with the magnetic field given by

$$B_\zeta = \frac{B_0 R_0}{R} = \frac{B_0}{[1 + r\cos\theta/R]} ,$$

and neglect the poloidal component B_θ. The density and temperature of the plasma are assumed to depend only on the radius r. We then consider the situation in which an electromagnetic wave at frequency range, $\omega_{ci} < \omega \ll \omega_{ce}$, is launched at the edge of the plasma along the magnetic field.

(i) We start from the dielectric tensor (7.3.8) for a cold plasma model. Show that in the present frequency range, $\chi_\sigma(\sigma = 1, 2, 3)$ can be written approximately as

$$\chi_1 = 1 + N_A^2 \left(m_e/m_i + \frac{1}{[1 - (\omega^2/\omega_{ci}^2)]} \right)$$

$$\chi_2 = -N_A^2 \frac{(\omega/\omega_{ci})}{[1 - (\omega^2/\omega_{ci}^2)]} , \quad \chi_3 = 1 - \omega_{pe}^2/\omega^2 ,$$

where $N_A^2 = \omega_{pi}^2/\omega_{ci}^2$. For an actual case of warm plasma with $\zeta_e = \omega/(k_\| v_{Te}) \sim 1$ or > 1, χ_3 has to be replaced by

$$\chi_3 = \left[\frac{\omega_{pe}^2}{(k_\| v_{Te})^2} \right] [1 + \mathrm{i}\sqrt{\pi}\zeta_e Z(\zeta_e)]$$

where $Z(\zeta_e)$ is the plasma-dispersion function.

(ii) Show that the transverse index of refraction $N_\perp \equiv k_\perp c/\omega$ can be written as

$$2\chi_1 N_\perp^2 = -(\chi_1 + \chi_3)\left(N_\|^2 - \chi_1\right) - \chi_2^2$$
$$\pm \left\{ \left[(\chi_1 + \chi_3)\left(N_\|^2 - \chi_1\right) + \chi_2^2 \right]^2 - 4\chi_1\chi_3 \left[\left(\chi_1 - N_\|^2\right)^2 - \chi_2^2 \right] \right\}^{1/2}$$

where $k_\perp^2 = k_r^2 + k_\theta^2$, with $k_\theta = m/r$ (m: integer) and $N_\| = k_\| c/\omega$ which is the longitudinal index of refraction. The upper sign corresponds to the slow mode of magnetosonic wave, and the lower sign to the fast mode. In the present frequency range, $|\chi_3| \gg |\chi_1|, |\chi_2|$. Show that in this case we have approximately

$$N_\perp^2 = \frac{\left(\chi_1 - N_\|^2\right)^2 - \chi_2^2}{\left(\chi_1 - N_\|^2\right)} \quad \text{(fast mode)} ,$$

$$= -\frac{\chi_3 \left(N_\|^2 - \chi_1\right)}{\chi_1} \quad \text{(slow mode)} .$$

Let us study the dependence of the index of refraction of the fast and slow modes on the density. Note that the density dependence on the major radius becomes important since the wave is launched in the horizontal plane with $\theta = 0$ and the wavelength is much shorter than the minor radius.

(iii) First consider the fast mode. Let N_r be the radial component of the index of refraction, show that the propagation is possible in the plasma core, i.e. $N_r^2 > 0$, up to the cut off point $r = r_{0f}$, where $N_r^2 = 0$, at which the following relation holds:

$$N_A^4 - N_A^2 \left[N_\theta^2 + 2\left(N_\parallel^2 - 1\right)\right]$$
$$+ \left[1 - (\omega^2/\omega_{ci}^2)\right] \left(N_\parallel^2 - 1\right) \left(N_\theta^2 + N_\parallel^2 - 1\right) = 0 \, ,$$

which gives

$$N_A^2 \doteq \left(N_\parallel^2 - 1\right) [(\omega/\omega_{ci}) + 1] \quad \text{when} \quad \left(N_\parallel^2 - 1\right) \gg N_\theta^2 \, ,$$
$$N_A^2 \doteq N_\theta^2 \quad \text{when} \quad \left(N_\parallel^2 - 1\right) \ll N_\theta^2$$

Show that at high density where $\chi_1 \sim N_A^2 \gg N_\parallel^2, N_\theta^2, N_\perp^2 = N_A^2$ corresponding to the Alfven phase velocity.

(iv) Consider next the slow mode. Show that the propagation is possible up to a very low density $r = r_{0s}$, where $N_A^2 \ll 1$ and $\chi_3 \doteq -N_\theta^2/(N_\parallel^2 - 1)$ or $\omega_{pe}^2/\omega^2 \doteq 1 + N_\theta^2/(N_\parallel^2 - 1)$, and that $N_\perp^2$ becomes infinite at the lower hybrid resonance point $r = r_{LF}$ where $\chi_1 = 0$ or $N_A^2 = (\omega^2/\omega_{ci}^2) - 1$ for $(\omega^2/\omega_{ci}^2) \ll m_i/m_e$.

(v) Show that at very small value of r, N_r^2 becomes negative for fast and slow modes. This implies that there exists another cut off or turning point of the electromagnetic wave at sufficiently small value of r. Draw curves showing $N_\perp^2$ and $N_r^2 = N_\perp^2 - N_\theta^2$ as functions of R.

A.15. Current Drive

Consider a one-dimensional kinetic equation along the magnetic field with a Lenard-Bernstein type collision term (6.8.30) and the quasilinear diffusion term for resonant electrons. Under the assumption that $v_{res} > v_{Te}$, where v_{res} is the resonant velocity, the kinetic equation can be written as

$$\frac{\partial F_e}{\partial t} = \frac{\partial}{\partial v} \left[\nu_e \left(\frac{v_{Te}^3}{v^3}\right) \left(v F_e + v_{Te}^2 \frac{\partial F_e}{\partial v}\right)\right]$$
$$+ \pi \left(\frac{e}{m}\right)^2 \frac{\partial}{\partial v} \int dk |E_k|^2 \delta(\omega - kv) \frac{\partial F_e}{\partial v} \, ,$$

where ν_e is the electron-ion collision frequency given by (3.5.30b), v and E are, respectively, the velocity and electric field component parallel to the magnetic field. The spectral energy density of high-frequency oscillations is given by $W_k = \varepsilon_0 |E_k|^2 (\omega_{pe}^2/\omega^2)$. The above equation is valid if the wave frequency spectrum is sufficiently broad that

$$\Delta(\omega/k) > eE/(m\omega) \, .$$

We consider the situation that an external source produces oscillations with the following spectrum in the plasma

$$W_k = W_0 > 0 \quad \text{at} \quad v_1 \leqq \omega/k \leqq v_2 \, ,$$
$$= 0 \quad \text{at} \quad \omega/k < v_1 \quad \text{and} \quad \omega/k > v_2 \, .$$

(i) Show that the stationary solution of the above kinetic equation is given by the Maxwellian distribution F_M in the region $v > v_2$ and $v < v_1$, and that it is given by the following expression in the region $v_1 \leqq v \leqq v_2$:

$$F_e = C \exp\left\{ -\nu_e v_{Te}^3 \int^v dv' / \left[\nu_e v_{Te}^5 / v' + \pi\omega^2 W_0 v' / (m n_e) \right] \right\} \, .$$

The integration constants are found from the continuity condition for F_e at $v = v_1$ and v_2 and the conservation of the total number of particles, $\int dv F_e = n_e$.

(ii) Show that for the case

$$\pi\omega^2 W_0/(m n_e) \gg \nu_e v_{Te}^5 / v'^2 \, ,$$

F_e is nearly a plateau in the region $v_1 \leqq v \leqq v_2$.

(iii) Under the above condition, calculate the current density $J = -e \int_{-\infty}^{\infty} dv v F_e$.

(iv) The power of the external source necessary to produce this current density is determined by electron collisions responsible for the energy flow from the plateau region to the thermal region

$$P = \int_{-\infty}^{\infty} dv (m v^2/2) \nu_e (v_{Te}/v)^3 (F_e - F_M) \, .$$

Show that the current drive efficiency is given by

$$J/P = e \left(v_2^2 - v_1^2 \right) / \left[m \nu_e v_{Te}^3 \log(v_2/v_1) \right] \, .$$

A.16. Wave Equation in an Inhomogeneous Cylindrical Plasma

Consider the propagation of a low-frequency wave in a cylindrical plasma in the presence of a static magnetic field $\boldsymbol{B}_0 = (0, B_{\theta 0}, B_0)$. In the frequency range $\omega < \omega_{LH}$ (the lower hybrid-wave frequency), electron mass effect is negligible in the equation of motion

$$\boldsymbol{E} + \boldsymbol{u} \times \boldsymbol{B}_0 + u_{\parallel 0} \boldsymbol{b} \times \boldsymbol{B} = 0 \, ,$$

where $\boldsymbol{E}$ and $\boldsymbol{B}$ are fluctuating fields and $u_{\|0}$ is the equilibrium electron flow velocity due to the magnetic field gradient being given by Ampere's law, i.e. $J_{\|0} = -en_{e0}u_{\|0} = (\nabla \times \boldsymbol{B}_0)/\mu_0$. Since we can neglect the electric field along the magnetic field, we do not need equation for $\boldsymbol{u} \cdot \boldsymbol{b} = u_{\|}$.

(i) By operating $\boldsymbol{b}\times$ on the above equation, derive the following expression for the perturbed electron current perpendicular to the magnetic field:

$$\begin{aligned} \boldsymbol{J}_{e\perp} &= -en_{e0}(\boldsymbol{u}_e - u_{\|}\boldsymbol{b}) \\ &= \omega_{pi}^2/(\omega\omega_{ci})\boldsymbol{b} \times \boldsymbol{E}/c^2 + (\nabla \times \boldsymbol{B}_0)_{\|}(\boldsymbol{B} - B_{\|}\boldsymbol{b})/(\mu_0 B_0) \,. \end{aligned}$$

(ii) For ions we can neglect equilibrium current and the equation of motion becomes (7.3.1). Derive the following relation for the perturbed ion current, see (7.2.13) and (7.3.9),

$$\boldsymbol{J}_i/\omega = \varepsilon_0 \left[-\mathrm{i}\boldsymbol{E}(\omega_{pi}/\omega_{ci})^2 - \boldsymbol{b} \times \boldsymbol{E}\omega_{pi}^2/(\omega\omega_{ci})\right] / \left[1 - (\omega/\omega_{ci})^2\right] \,.$$

(iii) Show that the dielectric tensor $\overleftrightarrow{\varepsilon}$ can be written as follows

$$\begin{aligned} \overleftrightarrow{\varepsilon}/\varepsilon_0 \cdot \boldsymbol{E} &= \boldsymbol{E} + \left(\mathrm{i}\mu_0 c^2/\omega\right)(\boldsymbol{J}_{e\perp} + \boldsymbol{J}_i) \\ &= \left(1 + \frac{(c/v_A)^2}{[1 - (\omega/\omega_{ci})^2]}\right)\boldsymbol{E} - \mathrm{i}\frac{(c/v_A)^2(\omega/\omega_{ci})}{[1 - (\omega/\omega_{ci})^2]}\boldsymbol{b} \times \boldsymbol{E} \\ &\quad + (\mathrm{i}c^2/\omega)(\nabla \times \boldsymbol{B}_0)_{\|}(\boldsymbol{B} - B_{\|}\boldsymbol{b})/B_0 \,, \end{aligned}$$

where v_A is the Alfven velocity which is a function of the radial position r since n_{i0} depends on r.

(iv) In the low-β, cylindrical tokamak model, the dependence of the cyclotron frequency on r is ignorable. Show the following relations

$$\begin{aligned} [\nabla \times (\nabla \times \boldsymbol{E})]_r &= \left(k_\perp^2 + k_\|^2\right) E_r + (\mathrm{i}k_\perp/r)d/dr(rE_\perp) \\ &\quad - ik_\| r d/dr[B_{\theta0}/(rB_0)]E_\perp \,, \end{aligned}$$

$$\begin{aligned} [\nabla \times (\nabla \times \boldsymbol{E})]_\perp &= k_\|^2 E_\perp - d/dr[-\mathrm{i}k_\perp E_r + (1/r)d/dr(rE_\perp)] \\ &\quad + ik_\| r d/dr[B_{\theta0}/(rB_0)]E_r \,, \end{aligned}$$

and

$$\begin{aligned} \left[(\omega/c)^2\overleftrightarrow{\varepsilon}/\varepsilon_0 \cdot \boldsymbol{E}\right]_r &= (\omega/c)^2\left\{1 + \frac{(c/v_A)^2}{[1 - (\omega/\omega_{ci})^2]}\right\}E_r \\ &\quad + \mathrm{i}\frac{(\omega/c)^2(c/v_A)^2(\omega/\omega_{ci})}{[1 - (\omega/\omega_{ci})^2]}E_\perp \\ &\quad + (1/r)d/dr(rB_{\theta0}/B_0)(-\mathrm{i}k_\|)E_\perp \,, \end{aligned}$$

$$\left[(\omega/c)^2 \overleftrightarrow{\varepsilon}/\varepsilon_0 \cdot \boldsymbol{E}\right]_\perp = (\omega/c)^2 \left\{1 + \frac{(c/v_A)^2}{[1-(\omega/\omega_{ci})^2]}\right\} E_\perp + \mathrm{i}\frac{(\omega/c)^2 (c/v_A)^2 (\omega/\omega_{ci})}{[1-(\omega/\omega_{ci})^2]} E_r + (1/r) d/dr (r B_{\theta 0}/B_0)(-\mathrm{i}k_\parallel) E_r \, .$$

(v) Noting Maxwell's equations:

$$\nabla \times (\nabla \times \boldsymbol{E}) = (\omega^2/c^2)\left(\overleftrightarrow{\varepsilon}/\varepsilon_0\right) \cdot \boldsymbol{E} \, ,$$

and the above equations, derive the following equations for $(E_r, E_\perp)$:

$$(\mathrm{i}k_\perp/r) d/dr (r E_\perp) = \left(A - k_\perp^2\right) E_r + \mathrm{i}G E_\perp \, ,$$

$$d/dr[(1/r) d/dr (r E_\perp) - \mathrm{i}k_\perp E_r] = \mathrm{i}G E_r - A E_\perp \, ,$$

where

$$A = (\omega^2/c^2) + \frac{(\omega/v_A)^2}{[1-(\omega/\omega_{ci})^2]} - k_\parallel^2 \, ,$$

$$G = \frac{(\omega/v_A)^2 (\omega/\omega_{ci})}{[1-(\omega/\omega_{ci})^2]} - (2k_\parallel/r)(B_{\theta 0}/B_0) \, .$$

Note that $A = 0$ represents the dispersion relation for the Alfven wave. The term (ω^2/c^2) is often negligible for $c \gg v_A$. Note also that $G = 0$ if $(\omega/\omega_{ci}) \to 0$ and equilibrium current term is neglected.

(vi) The MHD limit implies $(\omega/\omega_{ci}) \to 0$ and $c \gg v_A$. Then $A = (\omega/v_A)^2 - k_\parallel^2$ and $G = -(2k_\parallel/r)(B_{\theta 0}/B_0)$. Eliminating E_r from the above equations, derive

$$(d/dr)\left[\frac{A}{(A-k_\perp^2)} \frac{d}{dr}(r E_\perp)\right] + \left\{(A/r) - \frac{(G^2/r}{(A-k_\perp^2)} - \frac{d}{dr}\left[\frac{(k_\perp G/r)}{(A-k_\perp^2)}\right]\right\}(r E_\perp) = 0 \, .$$

For the MHD model in the slab geometry, this equation is reduced to the form,

$$(d/dx)\left[\frac{A}{(A-k_\perp^2)}\right](dE_\perp/dx) + A E_\perp = 0 \, .$$

A.17. Spectrum of Shear Alfven Wave in Toroidal Geometry

Shear Alfven waves exhibit interesting features in toroidal geometry. Consider a large aspect ratio tokamak which satisfies $a/R \ll 1$, where a and R are the minor and major radii, respectively. The toroidal coordinates are introduced through

$$R = R_0 - \Delta(r) + r\cos\theta ,$$
$$\phi = -\zeta/R_0 ,$$
$$Z = r\sin\theta ,$$

where (R, ϕ, Z) are the cylindrical coordinates. Here $\Delta(r)$ denotes the shift of center of the magnetic surface due to the toroidal effect and is called the *Shafranov shift.*

We describe the shear Alfven wave by fluctuating electric field. In the context of the ideal MHD, $E_{\parallel} = 0$ and $\boldsymbol{E}_{\perp} = -\nabla_{\perp}\Phi$.

(i) Using the Fourier-series representation $\Phi(r,\theta,\zeta) = \sum \Phi_{mn}\exp(im\theta - in\zeta)$ for the linearized MHD equations and neglecting $O(r^2/R_0^2)$, derive the following eigenvalue equation for the variable $E_m = \Phi_{mn}/r$:

$$\frac{d}{dt}\left[r^3\left(\frac{\omega^2}{v_A^2} - k_{\parallel m}^2\right)\frac{dE_m}{dr}\right] + \frac{d}{dr}\left(\frac{\omega^2}{v_A^2}\right)r^2 E_m$$
$$- r(m^2-1)\left(\frac{\omega^2}{v_A^2} - k_{\parallel m}^2\right)E_m$$
$$+ \frac{d}{dr}\left[r^3\left(\frac{\omega^2}{v_A^2}\right)\left(\frac{2r}{R_0} + \Delta'\right)\left(\frac{dE_{m+1}}{dr} + \frac{dE_{m-1}}{dr}\right)\right.$$
$$\left.-r^3 k_{\parallel m}\Delta'\left(k_{\parallel m+1}\frac{dE_{m+1}}{dr} + k_{\parallel m-1}\frac{dE_{m-1}}{dr}\right)\right] = 0 .$$

Here $\Delta' \equiv d\Delta/dr$, $k_{\parallel m}$ is the parallel wave number given by $k_{\parallel m} = (mB_\theta/B + nB_0/RB)$.

(ii) Consider the cylindrical limit, $r/R_0 \to 0$ and $\Delta \to 0$, and derive the dispersion relation $\omega^2 = k_{\parallel m}^2 v_A^2$, which corresponds to a continuous spectrum.

(iii) By including the toroidal effect, derive the following dispersion relation for the shear Alfven wave

$$|D_{mn}| = 0 ,$$

with

$$D_{mn} = \left(\omega^2/v_A^2 - k_{\parallel m}^2\right)\delta_{m,n}$$
$$+ \left[\left(\omega^2/v_A^2\right)(2r/R_0 + \Delta') - \Delta' k_{\parallel m}k_{\parallel n}\right](\delta_{m+1,n} + \delta_{m-1,n}) ,$$

where $\delta_{m,n}$ is Kronecker's delta. It should be noted that this dispersion relation resolves the degeneracy at radial locations where different shear Alfven waves intersect at $|k_{\parallel m}| = |k_{\parallel m+1}|$.

(iv) Solve the above dispersion relation for the case when two modes, $m = -1$ and $m = -2$, are present. Draw the (ω, r) diagram for the case of the safety factor $q(r) = 1 + (r/a)^2$, and confirm the existence of the frequency gap.

A.18. MHD Equilibrium with Anisotropic Pressure

When the particle distribution function is anisotropic with respect to $v_\parallel$ and $v_\perp$, the MHD equation has to be extended to include the pressure anisotropy, $P_\parallel \neq P_\perp$, where $\parallel$ and $\perp$ refer to the directions parallel and perpendicular to the magnetic field. By adding the gradient B drift current, the curvature drift current and the magnetization current, we have

$$\begin{aligned} \boldsymbol{J} &= \boldsymbol{J}_{\nabla B} + \boldsymbol{J}_R + \boldsymbol{J}_M \\ &= P_\perp \nabla(1/B) \times \boldsymbol{b} + (P_\parallel/B)(\nabla \times \boldsymbol{b})_\perp - \nabla \times (P_\perp \boldsymbol{b}/B) \,. \end{aligned}$$

(i) Show that

$$\boldsymbol{J} \times \boldsymbol{B} = \nabla_\perp P_\perp + (P_\parallel - P_\perp)\boldsymbol{b} \cdot \nabla \boldsymbol{b} \,.$$

(ii) If we define the pressure tensor $\overleftrightarrow{P}$ for the anisotropic plasma by

$$\overleftrightarrow{P} = P_\perp \overleftrightarrow{I} + (P_\parallel - P_\perp)\boldsymbol{b}\boldsymbol{b} \,,$$

show that the right hand side of the above equation is equal to $(\nabla \cdot \overleftrightarrow{P})_\perp$. Also show that the parallel force balance gives

$$(\nabla \cdot \overleftrightarrow{P})_\parallel = \nabla_\parallel P_\parallel + (P_\parallel - P_\perp)\boldsymbol{b} \cdot \nabla \boldsymbol{b} = 0 \,.$$

(iii) Assuming $P = P(\psi, B)$, where ψ is the flux function, derive the following relations

$$\begin{aligned} \partial P_\parallel / \partial B &= (P_\parallel - P_\perp)/B \,, \\ (\nabla \times \sigma \boldsymbol{B}) \times \boldsymbol{B}/\mu_0 &= (\partial P_\parallel / \partial \psi) \nabla_\perp \psi \,, \end{aligned}$$

where

$$\sigma = 1 - \mu_0 (P_\parallel - P_\perp)/B^2$$

A.19. Beta Limit

Stability criterion for the interchange mode is given by the Suydam criterion (10.3.36). In order to obtain the highest pressure for a given magnetic configuration, the pressure profile marginal to the interchange mode everywhere in the plasma column is selected. For the simplest model of a Heliotron E in the cylindrical limit, where the rotational transform is given by

$$\iota(r) = 0.51 + 1.69(r/a)^2 ,$$

r being the a radius corresponding to magnetic surface in the cylindrical model, estimate the highest beta value determined by the Suydam criterion.

In the toroidal geometry the Suydam criterion must be replaced by the Mercier criterion which takes into account the toroidal effect correctly [10.1].

A.20. Shafranov Shift

Consider an MHD equilibrium of tokamak plasma described by (10.2.21–27).

(i) Using (10.2.37, 42 and 49), show that the magnetic field components for $\varrho \geq a$ in the poloidal and radial directions are given by

$$B_\theta = B_\theta(a)\frac{a}{\varrho} - B_\theta(a)\frac{a}{2R}\left[\left(1+\frac{a^2}{\varrho^2}\right)\left(\Lambda(a)+\frac{1}{2}\right)+\ln\frac{\varrho}{a}-1\right]\cos\theta$$

and

$$B_\varrho = \frac{a}{2R}B_\theta(a)\left[\left(1-\frac{a^2}{\varrho^2}\right)\left(\Lambda(a)+\frac{1}{2}\right)+\ln\frac{\varrho}{a}\right]\sin\theta ,$$

respectively, where $B_\theta(a) = B_\theta(a, \theta = \pi/2)$ and

$$\Lambda(a) = \beta_p + \frac{l_i}{2} - 1$$

and a is the minor radius of toroidal plasma. Note that the above equations give

$$\begin{cases} B_\theta(a,\theta) = B_\theta(a)\left[1-\dfrac{a}{R}\Lambda(a)\cos\theta\right] \\ B_\varrho(a,\theta) = 0 . \end{cases}$$

(ii) For the MHD equilibrium with uniform toroidal current density $J_{\varphi 0}$ = const and parabolic pressure profile $P_0(r) = P_0(0)[1-(\varrho^2/a^2)]$ in the zeroth approximation, derive expressions for $B_\varrho(\varrho,\theta)$ and $B_\theta(\varrho,\theta)$ inside the plasma column. Also calculate the internal inductance and the Shafranov shift.

A.21 Stochastic Behavior of Magnetic Field Lines

Consider the equations for field lines of tokamak in the cylindrical coordinates (R, ϕ, Z) given by

$$\begin{cases} \dfrac{dR}{d\phi} = \dfrac{B_R}{B_\phi} \\ \dfrac{dZ}{d\phi} = \dfrac{B_Z}{B_\phi} \end{cases},$$

where B_ϕ is the toroidal field.

(i) Show that the above equations can be transformed into a form of canonical equations for a Hamiltonian system

$$\begin{cases} \dot{p} = -\dfrac{\partial\psi}{\partial q} \\ \dot{q} = \dfrac{\partial\psi}{\partial p} \end{cases},$$

where the toroidal angle ϕ corresponds to the time, p = const to the magnetic surface and q to the poloidal angle around the magnetic axis. The 'Hamiltonian' $\psi(p, q, \phi)$ is represented by the poloidal flux function.

(ii) As an example of tokamak equilibrium, consider the case

$$\frac{d\psi_0}{dp} = \frac{1}{1+5p^2},$$

where $0 \leq p \leq 1$ labels the equilibrium flux surfaces. This corresponds to the safety factor $q_0(p) = 1 + 5p^2$. Choose the perturbation $\tilde{\psi}$ in the following way, which is resonant in the confinement zone of $1 \lesssim q_0 \lesssim 2$,

$$\tilde{\psi} = \varepsilon \sum_{m=3}^{4} \frac{1}{(m+1)^2} \sin(mq - 2\phi) .$$

Here two basic driven modes (m, n) = (3, 2) and (m, n) = (4, 2) may be assumed to be of magnetohydrodynamic origin. Show the island half width Δp_{mn} is given by

$$\Delta p_{mn} = 2\sqrt{a_{mn} / \left(\frac{d^2\psi_0}{dp^2}\right)\Bigg|_{p=p_{mn}}}$$

and estimate Δp_{mn} for $\varepsilon = 0.025$, where $a_{mn} = \varepsilon/(m+1)^2$ and p_{mn} is the resonant surface satisfying $d\psi_0/dp|_{p=p_{mn}} = n/m$. The value of $\varepsilon = 0.025$ may correspond to $\delta B/B \simeq 10^{-3}$.

(iii) By preparing a computer program, follow 30 trajectories for 10^3 revolutions around the torus, starting at $q_i = 0$ and $p_i = 0.05 + (i-1) \times 0.031$ $(i = 1, \ldots\ldots, 30)$ and plot a phase diagram on (p, q) plane by using the periodicity in the ϕ direction with 2π. The following parameter is called the Chirikov parameter

$$S_{mn,m'n'} = \frac{\Delta p_{mn} + \Delta p_{m'n'}}{|p_{mn} - p_{m'n'}|} .$$

Calculate it for the above case of the perturbations of $(m, n) = (3, 2)$ and $(m, n) = (4, 2)$ and $\varepsilon = 0.025$.

A.22 Ion-temperature Gradient Driven Drift Wave

Consider the situation where an inhomogeneous plasma is immersed in the homogeneous magnetic field and the ion inertia is negligible for the perpendicular velocity component to the magnetic field with $\boldsymbol{b} = \boldsymbol{B}/B$.

$$\boldsymbol{v}_\perp^i = \left[\boldsymbol{b} \times \left(\nabla\varphi + \frac{\nabla P_i}{en}\right)\right] / B .$$

(i) By substituting the expression for $\boldsymbol{v}_\perp^i$ in the ion continuity equation, and in the equations for longitudinal ion motion and ion thermal transport, derive the following equations:

$$\frac{\partial n_i}{\partial t} + \frac{(\boldsymbol{b} \times \nabla\varphi)}{B} \cdot \nabla n_i + \boldsymbol{b} \cdot \nabla(n_i v_{\|i}) = 0 , \tag{A.19}$$

$$m_i n_i \left(\frac{\partial v_{\|i}}{\partial t} + v_{\|i}\boldsymbol{b} \cdot \nabla v_{\|i} + \frac{\boldsymbol{b} \times \nabla\varphi}{B} \cdot \nabla v_{\|i}\right)$$
$$= -\boldsymbol{b} \cdot \nabla P_i - e n_i \boldsymbol{b} \cdot \nabla\varphi , \tag{A.20}$$

$$\frac{\partial P_i}{\partial t} + \frac{\boldsymbol{b} \times \nabla\varphi}{B} \cdot \nabla P_i + v_{\|i}\boldsymbol{b} \cdot \nabla P_i + \frac{5}{3} P_i \boldsymbol{b} \cdot \nabla v_{\|i} = 0 . \tag{A.21}$$

It should be noted that for electrons

$$-\boldsymbol{b} \cdot \nabla P_e + en\boldsymbol{b} \cdot \nabla\varphi = 0 , \tag{A.22}$$

and $T_e = \text{const}$ are supplemented.

(ii) When n_0 and T_{0i} have gradients in the x direction, by linearizing (A.19–21) and using the perturbed quantities $\{\tilde{n}_i, \widetilde{v_{\|i}}, \widetilde{P}_i, \tilde{\varphi}\}\ \exp(-\mathrm{i}\omega t + \mathrm{i}k_y y + \mathrm{i}k_\| z)$, show that

$$\frac{\tilde{n}_i}{n_0} = -\frac{k_y}{B\omega n_0}\frac{dn_0}{dx}\tilde{\varphi} + \frac{k_\|\widetilde{v_{\|i}}}{\omega} ,$$

$$\widetilde{v_{\|i}} = \frac{k_\| e}{m_i \omega}\left(1 - \frac{\omega_{pi}^*}{\omega}\right)\tilde{\varphi} ,$$

$$\widetilde{P}_i = -\frac{k_y}{B\omega}\frac{dP_{0i}}{dx}\tilde{\varphi} ,$$

where k_y and $k_\parallel$ are wave numbers in y direction and along the magnetic field line, respectively, and $\omega_{pi}^* = (k_y/eBn_0)(dP_0^i/dx)$. Here incompressibility is assumed by neglecting the last term in (A.21).

(iii) Show the ion-density perturbation is given by

$$\frac{\tilde{n_i}}{n_0} = -\frac{k_y}{B\omega n_0}\frac{dn_0}{dx}\tilde{\varphi} + \frac{k_\parallel^2 e}{m_i\omega^2}\left(1 - \frac{\omega_{pi}^*}{\omega}\right)\tilde{\varphi}$$

and derive the dispersion relation

$$1 - \frac{\omega_e^*}{\omega} - \frac{k_\parallel^2 T_e}{m_i\omega^2}\left(1 - \frac{\omega_{pi}^*}{\omega}\right) = 0 \tag{A.23}$$

by using the Boltzmann relation $\widetilde{n_e}/n_0 = e\varphi/T_e$ obtained from (A.22) and assuming the charge neutrality $\widetilde{n_e} = \tilde{n}_i$. Here $\omega_{*e} = -(k_y T_e/eBn_0)(dn_0/dx)$ is the electron drift frequency.

(iv) Show that the dispersion relation (A.23) becomes

$$\omega^2 = -\frac{k_\parallel^2 T_{0i}}{m_i}(1 + \eta_i)$$

for $\omega \ll \omega_e^*$, where $\eta_i = d\ln T_{0i}/d\ln n_0$. In this case one of the roots indicates the instability for $\eta_i > -1$.

(v) For $\eta \gg 1$ and $\omega_{T_i}^* > \omega > \omega_e^*$, show the dispersion relation (A.23) becomes

$$\omega^3 = -\frac{k_\parallel^2 T_e}{m_i}\omega_{T_i}^* ,$$

which also includes the unstable oscillation, where $\omega_{T_i}^* = (k_y/eB_0)dT_{0i}/dx$.

(vi) When the last term of (A.21) is kept for compressible plasmas, derive the dispersion relation

$$1 - \frac{\omega_e^*}{\omega} - \frac{k_\parallel^2 T_e}{m_i\omega^2}\left(1 - \frac{\omega_{pi}^*}{\omega}\right) - \frac{5}{3}\frac{k_\parallel^2 T_{0i}}{m_i\omega^2}\left(1 - \frac{\omega_e^*}{\omega}\right) = 0 .$$

Also show that for $\omega \ll \omega_e^*$ and $T_e = T_{0i}$

$$\omega^2 = \frac{k_\parallel^2 T_{0i}}{m_i}\left(\frac{2}{3} - \eta_i\right) .$$

In this case the instabilities are excited for $\eta_i > 2/3$. Since η_i is the important parameter for the ion temperature gradient driven drift wave, this is also called η_i mode.

A.23 Transport Equations

Equations (14.1–5) describes global particle and energy balances of toroidal plasma without spacial dependence. However, in order to analyze experimental results, radial dependence of density and temperature becomes important. For such a study of tokamak plasmas the following balance equations are solved numerically

$$\frac{\partial n}{\partial t} = \frac{1}{r}\frac{\partial}{\partial r}\left(rD\frac{\partial n}{\partial r}\right) + S_I , \tag{A.24}$$

$$\begin{aligned} \frac{3}{2}\frac{\partial}{\partial t}(nT_e) &= \frac{1}{r}\frac{\partial}{\partial r}r\left(\kappa_e\frac{\partial T_e}{\partial r} + \frac{3}{2}T_e D\frac{\partial n}{\partial r}\right) + \frac{\eta}{\mu_0^2 r^2}\left[\frac{\partial}{\partial r}(rB_\theta)\right]^2 \\ &\quad -\frac{3m_e}{m_i}\frac{\eta}{\tau_{ei}}(T_e - T_i) - P_{\text{rad}} - P_I + P_A^e , \end{aligned} \tag{A.25}$$

$$\begin{aligned} \frac{3}{2}\frac{\partial}{\partial t}(nT_i) &= \frac{1}{r}\frac{\partial}{\partial r}r\left(\kappa_i\frac{\partial T_i}{\partial r} + \frac{3}{2}T_i D\frac{\partial n}{\partial r}\right) \\ &\quad +\frac{3m_e}{m_i}\frac{\eta}{\tau_{ei}}(T_e - T_i) - P_{cx} + P_A^i , \end{aligned} \tag{A.26}$$

$$\frac{\partial B_\theta}{\partial t} = \frac{1}{\mu_0}\frac{\partial}{\partial r}\left[\frac{\eta}{r}\frac{\partial}{\partial r}(rB_\theta)\right] . \tag{A.27}$$

Here we set $n = n_e = n_i$. In these equations S_I denotes the particle source due to ionization, P_I and P_{cx} denote energy losses due to ionization and charge exchange, respectively. The terms P_A^e and P_A^i denote additional heating of electrons and ions by neutral beams or electromagnetic waves. The term P_{rad} denotes the energy loss due to radiation. When impurity ions are disregarded, this term is simply the Bremsstrahlung given by (9.1.9). The classical resistivity η is given by (4.5.12).

There are many models for the electron and ion thermal conductivity, κ_e and κ_i, and the diffusion coefficient D.

First choose D, κ_e and κ_i appropriately from recent scientific papers. (Note that many papers give $\chi_e = \kappa_e/n$ and $\chi_i = \kappa_i/n$ instead of κ_e and κ_i.) Next develop a numerical code to solve (A.24–27) in a straight cylindrical plasma. Then study evolution of density and temperature profiles for an ohmically heated tokamak with $R = 1m$, $a = 0.2m$, $B = 2T$ and $I_p = 100KA$ from an initial low temperature plasma with $n_e(m^{-3}) = 2 \times 10^{19}[1 - (r/a)^2] + 2 \times 10^{18}$, $T_e(eV) = 40[1-(r/a)^2]+10$ and $T_i(eV) = 40[1-(r/a)^2]+10$. Finally calculate the particle and energy confinement at the steady state.

References

1.1 L. Tonks, I. Langmuir: Oscillations in Ionized Gases and Note on Oscillations in Ionized Gases, Phys. Rev. **33**, 195, 990 (1929)
1.2 A.A. Vlasov: Zh. Eksp. Teor. Fiz **8**, 291 (1938)
1.3 L.D. Landau: On the Vibration of the Electronic Plasma, J. Phys. (USSR) **10**, 25 (1946)
1.4 J.E. Mayer: The Theory of Ionic Solutions, J. Chem. Phys. **18**, 1426 (1950)
1.5 R. Balescu: Irreversible Processes in Ionized Gases, Phys. Fluids **3**, 52 (1960)
1.6 A. Lenard: On Bogoliubov's Kinetic Equation for a Spatially Homogeneous Plasma, Ann. Phys. **10**, 390 (1960)
1.7 R.L. Guernsey: Kinetic Equation for a Completely Ionized Gas, Phys. Fluids **5**, 322 (1962)
1.8 H. Alfven: *Cosmical Electrodynamics* (Oxford University Press 1951)
1.9 T.G. Cowling: *Magnetohydrodynamics* (Adam Hilger, Bristol 1976)
1.10 S.I. Braginskii: *Reviews of Plasma Physics*, ed. by M.A. Leontovich (Consultants Bureau, New York 1965) p. 205
1.11 R.Z. Sagdeev, A.A. Galeev: *Nonlinear Plasma Theory* (Benjamin, New York 1969)
1.12 R.C. Davidson: *Methods in Nonlinear Plasma Theory* (Academic, New York 1972)
1.13 V.N. Tsytovich: *Nonlinear Effects in Plasma* (Plenum, New York 1970)
1.14 V.I. Karpman: *Nonlinear Waves in Dispersive Media* (Pergamon, Oxford 1975)
1.15 J. Weiland, H. Wilhelmsson: *Coherent Nonlinear Interaction of Waves in Plasmas* (Pergamon, Oxford 1977)
1.16 C.F. Kennel, L.J. Lanzerotti, E.N. Parker (eds.): *Solar System Plasma Physics*, Vol.I, II, III (North-Holland, Amsterdam 1979)
1.17 D.B. Melrose: *Plasma Astrophysics* (Gordon and Breach, New York 1980)
1.18 Proc. of the Second United Nations Int. Conf. on the Peaceful Uses of Atomic Energy in Geneva: *Theoretical and Experimental Aspects of Controlled Nuclear Fusion*, Vol.31; *Controlled Fusion Devices*, Vol.32 (United Nations Publication, Geneva 1958)
1.19 Proceedings of IAEA International Conference on Plasma Physics and Controlled Thermonuclear Fusion Research, Novosibirsk 1968, Vol.1, p. 157

Chapter 2

2.1 M.N. Saha: On a Physical Theory of Stellar Spectra, Proc. Roy. Soc. A **99**, 135 (1921); W.B. Thompson, *An Introduction to Plasma Physics* (Pergamon, Oxford 1961) Sect. 2.3
2.2 N.A. Krall, A.W. Trivelpiece: *Principles of Plasma Physics* (McGraw-Hill, New York 1973) Chap. I, Part Two
2.3 H. Griem: *Spectral Line Broadening by Plasmas* (Academic, New York 1974)
2.4 G.Y. Marr: *Plasma Spectroscopy* (Elsevier, Amsterdam 1968)

Chapter 4

4.1 D.R. Nicholson: *Introduction to Plasma Theory* (John Wiley, New York 1983)
4.2 H.K. Wimmel: Consistent Guiding Center Drift Theories, Naturforschung **37a**, 985 (1982)

Chapter 5

5.1 K.V. Roberts, H.L. Berk: Nonlinear Evolution of a Two-Stream Instability, Phys. Rev. Lett. **19**, 297 (1967)
5.2 J.P. Freidberg, T.P. Armstrong: Nonlinear Development of the Two-Stream Instability, Phys. Fluids **11**, 2699 (1968)
5.3 T. O'Neil, J.H. Winfrey, J.H. Malmberg: Nonlinear Interaction of a Small Cold Beam and a Plasma, Phys. Fluids **14**, 1204 (1971)
5.4 A.B. Mikhailovskii: *Theory of Plasma Instabilities*, Vol. 2, Chap. 3 (Consultants Bureau, New York 1974)
5.5 D.W. Ross, S.M. Mahajan: Are Drift-Wave Eigenmodes Unstable?, Phys. Rev. Lett. **40**, 324 (1978)
5.6 K.T. Tsang, P.J. Catto, J.C. Whitson, I.Smith: Absolute Universal Instability is Not Universal, Phys. Rev. Lett. **40**, 327 (1978)
5.7 T.M. Antonsen, Jr.: Stability of Bound Eigenmode Solutions for the Collisionless Universal Instability, Phys. Rev. Lett. **41**, 33 (1978)
5.8 H. Ikezi, R.J. Taylor, D.R. Baker: Formation and Interaction of Ion-Acoustic Solitons, Phys. Rev. Lett. **25**, 11 (1970)
5.9 H. Ikezi: Experiments on Ion-Acoustic Solitary Waves, Phys. Fluids **16**, 1668 (1973)
5.10 M.Q. Tran: Ion Acoustic Solitons in a Plasma: A Review of Their Experimental Properties and Related Theories, Physica. Scripta **20**, 317 (1979)
5.11 V. Zakharov, A.B. Shabat: Exact Theory of Two-Dimensional Self-Focusing and One-Dimensional Self-Modulation of Waves in Nonlinear Media, Sov. Phys. –JETP **34**, 62 (1972)
5.12 T. Taniuti, N. Yajima: Perturbation Method for a Nonlinear Wave Modulation I., II., J. Math. Phys. **10**, 1369, 2020 (1969)

Chapter 6

6.1 L.D. Landau: On the Vibration of the Electron Plasma, J. Phys. (USSR) **10**, 25 (1946)
6.2 B.D. Fried, S.D. Conte: *The Plasma Dispersion Function* (Academic, London 1961)
6.3 T. Stix: *The Theory of Plasma Waves* (McGraw-Hill, New York 1962) Chap. 8.2
6.4 N.A. Krall: "Drift Waves", in *Advances in Plasma Physics* Vol. 1, ed. by A. Simon, W.B. Thompson (Interscience, New York 1968) p. 153
6.5 D.W. Ross, S.M. Mahajan: Are Drift-Wave Eigenmodes Unstable?, Phys. Rev. Lett. **40**, 324 (1978); K.T. Tsang, P.J. Catto, J.C. Whitson, I. Smith: Absolute Universal Instability is not Universal, Phys. Rev. Lett. **40**, 327 (1978); T.M. Antonsen, Jr.: Stability of Bound Eigenmode Solutions for the Collisionless Universal Instability, Phys. Rev. Lett. **41**, 33 (1978)
6.6 I.B. Bernstein, J.M. Greene, M.D. Kruskal: Exact Nonlinear Plasma Oscillations, Phys. Rev. **108**, 546 (1957)
6.7 F. Einaudi, R.N. Sudan: A Review of the Nonlinear Theory of Plasma Oscillations, Plasma Phys. **11**, 359 (1969)
6.8 R. Balescu: Irreversible Processes in Ionized Gases, Phys. Fluids **3**, 52 (1960); A. Lenard: On Bogoliubov's Kinetic Equation for a Spatially Homogeneous Plasma, Ann. Phys. **10**, 390 (1960); R.L. Guernsey: Kinetic Equation for a Completely Ionized Gas, Phys. Fluids **5**, 322 (1962)
6.9 A.A. Vedenov, E.P. Velikhov, R.Z. Sagdeev: Quasilinear Theory of Plasma Oscillations, Nuclear Fusion **1**, 82 (1962); Nuclear Fusion Suppl. Part II, 465 (1962); W.E. Drummond, D. Pines: Nonlinear Stability of Plasma Oscillations, Nuclear Fusion Suppl. Part III, 1049 (1969)
6.10 M.C. Wang, G.E. Uhlenbeck: On the Theory of Brownian Motion II, Rev. Mod. Phys. **17**, 323 (1945)
6.11 D.C. Montgomery, A. Tidman: *Plasma Kinetic Theory* (McGraw-Hill, New York 1964) Chap. 2

6.12 A. Lenard, I. Bernstein: Plasma Oscillations with Diffusion in Velocity Space, Phys. Rev. **112**, 1456 (1958)
6.13 V.I. Karpman: Singular Solutions of the Equation for Plasma Oscillations, Sov. Phys. JETP **24**, 603 (1967)

Chapter 7

7.1 K. Lonngren, D. Montgomery, I. Alexeff, D. Jones: Dispersion of Ion-Acoustic Waves, Phys. Lett. **25A**, 629 (1967)
7.2 Y. Yamashita, H. Ikezi, N. Sato, T. Takahashi: Propagation of Density Perturbation in a Plasma, Phys. Lett. **27A**, 79 (1968)
7.3 I. Alexeff, W.D. Jones, K. Lonngren: Excitation of Pseudowaves in a Plasma via a Grid, Phys. Rev. Lett. **21**, 878 (1968)
7.4 T. Stix: *The Theory of Plasma Waves* (McGraw-Hill, New York 1962) Chap. 2
7.5 B.B. Kadomtsev: *Plasma Turbulence* (Academic, London 1965) p. 22

Chapter 8

8.1 P.K. Kaw, W.L. Kruer, C.S. Liu, K. Nishikawa: "Parametric Instabilities in Plasma", in *Advances in Plasma Physics*, Vol. 6, ed by A. Simon, W.B. Thompson (Wiley, New York 1975) part I
8.2 K. Mima, K. Nishikawa: "Parametric Instabilities and Wave Dissipation in Plasmas", in *Basic Plasma Physics II*, ed. by A.A. Galeev, R.N. Sudan (North-Holland, Amsterdam 1984) Chap. 6.5
8.3 M. Abramowitz, I.A. Stegun: *Handbook of Mathematical Functions* (Dover, New York 1965) p. 724
8.4 K. Nishikawa: Parametric Excitation of Coupled Waves I, II, J. Phys. Soc. Japan **24**, 916, 1152 (1968)
8.5 J. Drake, P.K. Kaw, Y.C. Lee, G. Schmidt, C.S. Liu, M.N. Rosenbluth: Parametric Instabilities of Electromagnetic Waves in Plasmas, Phys. Fluids **17**, 778 (1974)
8.6 B.B. Kadomtsev, V.I. Karpman: Nonlinear Waves, Sov. Phys. (Uspekhi) **14**, 40 (1971)
8.7 B.N. Breizman, D.D. Ryutov: Powerful Relativistic Electron Beams in a Plasma and in a Vacuum (Theory), Nuclear Fusion **14**, 873 (1974)
8.8 K. Nishikawa, Y.C. Lee, C.S. Liu: Condensation and Collapse, Comments on Plasma Physics and Controlled Fusion II, 63 (1975)
8.9 V.E. Zakharov: Collapse of Langmuir Waves, Sov. Phys. JETP **35**, 908 (1972)
8.10 A.A. Galeev, R.Z. Sagdeev, Yu.S. Sigov, V.D. Shapiro, V.I. Shevchenko: Nonlinear Theory of the Modulation Instability of Plasma Waves, Sov. J. Plasma Phys. **1**, 5 (1975)
8.11 L.M. Degtyarev, V.E. Zakharov, L.I. Rudakov: Dynamics of Langmuir Collapse, Sov. J. Plasma Phys. **2**, 240 (1976)
8.12 L.M. Degtyarev, R.Z. Sagdeev, G.I. Soloviev, V.D. Shapiro, V.I. Shevchenko: One-Dimensional Langmuir Turbulence, Sov. J. Plasma Phys. **6**, 263 (1980)
8.13 J.M. Manley, H.E. Rowe: Some General Properties of Nonlinear Elements. Part I. General Energy Relations, Proc. of the IRE **44**, 904 (1956)

Chapter 9

9.1 D.J. Rose, M. Clark: *Plasma and Controlled Fusion* (Massachusetts Institute of Technology Press, Cambridge, Massachusetts 1961)
9.2 R.F. Post: "Experimental Base of Mirror-Confined Physics", in *Fusion*, Vol. 1, Part A, ed. by E. Teller (Academic, New York 1981) p. 358

9.3 T.K. Fowler: "Mirror Theory", *ibid.* p. 291

9.4 R.F. Post: The Magnetic Mirror Approach to Fusion, Nucl. Fusion **27**, 1579 (1987)

9.5 L.A. Artimovich: Tokamak Devices, Nucl. Fusion **12**, 205 (1972)

9.6 H.P. Furth: Tokamak Research, Nucl. Fusion **15**, 408 (1975)

9.7 W.B. Kunkel: "Neutral Beam Injection", in *Fusion*, Vol. 1, Part B, ed. by E. Teller (Academic, New York, 1981) p. 103

9.8 D.L. Jassby: Neutral Beam Driven Tokamak Fusion Reactors, Nuclear Fusion **17**, 309 (1977)

9.9 M. Porkolab: "Radio-Frequency Heating of Magnetically Confined Plasmas", in *Fusion*, Vol. 1, Part B, ed. by E. Teller (Academic, New York 1981) p. 151

9.10 H.A.B. Bodin, A.A. Newton: Reversed Field Pinch Research, Nucl. Fusion **20**, 1255 (1980)

9.11 D.A. Baker, W.E. Quinn: "The Reversed Field Pinch", in *Fusion*, Vol. 1, Part A, ed. by E. Teller (Academic, New York 1981) p. 438

9.12 M.N. Rosenbluth, M.W. Bussac: MHD Stability of Spheromak, Nucl. Fusion **19**, 489 (1979)

9.13 A.I. Morozov, L.S. Solov'ev: "The Structure of a Magnetic Field", in *Reviews of Plasma physics*, Vol.2 (Consultants Bureau, New York 1966) p. 1

9.14 J.L. Shohet: "Stellarators", in *Fusion*, Vol. 1, Part A, ed. by E. Teller (Academic, New York 1981) p. 243

9.15 K. Miyamoto: Recent Stellarator Research, Nucl. Fusion **18**, 243 (1978)

9.16 B.A. Carreras, G. Grieger, J.H. Harris, J.L. Johnson, J.F. Lyon, O. Motojima, F. Rau, H. Renner, J.A. Rome, K. Uo, M. Wakatani, M. Wobig: Progress in Stellarator/Heliotron Research: 1981–1986, Nuclear Fusion **28**, 1613 (1988)

9.17 H. Motz: *The Physics of Laser Fusion* (Academic, New York 1979)

9.18 W.L. Kruer: *The Physics of Laser Plasma Interactions* (Addison–Wesley, New York 1988)

9.19 M. Miyanaga, Y. Kato, Y. Kitagawa, T. Kouketsu, M. Yoshida, M. Nakatsuka, T. Yabe, C. Yamanaka: Efficient Spherical Compression of Cannonball Targets with 1.052μm Laser Beams, Jap. J. Appl. Phys. **22**, L551 (1983)

9.20 K. Okada, T. Mochizuki, M. Hamada, N. Ikeda, H. Shiraga, T. Yabe, C. Yamanaka: Spectrum-Resolved Absolute Energy Measurement of X-ray Emission in 0.17 – 1.6KeV Range from a 0.53μm Laser-Irradiated Au Target, Jap. J. Appl. Phys. **22**, L671 (1983)

Chapter 10

10.1 J.P. Friedberg: *Ideal Magnetohydrodynamics* (Plenum, New York 1987)

10.2 W.B. Thompson: *An Introduction to Plasma Physics* (Pergamon, Oxford 1962)

10.3 G. Schmidt: *Physics of High Temperature Plasmas* (Academic, New York 1966)

10.4 V.D. Shafranov: "Plasma Equilibrium in a Magnetic Field", in *Reviews of Plasma Physics*, Vol.2 (Consultants Bureau, New York 1966) p. 103

10.5 V.S. Mukovatov, V.D. Shafranov: Plasma Equilibrium in a Tokamak, Nuclear Fusion **11**, 605 (1971)

10.6 L.S. Solov'ev, V.D. Shafranov: "Plasma Confinement in a Closed Magnetic System", in *Reviews of Plasma Physics*, Vol.5 (Consultants Bureau, New York 1970) p. 1

10.7 M.D. Kruskal, R.M. Kulsrud: Equilibrium of a Magnetically Confined Plasma in a Toroid, Phys. Fluids **1**, 265 (1958)

10.8 J.M. Greene, J.L. Johnson, K.E. Weimer: Tokamak Equilibrium, Phys. Fluids **14**, 671 (1971)

10.9 J.F. Clark, D.J. Sigmar: High Pressure Flux-Conserving Tokamak Equilibria, Phys. Rev. Lett. **38**, 10 (1977)

10.10 B.B. Kadomtsev: "Hydromagnetic Stability of a Plasma", in *Reviews of Plasma Physics*, Vol.2 (Consultants Bureau, New York 1966) p. 153

10.11 W.A. Newcomb: Hydromagnetic Stability of a Diffuse Linear Pinch, Ann. Phys. (New York) **3**, 347 (1958)

10.12 I.B. Bernstein, E.A. Frieman, M.D. Kruskal, R.M. Kulsrud: An Energy Principle for Hydromagnetic Stability Problems, Proc. Roy. Soc. **A244**, 17 (1958)

10.13 J.A. Wesson: Hydromagnetic Stability of Tokamaks, Nuclear Fusion **18**, 87 (1978)
10.14 B.R. Suydam: Stability of a Linear Pinch, IAEA Geneva Conf. **31**, 157 (1958)
10.15 V.D. Shafranov, E.I. Yurchenko: Condition for Flute Instability of a Toroidal-Geometry Plasma, Sov. Phys-JETP **26**, 682 (1968)
10.16 G. Batemann: *MHD Instabilities* (Massachusetts Institute of Technology Press, Cambridge 1978)
10.17 V.D. Shafranov: Hydromagnetic Stability of a Current Carrying Pinch in a Strong Longitudinal Magnetic Field, Sov. Phys. Tech. Phys. **15**, 175 (1970)
10.18 J.P. Goedbloed: *Lecture Notes on Ideal Magnetohydrodynamics* (Rijnhuizen Report 83-145, FOM-Instituut Voor Plasmaphysica, Nederland 1983) p. 149

Chapter 11

11.1 H.P. Furth, J. Killeen, M.N. Rosenbluth: Finite-Resistivity Instabilities of a Sheet Pinch, Phys. Fluids **6**, 459 (1963)
11.2 J.M. Greene: *Introduction to Resistive Instabilities*, Centre de Recherches en Physique des plasmas report LRP 114/76 (Lausanne, Switzerland 1976)
11.3 B.B. Kadomtsev, O.P. Pogutse: "Turbulence in Toroidal Systems", in *Reviews of Plasma Physics*, Vol.5 (Consultants Bureau, New York 1967) p. 247
11.4 R.B. White: "Resistive Instabilities and Field Line Reconnection", in *Basic Plasma Physics I*, ed. by A.A. Galeev, R.N. Sudan (North-Holland, Amsterdam 1983) p. 611
11.5 H.R. Strauss: Dynamics of High β Tokamaks, Phys. Fluids **20**, 1354 (1977)
11.6 J.W. Connor, R.J. Hastie, J.B. Taylor: Shear, Periodicity, and Plasma Ballooning Modes, Phys. Rev. Lett. **40**, 396 (1978)
11.7 B. Coppi, A. Ferreira, J.J. Ramos: Self-Healing of Confined Plasma with Finite Pressure, Phys. Rev. Lett. **44**, 990 (1980)
11.8 M.S. Chance, "MHD Stability Limits on High-β Tokamaks", in *Proc. of 7th IAEA Conf. on Plasma Phys. and Controlled Nucl. Fusion Research*, Vol I (1978) p. 677
11.9 F. Troyon, R. Gruber, H. Saurenmann, S. Semenzato, S. Succi: MHD-Limits to Plasma Confinement, Plasma Phys. **26**, 209 (1984)
11.10 J.M. Greene, J.L. Johnson: Determination of Hydromagnetic Equilibria, Phys. Fluids **5**, 875 (1961)
11.11 H.R. Strauss: Stellarator Equations of Motion, Plasma Physics **22**, 733 (1980)
11.12 S. von Goeler, W. Stodiek, N. Sauthoff: Studies of Internal Disruptions and m=1 Oscillations in Tokamak Discharges with Soft-X-Ray Techniques, Phys. Rev. Lett. **33**, 1201 (1974)
11.13 J.D. Callen, G.L. Johns: Experimental Measurement of Electron Heat Diffusivity in a Tokamak, Phys. Rev. Lett. **38**, 491 (1977)
11.14 B.B. Kadomtsev: Disruptive Instability in Tokamaks, Sov. J. Plasma Phys. **1**, 389 (1975)
11.15 B.V. Waddel, M.N. Rosenbluth, D.A. Monticello, R.B. White: Nonlinear Growth of the m=1 Tearing Mode, Nucl. Fusion **16**, 528 (1976)
11.16 A. Sykes, J.A. Wesson: Relaxation Instabilities in Tokamaks, Phys. Rev. Lett. **37**, 140 (1976)
11.17 P.H. Rutherford: Nonlinear Growth of the Tearing Mode, Phys. Fluids **16**, 1903 (1973)
11.18 B.V. Waddel, B. Carreras, H.R. Hicks, J.A. Holmes: Nonlinear Interaction of Tearing Modes in Hightly Resistive Tokamaks, Phys. Fluids **22**, 896 (1979)
11.19 H.P. Furth, P.H. Rutherford, H. Selberg: Tearing Mode in the Cylindrical Tokamak, Phys. Fluids **16**, 1054 (1973)
11.20 B. Coppi, R. Galvao, R. Pellet, M.N. Rosenbluth, P.H. Rutherford, Resistive Internal Kink Modes, Sov. J. Plasma Phys. **2**, 533 (1976)
11.21 A.W. Edwards et al.: Rapid Collapse of a Plasma Sawtooth Oscillation in the JET Tokamak, Phys. Rev. Lett. **57**, 210 (1986)
11.22 A. Hasegawa, T. Sato: *Space Plasma Physics 1* (Springer, Heidelberg, Berlin 1989)
11.23 T. Sato, T. Hayashi: Externally Driven Magnetic Reconnection and a Powerful Magnetic Energy Convertor, Phys. Fluids **22**, 1189 (1979)
11.24 T. Sato: *Nonlinear Driven Reconnection*, HIFT-97, Institute For Fusion Theory (Hiroshima University Press 1984)

11.25 J.B. Taylor: Relaxation of Toroidal Plasma and Generation of Diverse Magnetic Fields, Phys. Rev. Lett. **33**, 1139 (1974)
11.26 R. Horiuchi, T. Sato: Self-Organization and Energy Relaxation in a Three-Dimensional Magnetohydrodynamic Plasma, Phys. Fluids **29**, 1161 (1986)
11.27 A. Reiman: Minimum Energy State of a Toroidal Discharge, Phys. Fluids **23**, 230 (1980)

Chapter 12

12.1 V.L. Ginzburg: *The Propagation of Electromagnetic Waves in Plasmas* (Pergamon, Oxford 1964) Chap. IV, V
12.2 A. Messiah: *Quantum Mechanics* (North-Holland, Amsterdam 1970) Vol.1, Chap. VI
12.3 T. Watanabe, H. Sanuki, M. Watanabe: New Treatment of Eigenmode Analyses for an Inhomogeneous Vlasov Plasma, J. Phys. Soc. Japan **47**, 286 (1979)
12.4 T. Watanabe, M. Watanabe, H. Sanuki, K. Iino, K. Nishikawa: Eigenmode Analyses of the Electrostatic Wave in the Nonuniform Vlasov Plasma. I. Integral Equation in the Wavenumber Space, J. Phys. Soc. Japan **50**, 1745 (1981)
12.5 P.A. Sturrock: Kinematics of Growing Waves, Phys. Rev. **112**, 1488 (1958)
12.6 R.N. Sudan: Classification of Instabilities from their Dispersion Relations, Phys. Fluids **8**, 1899 (1965)
12.7 L.D. Pearlstein, H.L. Berk: Universal Eigenmode in a Strongly Sheared Magnetic Field, Phys. Rev. Lett. **23**, 220 (1969)
12.8 T.H. Stix: *The Theory of Plasma Waves* (McGraw-Hill, New York 1962)
12.9 R.A. Cairns, O.N. Lashmore-Davies: The Absorption Mechanism of the Ordinary Mode Propagating Perpendicular to the Magnetic Field at the Electron Cyclotron Frequency, Phys. Fluids **25**, 1605 (1982)
12.10 N.J. Fisch, A.H. Boozer: Creating an Asymmetric Plasma Resistivity with Waves, Phys. Rev. Lett. **45**, 720 (1980)
12.11 W.L. Kruer: *The Physics of Laser Plasma Interactions* (Addision-Wesley, New York 1988)
12.12 C.E. Max: "Theory of Coronal Plasma in Laser Fusion Targets", in *Physics of Laser Fusion* Vol.1, Proc. Summer School of Theoretical Physics (Les Houches, France 1980)

Chapter 13

13.1 S.I. Bragniskii: "Transport Processes in a Plasma", in *Reviews of Plasma Physics*, Vol. 1 (Consultants Bureau, New York 1965) p. 205
13.2 F.L. Hinton, R.D. Hazeltine: Theory of Plasma Transport in Toroidal Confinement Systems, Rev. Mod. Phys. **48**, 239 (1976)
13.3 A.A. Galeev, R.Z. Sagdeev: "Theory of Neoclassical Diffusion", in *Reviews of Plasma Physics*, Vol. 7 (Consultants Bureau, New York 1975) p. 257
13.4 B.B. Kadomtsev, O.P. Pogutse: Trapped Particles in Toroidal Magnetic Systems, Nucl. Fusion **11**, 67 (1971)
13.5 P.C. Liewer: Measurements of Microturbulence in Tokamaks and Comparison with Theories of Turbulence and Anomalous Transport, Nucl. Fusion **25**, 543 (1985)
13.6 D. Montgomery: "Introduction to the Theory of Fluid and Magnet Fluid Turbulence", in *Nagoya Lectures in Plasma Physics and Controlled Fusion*, ed. by Y.M. Ichikawa, T. Kamimura (Tokai University Press 1989) p. 207
13.7 A. Hasegawa: Self-Organization Processes in Continuous Media, Adv. Phys. **34**, 1 (1985)
13.8 W.M. Tang: Microinstability Theory in Tokamaks, Nucl. Fusion **18**, 1089 (1978)
13.9 M.N. Rosenbluth, R.Z. Sagdeev, J.B. Taylor, G.M. Zaslavski: Destruction of Magnetic Surfaces by Magnetic Field Irregularities , Nucl. Fusion **6**, 297 (1966)
13.10 A.B. Rechester, M.N. Rosenbluth: Electron Heat Transport in a Tokamak with Destroyed Magnetic Surfaces, Phys. Rev. Lett. **40**, 38 (1978)

13.11 R.E. Waltz: "Turbulent Transport in Tokamaks", in *Nagoya Lectures in Plasma Physics and Controlled Fusion*, ed. by Y.H. Ichikawa, T. Kamimura (Tokai University Press 1989) p. 357
13.12 R.H. Kraichnan: Inertial Range in Two-Dimensional Turbulence, Phys. Fluids **10**, 1417 (1967)

Chapter 14

For reference on tokamaks:

14.1 D. Johnson, M. Bell, M. Bitter, K. Bol, K. Brau, D. Buchenauer, R. Budny, T. Crowley, S. Davis, F. Dylla, H. Eubank, H. Fishman, R. Fonck, R. Hawryluk, H. Hsuan, R. Kaita, S. Kaye, J. Manickam, D. Mansfield, E. Mazzucato, R. MacCann, D. McCune, K. Oasa, J. Olivain, M. Okabayashi, K. Owens, J. Ramette, C. Reverdin, M. Reusch, M. Sauthoff, G. Schilling, G. Schmidt, S. Sesnic, R. Slusher, C. Surko, J. Strachan, S. Suckewer, H. Takahashi, T. Tenny, P. Thomas, H. Towner, J. Valley: High-beta experiments with neutral beam injection on PDX, in *Proc. 9th IAEA Conf. on Plasma Physics and Controlled Nuclear Fusion Research* (Baltimore 1982) Vol. 1, p. 9
14.2 M. Okabayashi, K. Bol, M. Chance, P. Couture, H. Fishman, R.J. Fonck, G. Gammel, W.W. Heidbrinck, K. Ida, K.P. Jaehnig, G. Jahns, R. Kaita, S.M. Kaye, H. Kugel, B. LeBlanc, J. Manickam, W. Morris, G.A. Navratil, N. Ohyabu, S. Paul, E. Powell, M. Reusch, S. Sesnic, H. Takahashi: Stability and confinement studies in the Princeton beta experiment (PBX), in *Proc. 11th IAEA Conf. on Plasma Physics and Controlled Nuclear Fusion Research* (Kyoto 1986) Vol. 1, p. 139
14.3 R.D. Stambaugh, R.M. Moore, L.C. Bernard, A.G. Kellman, E.J. Strait, L. Lao, D.O. Overskei, T. Angel, C.J. Armentraut, J.F. Bauer, T.R. Blau, C. Bramson, N.H. Brooks, K.H. Burrell, R.W. Callis, R.P. Chase, A.P. Colleraine, G. Cottrell, J.C. Deboo, S. Ejima, E.S. Fairbanks, J. Fasolo, C.H. Fox, J.R. Gilleland, R. Groebner, F.J. Helton, R.M. Hong, C.L. Hsieh, G.J. Johns, C.L. Kahn, J. Kim, D. Knowles, J.K. Lee, P. Lee, A.J. Lieber, J.M. Lohr, D.B. McColl, C.H. Meyer, T. Ohkawa, N. Ohyabu, P.I. Schissel, J.T. Scoville, R.P. Seraydarian, B.W. Sleaford, J.R. Smith, R.T. Snider, R.D. Stav, H. St. John, R.E. Stockdale, J.F. Tooker, D. Vaslow, J.C. Wesley, S.S. Wojtowicz, S.K. Wong, E.M. Zawadski: Test of beta limits as a function of plasma shape in doublet III, in *Proc. 10th IAEA Conf. on Plasma Physics and Controlled Nuclear Fusion Research* (London 1984) Vol. 1, p. 217
14.4 T.S. Taylor, E.J. Strait, L.L. Lao, M. Mauel, A.D. Turnbull, K.H. Burrell, M.S. Chu, J.R. Ferrom, R.J. Groebner, R.J. La Haye, B.W. Rice, R.T. Snider, S.J. Thompson, D. Wroblewski, D.J. Lightly: Wall stabilization of high beta plasmas in DIII-D. Phys. Plasmas **2**, 2390 (1995)
14.5 M. Greenwald, M. Besen F. Camacho, C. Flore, M. Foord, R. Gandy, C. Gomez, R. Granetz, D. Gwinn, S. Knowlton, B. Labombard, B. Lipschultz, H. Manning, E. Marmar, S.C. McCool, J. Parker, R. Parker, R. Petrasso, P. Prinyl, J. Rice, D. Sigmar, Y. Takase, J. Terry, R. Watterson, S. Wolfe: Studies of the regime of improved particle and energy confinement following pellet injection into Alcator C, in *Proc. 11th IAEA Conf. on Plasma Physics and Controlled Nuclear Fusion Research* (Kyoto 1986) Vol. 1, p. 139
14.6 M. Keilhacker, G. Fussmann, G. von Gierke, G. Janeschitz, M. Kornherr, K. Lackner, E.R. Muller, P. Smuelders, F. Wagner, G. Becker, K. Bernhardi, U. Ditte, A. Eberhagen, O. Gehre, J. Gernhardt, E. Glock, T. Grove, O. Gruber, G. Haas, M. Hesse, F. Karger, S. Kissel, O. Kluber, G. Lisitano, H.M. Mayer, K. McCormick, D. Meisel, V. Mertens, H. Murmann, H. Niedermeyer, W. Poschenrider, H. Rapp, F. Ryter, F. Schineider, G. Siller, F. Soldner, E. Speth, A. Stabler, K.-H. Steuer, O. Vollmer: Confinement and beta-limit studies in ASDEX H-mode discharges, in *Proc. 10th IAEA Conf. on Plasma Physics and Controlled Nuclear Fusion Research* (London 1984) Vol. 1, p. 71
14.7 The ASDEX team: The H-mode of ASDEX. Nucl. Fusion **29**, 1959 (1989)
14.8 P.N. Yushmanov, T. Takizuka, K.S. Riedel, O.J.W.F. Kardaun, J.G. Cordey, S.M. Kaye, D.E. Post: Scalings for tokamak energy confinement. Nucl. Fusion **30**, 1999 (1990)
14.9 F.W. Perkins, C.W. Barnes, D.W. Johnson, S.D. Scott, M.C. Zarnstorff, M.G. Bell, R.E. Bell, C.E. Bush, B. Grek, K.W. Hill, D.K. Mansfield, H. Park, A.T. Ramsey, J. Schivell, B.C. Stratton, E. Synakowski: Nondimensional transport scaling in the Tokamak Fusion Test Reactor: Is tokamak transport Bohm or gyroBohm? Phys. Fluids B **5**, 477 (1993)

14.10 R. Aymar, V. Chuyanov, M. Huguet, R. Parker, Y. Shimomura and the ITER Joint Central Team and Home Team: ITER project: A physics and technology experiment, in *Proc. 16th IAEA Conf. on Fusion Energy* (Montreal 1996) Vol. 1, p. 3
14.11 S-I. Itoh, K. Itoh: Model of L- to H-mode transition in tokamak. Phys. Rev. Lett. **60**, 2276 (1988)
14.12 K.C. Shaing, E.C. Crume Jr.: Bifurcation theory of poloidal rotation in tokamaks: a model for the L-H transition. Phys. Rev. Lett. **63**, 2369 (1989)
14.13 M.G. Bell, S. Batha, M. Beer, R.E. Bell, A. Belov, H. Berk, S. Bernabei, M. Bitter, B. Breizman, N.L. Bretz, R. Budny, C.E. Bush, J. Callen, S. Cauffman, C.S. Chang, Z. Chang, C.Z. Cheng, D.S. Darrow, R.O. Dendy, W. Dorland, H. Duong, P.C. Efthimion, D. Ernst, H. Evenson, N.J. Fisch, R. Fisher, R.J. Fonck, E.D. Fredrickson, G.Y. Fu, H.P. Furth, N.N. Gorelenkov, V. Ya. Goloborod'ko, B. Grek, L.R. Grisham, G.W. Hammett, R.J. Hawryluk, W. Heidbrink, H.W. Hermann, M.C. Hermann, K.W. Hill, J. Hogan, B. Hooper, J.C. Hosea, W.A. Houlberg, M. Hughes, D.L. Jassby, F.C. Jobes, D.W. Johnson, R. Kaita, S. Kaye, J. Kesner, J.S. Kim, M. Kissick, A.V. Krasilnikov, H. Kugal, A. Kumar, N.T. Lam, P. Lamarche, B. LeBlanc, F.M. Levinton, C. Ludescher, J. Machuzak, R.P. Majeski, J. Manickam, D.K. Mansfield, M. Mauel, E. Mazzucato, J. McChesney, D.C. McCune, G. McKee, K.M. McGuire, D.M. Meade, S.S. Medley, D.R. Mikkelsen, S.V. Mirnov, D. Mueller, Y. Nagayama, G.A. Navratil, R. Nazikian, M. Okabayashi, M. Osakabe, D.K. Owens, H.K. Park, W. Park, S.F. Paul, M.P. Petrov, C.K. Phillips, M. Phillips, P. Phillips, A.T. Ramsey, B. Rice, M.H. Redi, G. Rewoldt, S. Reznik, A.L. Roquemore, J. Rogers, E. Ruskov, S.A. Sabbagh, M. Sasao, G. Schilling, G.L. Schmidt, S.D. Scott, I. Semerlov, T. Senko, C.H. Skinner, T. Stevenson, E.J. Strait, B.C. Stratton, J.D. Strachan, W. Stodiek, E. Synakowski, H. Takahashi, W. Tang, G. Taylor, M.E. Thompson, S. von Goeler, A. Von Halle, R.T. Walters, S. Wang, R. White, R.M. Wieland, M. Williams, J. R. Wilson, K.L. Wong, G.A. Wurden, M. Yamada, V. Yavorski, K.M. Young, L. Zakharov, M.C. Zarnstorff, S.J. Zweben: Deuterium-tritium plasmas in novel regimes in the Tokamak Fusion Test Reactor. Phys. Plasmas **4**, 1714 (1997)
14.14 A. Gibson and the JET Team: D-T plasmas in the Joint European Torus (JET): Behavior and implications. Phys. Plasmas **5**, 1839 (1998)
14.15 S. Itoh, Y. Nakamura, K. Nakamura, T. Fujita, K. Makino, E. Jotaki, S. Kawasaki, S.I. Itoh: High density and long duration current drive discharge with high frequency LH waves in TRIAE-1M, in *Proc. 14th IAEA Conf. on Plasma Physics and Controlled Nuclear Fusion Research* (Würzburg, Germany 1992) Vol. 1, p. 743
14.16 F.M. Levinton, M.C. Zarnstorff, S.H. Batha, M. Bell, R.E. Bell, R.V. Budny, C. Bush, Z. Chang, E. Fredrickson, A. Janos, J. Manickam, A. Ramsey, S.A. Sabbagh, G.L. Schmidt, E.J. Synakowski, G. Taylor: Improved confinement with reversed magnetic shear in TFTR. Phys. Rev. Lett. **75**, 4417 (1995)
14.17 E.J. Strait, L.L. Lao, M.E. Mauel, B.W. Rice, T.S. Taylor, K.H. Burrell, M.S. Chu, E.A. Lazarus, T.H. Osbone, S.J. Thompson, A.D. Turnbull: Enhanced confinement and stability in DIII-D discharges with reversed magnetic shear. Phys. Rev. Lett. **75**, 4421 (1995)
14.18 K.L. Wong, R.J. Fonck, S.F. Paul, D.R. Roberts, E.D. Fredrickson, R. Nazikian, H.K. Park, M. Bell, N.L. Bretz, R. Bundy, S. Cohen, G.W. Hammett, F.C. Jobes, D.M. Meade, S.S. Medley, D. Mueller, Y. Nagayama, D.K. Owens, E.J. Synakowski: Excitation of toroidal Alfvén eigenmodes in TFTR. Phys. Rev. Lett. **66**, 1874 (1991)
14.19 W. Heidbrink, E.J. Strait, E. Doyle, G. Sager, R.T. Snider: An investigation of beam driven Alfvén instabilities in the DIII-D tokamak. Nucl. Fusion **31**, 1635 (1991)
14.20 R. Nazikian, G.Y. Fu, S.H. Batha, M.G. Bell, R.E. Bell, R.V. Budny, C.E. Bush, Z. Chang, Y. Chen, C.Z. Cheng, D.S. Darrow, P.C. Efthimion, E.D. Fredrickson, N.N. Gorelenkov, B. Leblanc, F.M. Levinton, R. Majeski, E. Mazzucato, S.S. Medley, H.K. Park, M.D. Petrov, D.A. Spong, J.D. Strachan, E.J. Synakowski, G. Yaylor, S. Von Goeler, R.B. White, K.L. Wong, S.J. Zweben: Alpha-particle-driven toroidal Alfvén eigenmodes in the Tokamak Fusion Test Reactor. Phys. Rev. Lett. **78**, 2976 (1997)
14.21 A. Sykes, R. Akers, L. Applel, P.G. Carolan, N.J. Conway, M. Cox, A.R. Field, D.A. Gates, S. Gee, M. Gryaznevich, T.C. Hender, I. Jenkins, R. Martin, K. Morel, A.W. Morris, M.P.S. Nightingale, G. Ribeiro, D.C. Robinson, M. Tournianski, M. Valovic, M.J. Walsh, C. Warvick: High β performance of the START spherical tokamak. Plasma Phys. Control. Fusion **39**, B247 (1997)

14.22 M. Wakatani, S. Sudo: Overview of Heliotron E results. Plasma Phys. Control. Fusion **38**, 937 (1996)

14.23 S. Sudo, Y. Takeiri, H. Zushi, F. Sano, K. Itoh, K. Kondo, A. Iiyoshi: Scaling of energy confinement and density limit is stellarator/heliotron devices. Nucl. Fusion **30**, 11 (1990)

14.24 A. Iiyoshi, K. Yamazaki: The next large helical devices. Phys. Plasmas **2**, 2349 (1995)

14.25 F. Wagner, J. Baldzuhn, R. Brakel, R. Burhenn, V. Erckmann, T. Estrada, P. Grigull, H.J. Hartfuss, G. Herre, M. Hirsch, J.V. Hofmann, R. Jaenicke, A. Rudyj, U. Stroth, A. Weller and W7-AS Team: H-mode of W7-AS stellarator. Plasma Phys. Control. Fusion **36**, A 61 (1994)

14.26 R. Brake, M. Anton, J. Baldzuhn, R. Burhenn, V. Erckmann, S. Fielder, J. Geiger, H.J. Hartfuss, O. Heinvich, M. Hirsch, R. Jaenicker, M. Kick, G. Kühner, H. Maassberg, U. Stroth, F. Wagner, A. Weller, W7-AS Team, ECRH Group and NBI Group: Confinement in W7-AS and the role of radial electric field and magnetic shear. Plasma Phys. Control. Fusion **39**, B273 (1997)

14.27 J. Nührenberg, R. Zille: Quasi-helically symmetric toroidal stellarator. Phys. Lett. A **129**, 113 (1998)

14.28 C. Beidler, G. Grieger, F. Herrnegger, E. Harmeyer, J. Kisslinger, W. Lotz, H. Maassberg, P. Merkel, J. Nührenberg, F. Rau, J. Sapper, F. Sardei, R. Scardovelli, A. Schlüter, H. Wobig: Physics and engineering design for Wenderstein VII-X. Fusion Technol. **17**, 148 (1990)

14.29 P.R. Garabedian: Recent progress in the design of stellarator experiments. Phys. Plasmas **4**, 1617 (1997)

For reference on tandem mirror:

14.30 T. Cho, M. Ichimura, N. Inutake, K. Ishi, I. Katanuma, Y. Kiwamoto, A. Mase, S. Miyoshi, Y. Nakashima, T. Saito, K. Sawada, D. Tsubouchi, N. Yamaguchi, K. Yatsu: Studies of potential formation and transport in the tandem mirror Gamma 10, in *Proc. of 11th IAEA Conf. on Plasma and Controlled Nucl. Fusion Research* (Kyoto, 1986) Vol. 2, p. 243

For reference on inertial fusion:

14.31 H. Azechi, T. Jitsuno, T. Kanabe, M. Katayama, K. Mima, M. Miyanaga, M. Nakai, S. Nakai, H. Nakaishi, M. Nakatsuka, A. Nishigochi, M. Takagi, M. Yamanaka: High-density compression experiment at ILE, Osaka. Laser and Particle Beams **9**, 193 (1991)

14.32 Y. Kato, K. Mima, M. Miyanaga, S. Arinaga, Y. Kitagawa, M. Nakatsuka, C. Yamanaka: Random Phasing of high-power lasers for uniform acceleration and plasma-instability suppression. Phys. Rev. Lett. **53**, 1057 (1984)

14.33 M. Miyanga, H. Azechi, R.O. Stapf, K. Itoga, H. Nakaishi, H. Shiraga, M. Yamanaka, T. Yamanaka, R. Tsuji, S. Ido, K. Sakurai, K. Nishihara, T. Yabe, M. Takagi, M. Nakatsuka, Y. Izawa, S. Nakai, C. Yamanaka: Radiochemistry and secondry reactions for the diagnositcs of laser-driven fusion plasmas. Rev. Sci. Instrum. **57**, 1731 (1986)

14.34 S. Nakai, K. Mima, T. Yamanaka, Y. Izawa, Y. Kato, K. Nishikawa, H. Azechi, N. Miyanaga, H. Takabe, T. Sasaki, M. Nakatsuka, M. Yamanaka, T. Norimatsu, K.A. Tanaka, T. Jitsuno, M. Nakai, M. Takagi, M. Katayama, M. Kado, R. Kodama: Fermi degeneracy of high density imploded plasma and stability of hollow shell pellet implosion, in *Proc. 14th IAEA Conf. on Plasma Phys. and Controlled Nuclear Fusion Res.* (Würzburg, Germany 1992) Vol. 3, p. 13

14.35 H. Shiraga, N. Miyanaga, A. Heya, M. Nakasuji: Ultrafast two-dimensional x-ray imaging with x-ray streak cameras for laser fusion research. Rev. Sci. Instrum. **68**, 745 (1997)
H. Shiraga, M. Heya, M. Nakasuji, N. Miyanaga: One- and two-dimensional fast x-ray imaging of laser driven implosion dynamics with x-ray streak camera. Rev. Sci. Instrum. **68**, 828 (1996)

14.36 K. Mima, Y. Kato, H. Azechi, K. Shigemori, H. Takabe, N. Miyanaga, T. Kanabe, T. Norimatsu, H. Nishimura, H. Shiraga, M. Nakai, R. Kodama, K.A. Tanaka, M. Takagi, M. Nakatsuka, K. Nishihara, T. Yamanaka, S. Nakai: Recent progress of implosion experiments with uniformity-improved GEKKO XII Laser Facility at the Institute of Laser Engineering, Osaka University. Phys. Plasmas **3**, 2077 (1996)

14.37 See, for instance, The National Ignition Facility (UCRL-TB-118550 Rev 1, 1998)
14.38 M. Tabak, J. Hammer, M.E. Glinsky, W.L. Kruer, S.C. Wilks, J. Woodworth, E. M. Campbell, M.D. Perry: Ignition and high gain with ultrapowerful lasers. Phys. Plasmas **1**, 1626 (1994)
14.39 D. Strickland, G. Mourou: Compression of amplified chirped optical pulses. Opt. Commun. **56**, 219 (1985)
14.40 N.L. Tsintsadze: Possibility of parametric resonance in an electron plasma. Sov. Phys.-JETP **32**, 684 (1971)
14.41 B. Quesnel, P. Mora, J.C. Adam, A. Heron, G. Laval: Electron parametric instabilities of ultraintense laser pulses propagating in plasmas of arbitrary density. Phys. Plasmas **4**, 3358 (1997)
14.42 P. Gibbon, P. Monot, T. Auguste, G. Mainfray: Measurable signatures of relativistic self-focus in underdense plasmas. Phys. Plasmas **2**, 1305 (1995)

For reference on controlled fusion:

14.43 D.J. Rose, M. Clark: *Plasma and Controlled Fusion* (Massachusetts Institute of Technology Press, Cambridge, Massachusetts 1961)
14.44 W.M. Stacy, Jr.: *Fusion Plasma Analysis* (Wiley-Interscience, New York 1981)

Subject Index

Printed in the United States
98579LV00002BC/17/A

9 783540 652854